Summary: Transformative Education for Regeneration

Abstract The summary explains the meaning of transformative education and makes a case for a new form of thinking and practice to support social, economic and environmental regeneration that is relational in terms of its axiology. We need to consider the many social, political, cultural, biological, economic and environmental relationships that come into play when engaging in transformative thinking and practice. The volume makes an important contribution to critical, systemic thinking and practice by drawing on and extending the work of Freire and Foucault by adding a more explicitly gendered and multispecies focus that responds to the big issues of the day, namely poverty, climate change, displacement of species, hunger, pandemics and the digital divide.

Keywords Education • Critical • Systemic praxis • Transformation • Relationality re-generative education

Introduction: Re-conceptualising Education for Re-generative Development

The world has changed since Dewey (1916/2001) wrote about education but the basic premise, namely testing out ideas through participatory democracy, remains relevant in an increasingly globalised digital world in which social and environmental issues have no boundaries (Peters & Jandric, 2017).

'Power and knowledge are linked', and nowhere is Foucault's linkage more marked than in the biopolitical determination of *what* species are valued and *why* and how our values shape the curricula at all levels of education (see Foucault & Gordon, 1980; Foucault, 2008).

How we see the world matters, quite literally because our relationships with others and nature are shaped by our values. The lack of understanding of our relationships constitute what Coll (2022:1) calls 'a consciousness gap', which he argues is the result of the disconnect between human beings and nature (Coll, 2022:26) fostered by current neoliberal economics and governance systems. Coll advocates a change in values and praxis and draws on Buddhism and Taoism in his transformative approach to praxis.

Compassion, humility and wisdom are lacking in current approaches to the big challenges facing the world, namely climate change, displacement, pandemics, poverty and conflict (see McIntyre-Mills & Corcoran Nantes, 2021, McIntyre-Mills, 2021a, b, c, d, e, f, g, h), because the human rights of a *minority* cannot be protected at the expense of *the majority* of living systems *in this generation and the next* (Meadows & Randers, 1992; Stiglitz et al., 2010, 2011; IPPC, 2021).

This volume makes a case for a new relational approach to transformative education based on valuing relationships and engaging in pilots that prefigure change. Our critical pedagogy extends an understanding of oppression (Freire, 1980) to address *the importance of personal and group agency* for promoting both social and environmental justice. Our world view is shaped by honouring the ontology of kinship with nature as expressed by indigenous custodians underwritten by working across cultures and disciplines using a cross-cultural approach and mixed methodology. Axiologically, it is based on the notion of transformative research and education that promotes a non-anthropocentric approach; we place local custodianship centre stage, whilst also addressing and updating gender norms. All change begins with a change in values. An understanding of the dire impacts of climate change and 'business as usual' is indeed understood, and new governance structures, such as the proposed Ecocide Law (Higgins et al., 2013; Thunberg, 2021[1]), are urgently needed (see chapters "Regenerative Education and Safe Habitats for Diverse Species: Caterpillar Dreaming Butterfly Being", "Dis-Ability, Differently Abled and Sentient Beings: Transformative Regeneration: The Need for Public Education on Vulnerability and Interbeing" and "Systemic Praxis and Education to Protect the Commons" of this volume).

The volume strives to address the question: What should/could transformation entail? In the pathways to wellbeing prototype (McIntyre-Mills et al., 2014), people were asked to explore what they have, what they need and what they are willing to add or discard in order to move away from *Business as usual* towards supporting what Joseph Stiglitz et al. (2010) call 'Wellbeing Stocks'.[2] They are asked to express their thoughts and perceptions based on their own feelings and opinions. In chapters "Learning from Nature's Classroom: Reframing Economics, Accounting and Accountability", "AA!!Venda Women and Social Enterprise: Stepwise Process to Regenerative and Sustainable Living" and "Systemic Praxis and Education to Protect the Commons", we discuss the current design through which we hope to

[1] https://twitter.com/GretaThunberg/status/1456731487928205321 6 Nov 2021

[2] The definition is as follows: '1. Material living standards (income, consumption and wealth), 2. Health, 3. Education, 4. Personal activities including work, 5. Political voice and governance, 6. Social connections and relationships, 7. Environment (present and future conditions), 8. Insecurity, of an economy as well as a physical nature'. This definition of wellbeing stocks fits well with the way in which both Indigenous and Non-Indigenous Australians connect with country in Australia and elsewhere and the way in which critical systems thinkers and complexity theorists understand.

enable local residents to map and monitor local wellbeing stocks and, in particular, environmental wellbeing on which everything else depends. We need to learn how to live differently. Imagine if we could design schools, universities[3] and centres which support action learning in order to integrate teaching, research, community and professional service? The contributions created for this book can all be said to be resources that help us to advance our imaginations towards this end.

We address scenarios and implications each scenario poses for educational transformation, for example:

Scenario A: The implications of education as usual or neo-liberal *factory education* that poses a threat as it ***does not*** address the convergent social, economic and environmental challenges.

Senario B: Taking steps towards alternative forms of education at the primary, secondary and tertiary level whereby each level is linked with vocational education and training that addresses the socio-economic and environmental challenges to support changes by taking daily steps towards transformation, an approach that could be regarded as too little, too late and thus ineffectual given the extent of current challenges.

Scenario C: Transformative public education that focuses on relationships and balances individual and collective needs (social, economic and environmental), in order to enhance protection across rural and urban areas by protecting habitat, reducing emissions in order to prefigure transformative praxis in support of zero emissions.

Transformative education from the perspective of contributors to this volume advocates for critical and systemic approach plus applied learning. For us, the focus is on combining face-to-face learning in the community that draws on the local tacit knowledge and is supported by a community of practice that enables people to earn, while they learn and grow a future together in learning organisations and communities that build, sustain or re-generate respectful relationships with others. Our focus is on moving from dualism between human beings and other species to supporting relationships that protect living systems as detailed in the case studies in chapters "The Power of Story: Ocean View Organic Farmers Cooperative, Cape Town, South Africa: Growing Food with Heart and Soul", "Learning from Nature's Classroom: Reframing Economics, Accounting and Accountability", "Sacred Groves of the Tolon District of the Northern Region, Ghana: Where Spirituality Meets Education",

---

[3] International students have propped up Australian universities through subsidising local students, many hope to remain in Australia and many more hope to continue their careers as academics. The casualisation of work for many years helped to provide students the opportunity to make the first step towards becoming early career academics. The COVID-19 pandemic highlighted the challenges facing universities reliant on international students who were unable to return or to take up new positions as the borders closed. The casualisation of academic work has also attracted attention in the Australian news during the pandemic when universities were excluded from government subsidies to prop up the earnings of people at risk of losing their jobs. But the structural problems go far deeper, the question has to be asked, what is the point of teaching compartmentalised, discipline-based courses that perpetuate the status quo, namely an economy that exploits people plants, animals and the habitat on which we all depend?

"AA!!Venda Women and Social Enterprise: Stepwise Process to Regenerative and Sustainable Living" and "Women, Education and Socialisation: Transforming Gender and Power in the MST, São Paulo, Brasil" in which contributors explain the way they engage in public education to increase organic farming, to adapt and apply the Indonesian model of *one village one /or many enterprises*,[4] to support circular green economies and to enable people to earn, learn and grow together. This is supported by the theory chapters in "Combining Focus and Circumspection: An Education in Natural Inclusion" and "The γ – Principle of Natural Inclusion: From Competition Versus Cooperation to Heartfelt Relay" by Alan Rayner and chapter "Systemic Praxis and Education to Protect the Commons" on new architectures and forms of education to protect the commons and by theory chapters that make a case for compassion and recognition of rights for sentient beings based on an awareness that consciousness is not the preserve of human beings (chapters "Dis-Ability, Differently Abled and Sentient Beings: Transformative Regeneration: The Need for Public Education on Vulnerability and Interbeing" and "Public Education on the Rights of Sentient Beings to a Life and Death Free of Suffering"). This point has been made by Indigenous and spiritual leaders, physicists and plant and animal researchers.

Overall, the chapters address the potential to develop alternative ways of supporting education to address poverty through *hands-on intervention* to enable active learning in ways that protect and re-generate habitat. The themes addressed in the book include education to support non-anthropocentrism and *multispecies* relationships. Some of the contributors will focus on re-generative education in a range of contexts to support opportunities for the marginalised. The focus on indigenous people, young people and differently abled and displaced peoples is also central. Creating employment opportunities in the area of food, energy and water security is a key theme underpinning the volume which builds on previous research on pathways to wellbeing to address complex health, housing and social inclusion as well as mitigation and adaptation to climate change by considering 'if then' scenarios inspired by the Mont Fleur approach (Kahane, 1992).

The Context and Rationale for Transformative Education

Gildersleeve and Kleinhesselink (2019:3) explain that:

> … Our industrialization efforts have increased CO_2 in the atmosphere, changing weather patterns, the Earth's temperature, and the ocean's acidity. Looking forward, the Anthropocene's potential consequences grow progressively distressing. Human-caused climate change is demonstrated through carbon dioxide levels that are the highest in human recorded history, rising sea levels, increasingly extreme weather events, melting permafrost. These happenings necessarily change the habitats of animals and other nonhumans,

[4] (OVAP, Morihiko Hiramatsu—Governor of Oita prefecture, 1979, Yogyakarta, 2014) was applied by President Jokowi in 2008–2009.

> compromising their ability to live and thrive. Left unchecked, it is not unreasonable to contemplate that anthropogenic climate change may lead to a sixth mass extinction—and that we may be in the beginning of that extinction now.

The context and rationale for transformative education are poverty, pandemics and climate change, in which unemployment and hunger are exacerbated by lockdowns and the digital divide in many parts of the world, such as South Africa[5] (Ramaphosa, 2021,[6] as discussed in chapter "Nurdles and Food Security: Education for Wellbeing and Implications for Pan-African Social and Environmental Justice" of this volume, see, McIntyre-Mills, 2021a, b). The WHO[7] and UN[8] have also stressed the increased vulnerability of women to violence during Covid lockdowns.

Each of the projects discussed in this book contributes in a small way towards understanding the potential of transformative education in a bid to develop regenerative pathways to sustainable environments. Lessons are drawn from First Nations, such as the San, the Khoi of Southern Africa and other Indigenous heritages in Venda, Ghana, Aboriginal Australians and the Sundanese of Padjadjaran, West Java.

The San Bushman's traditional greeting is 'I saw you from afar and I am hungry' (Van der Post, 1958). The divide between haves and have-nots has never been greater (Stiglitz, 2011; Oxfam, 2021; UNDESA Report, 2020) and is linked with climate change, deforestation, the loss of habitat, displacement of many species and loss (Shiva, 2021). Goodall (2021) argues that pandemics are the result of habitat loss and the way animals that have never been in contact are thrust together through the way they are farmed, trafficked or contained. Gorbalenya et al. (2020) also stress that cross-species infections are the root of the corona pandemic.

It is worth remembering that in the quest for land, property and accumulation of assets we have lost our sense of awe for nature and our understanding that we are dependent on it. It is appropriate that the Nobel Peace Prize was awarded to the

---

[5] The high levels of unemployment and the poverty facing families have resulted in higher levels of domestic violence and child abuse. Citing a 2017 report Ramaphosa (2020) said the cost of domestic violence to the South African economy is more than 28 billion and possibly exceeding 42 billion. In the wake of Covid-19, this is much higher and protests to end violence against women in South Africa were led in August 2018.

During the week of action to raise awareness on the need to stem violence against women for 16 days, it was stressed that the number of violent crimes against women and children during lockdown tripled. In November, the UN 16 days of activism against gender-based violence.

[6] https://www.news24.com/news24/columnists/cyrilramaphosa/cyril-ramaphosa-gender-based-violence-it-is-time-to-turn-the-tide-on-the-scourge-20201123

[7] https://www.unwomen.org/en/news/in-focus/in-focus-gender-equality-in-covid-19-response/violence-against-women-during-covid-19?gclid=CjwKCAiAv_KMBhAzEiwAs-rX1KX1oZi4NUIEhwf80u3MKmCzo1gyZZ-ceOOEpZLqW5yNF_07UljjBRoCPPMQAvD_BwE

[8] https://www.unwomen.org/en/news/in-focus/in-focus-gender-equality-in-covid-19-response/violence-against-women-during-covid-19

WFF in 2020 as food security has never been more important,[9] but the role of nominee Chief Raoni Metuktire is equally important along with other Indigenous leaders who stressed the need to 'defend nature' and respect diversity (2020). It resulted in a four-page manifesto called the Piaraçu Declaration[10] representing 45 indigenous peoples in Brazil on 14 and 18 January 2020. The essence of the declaration is that forests sustain all life. The leaders condemn the ecocide committed by government. The current context of 'existential risk' (Bostom, 2011) requires an urgent response. We argue, transformative education needs to address social, economic and environmental issues in the interest of the common good and not in the narrow interests of a specific political party, social group or market segment. The opening question posed by Gildersleeve and Kleinhesselink (2019:1) is:

> What does it mean for higher education that the human is remaking its environment and consequently, remaking itself ? What future/now for the institution that was built to generate, harbor, share, and provide leadership for the knowledge that might support the human condition and its social experiments at living?

This volume suggests alternatives based on small pilots and ways in which to scale up a relational approach guided by *a priori* norms that underpin education for the common good and *a posteriori* indicators to measure performance – in terms of social, economic and environmental indicators of wellbeing to support the common good for current and future generations. In Systemic Ethics (McIntyre-Mills, 2014:27), I stress that:

> The axiom to guide and balance new forms of democracy and governance is that we can be free and diverse at a local, national or transnational regional level to the extent that we do not undermine the collective good … or the global commons. Initially, the concept of global commons was narrowly defined as 'assets outside the national frontiers such as oceans,

[9] World Food Programme, Nobel Peace Prize 2020 – https://www.nobelprize.org/prizes/peace/2020/press-release/#:~:text=The%20Nobel%20Peace%20Prize%202020%20was%20awarded%20to%20World%20Food,weapon%20of%20war%20and%20conflict.%22. The need for international solidarity and multilateral cooperation is more conspicuous than ever : 'The Norwegian Nobel Committee has decided to award the Nobel Peace Prize for 2020 to the World Food Programme (WFP) for its efforts to combat hunger, for its contribution to bettering conditions for peace in conflict-affected areas and for acting as a driving force in efforts to prevent the use of hunger as a weapon of war and conflict….The World Food Programme is the world's largest humanitarian organisation addressing hunger and promoting food security. In 2019, the WFP provided assistance to close to 100 million people in 88 countries who are victims of acute food insecurity and hunger. In 2015, eradicating hunger was adopted as one of the UN's Sustainable Development Goals. The WFP is the UN's primary instrument for realising this goal. In recent years, the situation has taken a negative turn. In 2019, 135 million people suffered from acute hunger, the highest number in many years. Most of the increase was caused by war and armed conflict….'

[10] Agrocultures.org: 'We are not alone. At this great meeting, we declare the resumption of the Forest People's Alliance, which includes the Caatinga, Pantanal, Cerrado, Atlantic Forest and the Amazon. We will be together defending the protection of our territories. This struggle is not only for indigenous peoples, but for all of us, it is a struggle for the life of the planet.'

space and the Antarctic'. OECD definition http://stats.oecd.org/glossary/search. asp. Our area of concern is the extent to which resilience could be achieved through attitudinal and behavioural change fostered through participatory democracy and systemic governance that incorporates structured 'if then' considerations.

The chapters in this volume offer some ideas as to how education systems could support transformation. They build on Gildersleeve (2016), who reflects in her paper on the way in which academics become their work and their work responds to market demands to be productive in terms of commodified measures of output applied to research, teaching, administration and service. The speed at which work needs to be performed takes its toll on staff, students and family. To engage carefully with students and research participants takes time, as does research based on connecting with people and place.

The Audience and Focus

The audience at whom we are aiming includes policymakers, academics and practitioners who can influence the design content, structure and process of education to respond to the convergent challenges we face, namely poverty, hunger, climate change, displacement of multiple species and pandemics.

The focus is on social, economic and environmental challenges facing educators striving to address lockdowns, capacity building and the digital divide. Contributors from diverse cultures and geographical areas suggest ways forward based on praxis and small case studies. Underpinning all these contributions is the importance of developing ways to learn in more inclusive, dynamic, organic and democratic ways to support living systems. The needs of those who are not currently protected by citizenship rights are centre stage in our examples of attempts to re-frame education in ways that extend 'the frontiers of justice'(Nussbaum, 2006) to include: young people, the differently abled, the voiceless and sentient beings *within a system spanning organic and inorganic life*.

Transformative education aims to foster values and modes of regenerative living which respect and strive to connect with 'all that exists' (as named by Chilisa, 2020), namely, the 'living and the nonliving, the land, the earth, the animals, and others beings' (p. 24), which are not adequately protected through current approaches to democracy and governance (see McIntyre-Mills & Christakis, 2020 and chapter "Systemic Praxis and Education to Protect the Commons" of this volume.

Taxonomies are constructs based on values that need to be carefully re-considered in terms of the biopolitical consequences of policy decisions.

Educators and policymakers need an 'ecology of mind' (Bateson, 1972) and a more relational and responsive approach to 'natural inclusion' as suggested by Alan Rayner.

The volume is based on the premise that 'A is better off when B is better off', to cite Von Foerster's (1995) well-known maxim, as such it:

- **Advocates for** teaching systemic ethics as a form of lifelong learning within nature's classroom to support social and environmental justice. Critical systemic thinking is both an individual and a collective responsibility through many ways of knowing spanning the arts and sciences to inspire creativity with nature.
- **Contributes to** theory and practice by making suggestions as to how to re-frame the content, structure and process of education for transformation.
- **Makes a case for**

  (a) a more relational understanding of human beings and other species
  (b) a more integrated curriculum where learners across primary, secondary and tertiary levels are given the opportunity to explore many ways of knowing and to do applied learning. In many contexts, unemployment is a family issue and thus enabling intergenerational learning needs to be explored through vocational training that enables people to earn, learn and grow a future. In this volume, we also explore green co-operatives linked with a learning community.

Regenerative Praxis: Safe Habitats for Diverse Species with Diverse Capabilities and Needs

The book links the displacement of people with the displacement of plants and animals and how we need one another to ensure a sustainable way of life that works with rather than against nature. Diversity in nature is vital, and the social contract (as it is currently defined) does not go far enough to protect living systems of which we are a strand.

According to Carrington (2020):

> Pandemics such as coronavirus are the result of humanity's destruction of nature, according to leaders at the UN, WHO and WWF International, and the world has been ignoring this stark reality for decades. The illegal and unsustainable wildlife trade as well as the devastation of forests and other wild places were still the driving forces behind the increasing number of diseases leaping from wildlife to humans….

Trees absorb carbon dioxide and provide habitat for a range of species. The importance of safe habitats for multiple diverse species is a key point. Once species lose their habitats they come into contact with other species that have no immunity to their diseases.

Just as colonisation led to human epidemics such as smallpox or the great war led to the so-called Spanish Flu, we need to realise the importance of habitat protection for living systems.

Making a Case for Critical Systemic Pedagogy and Praxis

The aim of the book is to develop critical systemic thinking and practice applied to education; as such, it extends the boundaries of critical pedagogy. It advocates for transformative, mixed methods (McIntyre-Mills & Romm, 2019) using action learning and action research within and beyond formal learning contexts (McIntyre-Mills, 2018).[11]

Re-generative, transformative education means working with nature and within nature's classroom and emphasises multispecies relationships (McIntyre-Mills & Corcoran Nantes, 2021).

We need to urgently 'grasp the nettle' and accept that business as usual needs to move towards taking urgent steps re-generating and transforming our environment to enable rural urban balance and protecting the wild areas and habitats for diverse species through addressing the concerns raised by the IPCC report (2020) and the plea to reduce emissions (United Nations, COP 2021).

A Relational Approach

As facilitators we need to remember that the observer and the object of the research are just two variables in a systemic relational context that needs to include the subjects, namely learners as an active part of the engagement process (McIntyre-Mills, 2021).

The subjective, objective and intersubjective domains are equally important when considering wicked complex problems (Rittel & Webber, 1982), such as poverty, climate change, conflict and pandemics. In many ways, the relationality of the subjective, objective and intersubjective domains shape policy in positive and negative ways; in other words, we can engage in education policy in ways that create turning points for the better or worse.

Our values and decisions shape the world in which we live, and thus public education and policy debates need to be informed by ecologists, zoologists, climatologists, physicists, economists, philosophers and social scientists who understand our interconnectedness if we are to have a hope of achieving policy transformations that re-generate (Wahl, 2010) and sustain the environment.

[11] The approach to research at this difficult time relied a great deal on communication via WhatsApp, Zoom, Skype and other digital platforms as well as some face-to-face engagement in South Africa in 2020 and in Indonesia during 2019 on which I draw in my contributions and joint chapters.

Part I: Education Praxis

This book includes authors from Africa, Middle East, Asia, Australia, North and South America and is based on working together with colleagues in a range of contexts. For some of the authors, English is not their first language; thus writing partnerships were considered to be preferable to enable co-creation. The project evolved through engaging with members of our networks and sending out invitations to potential authors, so the proposal was open-ended to enable creativity and flexibility. We engaged in zoom conversations to develop some of the chapters and combined zoom engagement with on-the-ground engagement where possible, given the need to remain safe during the Covid-19 pandemic. A reflection on the arts also helped to make sense of the pandemic whether the arts involve epic ballet in Jogjakarta or Jarusalema dance groups assembling in a minibus taxi rank near Masiphumelele in Cape Town, South Africa, to express solidarity by dancing during the pandemic or storytelling, drumming and community art.

The volume addresses the following three key themes, namely: (1) education thinking and practice, (2) extending the 'frontiers of justice' and (3) learning in 360 degrees, each of which will be summarised below.

The ability to connect with people in more remote areas is reliant on strong links with colleagues at NGOs or universities who in turn reach out to people within the community through face-to-face or facilitated conversations online. Working with graduates, long-standing colleagues and their networks made this book possible.

The pandemic has changed many aspects of teaching and learning. The digital divide has made the challenge of being able to afford to buy a phone or a computer out of reach for many who had relied on face-to-face engagement and minimal technology. The perceptions of how young people feel about the situation are summed up in *The Youth Submission Report* by South Africa's African Peer Review Mechanism (APRM) civil society working group (2020/2021:99) draws on Statistics South Africa and stresses that:

> According to Statistics South Africa, only 37% of South African households have constant access to the Internet, either by phones or computers. In provinces such as the North West or Limpopo, this figure drops to 3.6% and 1.6% respectively.

Being able to afford access to data bundles in South Africa is a burden that many find beyond their budget, given the high levels of unemployment and informal employment. The same ARP report (2020/21:47) stresses that:

> "South Africa's unemployment rate reached an all-time high of 32.5% in the fourth quarter of 2020" and cites Statistics South Africa, 'Quarterly Labour Force Survey.[12]

[12] Statistics South Africa, 'Quarterly Labour Force Survey (QLFS) – Q4:2020', http://www.statssa.gov.za/publications/P0211/Presentation%20QLFS%20Q4_2020.pdf

According to this report, the Gini Coefficient for poverty in South Africa[13] in 2021 is estimated as 0.63 and remains one of the highest internationally, despite dropping from 0.65 in recent years. For those who are unfamiliar with the measure: 0 represents complete equality in society, and 1 represents complete inequality. It is likely that the pandemic has reversed many of these gains. Relying on access to computing is unrealistic in South Africa *unless* people are able to learn, earn and grow a future together through vocational training linked with formal and informal sectors that enables them to escape the digital divide and to earn a sustainable living. It is time to re-think the structure, content and process of delivering education in ways that do not favour urban centres at the expense of the rural and regional areas.

Public education is needed to address social and environmental justice to create movements to support passing an Ecocide Law (Higgins et al., 2013) to protect food, water and energy security, which has implications for international relations. With this in mind, the structuring of the book is as follows:

Chapter "Regenerative Education and Safe Habitats for Diverse Species: Caterpillar Dreaming Butterfly Being" by Janet McIntyre-Mills includes contributions on educational philosophy by Veronica McKay to make a case for a more critical and systemic curriculum. The chapter explores transformative approaches to education within the context of climate change, pandemics and the challenge of ecocide raised by First Nations and the Ecocide Movement.

Chapter "Fluid, Organic Thinking and Relationality: Implications for Education and International Relations" by Janet McIntyre-Mills provides a context for the transformative thinking and practice and an example of the need for more public education and gives details of the way contamination of the food chain can be impacted by small plastic pellets called nurdles and how these impact food security. Education for wellbeing has implications for post-national regional approaches to support social and environmental justice. The systemic circular approach is different from a linear systematic approach. Systemic ethical approaches (McIntyre-Mills, 2014, 2017) support 'wellbeing stocks' (Stiglitz et al., 2010) and approaches that work with people and the natural environment.

Chapter "Nurdles and Food Security: Education for Wellbeing and Implications for Pan-African Social and Environmental Justice" discusses the impact of plastic pollution in Fish Hoek, Cape Town, and the practical community development education and intervention that strives to introduce a sustainable and re-generative approach. The author McIntyre-Mills then reflects in a day-to-day diary on life in Fish Hoek and the impact of the legacy of apartheid on our relationships. Chapters "Fluid, Organic Thinking and Relationality: Implications for Education and

[13] https://worldpopulationreview.com/country-rankings/gini-coefficient-by-country, The review explains that "The Gini coefficient ranges from 0 (0%) to 1 (100%), with 0 representing perfect equality and 1 representing perfect inequality. …. If every resident of a nation had the same income, the Gini coefficient would be zero. If one resident earned all of the income in a nation and the rest earned zero, the Gini coefficient would be 1… …. In South Africa, the richest 10% hold 71% of the wealth, while the poorest 60% hold just 7% of the wealth. Additionally, more than half of South Africa's population, about 55.5%, live in poverty, earning less than $83 per month."

International Relations" and "Nurdles and Food Security: Education for Wellbeing and Implications for Pan-African Social and Environmental Justice" focus on values and perspectives and ways to make a difference through addressing practical concerns, namely: food, energy and water security.

Chapter "The Power of Story: Ocean View Organic Farmers Cooperative, Cape Town, South Africa: Growing Food with Heart and Soul" is a vignette on Ocean View Organic Farm in Fish Hoek, South Africa, by Stephanie Swanepoel who discusses the innovative co-operative that provided inspiration for colleagues to start a co-operative in Venda (see chapter "AA!!Venda Women and Social Enterprise: Stepwise Process to Regenerative and Sustainable Living").

Chapter "Collective Action for Regeneration of the Web of Life in the Face of Disruptive Injustice" is a case study by Francis Akena Adyanga and Norma Romm, who address the way a community in Uganda became active agents to speak out against the operations of a foreign-owned factory. They organized several focus group sessions with participants from various villages to enable participants to 'reflect on the issues at stake' and the possibility of working together towards realising their power to initiate transformative thinking and action to address the injustice.

Part II: Extending the 'Frontiers of Justice'

Chapter "Learning from Nature's Classroom: Reframing Economics, Accounting and Accountability" with contributions by Janet McIntyre-Mills, Rudolf Wirawan and Ida Widianingsih makes a case for vocational education linked with primary, secondary and tertiary institutions that educate participants to understand that destruction of the environment is not an inevitable approach to development. It is possible to earn, whilst we learn and grow a future together. The volunteer ecovillage team enables participants to develop skill sets, build capacity and potentially learn how to map and model their own carbon footprints. It is linked with the following website on ecovillage: https://www.wirasoftfoundation.org/en_GB/web/bic-sig.

The chapter advocates an alternative way of doing things by discussing steps to develop a circular economy linked with a virtual engagement site, dedicated to enabling people to create a viable future for themselves, others and living systems on which we depend. The pathways to wellbeing away from business as usual towards the regeneration of living systems and a sustainable, liveable environment for diverse species require a change in values so that social, economic and environmental indicators of wellbeing are co-created at a community level. The regenerative interventions in degraded landscapes can be supported, mapped and measured through non-linear or web-like pathways to wellbeing. The organic farming movement could be supported by setting up knowledge management structures that support the concepts of ubuntu and explaining how the concept of sharing areas with people and nature is applicable.

Chapter "Transformative Education: Employing the Balanced Scorecard for Regenerative Development" by Saad Algraini discusses the need for a balanced scorecard to guide re-generative development.[14]

It is paired with chapter "Education and Gender Mainstreaming in a Patriarchal Setting: What Really Matters" in a patriarchal Bangladesh setting, where power is gendered as discussed by Shajeda Aktar and SM Mokhlasur Rahman.

In chapter "Gender Roles in Vietnam: A Metalogue on the Traditional and the New and Suggestions for Transformation" by Huong Nguyen, Janet McIntyre-Mills and Yvonne Corcoran-Nantes, Huong engages in a metalogue with Janet and Yvonne to explore the nature of gender roles in Vietnam and the way that they have been shaped. Together they explore the potential for further opportunities for transformation that build on the leadership roles that women have shown during war and the pandemic.

Chapter "Dis-Ability, Differently Abled and Sentient Beings: Transformative Regeneration: The Need for Public Education on Vulnerability and Interbeing" by Janet McIntyre-Mills addresses the rights of the disabled, differently abled and sentient beings and makes a case for transformative regeneration by extending Butler's approach to vulnerability. A case is made for extending the frontiers of justice, based on Nussbaum's (2006) contribution, in order to improve the curriculum content of education to address the rights of sentient beings and to improve engagement with differently abled.

Chapter "Public Education on the Rights of Sentient Beings to a Life and Death Free of Suffering" by Janet McIntyre-Mills addresses education on the *rights of sentient beings to a life free of suffering*.

Chapter "An Assault on Situational Complexity in the Arena of Education: The Potential of Structured Dialogic Design" addresses the challenge for education to respond to complexity in a paper titled: *Assault on Situational Complexity in the Arena of Education: The Potential of Dialogical Design* by Alexander N. Christakis and Jeff Diedrich in which an approach to respond to complex needs is detailed.

Chapter "Rethinking Teacher Education Practicum for Transformative and Regenerative University – Community Partnerships" by Veronica McKay and Pinkie Mabunda makes a vital contribution to the volume by discussing the challenges faced by learners during the pandemic. The chapter discusses the role of the University of South Africa which aims to address the needs of students, many of

[14] Despite the divides between genders and cultures and the need to comply with a state that does not tolerate dissent from civil society or the media and is in an ongoing state of tension with its neighbour Yemen. A proxy war has raged between USA and Iran and has resulted in extreme poverty and fracturing of Yemen into tribal sub groupings. An offer of a truce has been made, but the future of Yemen is fragile according to a Brookings review (Johnsen, 2021).

whom live in 'rural areas,[15] urban townships[16] and informal settlements[17] – environments with very high poverty rates. The majority of these students require financial assistance to study and approximately half of the Unisa students are supported by the Government's National Student Financial Assistance Scheme'.

Chapter "Online Learning and the Pedagogy of Resilience, Agency and Protest: Lessons from the COVID-19 Experience" by Pinkie Mabunda explores the literature on transformative education and its implications for South Africa. The chapter explores the way practicums help with transformative education by enabling student teachers to engage in action learning that enables them to obtain classroom experience. This chapter provides critical perspectives on transformative and regenerative discourses on teacher professional standards in practice. It also offers several case studies.

Chapter "Transformative Exhibitions and Artistic Practices: Vision and Culture in the Educational Environment of the Museum" by Rania Jaber and Yvonne Corcoran-Nantes discusses the importance of engaging with art through art galleries and museums which can be used to 'invite interaction between people from different ethnic backgrounds and socio-cultural contexts'. They consider case studies from different settings and question ideas of 'citizenship and belonging as well as thinking about heritage and multiculturalism from different geopolitical angles and demonstrate how museums can play a central role for disseminating knowledge and challenging notions of ourselves and others'.

Part III: Learning in 360 Degrees

Papers on this theme make the case for learning not only in classrooms, on line or face-to-face in educational institutions, but also through artistic expression and an understanding of from nature's classroom about our 'natural inclusion' to use Alan Rayner's concept detailed in this section – he stresses that relationality is 'an inescapable inference from our human experience of the relationship between softness and hardness – receptivity and resistance and we need to move towards a more relational educational approach' which embraces multiple species.

Chapter "Sacred Groves of the Tolon District of the Northern Region, Ghana: Where Spirituality Meets Education" by David Addae and Janet McIntyre-Mills discusses the need to undertake public education so that we can mobilise learning

[15] Areas usually without piped water or sanitation, electricity and tarred road infrastructure.

[16] Separate townships were established for Africans, Indians and people of mixed race. While the urban townships are underdeveloped when compared with urban suburbs where white people lived, they nevertheless offer more services than informal settlements and would be more organised than informal settlements.

[17] Informal settlements are unplanned and unauthorised residential areas. The informal dwellings are usually shacks made from corrugated iron sheets. These communities have poor quality housing, inadequate access to safe water and sanitation and insecure residential status.

together with practical intervention. The chapter discusses the potential of indigenous customs and Islamic faith to support government legislation to protect the sustainable development agenda. It is estimated that throughout Ghana sacred groves are protected by belief in the power of Jaagbo. The case study of Tolon in the Northern Region of Ghana is discussed with reference to Douglas' notion of sacred and profane and the implications for fostering Bateson's ecological mindset through public education that builds on indigenous values and customs and mainstreams the sense of awe and respect for the systemic role that forests play in protecting the wellbeing of living systems.

Chapter "AA!!Venda Women and Social Enterprise: Stepwise Process to Regenerative and Sustainable Living" by Patricia Lethole, Mphatheleni (Mphathe), Rudolf Wirawan and Janet McIntyre-Mills discusses the need to protect sacred areas such as the Fundudzi lake and Thathe Vondo Forest in Venda, Southern Africa. The chapter demonstrates the importance of environmental protection and sustainable employment options through setting up co-operatives that could be a way forward instead of destructive forms of development. The team explore ways to develop skill sets, build capacity and map and model their carbon footprints; it also enables them to engage with others to make better community decisions that support democracy, governance and a circular economy that enables them to earn, whilst they learn and grow a future together. Thus, it helps to pilot a scalable way forward.

Chapter "Women, Education and Socialisation: Transforming Gender and Power in the MST, São Paulo, Brasil" by Yvonne Corcoran Nantes discusses how women's transformative experience in the fight for land offers them knowledge and skills that see them develop individually and collectively revealing a different sense of self they had never possessed before. This has underwritten women's newly created individuality, autonomy and leadership in the construction of a crucible of empowerment that requires no power position. This chapter focuses on women's role in the movement, its development and longevity rooted in a transformative praxis that underwrote their philosophy and action.

Chapter "Encounters and Mis-Encounters: The Process of Political Pedagogy Between Educators and Families of the Landless Movement (MST) Settlement of Pirituba II, South-Eastern Sao Paulo, Brasil (1984–2006)" by Edvaneide Barbosa da Silva and Yvonne Corcoran-Nantes discusses the political agenda of the landless movement and the battle for a transformative curriculum underpinned by the Liberation Pedagogy of Paulo Freire. The chapter offers insights into the different teaching methods initially offered to the children of the MST and the relationship with their families. Moreover, it reveals the active role of the community in the school curriculum in efforts to narrow the political pedagogical gap between educators and the rural settlers in which pathways to the development of a transformative educational practice are created.

Chapters "Combining Focus and Circumspection: An Education in Natural Inclusion" and "The γ – Principle of Natural Inclusion: From Competition Versus Cooperation to Heartfelt Relay" by Alan Rayner elaborate the concept of *natural inclusion* which needs to be re-introduced to education through enabling students to understand this vital concept which he explains using a range of media.[18] He encourages learning within nature and explains relationality as:

> This flow of receiving, sustaining and passing on is at the heart of the ecological and evolutionary philosophy of 'natural inclusion' that I have been pioneering for the last 20 years.

Chapter "Systemic Praxis and Education to Protect the Commons" by Janet McIntyre-Mills in conversation with Tom Flannagan, Dennis Finlayson and Rudolf Wirawan sums up the main themes explored in the volume, namely, the need to protect the commons through transformative education and practice that helps to reframe the way in which human beings see themselves in relation to other living systems. It also stresses that technocratic approaches to poverty and climate change – without caring for sentient beings and nature – could lead to new forms of authoritarianism and exploitation. All decision-making needs to be guided by ethical decision-making on how to protect all the interdependent users of a shared habitat. The chapter makes the case that education, design and ethical interventions that undermine wellbeing – and contribute to pathology or the undermining of ecosystems – should be sanctioned in law.

Education to Enable Relationality: Finding a Voice With and for Living Systems

Overall, the volume advocates that the need for practical intervention is growing as a result of the way in which human beings treat the planet and other species. We are no longer top predator! (see McIntyre-Mills, 2021). Food, energy[19] and water security will become increasingly problematic as democracy and governance are increasingly under threat, as demonstrated in a range of international contexts from United States to Myanmar and Hong Kong, to Philippines and South Africa.

---

[18] Alan is sensitive to the language of 'connection', 'webs', 'hubs', etc. In his words: 'As you know, I am sensitive to this, and feel that if the move is to be towards a truly relational and fluid understanding of the nature of reality that language needs to be used very carefully and for many purposes avoided altogether. I generally find that the word "relate" works better for many purposes than "connect", and that the latter is best used only to describe tangible linkage. The purpose of a spider's web is to entrap life, not to facilitate its flow. As you also know, I also have reservations about the common reference to the "spaces in between", because in reality space is continuous – it does not stop and start at figural boundaries – a fact that has still not been recognised in conventional mathematics. These are subtle but crucial points that I make in my own chapters on "natural inclusion".'

[19] In South Africa, many people employed in mining and coal-based power generation are likely to lose their jobs. This makes it imperative to create social enterprise opportunities, see for instance: https://techcentral.co.za/upheaval-coming-over-south-africas-shift-to-renewables/105035/ downloaded 18/02/2021.

As the attempt to achieve a shift away from a carbon intensive economy continues, it will become important to demonstrate alternative forms of vocational education and employment to help *prefigure* change. In this spirit, Elizabeth Wathuti,[20] a young climate change activist in Kenya, reminded the audience of the work of Nobel Laureate Wangari Maathaai (2004) who planted over 30 million trees by creating a movement through which women were empowered to understand that forests, land and water systems need to be protected which requires a systemic approach.[21]

At the United Nations (2021), *Glasgow Climate Change Conference* by a young leader from Kenya, Elizabeth Wathuti stressed:

> By the time I am 50, 86 million people will be displaced in Sub Saharan Africa alone….

Maphateleni Makaulule (see Lethole et al., chapter "AA!!Venda Women and Social Enterprise: Stepwise Process to Regenerative and Sustainable Living" this volume) advocates for the greater protection of forests and the use of alternatives to wood, such as bamboo which in itself has multiple uses. To this end, they have set up a co-operative to enable public education and vocational training. For example, bamboo is a strong and resilient building material which has multiple uses such as making furniture, clothing, acting as pipes or fencing and providing biofuel.

Re-generative interventions include using bamboo in Indonesia (chapter "Learning from Nature's Classroom: Reframing Economics, Accounting and Accountability" of this volume) to protect steep landscapes from erosion and to provide an alternative to firewood – which is being addressed, for example, by the Tshidzivhe project in Venda, South Africa – in collaboration with the bamboo co-operative. This is important as the destruction of forests is one of the root causes of both climate change and pandemics.

In the wake of the United Nations Climate Change Summit in Glasgow, policy decisions are central. What kind of a future do we want? What sort of education will help to ensure that we achieve rapid transformation to address the necessity to lower carbon emissions as outlined in the recent IPCC (020) report?

Current social and economic structures are not only unsustainable but also the cause of the convergent crises: conflict, poverty and environmental collapse. Being able to earn, whilst we learn and grow a future together has never been more important. Water, food and energy insecurity have resulted in increased urbanization and loss of habitat, increased rural-urban imbalance and the need to re-connect with nature and over-reliance on a carbon-based economy as a result of inadequate governance and a lack of democratic representation and accountability which is focused on the increasing gap between rich and poor and the increased commodification of

[20] https://mobile.twitter.com/lizwathuti/status/1455518577327542273

[21] https://www.nobelprize.org/prizes/peace/2004/press-release/

living systems[22] (see McIntyre-Mills et al., 2018, 2019; McIntyre-Mills & Corcoran-Nantes, 2021).

The volume invites learning within and beyond formal organisations by inspiring people to learn not only by reading the books and articles by others, but by creating and growing their own futures, trying to re-generate their communities by connecting with people, plants and animals and treating the land, rivers and oceans with respect – not as infinite resources or dumping grounds.

Some of the most important actions and re-actions include the stance taken by young people to sue governments for destroying their future[23] or First Nations who wish to retain their history, written in the landscape which is increasingly at risk. The sacking of three CEOs for allowing the ancient Juukan Gorge[24] in Australia to be destroyed is just one recent example.

Donna Haraway insisted that those who are marginalised need to be the designers, not the object of other people's designs.

Gregory Bateson's (1972) 'ecology of mind' emphasises the positive and negative aspects of the way we think about the world. The pathology of assuming that algorithms can *predict* based on simple binary logic that is de-linked from embodied warm human beings is the greatest challenge. His daughter Nora Bateson (2021) therefore refers to the need for *warm data* based on stories.

If we spend too much time in front of computers in cyberspace (Gergen, 1991), it will have consequences on our ability to relate to one another face to face (Greenfield, 2008, 2015); nevertheless, online engagement can be used in creative ways. The potential for online engagement was explored by presenters.

Plenary speakers at the International Society for Systems Sciences conference (8–13 July 2021) such as Nora Bateson created enthusiasm and energy when she discussed the potential for learning to see differently through developing an 'ecology of mind'[25] and through 'side-by-side' learning in different contexts.

---

[22] 'Escaping the chicken coop' is the core theme in Adiga's (2008) debut novel ' White Tiger', a story of the life of a trapped so-called 'low caste' worker who escapes poverty in India. As a child he shows talent at school and is said to be exceptional by a school inspector who likens him to the rare 'White Tiger'. Adiga's protagonist uses the analogy of chickens lined up in cages facing their own death as they are systematically slaughtered and sold as commodities. He sees himself as justified in fighting his way out of unrewarded servitude by stealing money that was being used to corrupt politicians. Ironically calls his taxi business 'White Tiger' as a commentary on racism and a recognition that he had indeed fulfilled his dream to be exceptional by slaughtering his master.

[23] Taylor, M, Holden, E Collyns, D Standaert, M. *and* Kassam, A. https://www.theguardian.com/environment/2021/may/07/the-young-people-taking-their-countries-to-court-over-climate-inaction

[24] 'On Thursday 11 June 2020 The Senate referred the following inquiry to the Joint Standing Committee on Northern Australia for inquiry and report by 30 September 2020:The destruction of 46,000 year old caves at the Juukan Gorge in the Pilbara region of Western Australia.' https://www.aph.gov.au/Parliamentary_Business/Committees/Joint/Northern_Australia/CavesatJuukanGorge Inquiry into the destruction of 46,000 year old caves at the Juukan Gorge in the Pilbara region of Western Australia.

[25] https://www.youtube.com/watch?v=jJ4Rdkm1iBA

Engaging in dialogue helps to test out ideas, so the closest we can get to truth is through dialogue (McIntyre-Mills, 2000). All nature is in communication; it is the means by which living systems relate to one another, and this is a 'living dynamic', to use Nora Bateson's term (2011). In a plenary speech (2021), Nora Bateson shared memories of learning from her father and from her grandfather's legacy and elaborated on the importance of 'warm data' and the notion that telling stories can change the way people see their own stories. This resonated with the spirit of this book as did her notion of the importance of making sense of people's experiences in context by considering stories at different levels (micro to macro) and across species. Nora Bateson calls this process anticipatory as it addresses the unseen and through engagement brings unseen issues to the surface – a process she calls 'aphanipoesis' from the Greek words 'Afani', meaning 'unseen' and 'poesis' which is engaged forms of creativity.

All forms of life relate to their context. It is this relationality of consciousness in which I am interested as an action learner and which is perhaps different from what Maturana and Varela (1980) called 'autopoiesis' a form of self-generation which can be seen across many living systems. Perhaps human beings have the ability to generate a new higher level of consciousness?

In the documentary (Nora Bateson et al., 2011) about her father's notion of 'taking steps towards an ecology of mind', Gregory Bateson discusses the five fingers on our hand and the importance of *the spaces in between* which make it possible for the fingers *to relate to one another* and to the world in which we live. The flexibility of the fingers helped our creativity and our evolution as toolmakers and then as increasingly capable of conceptualisation. His understanding of the importance of relationships in context is the basis of an 'ecology of mind' that we need to foster everyday as we live our lives.

Creative thinking on alternatives[26] and syntheses need to be encouraged.[27] Places of learning need to avoid draining the fluidity of life into dry segmented disciplines.

---

[26] As a child I wanted to know why we could not count our fingers or toes and the spaces in between? It would have been useful to discuss conventions such as Dewey's digital system as well as other conventions. If we count the spaces in between we get 9, a recurring number; it seemed to me not a bad idea because we need to understand continuums! A more wholistic approach to understanding the world and how mathematics could help to understand, map or model could have addressed my concerns.

[27] Categories are constructs and clearly the values with which I was confronted seemed wanting. Living in South Africa raised daily questions about race, culture, language and religion, sanity and insanity. Gender rights were also an obvious concern. Why were there so few boys encouraged to pursue their talents as dancers, for example? The strength of a dancer equals the strength of any sports person. As a child I could also never understand why human rights did not include animal rights and why it was possible to love pets, but to abuse other animals. This led to early struggles about the nature of food. When presented with a thesis or argument, gaps and alternative antitheses sprang to mind but could not be explored until I reached university. I chose to do research on schizophrenia as the opportunity arose through an invitation extended by a psychiatrist at Valkenberg Hospital for mentally ill to the head of the sociology department at the University of Cape Town. It resulted in my following networks out of the hospital to learn how Xhosa speaking patients sought help from both indigenous healers and biomedically trained doctors and nurses. The Xhosa speaking staff were mostly nurses or social workers in the apartheid era with the exception of a psychiatrist who had learned to speak Xhosa as a child. At his invitation, I undertook research on how 'the calling by the ancestors' was interpreted within a context of limited resources and the need to access food.

This is why education policy needs to promote critical, systemic and applied learning within an integrated curriculum to support the common good (see McIntyre-Mills, 2010; McIntyre-Mills et al., 2019a, b).

The volume builds on and extends the public education implications of the lack of representation, accountability, regeneration and sustainability leading to the big challenges of the day (poverty, displacement, loss of habitat, cross-species infection (Goodall, 2020; Gorbalenya et al., 2020), conflict and pandemics), which could be addressed through re-framing educational structures, processes and content. The volume makes the case for lessons to be learned from nature's classroom and applied to protecting living systems. The capability to show compassion is vital for our shared survival and is the need to re-structure our education systems.

College of Education
University of South Africa
Pretoria, South Africa

J. J. McIntyre-Mills

References

Abram, D. Foreword in Kaza, S. (2020). *A wild love for the world: Joanna Macy and the work of our time*. Shambhala (pp. 1x–xv11).

Accra Agreement. (2008). https://www.afdb.org/fileadmin/uploads/afdb/Documents/AccraAgendaAaction-4sept2008-FINAL-ENG_16h00.pdf

Adiga, A. (2008). *The white tiger: A novel*. Free Press.

APRM. (2021). *Youth submission to the South Africa's African Peer Review Mechanism (APRM) in South Africa in 2020/2021.*

Bateson, G. (1972). *Steps to an ecology of mind*. Ballantine.

Bateson, N. (2021). *Plenary International Systems Sciences*. https://norabateson.wordpress.com/2017/05/28/warm-data/

Bostrom, N. (2011). *Existential risk prevention as the most important task for humanity*. Faculty of Philosophy & Oxford Martin School University of Oxford. http://www.existential-risk.org/

Coll, J. M. (2022). *Buddhist and Taoist systems thinking: The natural path to sustainable transformation*. Routledge.

Dewey, J. (1916/2001). *Democracy and education. An introduction to the philosophy of education*. The Pennsylvania State University.

Flood, R. L., & Romm, N. R. A. (2018a). A systemic approach to processes of power in learning organizations: Part I – literature, theory, and methodology of triple loop learning. *The Learning Organization, 25*(4), 260–272, https://doi.org/10.1108/TLO-10-2017-0101. Permanent link to this document: https://doi.org/10.1108/TLO-10-2017-0101

Flood, R. L., & Romm, N. R. A. (2018b). Systemic approach to processes of power in learning organizations: Part II – Triple loop learning and a facilitative intervention in the "500 Schools Project". *The Learning Organization, 25*(5) 344–352.

Forsetto, R. (2020, February 5). *Chief Raoni, Nobel Peace Prize nominee, presides over historic meeting with over 600 indigenous leaders in Brazil.*

Freire, P. (1970). *Pedagogy of the oppressed*. Penguin Books.

Foucault, M. (2008). The birth of biopolitics: Lectures at the College de France 1978–79. Palgrave Macmillan.

Foucault, M., & Gordon, C. (Eds.). (1980). Power/knowledge. Brighton.

Gergen, K. (1991). *The saturated self: Dilemmas of identity in contemporary life*. Basic Books.

Gildersleeve, R. E. (2016). The Neoliberal Academy of the Anthropocene and the Retaliation of the Lazy Academic. *Cultural Studies—Critical Methodologies, 17*(3), 286–293.

Gildersleeve, R. E., & Kleinhesselink, K. (2019, April). *Philosophy and theory in higher education Special Issue on the Anthropocene in the Study of Higher Education Guest Editors, University of Denver* (Volume 1 Issue 1, pp. 1–16). Peter Lang.

Greenfield, S. (2008). *ID: The quest for meaning in the 21st century*. Sceptre, Hodder and Stoughton.

Greenfield, S. (2015). *Mind change*. Random House.

Haraway, D. (1991). *Cyborgs, simians, and women: The reinvention of nature*. Free Association Books.

Haraway, D. (2011). Speculative fabulations for technoculture's generations: Taking care of unexpected country. *Australian Humanities Review, 50*.

IPCC. (2021). Summary for policymakers. In V. P. Masson Delmotte, A. Zhai & S. L. Pirani (Eds.), *Climate Change 2021: The physical science basis. Contribution of Working Group I to the Sixth Assessment Report of the Intergovernmental Panel on Climate Change*.

Kahane, A. (1992). The Mont Fleur scenarios: What will South Africa be like in the year 2002? *Deeper News*, Vol. 7 Number 1 republication of The Weekly Mail and Guardian Weekly.

Meadows, D., & Randers, J. (1992). *Beyond the limits: Global collapse of a sustainable future*. Earthscan Publications.

McIntyre-Mills, J. (2000). *Global citizenship and social movements: Creating transcultural webs of meaning for the new millennium*. Harwood/Routledge.

McIntyre-Mills, J. (2018). Policy design for non-anthropocentric pathways to protect biodiversity and regenerate the land. *International Education Journal in South Africa*. https://journals.co.za/content/journal/10520/EJC-117cea7615

McIntyre-Mills, J. J. (2021a). Communication and culture: A multispecies endeavor within a shared habitat. Year book for International Society for Systems Sciences, *Systems Research and Behavioral Science*, 1–14. https://doi.org/10.1002/sres.2810

McIntyre-Mills, J. J. (2021b). The importance of relationality: A note on co-determinism, multispecies relationships and implications for COVID-19. *Systems Research and Behavioral Science*, 1–15. https://doi.org/10.1002/sres.2817

McIntyre-Mills, J. (2021c). Chapter 3: From old to new taxonomies of rights, relationships and responsibilities to protect habitat. In *From polarisation to multispecies relationships in the age of mass extinctions* (pp. 31–54). Springer.

McIntyre-Mills, J. (2021d). Chapter 5: McIntyre-Mills, J: Consciousness for balancing individualism and collectivism. In J. J. McIntyre-Mills, & Y. Corcoran Nantes (Eds.), *From polarisation to multispecies relationships in the age of mass extinctions* (pp. 67–86). Springer.

McIntyre-Mills, J. (2021e). Chapter 8: McIntyre-Mills, J. 'Stewardship: An anthropocentric misnomer'. In *From polarisation to multispecies relationships in the age of mass extinctions* (pp. 119–140). Springer.

McIntyre-Mills, J. (2021f). Chapter 9: Social engagement to redress the banality of evil and the frontiers of justice: Limitations of the social contract to protect habitat and why an international law to prevent the crime of ecocide matters. In *From polarisation to multispecies relationships in the age of mass extinctions* (pp. 141–172). Springer.

McIntyre-Mills, J. (2021g). Chapter 11: Vignette: Why thinking matters: Constructivism, relationships and the performative universe. In *From polarisation to multispecies relationships in the age of mass extinctions* (pp. 213–222). Springer.

McIntyre-Mills, J. (2021h). Chapter 26: Biopolitics and food security to protect social and environmental justice. In *From polarisation to multispecies relationships* (pp. 575–602). Springer.

McIntyre-Mills, J. J., Romm, N. R. A., & Corcoran-Nantes, Y. (2018). *Balancing individualism and collectivism: Social and environmental justice*. Springer.

McIntyre-Mills, J. J., & Romm, N. R. A. (Eds.). (2019a). *Mixed methods and cross-disciplinary research: Towards cultivating ecosystemic living*. Springer. https://doi.org/10.1007/978-3-319-58014-2

McIntyre-Mills, J., Romm, N. R. A., & Corcoran Nantes, Y. (Eds.), (2019b). *Democracy and governance for resourcing the commons: Theory and practice on rural-urban balance.* Springer.

McIntyre-Mills, J., & Christakis, A. (2021). Chapter 14: Social and environmental justice: The legacy of structured democratic dialogue and the potential of pathways to wellbeing. In *From polarisation to multispecies relationships in the age of mass extinctions* (pp. 283–308). Springer.

Nora Bateson Plenary, July 12th International Systems Sciences https://zoom.us/rec/play/_oe07ngyK1jHXqZxIaT5nBr_c95ReZ2KWOwXCrYw7u1K6rKfJRUPtUH_urE-VNtlI9m1WDZbqZFxJ8k7i.po_IqnXsYX9swRJr?continueMode=true&_x_zm_rtaid=u3XXrLmPQCaoiZM3sR9LHA.1626085969999.920d3aa8d2c700ad3a46aaaca293852c&_x_zm_rhtaid=301

Odora-Hoppers, C. (2013), Community engagement, globalization, and restorative action: Approaching systems and research in the universities. *Journal of Adult and Continuing Education, 19*(2) Autumn 2013:94–102.

Oxfam. (2021). *5 shocking facts about extreme global inequality and how to even it up*. https://www.oxfam.org/en/5-shocking-facts-about-extreme-global-inequality-and-how-even-it

Peters, M. A., & Jandrić, P. (2017). Dewey's *Democracy and Education* in the age of digital reason: The global, ecological and digital turns. *Open Review of Educational Research, 4*(1), 205–218. https://doi.org/10.1080/23265507.2017.1395290

Paris Declaration. (2005). https://www.oecd.org/dac/effectiveness/parisdeclarationandaccraagendaforaction.htm

Stiglitz, J. J. (2011). Of the 1%, by the 1%, for the 1%. https://www.vanityfair.com/news/2011/05/top-one-percent-201105#

Stiglitz, J., Sen, A., & Fitoussi, J. P. (2010). *Mis-measuring our lives: Why the GDP doesn't add up.* The New Press.

Shiva, V. (2021). *The world after Covid with environmental activist Vandana Shiva* https://www.youtube.com/watch?v=kLO5rqEGRWw

Tretter, F. (2021). Bertalanffy's 'Organismic Biology' and its relevance for today's COVID-systems pathology, *International systems sciences plenary.*

United Nations Report. (2020). *Department of economic and social affairs social inclusion (UNDESA) world social report.* https://www.un.org/development/desa/dspd/world-social-report/2020-2.html

United Nations. (2021). *Glasgow climate change conference.* https://unfccc.int/process-and-meetings/conferences/glasgow-climate-change-conference

Wahl, C. (2019). Sustainability is not enough in. In G. D. Dabelko, & K. Conca (Eds.), *Green Planet Blues: Critical Perspectives on global environmental politics* (edited by Taylor & Francis Group). ProQuest Ebook Central, http://ebookcentral.proquest.com/lib/flinders/detail.action?docID=5892383.

Wahl, C. (2016) *Designing regenerative cultures*. Triarchy Press. https://upjournals.co.za/index.php/eac/article/view/6649/4329

Acknowledgements

The diagrams and artwork in this volume are conceptualised by the contributors or appear with the permission of the artist as in the case of chapters "An Assault on Situational Complexity in the Arena of Education: The Potential of Structured Dialogic Design" and "Rethinking Teacher Education Practicum for Transformative and Regenerative University – Community Partnerships".

Chapter "Systemic Praxis to Protect the Commons: The Potential of a Community of Practice to Inspire Agency and Hope?" by Janet McIntyre-Mills in conversation with Tom Flanagan, Dennis Finlayson and Rudolf Wirawan is adapted from an earlier version of an online conference paper delivered to the International Systems, July 2021, titled 'The art of the impossible'. A version appears in an online publication with a commons licence and appears as 'Architectures to protect the commons: a community of practice to inspire agency and hope'.

Contents

Editors and Contributors

About the Editors

Janet J. McIntyre-Mills (DLitt et Phil, Sociology) is Adjunct Professor at the University of South Africa in the College of Education[28] and Visiting Research Fellow at the Yunus Social Business Centre within the University of Adelaide Business School since December 2019. She was previously Associate Professor at Flinders University and currently holds affiliations with universities in Indonesia, such as Universitas Padjadjaran where she is affiliated with the Centre for Research and Participatory Development Research. She was nominated 'Sociologist of the Month' in August 2019 by the *Current Sociology Journal* in recognition of her paper: 'Recognising our hybridity and Connectedness'. In 2020, her research was successfully ranked by the National Research Foundation in South Africa.

Her research focuses on systemic representation, accountability and regeneration applied to social and environmental justice concerns and includes several articles in accredited journals as well as edited and sole-authored volumes such as *From Polarisation to Multispecies Relationships*, Springer (2021); *Planetary Passport: Re-presentation, Accountability and Regeneration* (2017); and *Systemic Ethics and Non-anthropocentric Stewardship* (2014), Springer, New York. She has supervised 34 PhD students to successful completion and continues to mentor and facilitates research with colleagues, graduates and higher degree students.

Yvonne Corcoran-Nantes (PhD Sociology and Latin American Studies) is an International Gender Consultant, Adjunct Associate Professor and Principal Research Fellow in Social Sciences at Flinders University, Associate, Gender Consortium at Flinders University and affiliated with the Centre for Research and Participatory Development, Universitas Padjadjaran, Indonesia. She has taught and researched in the field of women's studies and international relations/development

[28] https://www.unisa.ac.za/sites/corporate/default/Research-&-Innovation/News-&-events/Articles/Resilience-and-passion-result-in-quality-research

and gender mainstreaming education. She engages with some of the major issues facing women globally. Her specialism is gender-specific research in non-western countries in the field of development and international politics, dealing with gender issues such as gender-based violence, human rights and human security, sustainable development, popular social movements, terrorism and conflict. She has worked, and continues to do so, as an international gender consultant to most major INGOs such as UNESCAP, UN Women, UNDP, ADB and UNESCO in Central, East and Southeast Asia over a period of 25 years. She undertook the first national surveys on the status of women in post-communist society in Central Asia and latterly Mongolia. The former led to the writing and publication of a seminal work on women in Central Asia entitled *Lost Voices: Central Asian Women Confronting Transition*, which was published by Zed Books in 2006 and is in its 12th edition. She is a founding member of the UNESCO Asia Pacific Women's and Gender Studies Network which was established in 2003. She has undertaken fieldwork research and consultancies in countries such as Brazil, Uzbekistan, Kazakhstan, Kyrgyzstan, Mongolia, Indonesia, Timor Leste, Viet Nam and Mauritius. She continues to supervise PhD students and engages in collaborative research and publishing.

Contributors

David Addae is a lecturer in the Department of Adult Education and Human Resource Studies, University of Ghana, and an adjunct research fellow at the Department of Adult Basic Education at the University of South Africa. His research interests include adult learning, indigenous knowledge systems and transformative learning. He pursues these interests through an interdisciplinary approach with an interpretivist philosophical len.

Francis Akena Adyanga (PhD) is a senior lecturer at Kabale University in Uganda. He is also the dean of the Faculty of Education. Francis obtained his PhD from the University of Toronto. He is a passionate educator with keen teaching and research interest in indigenous knowledge, social and environmental justice education, and equity and diversity in education. Francis is currently (2020/2021) collaborating with Prof. Norma Romm (UNISA, South Africa) in a joint research project focussed on *Indigenous People's Social Movement Protests* in northern Uganda.

Shajeda Aktar (PhD, Public Policy) is currently a professor in the Department of Public Administration at the University of Rajshahi, Bangladesh, and focuses on grassroots politics in Bangladesh. She is currently involved in a number of research projects on women's agency, capacity building, maternal health, food security and gender power relations in Bangladesh.

Saad Algraini is a PhD graduate of the School of Social and Policy Studies, Flinders University, South Australia. His research examined public policy for human development in Saudi Arabia, qualitative research and strategic planning. Saad's recent publication is a book titled 'Qualitative Research: Strategies and Analysis'.

Edvaneide Barbosa da Silva was born in Sergipe, Brazil, but now resides in Sao Paulo and has 25 years of experience in the field of education, teaching and curriculum development. She is a historian and entered public service in 1980 when she graduated as a teacher. She became part of the education group in the Municipal Secretariat of Sao Paulo in 2001 and was the coordinator of the schools' educational network for the area until 2016. She received a Master's degree in Education in 2001 and went on to complete her PhD in history in 2008. Her Masters dissertation was turned into a book and published in 2004 entitled *Educação e Reforma Agrária – Práticas Educativas de Assentados do Sudoeste Paulista, Xamã: São Paulo*. She was also the co-director of the documentary *A Viola Te Esperar*. At present, she is a member of the Political-Education Working Group (GEPEA), Universidade de Sao Paulo and the working group of Feminist Agro-ecology, which is part of the Sempreviva Feminist organisation-SOF.

Alexander N. Christakis, founder of Institute for 21st Century Agoras, Heraklion, Crete, is the originator of Structured Dialogue. In 1968, he was cofounder of the Club of Rome. Since the 1970s, Christakis has served on the faculties of several universities including Yale University and University of Athens, and spent five years at George Mason University as the Director of the Center for Interactive Management. In 1989, Christakis founded CWA Ltd. as CEO to apply interactive management principles and has authored several sole and joint papers and books.

Yvonne Corcoran-Nantes is Associate Professor (Adjunct) and Principal Research Fellow in the Social Sciences at Flinders University. She is also an International Gender Consultant working with major INGOS and engages with some of the major issues facing women globally.

Jeff Dietrich is a member of the Institute for 21st Century Agoras and as the creator of the Logosofia platform for the application of the science of dialogic design.

D. Finlayson is a member of the International Systems Sciences with a long track record in international development

Tom Flanagan is an academic, public intellectual and member of Institute for 21st Century Agoras.

Rania Jaber works across anthropology and art history methods to research and write about contemporary art, translation and the movement of people, cultures, objects, words and ideas. She holds a PhD in Art History and World Art Studies from the University of East Anglia in the UK and has lectured in anthropology at the American University of Beirut in Lebanon. She is interested in the history of migration, multicultural identities, cross-cultural dialogue, the persistence of memory and reciprocal exchanges.

Patricia Vhahangwele Lethole is Higher Education Manager for distance learning and is currently engaged in a research project involving setting up a bamboo co-operative in Venda at the University of South Africa. She is also undertaking research towards her doctoral degree.

Pinkie Mabunda is Associate Professor at the University of South Africa; she is currently Deputy Executive Dean of the College of Education at the University of South Africa. As a Commonwealth scholar, she obtained a PhD from the University of Nottingham. Her research interests and work are in, inter alia, teaching, curricula studies, teaching practice and qualitative research.

Mphatheleni (Mphathe) Makaulule is Indigenous knowledge project coordinator of environmental heritage with communities around Vhembe District/Venda.

Janet J. McIntyre-Mills is Adjunct Professor at the University of South Africa; Visiting Research Fellow, University of Adelaide; and Adjunct Associate Professor at Flinders University. She has published widely in the areas of social and environmental justice.

Veronica I. McKay is Professor and Acting Vice Principal, Teaching and Learning, University of South Africa, where she is responsible for the implementation of programs for initial and continuous teacher development, from pre-school to post-schooling, including adult and community education, and has recently hosted a commonwealth youth project on Paulo Freire.[29]

Huong Nguyen is a PhD candidate in Women's Studies at Flinders University and a lawyer and advocate for women's rights with more than 17 years of experience serving NGOs and the public sector internationally and in Vietnam.

S. M. Mokhlasur Rahman is Associate Professor at the Department of Political Science, University of Rajshahi, Bangladesh. Before coming to the academic arena, he served in various government organizations in Bangladesh. His research interest includes child rights, child protection, political party system, ethnic rights, human rights, poverty, food security, health, nutrition, capacity building, governance and institutions.

[29] https://thecommonwealth.org/media/news/commonwealth-hosts-second-collaborative-virtual-youth-work-caf%C3%A9-celebrating-100-years

Alan Rayner is an evolutionary ecologist, writer and artist, as well as founder of natural inclusion network. He was a faculty member in biology at the University of Bath from 1985 to 2011 and has published widely in the field of biology. He was president of the British Mycological Society in 1998. He became a Fellow of the Royal Society of Arts in 2010.

Norma R. A. Romm is Research Professor in the Department of Adult, Community and Continuing Education, the University of South Africa. She has sole-authored four books, co-authored three books, co-edited five books and published over 100 research articles.

Reynaldo Trevino-Cisneros is an artist, educator, prospectivist, strategic planning and public policies evaluation expert and systems researcher who has worked actively with systems leaders such as Alexander Christakis, Hasan Ozbekhan and John Warfield on issues pertaining to democracy and sustainability. He is the co-author with Bethania Arango-Hisijara of the first monograph of a Series by the Institute for 21st Century Agoras, *Strategic Articulation of Actions To Cope with the Huge Challenges of Our World Today: A Platform for Reflection (2013).* He is also a co-author of *Guanajuato, Siglo XXI, a Regional Development Study* (ITESM, 1995).

Ida Widianingsih is Chairperson of Center for Decentralization and Participatory Development Research, Faculty of Social and Political Sciences, Universitas Padjadjaran, Bandung and VC of research.

Rudolf Wirawan is CEO of Wirasoft and higher degree researcher at Adelaide University.

List of Boxes

AA!!Venda Women and Social Enterprise: Stepwise Process to Regenerative and Sustainable Living

Systemic Praxis and Education to Protect the Commons

List of Figures

An Assault on Situational Complexity in the Arena of Education: The Potential of Structured Dialogic Design

Online Learning and the Pedagogy of Resilience, Agency and Protest: Lessons from the COVID-19 Experience

Systemic Praxis and Education to Protect the Commons

Nurdles and Food Security: Education for Wellbeing and Implications for Pan-African Social and Environmental Justice

AA!!Venda Women and Social Enterprise: Stepwise Process to Regenerative and Sustainable Living

Combining Focus and Circumspection: An Education in Natural Inclusion

List of Tables

Part I
Education Thinking and Practice

Regenerative Education and Safe Habitats for Diverse Species: Caterpillar Dreaming Butterfly Being

Janet J. McIntyre-Mills

Abstract Transformative education and practice are underpinned by the following axiom: human beings can be free and diverse to the extent that we do not undermine living systems in this generation and the next. This is in turn based on a recognition of our diversity and interconnectedness and that we are one of many species within a living system. The approach by the authors of the various chapters in this book is based on expanding pragmatism and rethinking the social, economic and environmental consequences of decisions using mixed methods matched to a specific area of concern through engaging with the relevant stakeholders. Social, economic and environmental challenges that are convergent can be perceived differently by diverse stakeholders, thus engaging with diverse groups is a vital for transformation. The focus of this volume is on values, perspectives and ways to make a difference through addressing a range of practical concerns, such as: food, energy and water security. The models of sustainable development across the volume emphasise education for wellbeing in line with the UNESCO approach outlined in 2021 that emphasises the systemic nature of education rooted in protecting the environment and supported by the participation of active global citizens.

The dreaming story of the caterpillar (mpantwe) and the butterfly (yeperenye) in Alice Springs is detailed in McIntyre-Mills (2003) as explained to me by Olive Veverbrandts as a metaphor for transformation.

Acknowledgment. I would like to acknowledge and thank Professor Veronica McKay for her contributions to this volume and this chapter based on emailed conversations and the box insert. Veronica is Acting Vice Principal: Teaching and Learning, University of South Africa.

J. J. McIntyre-Mills (✉)
College of Education, University of South Africa, Pretoria Central, Gauteng, South Africa

College of Professions, Adelaide University, Adelaide, Australia
e-mail: janet.mcintyre@adelaide.edu.au

J. J. McIntyre-Mills, Y. Corcoran-Nantes (eds.), *Transformative Education for Regeneration and Wellbeing*, Contemporary Systems Thinking,
https://doi.org/10.1007/978-981-19-3258-8_1

Keywords Regenerative education · Habitat · Bio diversity · Social and environmental justice

1 Introduction

The well-known transformation metaphor of the caterpillar and the butterfly has specific relevance for education (Banathy, 1996, 2000) since designers and practitioners need to enable and empower learners to find their wings.

In Alice Springs in the Northern Territory the dreaming story of the caterpillar (mpantwe) and the butterfly (yeperenye) has special significance as Alice Springs is named Mpantwe and the dreaming story is recorded in the landscape of the McDonald Ranges (shaped like caterpillars). The sacred site where ghost gums provide the breeding site for caterpillars is opposite The Gap between the McDonald Ranges. It is here that they create their cocoons from which the yeperenye emerge. I was privileged to spend time at this site to see the caterpillars early one morning. Olive Veverbrants, an Arrernte Indigenous leader mentored me whilst I was undertaking research. Protecting the sacred sites as a habitat for the caterpillars was highlighted as an issue, because the road along the Golf Course Estate cut through the sacred site and was known as 'Broken Promise Drive', because consultation with custodians underlined the sacredness of the site as they see the landscape and the entire ecosystem as a living being that needs to be carefully tended to ensure its regeneration. It is fitting that the Ecocide Movement[1] (building on the work of Higgins et al., 2013) and supported by young people's *Fridays for Change Movement* stresses the need to protect multispecies relationships and habitats.

Why is this important as an opening paragraph or story? We need to place the experience of 'existential risks' (Bostrom, 2011) associated with anthropocentrism and the protection of environmental habitat for diverse species at the heart of a critical and systemic curriculum.

For John Dewey (2018):

> 'experience' was the linch-pin to a broader theory of nature and humanity's axiological place within it. More than just another node in a system, experience also amounted to a *metaphilosophical method*, a way of doing philosophy.[2]

Public Education, policies and laws that support the protection and regeneration of the commons are vital for preventing ecocide as detailed by Higgins et al. (2013) when making a plea to extend and scale up the Ecocide Law so that it is post national in its reach with legal measures to ensure that the inhabitants of a specific region are protected. The phrasing of the law by an expert panel implies that all the species

---

[1] https://www.stopecocide.earth/press-releases-summary/top-international-lawyers-unveil-definition-of-ecocide

[2] https://plato.stanford.edu/entries/dewey/#SentMindCons

within a region (2021)[3] and requires an amendment to the Rome Statute of the International Criminal Court (ICC).

> The expert panel defines ecocide as: "unlawful or wanton acts committed with knowledge that there is a substantial likelihood of severe and either widespread or long-term damage to the environment being caused by those acts".[4]

Burke and Celermajer (2021) stress that so-called human progress has been at the expense of other species and underline that the current system is dysfunctional, because what is required is a fundamental shift in our relationships with others and nature. They make the point that to prevent:

> An international crime of ecocide …. should recognise that all forms of life, and the ecological systems that support them, have value for their own sake. This perspective is known as multispecies justice. It holds that human well-being is bound to flourishing ecosystems, which have an intrinsic value outside the human use for them.[5]

They emphasise that excuses such as building infrastructure for housing or roads should not be used as an excuse for environmental destruction.

This chapter expands on the themes mentioned in the summary and reflects on what Transformative Education entails from the perspective of the author of this chapter and contributors to the volume. It is in line with UNESCO's (2021) emphasis on 'sustainable development, global citizenship, health and well-being'[6] and details the ontological, epistemological and axiological approach of Transformative Education.

The authors in this volume draw on their experiences as academics, educators, participatory action researchers and administrators to reflect on the need for education to become more systemic and practical, in order to respond to the urgent needs highlighted by young people who will need to live with the results of current policy decisions. Unless climate warming is kept as close as possible to 1.5 degrees Celsius (Glasgow Summit, 2021), the world will face convergent social, economic and environmental risks which when compounded will result in a very grim scenario, namely "existential risk" (Bostrom, 2011). The urgency of the situation is underlined by Nobel Peace Prize Winner Muhammad Yunus (2020), who argues we are on a suicidal path and Covid-19 has offered humanity an opportunity to pause and to take stock. Transformation of our thinking and practice is overdue. Will Steffen (29 Jan

---

[3] Independent Expert Panel for the Legal Definition of Ecocide (2021) June COMMENTARY AND CORE TEXT June 2021. https://static1.squarespace.com/static/5ca2608ab914493c64ef1f6d/t/60d1e6e604fae2201d03407f/1624368879048/SE+Foundation+Commentary+and+core+text+rev+6.pdf

[4] https://www.weforum.org/agenda/2021/07/ecocide-environmental-harm-international-crime/

[5] Burke, A and Celermajer, D. (2021) Human progress is no excuse to destroy nature. A push to make 'ecocide' a global crime must recognise this fundamental truth. https://thefifthestate.com.au/columns/spinifex/human-progress-is-no-excuse-to-destroy-nature-a-push-to-make-ecocide-a-global-crime-must-recognise-this-fundamental-truth/

[6] sustainable development, global citizenship, health and well-being

2021, ABC news)[7] stressed that Australia is warming faster than anticipated and we need to cut emissions by 50% by 2030 to avert catastrophic climate change.

Transformation of our thinking and practice is overdue and calls for changing the narrative through our standing together and redesigning systems of education to prioritize a more holistic worldview that embraces the planet and all living things.

Ontologically our perspective is shaped by recognising kinship with nature, as expressed by Indigenous custodians. Epistemologically the editors and contributors to this volume explore ways to enhance education based on working across cultures and disciplines using a cross cultural approach and mixed methodology. Axiologically we support the notion of transformative research that promotes balancing non-anthropocentrism with an approach that draws on Indigenous wisdom whilst addressing patriarchal notions through gender mainstreaming.

Defending ourselves and nature (Chief Raoni, Nobel-peace prize nominee, 2020)[8] requires protecting the forests as these are the lungs of the world and vital for mitigating climate change. In a video of his meeting to develop a manifesto he speaks of defending a way of life,[9] cautioning that that deforestation is accelerating. He stresses that people and nature are part of one system and that people of all nations and cultures should stand together to protect the forests from logging and climate change. Similarly, Will Steffen (29 Jan 2021, ABC news)[10] warns that Australia is warming faster than anticipated and needs to cut emissions by 50% by 2030 to avert catastrophic climate change.

Chief Raoni's (2020) plea for people to stand together to prevent the destruction of the Amazon echoes similar threats to the forests in Asia and Africa as is detailed in "Sacred Groves of the Tolon District of the Northern Region, Ghana: Where Spirituality Meets Education for Sustainable Development" and "Aa!!Venda Women and Social Enterprise: Stepwise Process to Regenerative and Sustainable Living" which focus on the Tolon forest in Ghana and the sacred forests in Venda, South Africa, respectively.

[7] https://www.abc.net.au/radionational/programs/breakfast/much-deeper-emissions-cuts-needed/13097482. Much deeper emission cuts needed to keep global warming below 2 degrees. According to Steffen (2021) Lead author of the Paris Goals report by the Climate Targets Panel explains that: "An independent panel of senior climate scientists has reworked Australia's emission trajectories." He concludes "carbon output will need to be cut by 50% by 2030 if we are to play our part in keeping temperature rises below 2 degrees by the end of the century." Professor Will Steffen.

[8] https://news.mongabay.com/2020/02/chief-raoni-nobel-peace-prize-nominee-presides-historic-meeting-with-over-600-indigenous-leaders-in-brazil/

[9] https://www.youtube.com/watch?v=rgW2zq8_xu0

[10] Source: https://www.abc.net.au/radionational/programs/breakfast/much-deeper-emissions-cutsneeded/13097482 cites the work of Professor Steffen as follows: "much deeper emission cuts needed to keep global warming below 2 degrees. According to Steffen (2021) Lead author of the Paris Goals report by the Climate Targets Panel explains that: "An independent panel of senior climate scientists has reworked Australia's emission trajectories." He concludes "carbon output will need to be cut by 50% by 2030 if we are to play our part in keeping temperature rises below 2 degrees by the end of the century."

Indigenous custodians, such as the Makhadzi in Venda have a relational approach to the world which plays an essential role in protecting the environment. The relational worldview offers a harmonious ecological way of being, a way of connecting humans, animals and the natural environment and embracing the planet and its species (see "Aa!!Venda Women and Social Enterprise: Stepwise Process to Regenerative and Sustainable Living" this volume). Building on this relational approach the mission and vision for transformation resonates the ethic of Wangari Maathai (Nobel Peace Prize Winner, 2004 for her work on re-forestation), Vandana Shiva (Sydney Peace Prize Winner 2010 for her work on seed security and the rights of small farmers who resist the control of large corporations[11]), Gibson et al. (2015) for their *Manifesto for living in the Anthropocene*, Debra Bird Rose for her contribution to understanding the impact of exploitation on living Systems in her work 'Love in a time of extinction' (2010), and her many ethnographies drawing on the wisdom of her Indigenous mentors in order to change the narrative of power. Bart Williams (2020) makes the case in her Tedex talk that Africa needs to ensure that nature is not stripped for the benefit of some nations at the expense of others. Chilisa (2017: 814) sums up the issue that educators should take a more active role:

> ….to contest the role that colonisation, imperialism, and its new form of globalisation continue to play in suppressing, marginalising, and silencing knowledge systems of formerly colonised, historically marginalised, and oppressed groups.

Chilisa (2017: 814) in fact appeals to academia to challenge policy makers and corporations who benefit from such exploitation. As she argues:

> ….academia as a whole, need to contest the role that colonisation, imperialism, and its new form of globalisation continue to play in suppressing, marginalising, and silencing knowledge systems of formerly colonised, historically marginalised, and oppressed groups.

According to Greta Thunberg (2021):

> Many are asking what it'll take for people in power to wake up. But let's be clear – they're already awake. They know exactly what they're doing. They know exactly what priceless values they're sacrificing to maintain business as usual.[12]

2 Regeneration and Transformation

Regeneration, for the purposes of this volume refers to a non anthropocentric systemic approach of working with nature in ways that do not merely sustain, but help to re-generate current and future generations of living systems. *Transformative education* and *practice* are underpinned by the following axiom: human beings can be free and diverse to the extent that we do not undermine living systems in this

[11] https://sydneypeacefoundation.org.au/peace-prize-recipients/2010-dr-vandana-shiva/#:~:text=Dr%20Vandana%20Shiva%20was%20the,%2C%20social%20activist%2C%20and%20feminist

[12] https://twitter.com/GretaThunberg/status/1456731487928205321 6 Nov 2021

generation and the next. This is in turn based on a recognition of our diversity and interconnectedness and that we are one of many interdependent living systems.

According to Rees (2019: 240)[13] who coined the concept of a global footprint, sustainability is the process for ensuring that we do not overstep the boundaries and it thus goes hand in hand with resilience, but as Wahl (2016) explains this requires a new emphasis on re-generative culture so that we design a better way to protect living systems. Regeneration means living in ways that support nature, rather than depleting it.

Daniel Wahl (2016) draws on his book on regenerative systems, whilst Alan Rayner in conversation with Wahl[14] stresses that he had learned by observing nature and thinking about his own experiences which helped him to develop his theory of inclusionality about the way living systems relate to one another's mutual needs. Dauvergne (2019) stresses the problems of consumption and that addressing consumption:

> requires going beyond the assumptions, pathways, and measurements of an ecological modernization approach. It will require a questioning of our economic order.

Re-generative narratives draw on spirituality and First Nations. David Abram explains in his foreword to an edited edition (Kaza, 2020) on the work of Joanna Macy (2015),[15] that the notion of 'interbeing' can be understood very simply and directly when we comprehend that the air we breathe in, is the air plants breath out. We could add to this, that we are what we eat and what we consider constitutes food. By understanding the way *in which living systems relate dynamically* is the first step towards understanding mutual causation and the importance of dialogue in what Bohm (2002) describes as a 'participatory universe' which he explains in conversation with HH the Bohm and HH Dalai Lama (2018). By participatory universe he means that our thinking shapes the world we live in. If we compartmentalise human beings into body and mind we separate thinking from being. Anger, anxiety and depression impact our physical health. Our emotions have a direct effect on physical molecules in the body as explained by Candace Pert (1999) in her book "The Molecules of Emotion". In fact *the way we think* can influence *our physical health.*

Just as our thinking effects our physical health, it also effects the health of other species and the natural world, because our thinking shapes who and what we value.

---

[13] "From this perspective, sustainability implies avoiding critical boundaries and maintaining the combined SES within its historically viable stability domain, i.e., an operating range compatible with reasonable human demands. Resilience then becomes a theoretical construct for sustainability that: (a) guides against breaching unknown systems boundaries; (b) suggests that continuous changes in certain driving variables is inherently dangerous (e.g., continuously increasing fishing pressure, escalating GHG emissions, or constant material growth)and; (c) warns that surviving the breach of a major tipping point, whether human induced or natural, will require unprecedented levels of investment, cooperation and other forms of institutional and societal adaptation. Human-induced climate change will almost certainly validate all these assertions. In short, resilience thinking is a complement to sustainability, not a substitute."

[14] Daniel Wahl with you. Recorded last month. https://www.youtube.com/watch?v=IKBlGtzEJoo

[15] Joanna Macey discusses Buddhism and systems theory and underlines the similarities.

Our thoughts shape the landscape – as Indigenous first nations understand. If voters do not understand that our shared global future depends on limiting climate change to 1.5 degrees Celsius, then protecting forests will not be prioritised. For example, voters who support mining profits and cattle pastures to support the beef industry voted for the Brazilian President Bolsonaro at the expense of the Amazon.

Thus as Donna Haraway (1991) stresses, "we are the boundaries", we can recognise our solidarity with other species or we can regard them as "other". Haraway (2015, 2016a), Mitman (2019) stresses the importance of extending our solidary to include other species. By abandoning this relational approach (Chilisa, 2020; Harris & Wasilewski, 2004) as understood by First Nations and many spiritual philosophies – neoliberal economies are exposing humanity to existential risks (Bostrom, 2011), because we are living beyond sustainable limits (Meadows & Randers, 1992).

The Daoist philosophy of tai chi also stresses the notion of the energy flows of nature and that human beings need to work with nature rather than opposing it. The philosophy of interbeing (Macy, 2015; Macy et al., 2014) or as Alan Rayner calls it 'inclusionality' is also echoed in the philosophy of The Dalai Lama who follows the Mahayana Tradition of Buddhism.

The competition between us and them, the commodification of nature and the notion that systems are linear and that knowledge can be confined to specific disciplines have combined to limit our understanding that we are in fact *interbeings* reliant on one another and that all living systems need water, food and energy. These are the common threads that bind us.

To sum up, transformative education should be more problem based and focus on enabling people to live and regenerate social and environmental systems of which we are a strand in an ongoing cycle of life. Biology, geography, history, culture, arts language philosophy, science and politics could be taught in ways that address the evolutionary challenges whilst being open and responsive to the learning needs and talents of diverse students and their environments.

In line with Dewely's legacy the approach advocated is expanding pragmatism which means thinking of the social, economic and environmental consequences of decisions using mixed methods matched to a specific area of concern through engaging with the relevant stakeholders. One of the first steps for transformation is engagement to take into consideration the social, economic and environmental challenges as well as the way that diverse stakeholders perceive these convergent challenges.

At a time when anger and outrage has flared as a result of a lack of leadership to address social, economic and environmental challenges educators need to address what matters to those with whom they are engaging. The movements include, for example: Fridays for Change that supports the Ecocide Law and school protests against climate change.The #BlackLivesMatter Campaign in the wake of the brutal death of George Floyd in the USA that has sparked global movements and raised questions about social justice and the need for black voices to be heard globally. The UNDRIP (2007) which stresses the rights of Indigenous Peoples for socially inclusive decision making on habitat protection. The #Me Too movement to express women's rights and the *Fridays for Change Movement*, mentioned above (Box 1).

Box 1: Design Implications for Transformative Education

In order for the process of regeneration and educational redesign to be effective, it needs to be based on a recognition of our diversity and interconnectedness as part of one of many interdependent living systems. Transformative education involves the teaching and learning intended to empower learners to take informed decisions and actions at the individual, community and global levels. Education is critical for enabling the development of the requisite knowledge, skills and values for learners to participate in individual and collective decisions making. It is necessary to guide and evaluate actions, and how they might impact on the planet of the future.

This critical role of education has found expression 'education for sustainable development' (ESD) which is gaining prevalence as a means to address the challenges facing the planet. We recognise, however, that there is a need to go beyond 'sustainable development' and to accentuate regeneration holistically through a more holistic educational approach. In this chapter, we denote this expanded vision by referring to ESD-plus which we regard as embracing the focus on regeneration. We argue that ESD+ needs to be introduced and *mainstreamed across* every discipline to engender a mind shift and a transformative way of thinking and being. In order to focus on earth challenges holistically, ESD+ transcends traditional disciplinary boundaries and will strengthen the multi-disciplinary relationships between the natural and social sciences with a view to developing informed and critical students who are effective agents of change and stewards of the environment. Mabunda and McKay (2022, p. 377) in fact contemplate ESD+ in relation to the theory of pluralism which is based on the assumption of an interrelated, non-hierarchical coexistence of the human and nonhuman worlds, extending to the natural and spirit worlds which are interconnected in time and space.

They argue the import of ESD+ which needs to serve as a link between various species and holistic systems and they draw on African thought to advance their argument. African thought offers an ecosophical dimension that prioritises relatedness and interconnectedness – a wholeness that embraces the planet and its species, a collectivism and relatedness that connects humans, animals and the natural environment (Assié-Lumumba, 2017). They argue that the living and practising of Ubuntu, provides an epistemology and philosophical basis to embrace the universe in its entirety and in its "wholeness".

Global problems require global solutions and for ESD+ to be effective, it has to be offered on a global scale, with national curricula incorporating the principles of a healthy environment and social and environmental justice as principles that underpin learning and which prepares learners to pursue sustainable development (UNESCO, 2017). ESD essentially transcends disciplines. It cuts across the lifelong learning continuum of formal and non-formal education – from early childhood development, throughout schooling and

(continued)

Box 1 (continued)
beyond in pursuit of responsible global citizenship (Mabunda & McKay, 2021; Hanemann & McKay, 2015).

The United Nations 2030 Agenda for Sustainable Development's comprehensive plan for development, with its 17 Sustainable Development Goals and 169 targets, lays out a matrix of interlocking factors that are deemed to affect the development agenda. Goal 4.7 of the SDG stresses the importance of ESD stating that

> By 2030, ensure that all learners acquire knowledge and skills needed to promote sustainable development, including, among others, through education for sustainable development and sustainable lifestyles, human rights, gender equality, promotion of a culture of peace and nonviolence, global citizenship and appreciation of cultural diversity and of culture's contribution to sustainable development (United Nations, 2015).

If the world's nations hoping to meet the SDGs-scheduled by 2030 much needs to be done and education has an emergency role to play in driving the SDG agenda.

While the momentum for ESD emerged from the United Nations Decade of Education for Sustainable Development (DESD, 2005–2014) much of its historical lineage can be attributed to John Dewey (1859–1952), a philosopher, psychologist, pedagogue and public intellectual and the founding voice in pragmatism and student-centred experiential learning. Dewey's twentieth century work on education provides the early theoretical foundations for ESD and for the development of a skills-based and experiential eco-pedagogy which underlies contemporary ESD (Tarrant & Thiele, 2016).

Although the contemporary socio-ecological context differs from Dewey's era, Tarrant and Thiele (2016: 57) argue that "Dewey's contributions give the historical grounding for present-day sustainability pedagogy and present opportunities for its further development". They commend Dewey's early cautions about the deleterious impacts of scientific and technological innovation unchecked by the ethics of responsibility. As Dewey pointed out:

> We have displayed enough intelligence in the physical field to create the new and powerful instruments of science and technology. We have not as yet had enough intelligence to use this instrument deliberately and systematically to control its social operation and consequences (Dewey, 1931, p. 6: 60).

To mitigate these effects Dewey posited the deepening of democracy and the accentuation of communication and engagement with the relational processes and he emphasised the primacy of education and with the active learner-citizen and agent of change as central to eco-responsibility.

(continued)

Box 1 (continued)

The focus on education for transformation inevitably leads to Paulo Freire, who, like John Dewey, was critical of traditional approaches to education which see education as the one-way transmission of knowledge in a disconnected world. Both Freire and Dewey see education as dialogical and both stress the importance of dialogue, critical thinking and problem solving, attributing learning to the active role of students in envisioning alternative futures, and this informed by eco-literacy, provides the bedrock for eco-pedagogies. Freire's anti-banking education stance draws on generative themes that initiate the dialogue of education, as the practice of freedom, stimulating people's awareness for going beyond their existential experience – and herein lies the essence of transformative education (Aitchison & McKay, 2021).

By drawing on ecological thinking and praxis (within and beyond Western models) *balance* can be achieved. Perspectives that inform this volume and our approach to transformative education include Indigenous First Nations, spirituality, the liberative potential of critical systemic thinking and systemic ethics. Joanna Macy's (2015) integration of systems thinking, spirituality, practical engagement on *regeneration* and the need for re-connecting with one another and with nature also resonates with this approach.

Another source of inspiration is the legacy of critical systemic thinkers such as Churchman (1971, 1979, 1982)[16] who tried to address problems at the root by emphasizing the need to *re-design* our approach to the big issues of the day. He stressed the importance of unfolding values and 'sweeping in' the many social, cultural, political, economic and environmental aspects of an area of concern that need to be understood within the context of living systems of which we are a strand.

3 Power, Knowledge and Communication

Communication requires respect, patience and the will to listen, which is why in Chapters "Learning From Nature's Classroom: Reframing Economics, Accounting and Accountability", "Systemic Praxis and Education to Protect the Commons" the basic factors for decision making that are uncovered are engagement, salience and trust. Power imbalances make respectful communication difficult to achieve, which is why Chapter "Fluid, Organic Thinking and Relationality: Implications for Education and International Relations" details the importance of the proposed Ecocide Law (Higgins et al., 2013) to protect all the inhabitants within a region. The choice of the word 'inhabitants' is a deliberate attempt to protect multiple species.

[16] As well as his students, colleagues and those he inspired such as Werner Ulrich, Gerald Midgley, Mike Jackson and Robert Flood).

Post national laws (such as the proposed Ecocide Law (Higgins et al., 2013) are needed to provide the conditions for 'respectful dialogue' as Habermas (2001) has stressed in his essay on the need for *post national constellations* in which laws provide the necessary structures to support democracy and governance to protect living systems. All sentient beings have a right to a life worth living (Nussbaum, 2006) without the fear of discrimination on the basis of nationality, culture, gender, age, ability or species bias (see Chapters "Dis-ability, Differently Abled and Sentient Beings: Transformative Regeneration: The Need for Public Education on Vulnerability and Interbeing" and "Public Education on the Rights of Sentient Beings to a Life and Death Free of Suffering" in this volume). This requires access to habitat for animals in which they can express their full capabilities (Nussbaum, 2006, 2011; Donaldson & Kymlicka, 2011). Animals have the rights to live undisturbed lives, *simply because they are sentient*. Furthermore, animals that are healthy and living in appropriate environments do not become sources of pandemics. The right for plants to exist in biodiverse habitats also needs to be recognised if living systems are to have a hope of surviving (Higgins, 2012, 2018; Stephens et al., 2019). Non-anthropocentric researchers have stressed the need to recognise that many animal sentient beings have empathy, reciprocity and a sense of fairness (De Waal, 2009). Other researchers go further arguing that all living systems communicate in some way, from one-celled to the most complex organisms (Meijer, 2016; Rayner, 2010, 2017a; Stephens et al., 2019). Alan Rayner (2010), a botanist and former president of the British Mycological Society, refers to the collectivistic mechanisms of the mycelia beneath the earth that act as the neural networks of the forest, as the internet of life-feeding information and the sharing of nutrients.

In making a plea for an Ecocide Law and a recognition of an Earth Charter, both barrister Polly Higgins and physicist Vandana Shiva (1989, 2002, 2012a, b) stress that new ethical and legal boundaries need to be drawn to recognise justice for the environment. Just as the artificial human-animal barrier has been breached, so the artificial barrier between people and nature needs to be breached by recognising our hybridity and interconnectedness (McIntyre-Mills, 2018). We need systemic ethics (McIntyre-Mills, 2014) to underpin education policies as to who gets what, when, how and to what effect on multiple species.

4 Transformative Education and Design to Protect People and Nature by Fostering Multispecies Relationships and Standing Together

The volume responds to Odora Hoppers' (2013) plea for more experiential, systemic collaborative Mode 2 engagement to promote indigenous knowledge systems across the sectors and across disciplines, rather than Mode 1 approaches that are based on linear logic (Gibbons et al., 1994). Supporting a circular economy by working in ways that protect nature together with social enterprises that support " earning, learning and growing a future," is just one of the steps. In this volume, we link vocational training with other places of learning and make a plea for educators and learners at all educational levels to be more connected to the environment and to learn with nature by engaging beyond the school walls. Rantala et al. (2020) make a plea for a stronger relationship with the places where we live and the many species on which we rely. If our starting point is a concern for relationships with others (including other sentient beings) in a post anthropocentric world then we need to re-consider the nature of ethics.

If we were to ask *where do we draw the line of sentience*, this would be the wrong question as it assumes that dualistic relationships are a starting point for ethics. The volume makes the case for a non dualistic world in which our being is related to the being of other living systems.

Systemic ethical decisions need to be based on considering the consequences of our decisions. Public and vocational education needs to support a circular economy can be defined as working with, rather than against nature in such a way that sources of abundance are found or co-created so that resources are not depleted. Gunther Pauli (2010) explains a circular economy as eco-facturing. Making and creating a future in which human beings see themselves as part of a living system. Examples of eco-facturing include using the wind and sun as energy and appropriate local products such as bamboo for biochar, cassava for bioplastics and coffee grains for mushrooms. A circular economy could avoid building in obsolescence into digital technologies that rely on coltan for smart phones, for example or other minerals necessary for building solar batteries or wind turbines. Block chain could help to manage the careful use of materials.

The protests in the wake of social and environmental injustice need to be buttressed by a change in *a priori* laws, norms and narratives, along with indicators and measures to enhance representation and accountability to all species. This requires a new form of public education and new policies to ensure that schools support living systems.

4.1 Dualistic Thinking and Its Consequnces for People and the Planet

Dualistic thinking and justifications for the commodification of people, animals, plants and the environment are prevalent and supported in neoliberalism and new forms of surveillance capitalism (Zuboff & Schwandt, 2019) where people also become data (Harari, 2015, 2016). Harari engages with Kahneman (2021) and they agree that critical reflection on designs is vital, because technology and computer algorithms *are a product of design.*[17]

---

[17] It could also be argued that it is possible for programmers to take into *account quantum*, rather than binary logic (Bohm, 2002). Quantum thinking is relational, not binary.

By creating a sense of separation from nature our relationship with nature (recognized by many Indigenous philosophies and religions) is broken. This sense of alienation becomes an ongoing source of social, economic and environmental problems because once the relationship is broken, the natural world can be reified (from the Latin, meaning thing) and then treated as a commodity from which profit can be extracted, without regard to the long term effects on living systems. Furthermore, the dualistic mindset is extended through forms of education and development that divide areas of knowledge into disciplines that categorize knowledge into discrete subject areas, rather than teaching the ways that knowledge areas overlap and inform one another.

Dualistic premises justify exploitation based 'us versus them' philosophy rationalized through ethnocentrism, racism, sexism and speciesism. Thus categories such as People/Nature, Us/Them, Male/Female, Human/Animal are examples of polarisations based on categories which do not take into account that, for example others who appear strange can become familiar if we engage in dialogue to find out more about their worldviews, their religions, ethnic groups, languages and cultures. Gender and identity are far more complex than simply two oppositional categories but are enriched by a multifarious intersectionality (Crenshaw, 1991). For example, LGBTQ+(Lesbian, Gay, Bisexual Trans sexual, Questioning and more, Cherry, 2021) are identities that are now recognized as a rainbow of options along a diverse human gender continuum. Butler (whose work extends the thinking of Hannah Arendt) asks: what would the world be like if we failed to recognise one another's vulnerability? The next step is to state explicitly that by not recognising our shared vulnerability as human beings and importantly our connection with other species, we are denying the everyday injustice towards the differently abled. Thus policy can be said to deny the 'banality of evil' inherent in many current policies (Arendt, 1963; see McIntyre-Mills, 2019).

Furthermore, human beings along with all sentient beings ought to have rights, simply because they are sentient as detailed in Chapters "Dis-ability, Differently Abled and Sentient Beings: Transformative Regeneration: The Need for Public Education on Vulnerability and Interbeing" and "Public Education on the Rights of Sentient Beings to a Life and Death Free of Suffering" of this volume in which the literature is reviewed and in which a case is made for public education to extend the boundaries of solidarity beyond the anthropocentric to include recognition of the rights of the differently abled and all sentient beings.

Dualistic Western thinking underpinning neoliberalism is problematic and has shaped perceptions and values that have led to a misunderstanding and a misuse of living systems which prompted the economists Joseph Stiglitz, Amartya Sen and Fitoussi (2010) to write about how economists 'mismeasure' our lives (see Stiglitz et al., 2010). They stress the need to protect the fabric of life and so-called "wellbeing stocks" and to refrain from an exploitative economics that results in degeneration, rather than regeneration of people, society and nature.

4.2 Challenges and Opportunities for Education

Education should ideally support creativity and the ability of the next generation to 'design re-generative cultures' to cite Daniel Wahl (2016). Hope and inspiration can only be generated through active participation and social inclusion. Alan Rayner, botanist (see Chapters "Combining Focus and Circumspection: An Education in Natural Inclusion" and "The γ – Principle of Natural Inclusion: From Competition Versus Cooperation to Heartfelt Relay" in this volume) stresses the importance of recognising that non categorical thinking and natural inclusion should underpin education. Alan Rayner studied the relationships of fungi to trees and habitat has emphasised the notion that boundaries need to be re-thought as they are part of the problem. Trees survive because of relationships and networks of co-sharing. To assume that human beings are somehow separate from nature would be ludicrous if it were not so profoundly sad.

Some hope for the future rests in the way that young people in Australia are submitting law suits to argue that a lack of policy sensitivity to the implications of climate change will result in their lives and livelihoods being at risk.[18] A law student has focused on government bonds which she says cannot be relied upon, because they are based on unsustainable approaches to the environment.[19]

4.3 Hope and Inspiration Based on Systemic Praxis

The quantum physics of David Bohm and the Buddhist notion of human beings' interdependency on all sentient beings share much in common according to the Dalai Lama (2005) who stressed that the notion of consciousness is continuous and not the preserve of human beings (see McIntyre-Mills, 2021a). *Aspects of these ideas* are echoed by ecofeminist Donna Haraway (1991, 2015, 2016a, b), Anna Tsing (2013, 2015) and Barad (2003, 2007) who explain how relationships matter.

Thinking matters quite literally because it shapes how we see the world and our relationships with others, including people, animals and plants. Mahayana Buddhist modules on karma, samsara, the dharma path, mind and its potential underline that this way of knowing has been understood for centuries. Similarly, Indigenous ways of knowing in Australia, New Zealand, North America and Africa (Harris & Wasilewski, 2004; Romm & Lethole, 2021; Gibson et al., 2015), for example have made this point.

---

[18] https://www.theguardian.com/australia-news/2021/may/27/australian-court-finds-government-has-duty-to-protect-young-people-from-climate-crisis

[19] https://www.abc.net.au/news/2020-07-22/student-sues-australian-government-over-climate-change/12480612

4.4 Communication is a Multispecies Endeavour

The notion that communication transcends species (McIntyre-Mills, 2021a: 1) is summed up as follows:

> COVID-19 can be seen as feedback for anthropocentric social, economic and environmental decision-making that disrespects other living systems. The paper makes a case that respects for multiple species, and the onus of beneficence should be applied to all living systems of which we are a strand. Human beings are not exceptional insofar as they are able to communicate, make decisions, demonstrate a sense of community and show empathy or to make political calculations. What does this mean for the way we live our lives? Respect for persons needs to include animals, plants and the earth. It is time to rethink rights and responsibilities to protect habitat.

Communication is not the preserve of human beings. Alan Rayner (2010) has shown through his study of fungi that trees communicate and as Monica Gagliono (see Gagliano et al., 2018) has shown plants are capable of making decisions as do many other creatures. Meijer (2016) has detailed how squid make movements and change in colour in response to risks and birds use language and culture with intelligence and pass on cultural tropes to kith and kin. Mammals have a sense of fairness as De Waal (2009) demonstrates and Nagel (1974) stressed that consciousness is experienced differently by other species, but it is nevertheless a form of consciousness.

This has implications for education and the chapters in this volume make a case for transformation in the way we live and the way we educate current and future generations, to ensure that we try to reverse the harms we have perpetrated as a result of misdirected systems (Ackoff & Pourdehnand, 2001) that harm nature.

Our values matter quite literally as the questions we pose shape the problems and the policy 'solutions' that shape our world for better or worse. By finding ways to bridge the divided middle ground using dialogue, sharing knowledge, mapping it using pictures/art or diagrams or drama as guide to further decisions. Critical systemic praxis (McIntyre-Mills, 2004) provides the conceptual tools for helping people to work with complexity to help to ensure that people are able to see the bigger picture, realise that we are merely a strand within creation and hold in mind more variables. By expanding our vision through better public education we could make better decisions. Much has been written about the fact that human beings cannot manage more than about 7 variables (Miller, 1956) and arguably this is perhaps one of the reasons why human beings make such bad political decisions than jeopardize our futures. The other reason is probably narrow self-interest.

Furthermore, Crenshaw suggests that many categories interconnect and cannot be easily separated when looking at life opportunities. Intersectionality (Crenshaw, 1991; Dhamoon, 2011) refers to working across a range of factors, thus we also need to consider demographic, social, cultural, political and economic inhibitors when developing a more integrated curriculum linked with vocational training to support a green economy (see Chapters "The Power of Story: Ocean View Organic Farmers Cooperative, Cape Town, South Africa: Growing Food with Heart and Soul", "Learning from Nature's Classroom : Reframing Economics, Accounting and Accountability", "Aa!!Venda Women and Social Enterprise: Stepwise Process to Regenerative and Sustainable Living").

A conversation in December with colleagues in the International Society of Systems Sciences including James Rose, Isabelle Sebastian, Shankar Sakaran and Terence Love reminded me of how important communication is for education and regeneration. I mentioned the Dalai Lama's (2005) book on the convergence of science and spirituality and the important role that David Bohm had made in sharing the essence of how he saw the participatory potential of the universe and how he saw physics.[20] By this they mean that the questions we pose and our way of framing "problems" also shape our world. Human beings have the potential for positive or negative mutual causality. This applies to our relationships and to the way we see ourselves and others.

4.5 *Health and Education Policy Needs to Support Non Anthropocentrism*

Plants and animals communicate and are sentient.[21] Animals and the environment ought not to be treated as commodities unless we wish to experience the consequences of climate change detailed in the IPCC (2018, 2019, 2021) report and the full ramifications of an unsustainable factory based agriculture that brutally confines animals in conditions that are breeding grounds for viruses.

As stressed in 'From Polarisation to Multispecies Relationships' (McIntyre-Mills, 2021b, c, d)[22] an alternative hypothesis (Goodall, 2021; Gorbalena et al., 2020) about the root causes of the pandemic suggests that cross species infections could be a result of habitat loss and inappropriate forms of food production that commodifies and contains animals without regard for their wellbeing.

These are the important points we need to understand, in order to promote educational systems that will meet the needs of current and future generations of life.

[20] This version of Western Science was seen to have parallels with Buddhism according to the Dalai Lama. The main difference being the emphasis on testing based on perception, personal lived experience and thought experiments versus testing based on external empirical experiments.

[21] See McIntyre-Mills 2021 Communication and culture: a Multispecies endeavour within a shared habitat in International Systems Sciences Year Book to appear in Systems Research and Behavioural Science which develops this point. Also see McIntyre-Mills, J (2021a) Chapter 1 Communication and Culture: A Multispecies Endeavour Recognising Kinship with Multiple Species in From Polarisation to Multispecies Relationships which extends the argument.

[22] This key point has been made by Eva Meijer (2016 on animal communication), Frans De Waal (2009) explains how animals have a sense of empathy, fairness and ability to reciprocate, Susan Simard (2016), Simard and Defrenne (2019) explain how trees communicate as does the forester Peter Wohlleben (2016). The botanist Allen Rayner (2010, 2017a, b) explains how the mycelia underground create communication highways or internet systems whilst Gagliano et al. (2018) explain that plants are capable of 'remembering' and capable of decision making.

The focus needs to be on nurturing regeneration through respect for their sentience and desisting from the Cartesian fallacy of treating living systems as *objects.*

Three linked concerns underpin the risks – they include high rates of urbanisation, a lack of rural urban balance and habitat loss – impacting multiple species. The structure, content and process of education needs to take this into account by creating new structures for education linked with vocational education within community based learning organisations as discussed in Chapter "Learning from Nature's Classroom : Reframing Economics, Accounting and Accountability" of this volume.

Members of Homo Sapiens Sapiens, the so-called 'twice wise' (Banathy, 1996, 2000) are able to think about their thinking and to use this capability to create designs that are potentially positive or negative. If we accept that members of Homo Sapiens Sapiens are part of one living system then we are a step closer to understanding our interbeing and that species are part of a continuum of life, but unlike other species we are able to make decisions on a scale far beyond other species. As a result our categorical thinking has changed our environment. Categorical, dualistic thinking is the first hurdle to overcome. Teleonomy or the functional response of living systems is shaped by human decisions that have created the Anthropocene (Haraway, 2014, 2015, 2016a, b). The butterfly symbolizes transformation, but the caterpillar holds within it the potential to change and the potential for regeneration.

If we can understand that the caterpillar and the butterfly are merely different stages of one cycle then we are able to comprehend the notion of potential and interbeing (Abram, 2020). The pandemic has taught us that unless the vaccine is shared with everyone we will all suffer. Thus the notion of us-them, zero sum thinking is being challenged.

Learning about ourselves and others requires two kinds of knowing. Perceptive knowing based on *our perceived experience* and knowing based on *testing and measuring* aspects of the external world. Unfortunately the former aspect based on thinking and careful analysis of *the way we see the world* is in need of more attention so that we become more mindful and aware of mutual causation that begins with our attitudes and values.

Rayner (2020)[23] stressed that we have evolved with the binocular vision of a predator, rather that more peripheral vision of a buck who needs panoramic vision.

The notion of *us and them* or *self and others* need to be replaced with the notion that we can indeed bridge the divides by realising that 'we are the boundaries' (Haraway, 1991) and that we can re-integrate our view of the world.

[23] Daniel Wahl with you. Recorded last month. https://www.youtube.com/watch?v=IKBlGtzEJoo

5 Re-membering Relationships as Re-generative Design for Transformation

This volume draws on the wisdom of first nations, spirituality and religion as much as it draws on the social and natural sciences. As such it strives for interdisciplinarity and cross cultural awareness of the importance of interbeing and a recognition of our vulnerability as living systems that are mutually dependent. Our resilience flows paradoxically from a recognition of our vulnerability as Judith Butler has argued (see Stanescu, 2012).

'Us them' categorizing in recent months has also focused on cultural concerns, for example how to persuade people that wearing a mask and wearing it properly is about saving lives, not about politics or weakness. We stand and fall together as multiple species making up one tapestry of life. The axiom which underpins this work is that democracy and governance needs to foster freedom and diversity to the extent that we do not undermine the rights of others in this generation or the next.

The first step for thinking about the environmental context and the many interrelated aspects of wicked problems (Rittel & Webber, 1984) is the realization of our interdependency and interconnectedness. The second step is the need to work with nature, rather than trying to control mother nature.

The big picture challenges require many changes from reframing democracy and governance to protecting multiple species through post national regionalism (based on new architectures for democracy and governance).

Education policy needs to encourage problem based learning that spans disciplines and takes into account these big picture challenges by setting up new structures that take into account indigenous knowledge systems in the ways suggested in this volume.

User centric design needs to address these policy concerns that are relevant to educators in all sectors to give more attention to Indigenous ways of knowing as a means to *not merely* achieve the UN Sustainable Development Goals, but instead to enable *regeneration* as 'sustainability does not go far enough' (see Wahl, 2016).

Eilperin et al. (2020)[24] highlighted that a positive policy signal has been sent by Biden when he nominated an American Indian woman Deb Haaland to take a federal role in the new government and

> the first Black man to head the Environmental Protection Agency, as well as Obama administration veteran Brenda Mallory to serve as the first Black chair of the White House Council on Environmental Quality.

This could serve as a step towards achieving some of the goals set out by the UN Declaration of the Rights of Indigenous Peoples (2008). Eilperin et al. (2020) cite Deb Haaland's twitter comment:

[24] https://www.washingtonpost.com/climate-environment/2020/12/17/deb-haaland-interior-secretary-biden/

> "A voice like mine has never been a Cabinet secretary or at the head of the Department of Interior," Haaland tweeted Thursday night. "... I'll be fierce for all of us, our planet, and all of our protected land."

These ideals are vital within and beyond educational institutions. Thus praxis needs to be rooted in conceptualizations as well as practical pilots that address:

- User-centric approaches to address complex needs such as health, housing and social inclusion (McIntyre-Mills, 2008)
- Mitigation and adaptation to climate change through social, economic and environmental indicators of wellbeing. (McIntyre-Mills and De Vries 2011; McIntyre-Mills et al., 2014)

Demonstrating ways to de-couple economics, accounting and accountability from the carbon economy (McIntyre-Mills et al., 2019, 2021a, b)[25] and link it with advocacy for the ecocide law (Higgins et al., 2013) to ensure that individuals, organisations and nation states are held to account for the extensive damage or destruction of an ecosystem that "prevents the peaceful life of the inhabitants of a region."[26] Is vital. In line with this proposed law, engagement to address educational challenges also needs to address Indigenous knowledge systems (Chilisa & Mertens, 2021; Odora Hoppers, 2013; UNDRIP, 2007) and the protection of multiple species. This (related) volume makes a case for transforming education praxis. In Chapters "Learning from Nature's Classroom : Reframing Economics, Accounting and Accountability", "Women, Education and Socialisation: Transforming Gender and Power in the MST, São Paulo, Brasil", and "Systemic Praxis and Education to Protect the Commons") we explore potential pathways to promote opportunities to enhance education and wellbeing in line with the Paris Declaration (2005)[27] and the Accra Agreement (2008) on harmonising development goals, development needs to take into account the values of the participants, based on long term trust and in the interests of the planet (IPCC, 2018; Climate Change, 2022).

Bibliography

ABC News: Teenagers file Australian-first class action to stop planned coal mine extension. A class action launched on behalf of young people everywhere seeks an injunction to stop the Australian Government approving an extension to Whitehaven's Vickery coal mine, arguing it will harm young people by exacerbating climate change. HTTPs Appleby, G. 2019. https://theconversation.com/the-uluru-statement-is-not-a-vague-idea-of-being-heard-but-deliberate-structural-reform-142820

[25] See Chapter "Collective Action for Regeneration of the Web of Life in the Face of Disruptive Injustice" this volume.

[26] Published on 13 Mar 2018 cocide, the 5th Crime Against Peace – Polly Higgins.

[27] https://www.oecd.org/dac/effectiveness/parisdeclarationandaccraagendaforaction.htm

ABC News: Student files climate change class action against Australian Government over 'risks' to her money. Student files climate change class action against Australian Government over 'risks' to her money 23-year-old Melbourne law student is suing the Australian Government for failing to disclose the risk climate change poses to Australians' super and other safe investments. Https.

Abram, D. (2020). Foreword in Kaza, S. 2020 *A wild love for the world: Joanna Macy and the work of our time*. Shambhala.

Ackoff, R. L., & Pourdehnand, J. (2001). On misdirected systems. *Systems Research and Behavioural Science, 18*(3), 199–205. AEST. https://theconversation.com/as-the-federal-government-debates-an-indigenous-voice-state-and-territories-are-pressing-ahead-120411

Aitchison, J., & McKay, V. (2021). A historical remembrance of Paulo Freire in South Africa: A tale of two activists. *Australian Journal of Adult Learning., 61*(3), 334–354.

Arendt, H. (1963). *Eichmann in Jerusalem: A report on the Banality of evil*. Viking Press.

Assié-Lumumba, N. T. (2017). The Ubuntu Paradigm and comparative and international education: Epistemological challenges and opportunities in our field. *Comparative Education Review, 61*(1), 1–21. https://doi.org/10.1086/689922

Banathy, B. (1996). *Designing social systems in a changing world*. Plenum.

Banathy, B. (2000). *Guided evolution of society: A systems view*. Kluwer/Plenum.

Barad, K. (2003). Posthumanist performativity: Toward an understanding of how matter comes to matter. *Signs, 28*, 801–831. https://doi.org/10.1086/345321

Barad, K. (2007). *Meeting the universe halfway: Quantum physics and the entanglement of matter and meaning*. Durham.

Bart Williams, M. (2020). *TEDxBerlinSalon | Why the wealth of Africa does not make Africans wealthy?* https://www.youtube.com/watch?v=SPqMxRO8_2U

Bateson, G. (1972). *Steps to an ecology of mind*. Ballantine.

Bateson, N., & Guddemi, P. (2013). Systems, tools and Bateson's cybernetics: A metalogue. In D. Arnold, & R. King (Eds.), *Traditions of systems theory: Major figures and contemporary developments*. ProQuest Ebook Central. https://ebookcentral-proquest-com.ezproxy.flinders.edu.au

Bateson, N., Sieburg, D., Bateson, G., & Films, B. (2011). *An ecology of mind a daughter's portrait of Gregory Bateson*. Bullfrog Films.

Berger, P. L. (1977). *Facing up to modernity*. Penguin.

Bohm, D. (2002). *Wholeness and the implicate order*. Routledge/Routledge Classics.

Bohm, D., & HH Dalai Lama. (2018). *From fragmentation to wholeness: David Bohm & Dalai Lama life and mind enfolded in everything*. https://www.youtube.com/watch?v=V2He4oU32sg

Bostrom, N. (2011). *Existential risk prevention as the most important task for humanity*. Faculty of Philosophy & Oxford Martin School University of Oxford. http://www.existential-risk.org/

Brigg, M., & Graham, M. (2020). *The COVID-19 pandemic demonstrates the relevance of Aboriginal political concepts*. https://www.abc.net.au/religion/coronavirus-and-aboriginal-political-concepts-morgan-brigg-and/12281304

Burke, A., & Celermajer, D. (2021). *Human progress is no excuse to destroy nature. A push to make 'ecocide' a global crime must recognise this fundamental truth*. https://thefifthestate.com.au/columns/spinifex/human-progress-is-no-excuse-to-destroy-nature-a-push-to-make-ecocide-a-global-crime-must-recognise-this-fundamental-truth/

Carrington, D. (2020). *Pandemics result from destruction of nature, say UN and WHO. Experts call for legislation and trade deals worldwide to encourage green recovery*. https://www.theguardian.com/world/2020/jun/17/pandemics-destruction-nature-un-who-legislation-trade-green-recovery

Cherry, K. (2021). https://www.verywellmind.com/what-does-lgbtq-mean-5069804#:~:text=Negative%20and%20positive%20factors%20associated,and%20questioning%20(LGBTQ)%20youth

Chilisa, B. (2017). Decolonizing transdisciplinary research approaches: An African perspective for enhancing knowledge integration in sustainability science. *Sustainability Science, 12*(5), 813–827.

Chilisa, B. (2020). *Indigenous research methodologies* (2nd ed.). Sage.

Chilisa, B., & Mertens, D. M. (2021). Indigenous made in Africa evaluation frameworks: Addressing epistemic violence and contributing to social transformation. *American Journal of Evaluation, 42*(2), 241–253. https://doi.org/10.1177/1098214020948601

Climate Change. (2022, February). *Impacts, adaptation and vulnerability*. https://report.ipcc.ch/ar6wg2/pdf/IPCC_AR6_WGII_SummaryForPolicymakers.pdf

Christakis, A., & Bausch, K. (2006). *How people harness their collective wisdom and power to construct the future in co-laboratories of democracy*. Information Age.

Churchman, C. W. (1971). *The design of inquiring systems*. Basic Books.

Churchman, C. W. (1979). *The systems approach and its enemies*. Basic Books.

Churchman, C. W. (1982). *Thought and wisdom*. Intersystems Publications.

Clack, B. (1999). René Descartes 1596–1650. In B. Clack (Ed.), *Misogyny in the Western philosophical tradition*. Palgrave Macmillan. https://doi.org/10.1057/9780230212800_8

Crenshaw, K. (1991). Mapping the margins: Intersectionality, identity politics and violence against women of color. *Stanford Law Review, 43*(6), 1241–1299.

Dabelko, G. D., & Conca, K. (Eds.). (2019). *Green planet blues: Critical perspectives on global environmental politics*. ProQuest Ebook Central. https://ebookcentral-proquest-com.ezproxy.flinders.edu.au

Dalai Lama. (2005). *The universe in a single atom. The convergence of science and spirituality*. Harmony.

Dauvergne, P. (2019). *The problem of consumption. Green planet blues: Critical perspectives on global environmental politics*. ProQuest Ebook Central. https://ebookcentral-proquest-com.ezproxy.flinders.edu.au

De Waal, F. (2009). *The age of empathy: Nature's lessons for a kinder society*. Harmony Books.

Descartes, R. 1596–1650. (1986). *Discourse on method*. Macmillan/Collier Macmillan.

Descartes, R. (2017). *Meditations on first philosophy* (J. Veitch).

Dewey, J. (1931). Science and society. In *Philosophy and civilization* (pp. 318–330). Minton, Balch, and Co.

Dhamoon, R. K. (2011). Considerations on mainstreaming intersectionality. *Political Research Quarterly, 64*(1), 230–243. https://doi.org/10.1177/1065912910379227

Donaldson, S., & Kymlicka, W. (2011). *Zoopolis: A political theory of animal rights*. Oxford University Press.

Eder, K. (1996). Part 3: The politics of the environment: Exhaustion or renewal. The institutionalisation of environmentalism: Ecological discourse and the second transformation of the public sphere. In S. Lash, B. Szerzynski, & B. Wynne (Eds.), *Risk, environment and modernity: Towards a new ecology* (pp. 203–223). Sage.

Eilperin, J., Grandon, D. I., & Brady, D. S (2020). With historic picks, Biden puts environmental justice front and center. In *The selection of the first Native American interior secretary and first Black male EPA chief highlights pollution disparities*. https://www.washingtonpost.com/climate-environment/2020/12/17/deb-haaland-interior-secretary-biden/. Accessed 11 Jan 2021.

Feketha, S., & Mavuso, S. (2020, October 13). Mbalulu hits back at MKMVA. *Cape Times*, p. 1.

Freire, P. (1972). *Pedagogy of the oppressed*. Penguin.

Gagliano, M., Abramson, C. I., & Depczynski, M. (2018). Plants learn and remember: Let's get used to it. *Oecologia, 186*, 29–31. https://doi.org/10.1007/s00442-017-4029-7

Gaventa, J. (2001). Global citizen action: Lessons and challenges. In M. Edwards & J. Gaventa (Eds.), *Global citizen action*. Lynne Rienner.

Geertz, C. (1973). Thick description: Towards an interpretive theory of culture. In C. Geertz (Ed.), *Of cultures selected essays* (pp. 3–32). Basic book.

Geertz, C. (1980). *Negara: The theatre state in nineteenth century Bali*. Prinversity Press.

Geertz, C. (2004). What is a state if it is not sovereign? *Reflections on Politics in Current Anthropology, 45*, 577–593.

Getano, L. (1993). Voices from the land. Boyer Lecture Series.

Gibbons, M., Limoges, C., Nowotny, H., Schwartzman, S., Scott, P., & Trow, M. (1994). *The new production of knowledge: The dynamics of science and research in contemporary societies*. Sage.
Gibson, K., Rose, D. B., & Fincher, R. (Eds.). (2015). *Manifesto for living in the Anthropocene* (pp. 17–22). Punctum books.
Giddens, A. (2009). *The politics of climate change*. Polity Press.
Gildersleeve, R. E. (2016). The neoliberal academy of the Anthropocene and the retaliation of the lazy academic. *Cultural Studies – Critical Methodologies, 17*(3), 286–293.
Gildersleeve, R. E., & Kleinhesselink, K. (2019). *Philosophy and theory in higher education Special Issue on the Anthropocene in the study of higher education* (pp. 1–16). Guest Editors, University of Denver, Vol. 1, Issue 1, April 2019, Peter Lang.
Goldson, R. (2020). *Jah in the flesh: An examination of spirit, power, and divine envesselment in Rastafari.* ProQuest Dissertations Publishing. https://www.researchgate.net/publication/321076086_
Goodall, J. (2021). https://www.theguardian.com/environment/2021/jan/03/jane-goodall-change-is-happening-there-are-many-ways-to-start-moving-in-the-right-way?
Gorbalenya, A. E., Baker, S. C., Baric, R. S., et al. (2020). The species severe acute respiratory syndrome-related coronavirus: Classifying 2019-nCoV and naming it SARS-CoV-2. *Nature Microbiology, 5*, 536–544. https://doi.org/10.1038/s41564-020-0695-z
Gore, A. (2007). *Assault on reason*. Bloomsbury.
Graver, M. (2007). *Stoicism and emotion*. University of Chicago.
Greenfield, S. (2000). *The private life of the brain: Emotions, consciousness self.* Wiley.
Greenfield, S. (2003). *Tomorrow's people how 21st century technology is changing how we think and feel*. Penguin.
Greenfield, S. (2008). *ID: The quest for meaning in the 21st century*. Stoughton.
Guddemi, P. (2006). Human views of ecological nature. *Cybernetics and Human Knowing, 12*(1), 86–90.
Habermas, J. (2001). *The postnational constellation. Political essays*. Polity Press.
Hanemann, U., & McKay, V. (2015). Lifelong literacy: Towards a new agenda. *International Review of Education, 61*, 265–272. https://doi.org/10.1007/s11159-015-9497-6
Harari, Y. N. (2015). *Sapiens: A brief history of humankind*. Harper.
Harari, Y. N. (2016). *Homo Deus*. Harvill Secker.
Harari, Y. N. (2020, June 14). *Organisms are algorithms. Body is calculator*. Answer = Sensation~Feeling~Vedanā. https://www.youtube.com/watch?v=GrQ7nY-vevY
Harari, Y. N. (2021, May 16). *How human can be hacked.* https://www.youtube.com/watch?v=eZuJ4WyJNGQ
Harari, Y. N., Kahneman, D., & Swisher, K. (2021). *Daniel Kahneman and Yuval Noah Harari in conversation*.
Haraway, D. (1991). *Cyborgs, simians, and women: The reinvention of nature*. Free Association Books.
Haraway, H. (2014, September 5). *Anthropocene, Capitalocene, Chthulucene: Staying with the trouble.* https://vimeo.com/97663518
Haraway, D. J. (2015). Anthropocene, Capitalocene, Plantationocene, Chthulucene. *Environmental Humanities, 6*(1), 159–165. https://doi.org/10.1215/22011919-3615934
Haraway, D. J. (2016a). *Staying with the trouble: Making Kin in the Chthulucene*. Duke University Press.
Haraway, D. J. (2016b). *Anthropocene, Capitalocene, Chthulucene: Making string figures with biologies*. Arts, Activisms – Aarhus University YouTube. https://m.youtube.com/watch?v=CHwZA9NGWg0
Harper, P. (2020). Free state branches gun for Ace. *Mail and Guardian*, pp. 23–29, p. 3.

Harris, L. D., & Wasilewski, J. (2004). Indigeneity, an alternative worldview: Four R's (relationship, responsibility, reciprocity, redistribution) vs. two P's (power and profit). Sharing the journey toward conscious evolution. *Systems Research and Behavioral Science, 21*(5), 489–503.
Higgins, P. (2012). *Ecocide*, the 5th crime against peace: https://www.youtube.com/watch?v=8EuxYzQ65H4 see the link in this lecture to Stop Ecocide, Change the Law https://www.youtube.com/channel/UCCYoIf880oaM419JO8XUhEA
Higgins, P. (2018). *The 2018 Hague freedom lecture by Polly Higgins*, https://www.youtube.com/watch?v=kQn8oA6e9To
Higgins, P., Short, D., & South, N. (2013). 2013 *Protecting the planet: a proposal for a law of ecocide. Crime Law Soc Change, 59*, 251–266. https://doi.org/10.1007/s10611-013-9413-6
Hoob, H.,Whittaker, A., & Coombes, L. (2019). *As the federal government debates an Indigenous Voice, state and territories are pressing ahead July 16, 2019 4.42 pm.*
IPCC. (2018). https://www.ipcc.ch/site/assets/uploads/2018/11/SR1.5_SPM_Low_Res.pdf
IPCC. (2019). Summary for policymakers. In P. R. Shukla, J. Skea, E. C. Buendia, V. Masson-Delmotte, H.-O. Pörtner, D. C. Roberts, P. Zhai, R. Slade, S. Connors, R. van Diemen, M. Ferrat, E. Haughey, S. Luz, S. Neogi, M. Pathak, J. Petzold, J. P. Pereira, P. Vyas, E. Huntley, et al. (Eds.), *Climate change and land: An IPCC special report on climate change, desertification, land degradation, sustainable land management, food security, and greenhouse gas fluxes in terrestrial ecosystem.* Cambridge University Press.
IPCC. (2021). Summary for policymakers. In V. Masson Delmotte, P. Zhai, A. Pirani, S. L. Connors, C. Péan, S. Berger, N. Caud, Y. Chen, L. Goldfarb, M. I. Gomis, M. Huang, K. Leitzell, E. Lonnoy, J. B. R. Matthews, T. K. Maycock, T. Waterfield, O. Yelekçi, R. Yu, & B. Zhou (Eds.), *Contribution of working group I to the sixth assessment report of the intergovernmental panel on climate change. Climate change 2021: The physical science basi.* Cambridge University Press.
Kaza, S. (2020). *A wild love for the world: Joanna Macy and the work of our time*. Shambhala. Is in chapter 2.
Kelly, L. (2016). *The memory code*. Allen and Unwin.
Kruger, J. (2020). Self-directed education in two transformative pro-environmental initiatives within the eco-schools programme, a South African case study. *Education as Change, 24*(6649), 1–23.
Laouris, Y., Economidou, A., Georgiou, M., Philippou, E., & 16 others. (2017). *Reinventing democracy in the digital era – African Initiative*. Future Worlds Center, Nicosia, Cyprus. This report has been developed in the context of the "Reinventing Democracy in the Digital Era" Project (http://reinventdemocracy.info), which was funded by the United Nations Democracy Fund (UNDEF),and implemented by Future Worlds Center (FWC) with the support of Regional Coordinators, and many partners and liaisons, who continue to support local, national, regional and/or global activities.
Le Duc, A. (2015). *The doctrine of dependent origination as basis for a paradigm of human-nature relationship of pesponsibility and accountability?*
Lovelock, J. (2020). *Beware: Gaia may destroy humans before we destroy the Earth Covid-19 may well have been one attempt by the Earth to protect itself. Gaia will try harder next time with something even nastier Guardian Australia.* https://apple.news/AWWBgfhQST_ugWXFUYlBhFw
Mabunda, P., & McKay, V. (2021). Education curriculum and multispecies relations. In J. McIntyre-Mills & C. Corcoran-Nantes (Eds.), *From polarisation to multipspecies relationships*. Springer.
Mabunda & McKay. (2022). *Chapter 13 Online learning and the pedagogy of resilience, agency and protest: The COVID-19 experience.*
Macy, J. (2015). *Mutual causality in Buddhism and general systems theory*. https://www.sunypress.edu/p-1176-mutual-causality-in-buddhism-an.aspx
Macy, J., Young-Brown, M., & Fox, M. (2014). *Coming back to life: The updated guide to the work that reconnects*. New Society Publishers.

Maliba, A. (2020). Mixed feelings about bail for suspect. *Weekend Argus*, 25th October.
Maturana, H., & Varela, F. (1980). *Autopoeisis and cognition: The realization of living*. Reidel.
Maxwell, N. (2019). Chapter 5. The scandal of the irrationality of academia. In *Philosophy and theory in higher education Special Issue on the Anthropocene in the Study of Higher Education* (pp. 105–128). Guest Editors, University of Denver, Vol. 1, Issue 1, April 2019 Peter Lang. https://doi.org/10.3726/ptihe.2019.01.06
McCall Smith, A. (2019). *To the land of long lost friends*. Anchor/Penguin.
McIntyre-Mills, J. (2004). Facilitating critical systemic praxis (CSP) by means of experiential learning and conceptual tools. *Systems Research and Behavioural Science, 21*, 37–61. (A ranking ABDC Australian Business Deans Council Journal Quality List).
McIntyre-Mills, J. J. (2008). *User centric policy design to address complex needs*. Nova Science.
McIntyre-Mills, J. (2010). Wellbeing, mindfulness and the global commons. *Journal of Consciousness Studies, 17*(7–8), 44–72.
McIntyre-Mills, J. (2014). *Systemic ethics and non-anthropocentric stewardship: Implications for transdisciplinarity and cosmopolitan politics* (Contemporary systems series). Springer.
McIntyre-Mills, J. (2017). *Planetary passport for representation, accountability and re-generation* (Contemporary systems series). Springer.
McIntyre-Mills, J. (2018). Policy Design for Non-Anthropocentric Pathways to Protect Biodiversity and Regenerate the Land. *International Education Journal in South Africa., 4*(1), 1–28. https://journals.co.za/content/journal/10520/EJC-117cea7615
McIntyre-Mills, J. (2019) Social engagement to redress the banality of evil and the frontiers of justice: Limitations of the social contract to protect habitat and why an international law to prevent the crime of ecocide matters. In *Proceedings of International Systems Sciences The 63nd Annual Meeting of ISSS*, Oregon State University, Corvallis, Nature's Enduring Patterns: A Path to Systems Literacy (2nd-5th July). http://journals.isss.org/index.php/proceedings63rd/article/view/3642
McIntyre-Mills, J. (2021a). Chapter 5: McIntyre-Mills, J: Consciousness for Balancing Individualism and Collectivism. In J. J. McIntyre-Mills & Y. Corcoran Nantes (Eds.), *From polarisation to multispecies relationships in the age of mass extinctions* (pp. 67–86). Springer.
McIntyre-Mills, J. (2021b). Chapter 3 'From old to new taxonomies of rights, relationships and responsibilities to protect habitat'. In *From polarisation to multispecies relationships in the age of mass extinctions* (pp. 31–54). Springer.
McIntyre-Mills, J. (2021c). Chapter 10: From polarisation to multispecies relationships: Re-membering narratives. In *From polarisation to multispecies relationships in the age of mass extinctions* (pp. 173–212). Springer.
McIntyre-Mills, J. (2021d). Chapter 26: McIntyre-Mills, J. Biopolitics and food security to protect social and environmental justice. In *From polarisation to multispecies relationships in the age of mass extinctions* (pp. 575–602). Springer.
McIntyre-Mills, J. J., & De Vries, D. (2011). *Identity, democracy and sustainability: Facing up to convergent social, economic and environmental challenges* (p. 331). ISCE/ Emergence / Complexity and Organization. ISBN 97 80984 216536.
McIntyre-Mills, J., & Romm, N. (2019). *Mixed methods and cross disciplinary research: Towards cultivating eco-systemic living*. Springer.
McIntyre-Mills, J., De Vries, D., & Binchai, N. (2014). *Transformation from wall street to well-being: Joining up the dots through participatory democracy and governance to mitigate the causes and adapt to the effects of climate change* (p. 253). Springer.
McIntyre-Mills, J. J., Romm, N. R. A., & Corcoran-Nantes, Y. (Eds.). (2019). *Democracy and governance for resourcing the commons: Theory and practice on rural-urban balance*.
McKie, R. (2020). *Rampant destruction of forests 'will unleash more pandemics' Researchers to tell UN that loss of biodiversity enables rapid spread of new diseases from animals to humans*. Sun 30 Aug 2020 16.03 AEST. https://www.theguardian.com/environment/2020/aug/30/rampant-destruction-of-forests-will-unleash-more-pandemics

Meadows, D., & Randers, J. (1992). *Beyond the limits: Global collapse of a sustainable future*. Earthscan Publications.
Meijer, E. (2016, Translated, 2019). *Animal languages: The secret conversations of the living world.* John Murray.
Melbourne Law School. (2018). *Uluru statement from the heart: Information booklet.* https://law.unimelb.edu.au/__data/assets/pdf_file/0010/2764738/Uluru-Statement-from-the-Heart-Information-Booklet.pdf
Miller, G. A. (1956). The magical number seven plus or minus two: Some limits on our capacity for processing information. *Psychological Review, 63*(2), 81–97.
Mitman, G. (2019). Reflections on the Plantationocene: A conversation with Donna Haraway and Anna Tsing June 18, 2019 updated October 12, 2019. https://edgeeffects.net/haraway-tsing-plantationocene/
Nagel, T. (1974). What is it like to be a bat? *The Philosophical Review, 83*(4), 435–450. https://doi.org/10.2307/2183914.JSTOR2183914
Nussbaum, M. (2006). *Frontiers of justice*. Harvard University Press.
Nussbaum, M. (2011). *Creating capabilities: The human development approach*. Belknap Press of Harvard University Press.
Odora-Hoppers, C. (2013). Community engagement, globalization, and restorative action: Approaching systems and research in the universities. *Journal of Adult and Continuing Education, 19*(2), 94–102.
Pathways to Wellbeing. https://ia801606.us.archive.org/20/items/pathway_DEMO_1/pathway_DEMO_1.mp4
Pauli, G. (2010). *The Blue Economy: Report to the Club of Rome*. Paradigm Publications.
Pert, C. (1999). *The molecules of emotion: Why you feel the way you feel*. Simon and Schuster ebook.
Pre Greenberg, G., & Tobach, E. (1987). *Cognition, language and consciousness* (Vol. 2).
Rantala, O., Salmela, T., Valtonen, A., & Höckert, E. (2020). Envisioning tourism and proximity after the Anthropocene. *Sustainability, 12*(10), 3948.
Rayner, A. D. M. (2010). Inclusionality and sustainability – Attuning with the currency of natural energy flow and how this contrasts with abstract economic rationality. *Environmental Economics, 1*, 98–108.
Rayner, A. (2017a). *The origin of life patterns in the natural inclusion of space in flux.*: Springer. https://en.wikiversity.org/wiki/Natural_Inclusion
Rayner, A. (2017b). *Natural inclusion*. In J. J. McIntyre-Mills, N. Romm, & Y. Corcoran-Nantes (Eds.), *Balancing individualism and collectivism: Social and environmental justice* (pp. 461–470). Springer.
Rayner. (2020). *Rayner in conversation with Daniel Wahl.* See Utube link https://www.youtube.com/watch?v=IKBlGtzEJoo
Rees, W. (2019). *Sustainability versus Resilience* In G. D. Dabelko & K. Conca (Eds.), *Green planet blues: Critical perspectives on global environmental politics*. Taylor & Francis Group. ProQuest Ebook Central, http://ebookcentral.proquest.com/lib/flinders/detail.action?docID=5892383
Rittel, H., & Webber, M. (1984). *Planning problems are wicked problems, developments in design methodology*. Wiley.
Romm, N. R. A., & Lethole, P. V. (2021). Prospects for sustainable living with focus on interrelatedness, interdependence and mutuality: Some African perspectives. In *Chapter 6 in from polarisation to multispecies relationships in an era of mass extinctions*. Springer.
Shiva, V. (1989). *Staying alive: Women, ecology and survival in India*. Zed Books.
Shiva, V. (2002). *Water wars: Privatization, pollution and profit*. Pluto Press.
Shiva, V. (2012a). *Making peace with the Earth*. Fernwood Publishing.
Shiva, V. (2012b). *Monocultures of the mind*. Third World Network.
Simard, S. (2016). *How trees talk to each other*. https://www.youtube.com/watch?v=Un2yBgIAxYs
Simard, S., & Defrenne, C. (2019). *The secret language of trees*. https://www.youtube.com/watch?v=V4m9SefyRjg

Stanescu, J. (2012). Species trouble: Judith Butler, mourning, and the precarious lives of animals. *Hypatia, 27*, 563–582.

Stephens, A., Taket, A., & Gagliano, M. (2019). Ecological justice for nature in critical systems thinking. *Systems Research and Behavioral Science, 36*, 3–19. https://doi.org/10.1002/sres.2532

Stiglitz, J., Sen, A., & Fitoussi, J. P. (2010). *Mis-measuring our lives: Why the GDP doesn't add up*. The New Press.

Synot, E. (2019). https://theconversation.com/listening-with-our-ears-and-our-eyes-ken-wyatts-big-promises-on-indigenous-affairs-120165

Tariq Mellet. (2020). *Camissa people at camissa people*. wordpress.com.

Tarrant, S., & Thiele, L. (2016). Practice makes pedagogy – John Dewey and skills-based sustainability education. *International Journal of Sustainability in Higher Education, 17*(1), 54–67. https://doi.org/10.1108/IJSHE-09-2014-0127

Trump is holding a News Conference today. https://www.youtube.com/watch?v=_4b2v96W5Nk. https://www.youtube.com/watch?v=JTVe0vZjXvg

Tsing, A. (2013). More-than-human sociality: A call for critical description. In K. Hastrup (Ed.), *Anthropology and nature* (pp. 27–42). Routledge.

Tsing, A. L. (2015). *The Mushroom at the end of the world: On the possibility of life in capitalist ruins* (p. 352). Princeton University Press.

Turchin, P. (2016). *Ultra society: How 10,000 years of war made humans the greatest cooperators on earth*. Beresta Books, LCC.

UNDRIP. (2007). *United Nations Declaration of the rights of Indigenous peoples*. https://www.humanrights.gov.au/publications/un-declaration-rights-indigenous-peoples-1.

UNESCO. (2017). https://www.sdg4education2030.org/education-sustainable-development-goals-learning-objectives-unesco-.

UNESCO. (2021). *Education for sustainable developmnt*. https://en.unesco.org/themes/education-sustainable-development. https://en.unesco.org/news/5th-unesco-forum-transformative-education-sustainable-development-global-citizenship-health

United Nations. (2015). https://www.un.org/en/chronicle/article/goal-4-education-post-2015-sustainable-development-agenda

Urry, J. (2010). Consuming the planet to excess. *Theory, Culture and Society, 27*(2–3), 191–212. https://doi.org/10.1177/0263276409355999

Von Foerster, H. (1995). *Cybernetics of cybernetics* (2nd ed.). Future Systems, Inc.

Wahl, C. (2016). *Designing regenerative cultures*. Triarchy Press.

Wahl, C. (2019). Sustainability is not enough in. In G. D. Dabelko, & K. Conca (Eds.), *Green planet blues : Critical perspectives on global environmental politics*. Edited by Taylor & Francis Group. ProQuest Ebook Central. http://ebookcentral.proquest.com/lib/flinders/detail.action?docID=5892383

Watts, J. (2021). *Jane Goodall: Change is happening. There are many ways to start moving in the right way.*

Wohlleben, P. (2016). *The hidden life of trees: What they feel, how they communicate: Discoveries from a secret world*. Greystone Books.

Wohlleben, P., & Simard, S. (2016). https://www.intelligent-trees.com/. https://vimeo.com/ondemand/intelligenttrees

Woo, Y. L., Mokhtar, M., Komoo, I., & Azman, N. (2012). Education for sustainable development: A review of characteristics of sustainability curriculum. *OIDA International Journal of Sustainable Development, 3*(8), 33–44. Available at SSRN https://ssrn.com/abstract=2031102

Yunus, M. (2020). Pandemic has given us an opportunity to protect the world from its suicidal path – Muhammad Yunus The interview was published in Brazilian newspaper 'Estado de Sao Paulo' on 18 December 2020.

Zuboff, S., & Schwandt, K. (2019). *The age of surveillance capitalism: The fight for a human future at the new frontier of power*. Profile Books.

Web Links Cited

https://www.capetownetc.com/news/nurdles-wash-up-on-cape-town-beaches/
https://www.news24.com/news24/columnists/cyrilramaphosa/cyril-ramaphosa-gender-based-violence-it-is-time-to-turn-the-tide-on-the-scourge-20201123
https://www.news24.com/news24/southafrica/news/just-in-president-cyril-ramaphosa-to-face-motion-of-no-confidence-next-week-20201126
https://www.aquarium.co.za/blog/entry/the-great-nurdle-disaster-what-to-do-if-you-find-nurdles

Fluid, Organic Thinking and Relationality: Implications for Education and International Relations

Janet J. McIntyre-Mills

Abstract The chapter aims to highlight the importance of subjective, objective and intersubjective domains when engaging in public education and research on wicked problems such as poverty, climate change, loss of habitat and pandemics. The case is made that public education and awareness of critical systemic thinking and practice underpinned by a more relational understanding of complexity and the implications of convergent social, economic and environmental challenges is vital for addressing 'existential risks' through transformative thinking and practice.

Keywords Systemic interrelationships · Relationality · Co-evolution · Existential risk · Ecology of mind

1 Engagement, Scenarios and Support for Global Action and an Ecocide Law

Educational philosophy needs to support more interdisciplinary, cross cultural understanding and a willingness to rely on diverse ways of knowing that help to ensure that the global citizens of tomorrow develop an ecological mindset, to use Gregory Bateson's (1972) term.

The notion of shifting blame to specific nations will not help and is not inevitable (Peters et al., 2020). In the long run, the root cause for the escalating risks is an unsustainable way of life supported by all nation states who are "living beyond the limits" (Meadows et al., 1972).

J. J. McIntyre-Mills (✉)
College of Education, University of South Africa, Pretoria Central, Gauteng, South Africa

College of Professions, Adelaide University, Adelaide, Australia
e-mail: janet.mcintyre@adelaide.edu.au

J. J. McIntyre-Mills, Y. Corcoran-Nantes (eds.), *Transformative Education for Regeneration and Wellbeing*, Contemporary Systems Thinking,
https://doi.org/10.1007/978-981-19-3258-8_2

A shortcoming of the original Club of Rome Report and approach according to the participants such as Alexander Christakis and C. West Churchman was the lack of engagement with the broader public, who needed to be engaged to understand the implications of the modelling (Christakis, 2006).

The use of scenarios could help young people to move towards becoming global citizens who support new architectures and laws to protect the environment.

If the proposed Ecocide Law were to be implemented it could result in the earth, air and waters being less vulnerable to pesticides, pollution by plastic nurdles or carbon emissions so that food, energy and water security becomes a post national regional goal.

The tracking of containers using block chain could at the very least ensure that companies responsible for polluting the oceans are held to account as a way to dissuade risk takers. Currently the costs are externalised or shifted from the companies to the oceans, coastlines, creatures and people who consume the toxic nurdles as discussed in chapter "Nurdles and Food Security: Education for Wellbeing and Implications for Pan-African Social and Environmental Justice" of this volume.

When we face the challenge of rising temperatures, rising sea levels and shared fears of climate change and pandemics then human beings could learn that competitive politics of Thucydides[1] is not inevitable (see Peters et al., 2020) and that the world today with convergent social, economic and environmental challenges is different from the world of ancient Greece. Covid 19 has demonstrated that we stand and fall together.

Nevertheless the patterns of history should not be forgotten. This is why curricula at school should make space for humanities, history and contemporary issues that require the ability to think in terms of the area of concern, the lenses through which it can be assessed, the way it is framed, the context or environment of the problem and the different methods that ought to be brought to bear in order to address the problem.[2]

Good post national regional governance will need an architecture that protects natural resources as outlined by Higgins et al. (2013) that focuses on planetary wellbeing and that could draw on vision for a common future based on a sense of shared community.

In some respects this requires a new form of governance and democracy that balances rights and responsibilities to ensure that decisions are made to protect regional biospheres and all the inhabitants.

Circular economies that work with, rather than against nature are vital for the future wellbeing of post national communities.

The pursuit of business as usual in a carbon economy is no longer viable as it threatens our shared habitat.

---

[1] The Internet Classics Archive | The History of the Peloponnesian War by Thucydides (mit.edu)

[2] The Covid-19 pandemic has placed further challenges for international relations.

For Australia to renegotiate cordial relationships with her powerful neighbour it will be vital to stress our *common* goals and *common* future based on food, water and energy security to ensure that the convergent challenges of climate change and pandemics are averted.

2 Relationality and Co-evolution: Implications for Education Policy

Carlo Rovelli (2021)[3] is a physicist and scholar who discussed philosophical principle that when we try to understand reality we need to look at the relationships across variables (McIntyre-Mills, 2021a, b, c). The papers in this volume make the case that the problems underpinning pandemics are relational and that we need greater recognition of our hybridity and interconnectedness (McIntyre-Mills, 2017a, b). This has implications for public education, economics, governance, ethics and democracy as we can no longer pursue business as usual (McIntyre-Mills, 2020).

The ethics of production, marketing and consumption needs to be revisited as we reflect on virus. David Attenborough (2020)[4] reminds us that human beings are superfluous on the planet but insects are not!

The roots of the problems lie in anthropocentrism (see Goodall, 2020) and approaches to economics that work *against*, rather than *with* nature. Donaldson and Kymlica (2011) stress the need for safe habitats for animals and the implications this has on human security. The big issues of the day, namely our relationships with other species and the environment are addressed in "From Polarisation to Multispecies relationships" (McIntyre-Mills & Corcoran-Nantes, 2021) in which a case is made for re-membering our place in living systems through embracing a non- anthropocentric approach to ethics and governance.

Stewardship and solidarity ought to be core public education themes and at all educational levels: primary, secondary and tertiary.

The next step is to develop a governance system to enable the monitoring of wellbeing stocks as well as the application of positive and negative sanctions. For this to occur more public education on the need for global solidarity to support the common good is overdue. The role played by Greta Thunberg and the movement by school children is significant (McIntyre-Mills, 2021, chapters "Chapter 8 Education and gender mainstreaming in a patriarchal setting: What Really Matters" and "Chapter 20 Combining Focus and Circumspection: An Education in Natural Inclusion"). Public education for transformation involves responsiveness. Groups of people are capable of working together in new configurations for the common good once we recognise we are 'no longer top predator' (McIntyre-Mills, 2021).

---

[3] 'Relational Interpretation of Quantum Mechanics and Alexander Bogdanov's Worldview' by Carlo Rovelli. https://stream.syscoi.com/2021/05/05/the-centre-for-systems-studies-two-even-celebrating-the-life-and-contribution-of-alexander-bogdanov-the-annual-mike-jackson-lecture-2-june-with-carlo-rovelli-and-seminar-on-3-june-both-online/. In this series Rovelli pays tribute to Werner Heisenberg, who discovered the fundamental equations of quantum mechanics. The other is Alexander Bogdanov.

[4] https://www.abc.net.au/classic/the-margaret-throsby-interviews/sir-david-attenborough-interviewed-by-margaret-throsby/11757192

'Nature is a Great Experimenter' (Sam Sen Kim, 2021)[5] it discards the species that are not "contributing to the whole" she lists the species that have been lost and she asks:

> how sure are we about the success of our human species? If we are not benefiting nature, it will discard us! She continues by reflects on what the planet could be telling us : "We are damaging and cruel to the planet.... we kill ... not for survival but to prove our ....dominance... and for pleasure....What if Covid 19 is nature's way of eliminating the human virus? ....We are not separate from other species...

She makes the point that the problem of Covid is not only the fault of one nation and ends by summing up:

> It is a problem of our consciousness..... we need transformation .... It is already too late...

The pandemic has taught us that sharing vaccines, as opposed to hoarding them is the first cosmopolitan lesson that the global community will need to learn Safi (2021). This illustrates the point that 'A is indeed better off when B is better off' (Von Foerster, 1995) and that international commerce needs to understand that the *zero sum* approach is not only misguided it poses an 'existential risk' or in the frank terminology of Mohammad Yunus, it is suicidal (2021). Social, economic and environmental challenges are convergent which is why transformative praxis requires actions at the local level that can be quickly scaled up by drawing together existing laws to help foster a post national regional approach to governance that supports not only sustainable, but most importantly also re-generative living (Wahl, 2016, 2019; McIntyre-Mills, 2017a, b) by drawing on the following conventions, covenants, policies and bills which could help to develop alternative architectures for democracy and governance that have been detailed elsewhere (see McIntyre-Mills, 2017a, b, McIntyre-Mills and Romm 2019, McIntyre-Mills et al. 2019, McIntyre-Mills 2019a, b, 2021) in which the case has been made for local and post national regional governance structures that build on:

- The Aarhus Convention (1998), UN Sustainable Development Goals,
- The UN Sundai Risk platform (2030) as well as by policy theorists such as Anne Florini (2003) who makes a case for scaling up the potential of EU federalism. This is now possible if we apply the ideas of cyberneticists such as Stafford Beer (1974, 1994) and physicists, such as Alexander Christakis and Kakoulaki (2021) who stress the potential of structured dialogue to enable better decision making in regional and global Agoras.
- Participatory democracy needs to be informed by local indigenous knowledge (McIntyre-Mills et al., 2014, 2017a, b) and to protect it also needs to be buttressed through The United Nations Declaration on the Rights of Indigenous People (2008) and supported by global international laws to protect the common good, such as the Ecocide Law (Higgins et al., 2013) that could make destruction

[5] https://www.youtube.com/watch?v=v9De4eYk1ls&ab_channel=SamSSenSKim%E5%90%89%E7%8C%AB

of habitat illegal and the companies and nations liable to prosecution in an International Criminal Court.

Hopefully processes such as the above could prevent damage to the environment[6]so that innovative remedial action, such as self-sustaining floating continents that clean the ocean as described by the award winning designer can be accompanied by innovative preventative actions. Hauser et al. (2014: 3) in co-operating with the future stress:

> More generally, our results emphasize the importance of institutional designers moving away from the assumption of universal self-interest...by demonstrating how voting can allow a majority of pro-social individuals to override a purely selfish minority,

3 From Dualism to Relationality

Human security is dependent on the wellbeing of post national regions, *not* nationalistic patriotism that inspires 'us them' zero sum competitive thinking. Instead education needs to foster collaboration across conceptual and spatial boundaries to protect the common good.

Vandana Shiva (2012, 2020a, b) makes the point, that no matter how hard local people work on protecting the commons we still need laws and governance systems to protect the majority from the small minority who control enormous assets and sway states to support the markets, rather than the common good. Stafford Beer (1974, 1994) argued for governance supported by cybernetic networks. He demonstrated the viability through using a geodesic dome to model the interactions as explained by Allena Leonard (1994, 2005). Beer argued that governance could span nodes through being supported by cybernetic networks and demonstrated the viability through using a geodesic dome to model the interactions. Each node represents a nation state/organisation/cell and each strut a connection based on cybernetic energy flows and feedback loops. Beer's approach to governance based on nodes and webs is now increasingly relevant to balance individual and collective needs. The architecture of a Buckminster Fuller's geodesic governance 'dome' that protects through multiple governance flows could provide a means to monitor and thus support social, economic and environmental regeneration and sustainability *from below* and not only *from above*. To draw on and paraphrase the frequently cited warning by Kenneth Boulding (1966): *we are not in separate life boats, we are on one space ship earth.*

Whilst closed borders have been used as a response – as a result of the pandemic, in order to protect populations from Covid-19 – the need for cosmopolitanism and a concern for our common fate has never been more urgent. This is a core

[6]Zaha Hadid, Lenka Petráková, developed the idea for her student master thesis on ocean pollution. https://www.euronews.com/living/2021/02/04/this-floating-continent-is-self-sustainable-and-cleans-the-ocean

consideration for transformative education praxis rooted in organic contextual situations. Nora Bateson (2021) sums up an approach that resonates, she calls it a 'transcontextual' and dynamic approach to working with people and warns (Bateson, 2016) of systems approaches that become top down and controlling. Thus the delicate balancing of the individual and the collective is an ongoing dance to sustain the common good by learning about 'natural inclusion (Rayner, 2010, 2017a, b, 2021 and chapters "The γ – Principle of Natural Inclusion: From Competition Versus Cooperation to Heartfelt Relay" and "Systemic Praxis and Education to Protect the Commons", this volume).

3.1 Nationalism is Harmful if It Supports Us/Them Polarisation

Nationalism is an ideology that – although outdated and ill-suited to addressing the big challenges of the day, namely poverty and climate change – is on the rise.[7] Nationalism is unable to protect the biospheres on which we depend for water, food and energy security. If we are to ensure that the global commons is protected then we need a global ecocide law (Higgins et al., 2013[8]) to ensure that over fishing, dumping of waste, mismanagement of cargo leading to spillage and pollution do not continue to threaten oceans[9] and rivers.

Small face-to-face communities understand the implications of damaging local food sources, solidarity needs to be scaled up through laws that give power to implement a Global Covenant, as suggested by David Held (2004).

The preamble to the concept paper for the *Chair of the Earth Community Trust* stresses the need to extend Rome Statute Article 8.2.b(iv) because, although it is already decreed that it is a crime to destroy the environment during war (in the wake

[7] One example is the nationalist group called "The Proud Boys" who are accused of leading the attempted subversion of democracy in USA and have recently been declared a terrorist group in Canada, according to Newton (2021), who explains that the organisation was founded in 2016 and that:

> the Proud Boys lists among its central tenets a belief in "closed borders" and the aim of "reinstating a spirit of Western chauvinism." In online statements, Proud Boys have claimed they have used violence only in self-defence. But members … have been convicted of crimes against anti-fascist protesters.

[8] Polly Higgins et al., (2013, 3–4), Protecting the Public Interest) Bill 2014 Submission 93 – Attachment 1. Closing the door to dangerous industrial activity A concept paper for governments to implement emergency measures Trade and Foreign Investment (Protecting the Public Interest) Bill 2014 Submission 93 – Attachment 1.

[9] https://www.themarinediaries.com/tmd-blog/stop-ecocide-change-the-law

of the use of Agent Orange in Vietnam (paradoxically) in peace time extensive damage daily, without any global laws to prevent the destruction:

> It is already an international crime to destroy the environment in wartime: ... By making an international law that effectively ends extensive damage and destruction to the environment during peace time, humanmade Ecocide can be outlawed forever.

The concept paper continues by arguing that

> ... by creating a law of Ecocide ...[j]ob creation will accelerate, clean energy will be prioritised, cities will benefit from both, food will be grown using regenerative agriculture principles, water will become less polluted and water poverty will be abated, oceans will be protected from mass damage, and disasters will be better managed.

4 Innovation for the Future: Wellbeing Is Possible

According to Khimba and Dyson (2020),[10] the aim of the organisation to support passing a law to prevent ecocide is to:

> get an amendment to the Rome Statute, so the International Law against Ecocide is added as the fifth crime against peace....

They explain that the law aims to hold organisations to account and that their organisation was founded by Polly Higgins who:

> .... recognised that the ICC could be used as a legal fast track to protect the Earth, its oceans and all its inhabitants ... for future generations.

Greene (2019) cites Higgins et al.'s (2013)[11] definition of ecocide as follows:

> 1. Acts or omissions committed in times of peace or conflict by any senior person within the course of State, corporate or any other entity's activity which cause, contribute to, or may be expected to cause or contribute to serious ecological, climate or cultural loss or damage to or destruction of ecosystem(s) of a given territory(ies), such that peaceful enjoyment by the inhabitants has been or will be severely diminished. 2. To establish seriousness, impact(s) must be widespread, long-term or severe.

Importantly inhabitants is a generic term that refers to all species and despite Greene's (2019) reservations, it is likely that the context for supporting laws to prevent the destruction of habitat may *indeed gain traction* when the cascading consequences are understood (see McIntyre-Mills, 2020).Given that both USA and China have to face up to being the world's greatest emitters (along with Australia) and most at risk as a result of habitat destruction and the tensions between USA and her allies and China as a result of the COVID-19 pandemic, the questions raised by the Australian government, the unfolding tensions caused by the cancellation of

---

[10] Khimba, J and Dyson, C (2020) Stop ecocide: change the law. https://www.themarinediaries.com/tmd-blog/stop-ecocide-change-the-law

[11] *Ecocide Law*, MISSION LIFEFORCE. https://www.missionlifeforce.org/ecocide-law [https://perma.cc/L326-S4KA].

markets for Australian products, it is increasingly important to look beyond us/them nationalism to the more important issues of solidarity to protect humanity from escalating risks described by Bostrom (2011) as an 'existential risk'.

Allison (2017) suggests that when two nations with differing ideologies are in competition there is a high chance of conflict, particularly if they fear each other or are driven by a sense of honour/outrage or self-interest. Peters et al. (2020) cite an article in the Atlantic by Allison (2015) in which Xi Jinping is cited warning that avoiding the so-called *'Thucydides Trap'* caused by *'strategic miscalculations.'*[12]

"Us/them" dualistic thinking has been responsible for many problems as detailed in other chapters of this book, but it is not inevitable.

Whilst I visited Beijing in 2011 as an academic researcher at the invitation of the Chinese Academy of Social Sciences to a Research Conference on 'Sustainable Environmental Policy and Global Governance' – our delegation walked in Tiananmen Square close to the hotel where the CASS conference was held. It was clear that a national sanctioned protest was supported by China as posters were draped over buildings around the square, declaring the importance of Chinese territorial rights.[13]

The Chinese Academy of Sciences meeting in 2012 which I attended and presented a paper,[14] was opened with a quote from Buckminster Fuller recognising that we are part of one space ship earth and that ultimately co-operation to protect the planet is important.

The potential of systemic thinkers such as Buckminster Fuller could be extended to develop governance systems that enable balancing individual and collective concerns that protect the common good. For this to occur public education needs to bolster the ability of people to think beyond simplistic us/them dualism by supporting the ability of primary, secondary and tertiary education to teach critical systemic thinking skills.[15]

---

[12] Allison, G 2015. The Thucydides Trap: Are the U.S. and China Headed for War? In 12 of 16 past cases in which a rising power has confronted a ruling power, the result has been bloodshed, SEPTEMBER 24. https://www.theatlantic.com/international/archive/2015/09/united-states-china-war-thucydides-trap/406756/

[13] The handover of Hong Kong back to China by Britain is not a direct parallel with the takeover of neighbouring countries of Germany, nevertheless history was also cited as a justification by Germany for taking over neighbouring countries. The dispute over the Spratly islands in the South China sea (Lewis and Letail, 2016) are claimed by the Philippines, Malaysia, China and Taiwan. Whilst the Senkaku Islands in the East China Sea (known by China as the The Diaoyu archipelago) are also a source of tension around fishing rights, potential oil supplies and are claimed by China, Taiwan and Japan. Lewis and Letail (2016) cite Holloway and Cui (2014) and explain that the Senkaku Islands were annexed by Japan when Japan defeated areas of China, but at the end of the second World War when Japan was defeated, these islands were left out of negotiations.

[14] McIntyre-Mills, J. (2012). Wellbeing and existential risk: engagement to address the policy and governance challenges of mitigation and accountability', paper presented to the Chinese Academy of Social Sciences, CASS Research Conference Sustainable Environmental Policy and Global Governance, Beijing, China, 18–19 September.

[15] China has a growing economy and the belt and road approach could be said to make old fashioned conflicts obsolete, even if the build-up in the South Pacific and the mounting trade war seems to belie this argument. Another plus is that the election of Biden could result in more diplomatic

Mind traps are like lobster traps, according to Geoffrey Vickers (1968, 1983) a critical systemic thinker who coined the term. If we view the way problems are framed and then overturn the frame, we have a hope of escaping.

The way out requires looking at the situation from many angles and realising. Creating or co-creating a way. Often problem solving involves the ability to hold in mind many variables and to realise that *"we are the boundaries"* (to draw on and adapt Donna Haraway's 1991 phrase). The boundaries are based on values.

We can re-frame situations and find ways of working towards solutions that require new forms of global citizenship to protect he common good.

The successful resolution[16] of the Brexit talks between Britain and the EU has been approved and underlined that both sides needed to give ground and to be prepared to engage in respectful dialogue for the common good and to agree on fishing quotas[17] and has been approved by the British Parliament (Murphy & O'Carroll, 2020).

The European Commission President Ursula von der Leyen stressed that finding a solution was the right thing to do as being good neighbours is vital in a world where boundaries and borders need to be set on the basis of common interests.

The pandemic has taught us that closing borders can sometimes be vital to protect health, but in the long term we are all interdependent and part of one living system.[18] Sharing vaccines, as opposed to hoarding them is the first lesson that the global community will need to learn in 2021.

5 Relationality and the Pandemic

Paradoxically, the virus lock downs have resulted in the lowering of carbon emissions (Abram, 2020) Another paradox is that human beings strive to be individuals and part of a group (Berger, 1977) but:

> "...Covid-19 is teaching us that "A is better off when B is better off", to cite the cybernetics axiom popularized by Von Foerster (1995). The caveat "as long as B is not undermining the rights of current and future generations of living systems" could be added. (See McIntyre-Mills, 2021b)

---

relations, but even during the Obama administration, Hilary Clinton made statements about the importance of territorial waters being respected and clearly this approach was unpopular as stressed by my hosts at the Chinese Academy of Social Sciences (CASS) Conference.

[16] Murphy and O'Carroll (2020) Johnson to hail 'historic resolution' as Brexit bill comes before Commons. https://www.theguardian.com/politics/2020/dec/29/brexit-trade-deal-wins-support-of-conservative-erg-mps

[17] Brexit: Boris Johnson set to unveil trade deal with EU. https://www.bbc.com/news/uk-politics-55433410

[18] The disparities within and across nation states will need to be addressed to ensure that everyone has equal access so that at least 50–70% of the population can become immune. This will help to ensure that the pandemic is slowed, vaccines can be updated and eventually when enough people achieve immunity the pandemic can be halted.

This axiom is inherently cosmopolitan as stressed in "Systemic Ethics" (McIntyre-Mills, 2014: 48) and it requires laws such as the bill for an Ecocide Law (Higgins et al., 2013) to make a practical difference. From this basic precept, many ethical principles follow. For example, if viable vaccines are not widely shared we will all suffer, because no one is safe until we are all safe'.

According to Markson's (2021) contentious lead article in the Australian, Fauci was aware of the risks associated with 'gain of function' corona virus research that was funded by grants from the USA, but considered the institute to be of a high standard and worthy of funding.

The social, economic environmental and political context all play a role in the creation of the perfect storm (McIntyre-Mills, 2021d, e) as governments hastened to point the finger of blame; instead we need to look at the way current systems of global food production are potential contributors to the multiple interrelated causes of pandemics – aided and abetted by industrialised production which harms biodiversity and small local producers. The political context of the origins and spread of the virus have also led to a cooling of international relations and escalating tensions expressed in terms of trade, whilst border closures have also impacted the higher education and research sectors.

6 Integrating the Subjective, Objective and Intersubjective Domains

Generosity and compassion help us to turn strangers into friends and to take the first step towards making friends with enemies (Yu, 2006). This echoes C. West Churchman who reflects that "religion, morality, politics and aesthetics" are the so-called 'enemies within' through which we filter the way we see the world.

All human beings view the world through value filters – but despite our differences all human beings share the need for food, water and safe habitat. These basic needs are shared by many other species. Once we regard these as basic elements of a shared future we could develop the will to foster better relationships and to engage more respectfully with one another, based on the right of *all inhabitants* to live peacefully within a shared habitat (Higgins et al., 2013).

Bibliography

Aarhus Convention. (1998). *On access to information*. Public participation and access to justice in environmental matters, Denmark, 25 June. http://ec.europa.eu/environment/aarhus/

Abram, S. (2020) Coronavirus: world's response has slashed CO_2 emissions – Here's how to keep them down. https://theconversation.com/coronavirus-worlds-response-has-slashed-co-emissions-heres-how-to-keep-them-down-134094

Allison, G (2015) The Thucydides Trap: Are the U.S. and China headed for war? In 12 of 16 past cases in which a rising power has confronted a ruling power, the result has been

bloodshed, September 24. https://www.theatlantic.com/international/archive/2015/09/united-states-china-war-thucydides-trap/406756/
Allison, G. (2017). *Destined for war: Can America and China escape Thucydides's trap?* Harcourt Press.
Ashby, W. R. (1956). *An introduction to cybernetics*. Chapman and Hall.
Attenborough, D. (2020). Life on our planet documentary. https://www.youtube.com/watch?v=mU4wZhFCsGs
Bai, Z., & Liu, J. (2020). China's governance model and system in transition. *Journal of Contemporary East Asia Studies, 9*(1), 65–82. https://doi.org/10.1080/24761028.2020.1744229
Barad, K. (2003). Posthumanist performativity: Toward an understanding of how matter comes to matter. *Signs, 28*, 801–831. https://doi.org/10.1086/345321
Barad, K. (2007). *Meeting the universe halfway: Quantum physics and the entanglement of matter and meaning*. Durham.
Bateson, N. (2016). *Small arcs of larger circles framing through other patterns*. Triarchy Press.
Bateson, N. (2021). Plenary International Systems Sciences. https://norabateson.wordpress.com/2017/05/28/warm-data/. Nora Bateson Plenary, July 12th International Systems Sciences. https://zoom.us/rec/play/_oe07ngyK1jHXqZxIaT5nBr_c95ReZ2KWOwXCrYw7u1K6rK-fJRUPtUH_urEVNtlI9m1WDZbqZFxJ8k7i.po_IqnXsYX9swRJr?continueMode=true&_x_zm_rtaid=u3XXrLmPQCaoiZM3sR9LHA.1626085969999.920d3aa8d2c700ad3a46aaaca293852c&_x_zm_rhtaid=301
Beer, S. (1974). *Designing freedom*. Wiley.
Beer, S. (1975). *Platform for change*. Wiley.
Beer, S. (1994). Governance or government. In S. Beer (Ed.), *Beyond dispute: The invention of team syntegrity*. Wiley.
Berger, P. L. (1977). *Facing up to modernity*. Penguin.
Bohm, D. (2002). *Wholeness and the implicate order*. Routledge, Routledge Classics.
Bohm, D, & Dalai Lama, HH. (2018). *From fragmentation to wholeness: David Bohm & Dalai Lama Life and mind enfolded in everything* https://www.youtube.com/watch?v=V2He4oU32sg
Bostrom, N. (2011). *Existential risk prevention as the most important task for humanity*. Faculty of Philosophy & Oxford Martin School University of Oxford. http://www.existential-risk.org/
Boulding, K. (1966). Space ship earth the economics of the coming spaceship earth. Kenneth E. Boulding. In H. Jarrett (Ed.), *Environmental quality in a growing economy* (pp. 3–14). Resources for the Future/Johns Hopkins University Press.
Buckminster Fuller, R. (1978). *Operating manual for space ship earth*. Dutton.
Capra, F. (1996). *The web of life: A new synthesis of mind and matter*. Harper Collins.
Christakis, A., & Bausch, K. (2006). *How people harness their collective wisdom and power to construct the future in co-laboratories of democracy*. Information Age.
Christakis, A., & Kakoulaki, M. (2021). Chapter 35 objectifying intersubjectivity through inclusion for a scientific [r] evolution: avoiding polarization by engaging stakeholders for saliency, priority and trust. In *From polarisation to multispecies relationships: Re-generation of the commons in an era of mass extinctions* (pp. 699–728). Springer.
Churchman, C. W. (1971a). *The design of inquiring systems: Basic concepts of systems and organization*. Basic Books.
Churchman, C. W. (1971b). *The design of inquiring systems: Basic concepts of systems and organisations*. Basic Books.
Churchman, C. W. (1979). *The systems approach and its enemies*. Basic Books.
Diedrich, J., & Christakis, A. N. Chapter 19 The co-laboratory of democracy archetypes: Engaging stakeholders in deliberative democracy to respond proactively to diversity. In *From polarisation to multispecies relationships: Re-generation of the commons in an era of mass extinctions* (pp. 451–472).
Diehm, C. (2007). Identification with nature: What it is and why it matters. *Ethics and the Environment, 12*(2), 1–22. Retrieved June 12, 2021, from http://www.jstor.org/stable/40339138

Donaldson, S., & Kymlicka, W. (2011). *Zoopolis: A political theory of animal rights*. Oxford University Press.
Foucault, M. (2008). *The birth of biopolitics: Lectures at the college de France 1978–79*. Palgrave Macmillan.
Foucault, M., & Gordon, C. (Eds.). (1980). *Power/knowledge*. Brighton.
Freitas, E. (2017). Karen Barad's quantum ontology and Posthuman ethics: Rethinking the concept of relationality. *Qualitative Inquiry, 23*(9), 741–748.
Gillespie, A. (2021). The COVID-19 lab leak theory highlights a glaring lack of global biosecurity regulation. https://theconversation.com/the-covid-19-lab-leak-theory-highlights-a-glaring-lack-of-global-biosecurity-regulation-162419
Goodall, J. (2020). https://www.cnn.com/videos/us/2020/03/19/coronavirus-jane-goodall-acfc-full-episode-vpx.cnn
Green, A. (2019). Campaign to make ecocide an international crime: Quixotic quest or moral imperative? *Fordham Environmental Law Review, 30*(3) 2019 Article 1,SPRING 2019. https://ir.lawnet.fordham.edu/elr
Haraway, H. (2014). Anthropocene, Capitalocene, Chthulucene: *Staying with the Trouble*, May 9, 2014. https://vimeo.com/97663518
Haraway, D. J. (2015). Anthropocene, Capitalocene, Plantationocene, Chthulucene. *Environmental Humanities*. 1 May 2015, *6*(1), 159–165. https://doi.org/10.1215/22011919-3615934
Haraway, D. J. (2016a). *Staying with the trouble: Making kin in the Chthulucene*. Duke University Press.
Haraway, D. (2016b). *'Anthropocene, Capitalocene, Chthulucene: Making string figures with Biologies', arts, activisms – Aarhus University*. YouTube. https://m.youtube.com/watch?v=CHwZA9NGWg0
Harris, L. D., & Wasilewski, J. (2004). Indigeneity, an alternative worldview: Four R's (relationship, responsibility, reciprocity, redistribution) vs. two P's (power and profit). Sharing the journey toward conscious evolution. *Systems Research and Behavioral Science, 21*(5), 489–503.
Hauser, O., Rand, D., Peysakhovich, A., et al. (2014). Cooperating with the future. *Nature, 511*, 220–223. https://doi.org/10.1038/nature13530
Held, D. (2004). *Global covenant: The social democratic alternative to the Washington consensus*. Polity.
Higgins, P., Short, D., & South, N. (2013). Protecting the planet: A proposal for a law of ecocide. *Crime, Law and Social Change, 59*, 251–266. https://doi.org/10.1007/s10611-013-9413-6
Holloway, D., & Cui, L. (2014). US–China relations in the shadow of the future. *Dynamics of Asymmetric Conflict, 7*, 137–149. https://doi.org/10.1080/17467586.2014.977316
IPCC. (2019). In P. R. Shukla, J. Skea, E. C. Buendia, V. Masson-Delmotte, H.-O. Pörtner, D. C. Roberts, P. Zhai, R. Slade, S. Connors, R. van Diemen, M. Ferrat, E. Haughey, S. Luz, S. Neogi, M. Pathak, J. Petzold, J. P. Pereira, P. Vyas, E. Huntley, K. Kissick, M. Belkacemi, & J. Malley (Eds.), *Summary for policymakers*. Climate Change and Land: an IPCC special report on climate change, desertification, land degradation, sustainable land management, food security, and greenhouse gas fluxes in terrestrial ecosystems. In press.
IPPC. (2018). https://www.ipcc.ch/sr15/chapter/spm/
Kaza, S. (2020). *A wild love for the world: Joanna Macy and the work of our time*. Shambhala.
Khimba, J., & Dyson, C. (2020). Stop ecocide: Change the law. https://www.themarinediaries.com/tmd-blog/stop-ecocide-change-the-law
Kim, S. S. (2021). https://www.youtube.com/watch?v=v9De4eYk1ls&ab_channel=SamSSenSKim%E5%90%89%E7%8C%AB
Leonard, A. (1994). The very model of a modern system-general. In *Beyond dispute*. Beer, S. Wiley and Sons.
Leonard, A. (2005). *Between momentum and control: A dynamic of democracy*. Emerald Group Publishing Limited.
Lewis, J. L., & Litai, X. (2016). China's security agenda transcends the South China Sea. *Bulletin of the Atomic Scientists, 72*(4), 212–221. https://doi.org/10.1080/00963402.2016.1194056

Lovelock, J. (2000). *Gaia: The practical science of planetary medicine*. Oxford.
Lovelock. (2006). *The revenge of Gaia*. Penguin.
Maathai, W. (2010). *Replenishing the earth: Spiritual values for healing ourselves and the world*. Doubleday.
Macy, J (2017). Mutual Causality in Buddhism and General Systems Theory. https://www.sunypress.edu/p-1176-mutual-causality-in-buddhism-an.aspx
Macy, J. (1995). Working through environmental despair. In T. Roszak, M. E. Gomes, & A. D. Kanner (Eds.), *Ecopsychology: Restoring the earth, healing the mind*. Sierra Club Books.
Markson, S. (2021). PLA's US Cash to engineer coronaviruses. *The Australian*, 5th–6th June.
Maturana, H. R. (2002). Autopoiesis, structural coupling and cognition. *Cybernetics and Human Knowing, 9*(3–4), 5–34.
Maturana, H., & Varela, F. (1973). *Autopoeisis: The organization of living*. Reidel.
McIntyre-Mills, J. (2008a). *User-centric design to meet complex needs*. Nova Science. 423 page book. Appendix on software design by Denise de Vries.
McIntyre-Mills, J. (2008b). Mobius bands and Mandelbrot sets as metaphors for systemic praxis. *Systems Research and Behavioural Science, 25*(2), 323–330.
McIntyre-Mills, J. (2012). *Wellbeing and existential risk: Engagement to address the policy and governance challenges of mitigation and accountability*. Paper presented to the Chinese Academy of Social Sciences, CASS Research Conference Sustainable Environmental Policy and Global Governance. 18–19 September.
McIntyre-Mills, J. (2014). *Systemic ethics and non-anthropocentric stewardship: Implications for transdisciplinarity and cosmopolitan politics. Contemporary systems series*. Springer.
McIntyre-Mills, J. (2017a). *Planetary passport: Re-presentation, accountability and re-generation*. Springer International Publishing AG. eBook ISBN, 978-3-319-58011-1, 10.1007/978-3-319-58011-1.
McIntyre-Mills, J. (2017b). Recognizing our hybridity and interconnectedness. *Current Sociology, 66*, 1–24. https://doi.org/10.1177/0011392117715898
McIntyre-Mills, J. J. (2019a). Summary and key themes: We are the land and the waters. In J. McIntyre-Mills & N. R. A. Romm (Eds.), *Mixed methods and cross disciplinary research: Towards cultivating eco-systemic living* (pp. 1–13). Springer.
McIntyre-Mills, J. J. (2019b). Dynamic weaving together strands of experience: Multiple mixed methods approaches to resilience and re-generation based on intra-, inter- and cross-disciplinary approaches. In J. McIntyre-Mills & N. R. A. Romm (Eds.), *Mixed methods and cross disciplinary research: Towards cultivating eco-systemic living* (pp. 15–58). Springer.
McIntyre-Mills, J. (2020). The Covid-19 era: No longer business as usual. *Systems Research and Behavioral Science, 37*, 827–838. https://doi.org/10.1002/sres.2745
McIntyre-Mills, J. (2021). Chapter 20: The Greta factor: Turning point and need for transformative research. In *Polarisation to multispecies relationships in the age of mass extinctions* (pp. 473–484). Springer Nature.
McIntyre-Mills, J. J. (2021a). The importance of relationality: A note on co-determinism, multispecies relationships and implications for COVID-19. *Systems Research and Behavioral Science, 1–15*. https://doi.org/10.1002/sres.2817
McIntyre-Mills, J. J. (2021b). Prologue: No longer top predator. In *McIntyre-Mills and Corcoran Nantes from polarisation to multispecies relationships*. Springer.
McIntyre-Mills, J. (2021c). Chapter 8: 'Stewardship: An anthropocentric misnomer'. In *From polarisation to multispecies relationships in the age of mass extinctions* (pp. 119–140). Springer Nature.
McIntyre-Mills, J. J. (2021d). From old to new taxonomies of rights, relationships and responsibilities to protect habitat. In J. J. McIntyre-Mills & Y. Corcoran Nantes (Eds.), *From polarisation to multispecies relationships in the age of mass extinctions* (pp. 31–55). Springer International.
McIntyre-Mills, J. J. (2021e). From Polarisation to multispecies relationships: Re-membering narratives. In J. J. McIntyre-Mills & Y. Corcoran Nantes (Eds.), *From polarisation to multispecies relationships in the age of mass extinctions* (pp. 173–213). Springer.

McIntyre-Mills, J. J., & Corcoran Nantes, Y. (2021). *From polarisation to multispecies relationships in the age of mass extinctions*. Springer International.

McIntyre-Mills, J. J., & De Vries, D. (2011). *Identity, democracy and sustainability: Facing up to convergent social, economic and environmental challenges*. ISCE/Emergence/Complexity and Organization. ISBN 97 80984 216536. 331 pp.

McIntyre-Mills, J. J., & Romm, N. R. A. (Eds.). (2019). *Mixed methods and cross disciplinary research: Towards cultivating ecosystemic living*. https://doi.org/10.1007/978-3-319-58014-2.

McIntyre-Mills, J., Vries, D., & Binchai, N. (2014). *Transformation from wall street to wellbeing: Joining up the dots through participatory democracy and governance to mitigate the causes and adapt to the effects of climate change* (p. 253). Springer. Second and third authors provided the appendix on software user guide (pages 193–198).

McIntyre-Mills, J., Romm, N. R. A., & Corcoran Nantes, Y. (Eds.). (2019). *Democracy and governance for resourcing the commons: Theory and practice on rural-urban balance*. Springer. https://doi.org/10.1007/978-3-319-58014-2

Meadows, D., & Randers, J. (1992). *Beyond the limits: Global collapse of a sustainable future*. Earthscan Publications.

Meadows, Donella H. . Meadows Dennis L, Randers, J. Behrens, W.W III (1972) A report for the Club of Rome's project on the predicament of mankind universe books. .

Murphy, S., & O'Carroll, L. (2020). Johnson to hail 'historic resolution' as Brexit bill comes before Commons. https://www.theguardian.com/politics/2020/dec/29/brexit-trade-deal-wins-support-of-conservative-erg-mps

Naess, A. (1995). *Self-realization: In ecological approach to being-in-theworld* (pp. 225–239). Deep Ecology for the Twenty-First Century.

Newton, P. (2021). Canada will list the Proud Boys movement as a terrorist group, February 4, 2021.

Nussbaum, M. (2006). *Frontiers of justice: Disability, nationality, species membership*. Belknap Press.

Peters, M. A., Green, B., Mou, C., Hollings, S., Ogunniran, M. O., & Rizvi, F. (Open Review), Rider S. (Open Review) & Tierney, R. (Open Review). (2020). US–China rivalry and 'Thucydides' trap': Why this is a misleading account. *Educational Philosophy and Theory*. https://doi.org/10.1080/00131857.2020.1799739

Raoni (2020) see https://news.mongabay.com/2020/02/chief-raoni-nobel-peace-prize-nominee-presides-historic-meeting-with-over-600-indigenousleaders-in-brazil/

Rayner, A. (2010). Inclusionality and sustainability – Attuning with the currency of natural energy flow and how this contrasts with abstract economic rationality. *Environmental Economics, 1*(1), 98–108.

Rayner, A. (2017a). https://en.wikiversity.org/wiki/Natural_Inclusion

Rayner, A. (2017b). Natural inclusion. In J. J. McIntyre-Mills, N. Romm, & Y. Corcoran-Nantes (Eds.), *Balancing individualism and collectivism: Social and environmental justice* (pp. 461–470). Springer.

Rayner. (2021). Chapter 36 The natural inclusion of diversity in community: Swirls within swirls and variations around a central theme and variations around a central theme. In *From polarisation to multispecies relationships: Re-generation of the commons in an era of mass extinctions* (pp. 729–748).

Rittel, H., & Webber, M. (1984). *Planning problems are wicked problems, developments in design methodology*. Wiley.

Rovelli, C. (2021). Relational Interpretation of Quantum Mechanics and Alexander Bogdanov's Worldview' by Carlo Rovelli. https://stream.syscoi.com/2021/05/05/the-centre-for-systems-studies-two-even-celebrating-the-life-and-contribution-of-alexander-bogdanov-the-annual-mike-jackson-lecture-2-june-with-carlo-rovelli-and-seminar-on-3-june-both-online/

Safi, M. (2021). Hoarding Covid Vaccine. https://www.theguardian.com/world/2021/jan/25/hoarding-covid-vaccines-could-cost-wealthy-countries-45tn

Scott, J. (1998). *Seeing like a state*. Yale University Press.

Shiva, V. (1989). *Staying alive: Women, ecology and survival in India*. Zed Books.
Shiva, V. (2012). *Making peace with the earth*. Fernwood Publishing.
Shiva, V. (2012a). *Monocultures of the mind*. Third World Network.
Shiva, V. (2012b). *Making peace with the earth*. Fernwood Publishing.
Shiva, V. (2020a). Ecofeminism and the decolonization of women, nature and the future. https://www.youtube.com/watch?v=hVbbov9Rfjg Mar 25, 2020 – Uploaded by San Telmo Museoa.
Shiva, V. (2020b). *Two Paths to the Future of Food and Farming – EcoFarm* Keynote. https://www.youtube.com/watch?v=9vNPGwqR4TQ
Shiva, V. (2020c). Ecofeminism and the decolonization of women, nature and the future. https://www.youtube.com/watch?v=hVbbov9Rfjg Mar 25, 2020 – Uploaded by San Telmo Museoa.
Shiva, V. (2020d). *Two Paths to the Future of Food and Farming – EcoFarm* Keynote. https://www.youtube.com/watch?v=9vNPGwqR4TQ
Simard, S. (2016). *How trees talk to each other*. https://www.youtube.com/watch?v=Un2yBgIAxYs
Simard & Defrenne. (2019). The secret language of trees – Camille Defrenne and Suzanne Simard. https://www.youtube.com/watch?v=V4m9SefyRjg
South African Statistics. https://www.statista.com/statistics/455931/urbanization-in-south-africa/
Spash, C. L. (2009). The development of environmental thinking in economics. In C. L. Spash (Ed.), *Ecological economics* (pp. 62–83). Routledge.
Stempel, J.L. (2020). https://unherd.com/2020/11/factory-farming-will-kill-us-all/
Stiglitz, J. J. (2011) Of the 1%, by the 1%, for the 1%. https://www.vanityfair.com/news/2011/05/top-one-percent-201105#
Stiglitz, J., Sen, A., & Fitoussi, J. P. (2010). *Mis-measuring our lives: Why the GDP doesn't add up*. The New Press.
Tierney (Open Review). (2020). US–China rivalry and 'Thucydides' trap': Why this is a misleading account. *Educational Philosophy and Theory*. https://doi.org/10.1080/00131857.2020.1799739
Tsing, A. L. (2015). *The mushroom at the end of the world: On the possibility of life in capitalist ruins* (p. 352). Princeton University Press.
UN Declaration of the Rights of Indigenous Peoples. (2008). http://www.ohchr.org/
United Nations. (2003). Human Development Index. In *A compact among nations to end poverty*. Oxford University Press.
United Nations. (2007). *Declaration of the rights of Indigenous peoples*. https://www.humanrights.gov.au/publications/un-declaration-rights-indigenous-peoples-1
United Nations. (2014). World urbanisation prospects: The 2014 revision. https://esa.un.org/unpd/wup/publications/files/wup2014-highlights.Pdf
United Nations. (2017). Sustainable development goals. https://www.un.org/development/desa/publications/sdg-report-2017.html
United Nations Office for Disaster Risk Reduction. (2015–2030). Sendai framework. http://www.preventionweb.net/drr-framework/sendai-framework/
United Nations Office for Disaster Risk Reduction. (2017). http://www.preventionweb.net/files/55465_globalplatform2017edings.pdf
United Nations Paris Climate Change. (2015). Conference of the parties twenty-first session Paris, 30 November to 11 December 2015. https://unfccc.int/resource/docs/2015/cop21/eng/l09r01.pdf
United Nations Sustainable Development Goals. (2017). https://www.un.org/development/desa/publications/sdg-report-2017.html
Vickers, G. (1968). *Value systems and social process*. Tavistock Publications.
Vickers, G. (1983). *Human systems are different*. Harper and Row.
Von Foerster, H. (1995). *Cybernetics of cybernetics* (2nd ed.). Future Systems.
Wahl, C. (2016). *Designing regenerative cultures*. Triarchy Press.
Wahl, C. (2019). Sustainability is not enough. In G. D. Dabelko & K. Conca (Eds.), *Green planet blues: Critical perspectives on global environmental politics*. edited by Taylor & Francis Group, 2019. ProQuest Ebook Central. http://ebookcentral.proquest.com/lib/flinders/detail.action?docID=5892383

Xi, J. (2015). *Towards a Community of Common Destiny and a new future for Asia*. Xinhua. March 29. http://news.xinhuanet.com/english/2015-03/29/c_134106145.htm

Yong, E. (2021). Where year 2 of the Pandemic will take us? As vaccines roll out, the U.S. will face a choice about what to learn and what to forget. https://www.theatlantic.com/health/archive/2020/12/pandemic-year-two/617528/

Young-Brown, M an Fox, M. (2014.) Coming Back to Life: The Updated Guide to the Work That Reconnects New Society Publishers. USA.

Yu, J. (2006). Making friends of enemies. In V. Gigch & J. McIntyre-Millls (Eds.), *Wisdom, knowledge, and management* (pp. 305–329). Springer. https://doi.org/10.1007/978-0-387-36506-0_16

Web Links Cited

https://www.capetownetc.com/news/nurdles-wash-up-on-cape-town-beaches/

https://www.news24.com/news24/columnists/cyrilramaphosa/cyril-ramaphosa-gender-based-violence-it-is-time-to-turn-the-tide-on-the-scourge-20201123

https://www.news24.com/news24/southafrica/news/just-in-president-cyril-ramaphosa-to-face-motion-of-no-confidence-next-week-20201126

https://www.aquarium.co.za/blog/entry/the-great-nurdle-disaster-what-to-do-if-you-find-nurdles

Nurdles and Food Security: Education for Wellbeing and Implications for Pan-African Social and Environmental Justice

Janet J. McIntyre-Mills

Abstract The chapter discusses an NGO in Cape Town that works successfully with a local community to provide education and opportunities to re-generate the community through sustainable organic farming and a range of associated activities to address employment and food security through supporting aspects of a circular economy at the community level. The case study provides an example of how 'business as usual', could be replaced with a desire to achieve a better non- anthropocentric way of engaging with living systems. Local initiatives to support organic, non anthropocentric food production needs to be buttressed by educational policy and laws at the local, national and international level to protect food security by protecting the environment and multispecies relationships, because no matter how well local communities protect local environments they are placed at risk if national and international laws are not set in place to protect local environments.

Keywords Wellbeing · Regeneration · National and international law · Circular economy and food security

1 Nurdles and Food Security: Telling a New Narrative to Promote Wellbeing

I wrote this chapter whilst in Fish Hoek in the Western Cape, known as Camissa, or 'The Place of Sweet Waters' (Mellet, 2020) by the Khoi who traded with the Dutch and supplied them with food and fresh water, essential for the voyage to Batavia and Malaysia. In the Fish Hoek Valley the beach has been strewn with nurdles washed

J. J. McIntyre-Mills (✉)
College of Education, University of South Africa, Pretoria Central, Gauteng, South Africa

College of Professions, Adelaide University, Adelaide, Australia
e-mail: janet.mcintyre@adelaide.edu.au

J. J. McIntyre-Mills, Y. Corcoran-Nantes (eds.), *Transformative Education for Regeneration and Wellbeing*, Contemporary Systems Thinking,
https://doi.org/10.1007/978-981-19-3258-8_3

up from a container. Local volunteers were working to try to pick up the tiny plastic balls and flakes.

The fisherman were out pulling in their nets of Yellow Tail, feeding in the same tidal waters. Some of the fisherman are also organic farmers at the Ocean View Organic farm but have to earn a living from diverse sources of income. Eating fish caught locally results in chewing little plastic pellets that have contaminated the ocean and entered the food chain. If the United Nations Sustainable Development Goals and United Nations Local Agenda 21 are to have a hope of being implemented then they will need to be buttressed by laws not only locally, but also regionally and internationally.

The contamination of the ocean by nurdles is a result of spillage from a container ship. According to Dordley (2020) who cites a spokesperson from South African National Parks as follows:

> Unfortunately, there are occasional incidents at sea where containers are lost overboard and the nurdles are released, causing a major environmental disaster

The further contamination of the fishing and bathing area with e-coli is the result of inadequate management of sewage, according to Kotze and Watkins (2020):

> In the Coastal Water Quality Report released by the City on March 20, Muizenberg, Fish Hoek, Monwabisi and Beta Beach are listed among 31 polluted sites along Cape Town's 307 km of coast-line. Almost half (45%) of the 49 testing sites in False Bay failed the minimum water quality guidelines in 2019.... committee member for water and waste Xanthea Limberg said routine monthly sampling in the Kirstenhoff area indicated periods of elevated E coli levels linked to sewage overflows in the general catchment…

As discussed in chapters "The Power of Story: Ocean View Organic Farmers Cooperative, Cape Town, South Africa: Growing Food With Heart and Soul" and "Learning from Nature's Classroom: Reframing Economics, Accounting and Accountability" of this volume, local organic farming and training programs to support wellbeing in this coastal area is growing, but public health and education policy needs to be supported not only at the local level, but also at the national and international level to support social, economic and environmental wellbeing. The big issues of the day are systemic and include poverty, climate change and pandemics that will impact the health of all living systems. Food security can only be protected locally when the environment of food protection is globally supported. Nakamori et al. (2011) have developed Nonaka and Takuchi's (1995) work and stressed that it is more than information management. This is the point of the argument developed in *SystemicGovernance and Accountability* (McIntyre-Mills, 2006) in which I stressed the importance of extending democracy to improve governance and accountability to protect the environment which could include monitoring to protect the food chain from pollution and contamination. This is discussed further in chapters "Learning from Nature's Classroom: Reframing Economics, Accounting and Accountability" and "Systemic Praxis and Education to Protect the Commons" of this volume which explain how positive cascades can be created by working with nature in a circular economy, instead of exploiting nature.

2 Praxis and Research Approach

2.1 Assumptions and Values

The pandemic has posed opportunities and threats. The need to be creative during lock down has paradoxically brought us closer together.[1] I 'skyped, zoomed, phoned and WhatsApped' in order to supplement face to face engagement to extend the face to face engagement that was hard won during the pandemic when lockdowns required strict social distancing and it was possible to engage wearing masks (for a short window of time from Oct to early Dec in South Africa).

3 Case Study of Transformative Praxis

The community project in Ocean View (see chapter "The Power of Story: Ocean View Organic Farmers Cooperative, Cape Town, South Africa: Growing Food with Heart and Soul" of this volume) is a beacon of hope in a community that has a high crime rate linked with unemployment, poverty, drug gangs resulting in violence.[2] The project served to inspire others as detailed in chapter "Aa!! Venda Women and Social Enterprise: Stepwise Process to Regenerative and Sustainable Living" of this volume on the Venda co-operative supported by a digital community of practice including colleagues in Indonesia and elsewhere as detailed in chapter "Transformative Education: Employing the Balanced Scorecard for Regenerative Development" of this volume.

---

[1] We need to enable people to do creative research that enables both the generation of unique contextual ideas as well as to seek patterns based on shared meanings which we co-create through dialogue. It is possible to find a way through the structural challenges by co-creating meaningful research outcomes that address the convergent social, economic and environmental challenges. This requires long term engagement with the academic community and with members of the community to enable people to learn and manage the development of re-generative and sustainable livelihoods. It can involve setting up one's own engagement processes in one's own time and /or working closely with industry partners who develop an ongoing relationship and are then willing to work with one for the common good. If we enjoy our research and our research relationships it can become a way of life that gives meaning (and hopefully a re-generative lifestyle). Our research is still in process and so we will find out how successful this approach is down the track as we map and monitor the process. It is possible to find a way through the structural challenges embedded in the academic system by creating one's own research environment and working through a digital environment to co-create meaningful research outcomes that address the convergent social, economic and environmental challenges.

[2] I visited these non-government organisations (Oct–Dec, 2020) and met with the organisers and facilitators at the organic farm and the nearby a larger organisation providing integrated care. Both NGOs are situated alongside each other and serve the surrounding community of Ocean View (comprising formal social housing and Masimuphelele an informal settlement.

3.1 Connecting with NGOs and Reflection on Engagement and Setting up a Community of Practice

The talk given by Sophie on Cape Talk the 25th October inspired me to connect with the group at the Ocean View Organic Farm and to do some research on the 'Kos System' that provides hands on training for farmers. Their mission is to be organic and to set a positive example for the community, using a gender mainstreaming approach that enables women to find their voice and opportunities for men who want to do the work as partners, rather than as co-op members. The site of the farm used to be a tomato farm. Now they have several sources of income or profit bases and their restaurant supports local vendors, they buy roti and wraps, but they supply their own vegetables, in the words of one of the women: "The mission is to grow a future that is not based on drug money and to foster a bloodless revolution …. a transformation." One of the facilitators explained her Khoi heritage and the importance of storytelling and the analysis of dreams.[3] She runs workshops with a local Khoi chief so that young people can learn their history and re-connect with the land which helps them to resist the drug gangs and become 'food gangsters' who grow and restore the land which is part of their rich heritage.

The members of Ocean View are fisherman and tailors, many members of the community are talented story tellers, musicians and artists. Some of the residents living in the valley were re-located as a result of apartheid and some have forgotten their own stories; thus the Khoi chief is trying to ensure that they learn about their identity. We discussed the importance of positive stories at a time when there was so much sadness in the valley. She then explained that the remembrance garden is important for those who have lost family.

The task of telling history is led by the Khoisan chief who teaches young people who are the future stakeholders of the Organic farm. When the African National Congress was elected to power many of the indigenous people felt that they had been 'forgotten'. The Ocean View project helps to foster a sense of community empowerment.

The original traders of the Cape called it the place of sweet waters. Today the pollution of the beaches is from plastic and storm water runoff. Nurdles washed up on the beach and needless to say are ingested by the fish and the people who eat the line fish of the day sold by the local fishermen who rely on the income to feed their families.

Some of the fishermen also work in the Ocean View Organic Farm, one of at least four neighbourhood farms set up in the valley to help feed people and boost food security during the pandemic which led to many people losing their jobs in tourism and other service industries.

[3] Dreams are vitally important in Indigenous cultures. Vera Burmann (a psychiatrist) whose work I read when doing fieldwork with Xhosa Indigenous Healers discussed the Jungian symbolism in dreams that are shared during healing ceremonies. In some ways this book is a dream of how education could be. When I met Sophia at Ocean View Organic Farm she said that one of the ways they empower young people is to ask them to share their dreams and to tell stories. She said that she had been inspired by her own school nearby and by a Montessori course she had attended.

The leadership team have dreams of scaling up the farm and to make it a tourism destination. Story telling builds on the rich oral history of the people.

They supply fresh vegetables and bread to members of the community as a form of community co-operative. Those who cannot afford food are given food if they volunteer their time. This is one of the options for reciprocal benefit. Another is to join the co-operative and to work full time or to join as partners who work flexible hours so that they can combine farming with other sources of income. Some of the fisherman are part of the organic farm where plastic from a range of sources is re-purposed for building houses, planting or drip watering systems using ingenuity and creativity. The chicken coop is now used as a restaurant and meeting place where organic food is served.

Nearby a container has been used as a bicycle recycling shop. New bicycles are created from scraps and old cycles are repaired. Nearby is a train the trainer program is linked and NGO focusing on integrated health care.

Innovative approaches to agriculture were also being shared whilst I was there. A pipe fixed to a wall with holes drilled into it into which plastic bottles are slotted make neat wall containers for a vertical, irrigated garden. Some of the fishermen also work in the Ocean View Organic Farm, one of at least four neighbourhood farms set up in the valley to help feed people and boost food security during the pandemic which led to many people losing their jobs in tourism and other service industries. The facilitators at the farm have dreams for scaling up the farm and to make it a tourism destination. The Community Organic Farm and Organic Groceries[4] is based on piece of land (6000 square metres rented from the Ocean View High School and further land is allocated near the school's soup kitchen. The five women who are members of the co-op are all graduates of the school (Photos 1, 2, 3 and 4).

Photos 1 and 2 The chicken coop restaurant

[4] https://www.facebook.com/kosgangsters/

Photo 3 Plastic bottles become building material

Photo 4 Bicycle repair shop. (Sources: Authors own photographs)

4 A New Narrative

The Khoisan Chief teaches Khoi and San language to the children as well as stories of their heritage. He has taught across the Cape and in the Kalahari and explained that the Khoi were pastoralist and the San hunter gatherers and his narrative emphasised a shared identity and did not address the way that the Khoi had competed with the San along with the African tribes that came from the north. His narrative emphasised the potential of the rainbow nation[5] (a metaphor used by Nelson Mandela) to enable people to live together in harmony despite their differences and that the farm

[5]This was an important statement because the South African economy was described as 'broken rainbow', according to a commentator on SABC discussion of the mini budget on the 29th November, 2020. The corruption scandals span all sectors of the economy and in the OFS mass action protests have expressed the need for justice, according to Harper (2020: 3):

> Mlamleli, an ally of Magashule, was arrested last month, together with Edwin Sodi, Chief Executive of Blackhead Consulting, which was paid for the 2014 tender to replace asbestos roofs. The work was never carried out.

teaches people to grow. By this he meant food as well as spiritual growth.[6] Some of the participant elders wore the dreadlocks of Rastafarianism and espouse the ideas of the spiritual movement of Ras Tafari and believe that they embody the spirit of the Lion of Juda. Their knowledge of herbs is passed on in workshops which anyone is welcome to attend.

The descendants of the dissenters that they brought as slaves to the Cape live in the Fish Hoek Valley. The call to prayers from a mosque near the Ocean View Organic Farmers Market next to a Christian NGO and the story workshops by Khoi chief and herbal medicine workshop by a Rasta, are an illustration of the rainbow nation at its best.

Story workshops along with herbal healing workshops are part of an alternative curriculum taught by the members. They ask the children to discuss their dreams and try to empower them. Story telling builds on the rich oral history of the people. Aspects of the Rastafarian movement fits well with the mission to foster a pride in African heritage and a re-connection with nature. Goldston (2020) explains the philosophy of believing in the spirit of the lion of Judah embodied by believers. But the women's co-op stresses that they are emphasising the kos (food) system as a counterpoint to the drug culture that is prevalent in the Cape and in the Ocean view area,

The high levels of unemployment and the poverty facing families has resulted in higher levels of domestic violence and child abuse. Citing a 2017 report Ramaphosa (2020) said the cost of domestic violence to the South African economy is more than 28 billion and possibly exceeding 42 billion. In the wake of Covid-19 this is much higher and protests to end the violence in South Africa were led in August 2018.[7] In the motion of no confidence by the African Transformation Movement it was stressed that more than ten million are unemployed and that 30% and that in 2018 300 women were murdered,

> while the number of sexual crimes including rape and sexual assault jumped by 4.6% to more than 50,000.

---

[6] By this he meant food as well as spiritual growth. Some of the participant elders wore the dreadlocks of Rastafarianism and espouse the ideas of the spiritual movement of Ras Tafari and believe that they embody the spirit of the Lion of Juda. Their knowledge of herbs is passed on in workshops which anyone is welcome to attend. Visited the Ocean View Organic Farm run by descendants of the first inhabitants of the Cape, the Khoisan people and the displaced slaves from Batavia and Malaysia. The call to prayer from the Mosque was evocative of the sounds of Indonesia. Nearby the Christian Church mission Living Hope is part of the rainbow nation along with the Rastafarian members and those who uphold the traditional beliefs held by the Khoi.

[7] https://www.unwomen.org/en/news/in-focus/end-violence-against-women?gclid=Cj0KCQiAh4j-BRCsARIsAGeV12D93zaolGtaTb7VB4uXomcoNt9G-zurpj85oOuVm1RAEaPKZu

In South Africa, women call for #TotalShutdown of gender-based violence. Date: Friday, August 3, 2018

During the week of action to raise awareness on the need to stem violence against women for 16 days, it was stressed that the number of violent crimes against women and children during lockdown tripled. In November the UN 16 Days of Activism against Gender-based Violence.

The high rate of crime, domestic violence and poverty is linked with gang culture which is being replaced with a love of gardening and Khoisan culture combined with some of the Rastafarian principles of being filled with a sense of empowerment and the spirit of the lion of Juda or Africa. Sophia's motto is that the participants in the project are now no longer drug gangsters they are food gangsters. Thus, a strong emphasis on women's rights and gender mainstreaming puts in place the mission Reggae music, dreadlocks resonate and it is interesting that the Reggae movement has also mobilised the young farmers in Indonesia in the Bandung region. During my visit I connected the facilitator who works as the farm manager with a senior academic in Indonesia who works closely with both Indigenous Sundanese First Nations and with the Reggae farmers near Bandung. The Organic Farm in Cape Town and regional areas of Bandung West Java both strive towards displacing colonial culture and restoring the Indigenous spirit and indigenous knowledge systems. The Africa Indonesia Conference (1955) held in Bandung emphasised the importance of displacing oppressive systems.

Indonesia and South Africa need to pay more attention to Indigenous first nations. This group in the Cape think that the ANC has not given the so-called Cape Malays and Khoisan Camissa People enough voice and in Indonesia the Sundanese Indigenous people want the keep their traditions and their pristine forests that they have nurtured for centuries. The UN Declaration of the Rights of Indigenous people is very important in this context.

The strength of Organic Farm is the ability of the participants to see the big picture by building on the Indigenous understanding of our human connectivity which has been lost in the process of rigorously imposing us/them thinking. The potential of a digital community of practice to share lessons of inspiration cannot be underestimated.

The story telling[8] is part of the healing process and the Khoi chief shares some of their history and stories of origin. The young children were asked to explore the

[8] Whilst I was not privy to these stories I have read Van der Post (1961: 152–163) whose fabricated versions re-tell the stories of origin of Dxui who created the world and who understood that they were part of a creation in which they participated. Dxui takes many forms as part of the living system. He becomes male and female, water, a lizard, is rescued by a bee and deposited upon a flower and become Mantis, the insect who bows and who has a face very like the face of the San. The bee is sacred and creates honey, the lizard is sacred as it re-generates when it loses a tail. Peter Ammann gave a zoom talk for the Jung society in Cape Town. My friend is the librarian for this society. 'The lost world of the Kalahari' (1957) and the sequel 'The Heart of the Hunter' (1961) by Laurens Van der Post provide insights into oral history of the San of the Kalahari. According to Van der Post (1958) they were the first people of South Africa and were probably driven South from Egypt and perhaps crossed land bridges from Asia when Gondwana land was one. They were driven out of the Western Cape by black and white pastoralists who believed in domesticating flocks and owning property. The themes in their stories have universal appeal and speak to the

organic garden and to tell what they had seen, heard, tasted, smelled or perceived with any of their senses. They were very shy when they told their stories about the butterflies, the vegetables, their enjoyment of being out in the sun with their friends and looking forward to the pizza that they could smell being cooked for their lunch. They said they had heard one of the baboons calling and the sound of the birds. They were much more confident when they sang as a group and when they danced and some took turns at drumming and storytelling.

The high rate of crime, domestic violence and poverty is linked with gang culture which is being replaced with a love of gardening and Khoisan culture combined with some of the Rastafarian principles. The motto is that the participants in the project are now no longer "drug gangsters", they are "food gangsters". Reggae music, dreadlocks resonate and it is interesting that the Reggae movement has also mobilised the young farmers in Indonesia in the Bandung region.

The Africa Indonesia Conference held in Bandung emphasised the importance of displacing oppressive systems. Indonesia and South Africa need to pay more attention to Indigenous first nations. The so-called Cape Malays and Khoisan[9] and the Sundanese Indigenous people want the keep their traditions and their pristine forests that they have nurtured for centuries. The UN Declaration of the Rights of Indigenous people (2007) is very important in this context.

A sense of being connected with the land and ocean has long been understood by Khoisan and the philosophy of Rastafarianism has been embraced by some of the male participants in particular at the Organic Farm. They believe that as Rastafarians they hold the spirit of the lion of Judah or embodied by the progressive Ethiopian leader Hailie Salasi or Ras Tafari (Goldston, 2020). The patriarchal elements of the approach seem to be less evident in the context of the Organic Farm where five

---

human spirit of the first peoples who saw themselves as part of the land and sky. The baboons in the Cape were known as 'the people who sit on their heels' (Van der Post, 1961: 195). The threat to the habitat of the baboons and their displacement by houses is a threat to the biodiversity of Table Mountain. The baboons are the gardeners who fertilise the fynbos. The distinction between human and animal taxonomy was blurred by the San. Their stories of hunters in the sky, the origin of fire stolen by their Mantis God (the trickster) from the Ostrich are just some of the stories that have layers of meaning that are unfolded by Jungian analysts such as Peter Amann in his lecture on early African stories. The Mantis is central and plays a role as the re-generator of the slaughtered Hartebees who symbolises the San culture because he lives in small groups like the San. Mantis suffered for tricking the Ostrich (the largest bird) by stealing (some of the)fire from under her wing. As a result the ostrich lost the ability to fly. The Mantis also suffers and dies when he has fulfilled the creative act of procreating and his wife uses his body to provide food for her young. So there is a parallel with the crucifixion and resurrection story. The trickster gave the San fire, which is life, culture and safety and willingly sacrifices his life to through ensure the safety of his people. Mantis is s a god for his people. His little face looks oriental and he is always praying. Sadly the Dutch could not tell the difference between the Khoi (Hottentots) and the San (Bushman) so they mistakenly called Mantis (Hotnotsgod, the Hottentots God).

[9]The first people of South Africa have been neglected by all South Africans. The divide between the pastoralists seeking territory and the hunter gatherers was drawn early in the history of South Africa. We need to regain the soul of the first people who walked gently on the land and took no more than was necessary.

women have managed to form a co-operative in which they work full time. They invite those who wish to work part time to form a partnership without fixed hours, so that they are able to earn a living by fishing or other forms of employment. Currently gender mainstreaming seemed to be along the lines of men preferring a more flexible arrangement and woman being prepare to work full time in the co-operative. Ad hoc volunteering is also encouraged so that members of the Ocean View community can spend a day working on the farm and they will receive a box of vegetables.

5 New Narratives to Inspire Alternative Forms of Organisation

A new narrative is needed to enable people to ensure their food security and to help us to realise that we are part of this food system not separate from it. Vandana Shiva uses the concept 'ecological apartheid' (Shiva, 2020 a, b) which accurately sums up the challenge of 'making peace with earth' (Shiva, 2012) which involves re-connecting through telling new stories and engaging with one another in *real* time, not only on digital screens (Greenfield, 2015) which is alienating and paradoxically become increasingly important during pandemics.

New Zealand has supported the notion of a 'Wellbeing Budget'[10] which resonates with some of the Māori notions of wholistic wellbeing (New Zealand Dept of Public Health, 2007). In 2019 Jacinda Ardern supported the notion that wellbeing indicators should be measured by government departments:

> clean air and water, access to housing and health care, education standards, economic mobility, social harmony and community safety, and a safe climate, is the core of work of government. (Field, 2019: 32)

This policy provides a forward and is in line with aspects of the work by Stiglitz et al. (2010) who stress the importance of taking *wellbeing* into account, rather than productivity and profit. This, however cannot be achieved unless it is supported globally as pollution, carbon emissions and climate change and commodification of the environment have no barriers.

The award of the Nobel Peace Prize to honour the work of the World Food Programme (29th October 2020)[11] makes an important point about the link between food security and world peace. It was poignant that on that day the plastic nurdles washed up along the Cape Coast.

Society needs to face up to the implications of "no longer being top predator" (McIntyre-Mills, 2021) and learning from the implications of the Covid-19

[10] https://www.weforum.org/agenda/2019/05/new-zealand-is-publishing-its-first-well-being-budget/

[11] 29 October 2020, The Nobel Peace Prize for 2020. NobelPrize.org. Nobel Media AB 2020. Thu. 29 Oct 2020. https://www.nobelprize.org/prizes/peace/2020/press-release/

pandemic that has required us to pause and to reflect on what we value and why. We could choose to maximise social and environmental factors and meaningful jobs, rather than focusing on business as usual, where a minority wins at the expense of current and future generations of life. Clearly nation states need to pivot quickly to mitigate and adapt to the pandemic. Australia has demonstrated that society is capable of rapid transformation to flatten the curve of COVID-19, but it needs to apply this zeal to flattening our carbon footprint to minimize the risks of further loss of life through climate related disasters that result death and displacement on an unprecedented scale. Lowering our dependency on carbon is linked with eliminating the use of plastics from the food chain. The potential for alternative bioplastics is discussed in chapter "Learning from Nature's Classroom: Reframing Economics, Accounting and Accountability" which suggests using ecofacturing of cassava or other products to create employment in a circular economy.

The lack of non-anthropocentric emphasis on the way in which nature has been controlled and mastered needs to be remedied. Alternative forms of organisation are possible to support 'wellbeing stocks' but this needs to be achieved through creating a change in values through reaching out and supporting one another. New forms of education need to de-centre colonialism The Rhodes Must fall campaign, The Fees Must Fall campaign (see chapters "Rethinking Teacher Education Practicum for Transformative and Regenerative University–Community Partnerships" and "Online Learning and the Pedagogy of Resilience, Agency and Protest: Lessons from the Covid-19 Experience" by McKay and Mabunda) and The Black Lives Matter Campaign in the wake of the brutal death of George Floyd (when he was arrested by police in the USA) has sparked a global movements that have raised questions about social justice and the need for black voices to be heard. In South Africa the high cost of education and the high rate of unemployment requires finding ways to ensure that people remain on the land and that farming is given the respect that it deserves through enabling land sharing and co-operatives based on social and environmental justice principles that support a sustainable and cyclical economy.

In order to manage the commons, mutual agreements need to be negotiated and records need to be kept, in order to protect the interests of stakeholders.[12]

---

[12] User centric design and participation by those who are to be at the receiving end of health and education needs to inform the process(McIntyre-Mills, 2008). In order to manage the commons, mutual agreements need to be negotiated and records need to be kept, in order to protect the interests of stakeholders. The commons needs to be theorized as a legal concept (Marella, 2017) and as a transformative governance concept (see Planetary Passport (PP), McIntyre-Mills, 2017, Systemic Ethics, McIntyre-Mills, 2014, Wirawan, and McIntyre-Mills, 2019). Covid-19 provides a once in a lifetime opportunity to re-think our place in the web of life and to think seriously about the consequences of our decisions.

Furthermore, in order to have a hope of achieving the UN Sustainable Development Goals (2030) this volume proposes a non anthropocentric approach that recognizes humanity as a strand in the web of life. Protecting habitat for diverse species is vital for human security.

Indigenous people the world over have placed their sense of connection as part of nature at the centre of their beliefs and this philosophy and its political implications have been buttressed by the UN Declaration of Human Rights (2007) that stresses that indigenous rights are linked with culture and land. In many parts of the world these rights are ignored or not placed centre stage, this is true for Indonesia where the Sundanese strive to keep their ancient forests intact and in South Africa the Khoi and San are not given a voice, whilst in Chile they are also excluded.

The Africa-Indonesia Bandung Conference in 1955 on development stressed the importance of de-colonisation and local voices, but indigenous people and many of the rural and urban poor feel left behind in South Africa and West Java. The reggae movement and music, along with dreadlocks and trendy colourful clothing is considered important for young farmers in It is important to note that in a conversation with the Khoisan chief, he explained that they also have close familial links with the Xhosa and other groups in Botswana and in the Kalahari (where they were displaced) so they are not trying to emphasise separation, rather their right to be recognised and integrated as part of the rainbow nation of South Africa who have strong links with neighbouring regions and a sense of the need for local ideas to be emphasised through knowledge management that honours local traditions, shares and applies the tacit and applied knowledge that is then institutionalised (Nonaka and Takuchi 1995, McIntyre-Mills et al., 2014). Improving health and wellbeing is dependent on re-framing social, historical, economic and environmental narratives.

6 Public Education and Post National Policy to Protect the Commons

The Paris Climate Change agreement (2015) and the forthcoming Glasgow summit (2021) could be buttressed in international law to ensure that neoliberal economy is transformed so that it no longer relies on fossil fuels and plastics that discount the future in the bid to make short term profits.[13]

[13] A further policy concern that needs to be born in mind to ensure food security – is the UN Report on Urbanisation (2014) that stressed that the majority of the worlds urban populations will be in Africa and Asia by 2050 and that the risks associated climate change will exacerbate the convergent social, economic and environmental challenges associated with urban living. This has also been stressed by the Sendai Risk Platform and associated reports (2015–2030)). Importantly, by 2020, 66% of South Africans live in urban areas which has implications for food security and the need to balance rural and urban living. This requires implementing new architectures for governance, economics and education.

An added problem is a lack of understanding that all living systems are sentient (Rayner, 2021, pers. comm) and that they re-generate they are *not* merely 'building blocks' to be manipulated by human architects. The fear of Covid-19 has brought the global economy to a halt in many parts of the world. The desire to resume 'business as usual' (McIntyre-Mills, 2020) perhaps should be replaced with a desire to achieve a better way of engaging with living systems of which we are a strand. Bollier and Helfrich (2012: xii) stress that as commoners we need to:

A further policy concern that needs to be born in mind to ensure food security – is the UN Report on Urbanisation (2014) that stressed that the majority of the worlds urban populations will be in Africa and Asia by 2050 and that the risks associated climate change will exacerbate the convergent social, economic and environmental challenges associated with urban living. This has also been stressed by the Sendai Risk Platform and associated reports (2015–2030)). Importantly, by 2020, 66% of South Africans live in urban areas which has implications for food security and the need to balance rural and urban living. This requires implementing new architectures for governance, economics and education.

An added problem is a lack of understanding that all living systems are sentient (Rayner, 2021, pers. comm) and that they re-generate they are *not* merely 'building blocks' to be manipulated by human architects. The fear of Covid-19 has brought the global economy to a halt in many parts of the world. The desire to resume 'business as usual'(McIntyre-Mills, 2020) could be replaced with a desire to achieve a better way of engaging with living systems of which we are a strand. Bollier and Helfrich (2012: xii) stress that as commoners we need to:

> start seeing ourselves as commoners in relation to others, with a shared history and shared future" to create "a culture of stewardship and co-responsibility for our common resources while at the same time defending our livelihoods… It asks us to think about the world in more organic, holistic and long-term ways…

In order to protectwellbeing we need *both* a transformation of *a priori* norms and values and *also a-posteriori measures* of wellbeing to protect and re-generate living systems by addressing 'wellbeing stocks' a concept adapted from Stiglitz et al. (2010, 15) to refer to a multidimensional measure of wellbeing as detailed in Planetary Passport (McIntyre-Mills, 2017). Local, national and international laws

---

> start seeing ourselves as commoners in relation to others, with a shared history and shared future" to create "a culture of stewardship and co-responsibility for our common resources while at the same time defending our livelihoods… It asks us to think about the world in more organic, holistic and long-term ways…

In order to protect wellbeing we need *both* a transformation of *a priori* norms and values and *also a-posteriori measures* of wellbeing to protect and re-generate living systems by addressing 'wellbeing stocks' a concept adapted from Stiglitz et al. (2010: 15) to refer to a multidimensional measure of wellbeing as detailed in Planetary Passport (McIntyre-Mills, 2017). Local, national and international laws are needed to protect the environment spanning local engagement to protect the environment such as the Aarhus Convention (1998) that provides the right for all residents of the European Union to have access to information, to speak out and be heard at the local government level. If they are not heard then they have the right to take their concerns to the next level, namely the European Parliament and the European Criminal Court. Florini (2003) suggests that this convention could be usefully applied in other post national contexts. Social and environmental justice in turn needs to be buttressed by supporting an international law of Ecocide (Higgins et al., 2013) that stresses the need to protect the wellbeing of an ecosystem and *all the inhabitants* of a region. It is couched in deliberately non- anthropocentric language and goes on to stress that organisations and companies need to be held to account.

are needed to protect the environment spanning local engagement to protect the environment such as the Aarhus Convention (1998) that provides the right for all residents of the European Union to have access to information, to speak out and be heard at the local government level. If they are not heard then they have the right to take their concerns to the next level, namely the European Parliament and the European Criminal Court. Florini (2003) suggests that this convention could be usefully applied in other post national contexts. Social and environmental justice in turn needs to be buttressed by supporting an international law of Ecocide (Higgins et al., 2013) that stresses the need to protect the wellbeing of an ecosystem and *all the inhabitants* of a region. It is couched in deliberately non- anthropocentric language and goes on to stress that organisations and companies need to be held to account. Only by introducing international law could an event such as the spillage of nurdles or oil along beaches be prevented and thus end the contamination of the ocean and the food chain.

A further law that could help to protect the rights of the Indigenous people internationally (such as the San and the Khoi) is the UN Declaration of the Rights of Indigenous Peoples (2008) that stresses that identity and wellbeing are linked with culture and access to the land that they have nurtured for centuries and from which they have been dispossessed. By re-centring indigeneity and changing the narrative some of the issues raised by the neo-liberal carbon economy could be addressed.

Pan Africanism could usefully be applied to support a version of the Aarhus Convention (1998) which requires that all local residents who are citizens of a post national regional federation be given the right to access environmental (and related) information, a voice and a right to be heard. This governance architecture of supporting voices at the local government level and post national level through regional parliaments and courts has been described in detail by Ann Florini (2003) in her book on alternative forms of democracy and could be usefully adapted and applied in the African continent. Social and environmental justice in turn needs to be buttressed by supporting an international law of Ecocide (Higgins et al., 2013) that stresses the need to protect the wellbeing of an ecosystem and *all the inhabitants* of a region. It is couched in deliberately non- anthropocentric language and goes on to stress that organisations and companies need to be held to account. Only by introducing international law could an event such as the spillage of nurdles or oil along beaches be prevented and thus end the contamination of the ocean and the food chain.

7 Conclusion

In order to re-generate wellbeing and prevent pandemics we need transformation of *a priori* norms and values, *a-posteriori measures* of wellbeing to protect and re-generate living systems locally, nationally and at a post national level and laws to buttress multispecies protection through international law, such as the proposed Ecocide Law (Higgins et al., 2013), so that 'all the inhabitants' of a region can live peacefully.'

A case is made for (a) expanding pragmatism so people think about the consequences of their actions and (b) enabling people at the local level to create an alternative future whilst (c) networking with wider post national groups. Nakamori et al. (2011) have developed Nonaka and Takuchi's work and stressed that it is more than information management.

We face a dismal future if 'Business as Usual' prevails (Beck, 2010; Bostrom, 2011; IPPC, 2018, 2021; McIntyre-Mills, 2020). Facing up to 'existential risks' (Bostrom, 2011) can take the form of community projects such as Ocean View organic farm as a practical way to engage in action learning and through protecting current and future generations (see Thunberg, 2021[14]). Education and research needs to be re-conceptualised to take into account many ways of knowing including the arts, sciences, spirituality, an appreciation of diverse species intelligences and a willingness to learn from and adapt to the complexities of nature.

'Earn while we learn and grow a future together' is the slogan for working together with rural communities supported by vocational education and training through a community of practice as detailed in chapter "Learning from Nature's Classroom: Reframing Economics, Accounting and Accountability" of this volume.[15] The diary below gives a sense of the trans contextuality of this research (Box 1).

Box 1: COVID Diary and Self Reflection
5th of September, 2020

I watched Ramayana a livestreamed ballet from Jogjakarta. The heroic ballet is a timeless story of good and evil, love and perseverance as well as the way human beings rely on nature. The princess abducted to the lower realms is rescued by her prince with the help of an army of white monkeys who help to defeat evil. The performance was shared as a gift and as an act of cultural diplomacy by the Indonesian government with the Indian government. The performance was danced with transparent face shields and cloth face coverings to protect the dancers.

(continued)

[14] School strike week 166. Global climate strike in Stockholm #Uproot the system and #FridaysForFutyre#ClimateStrke.

[15] A paper entitled: McIntyre-Mills, J. with Wirawan, R. (2020) De-coupling social enterprise from neo-liberalism: the potential of ecovillages 9th Social Business Academia Conference 2020, November 5th as part of the program from 4 to 6, 2020, SBAC 2020

Box 1 (continued)

The rate of increase in Covid-19 infections in Indonesia[16] at the time had reached 846,765, whilst in India[17] it was beginning to peak at about 10.5 million, according to the World Health Organisation at that time.

The message about the importance of cultural heritage and the Buddhist notions of respect for all sentient beings was timely and it underlined the way in which education could be enhanced through cultural messages.

Without protecting naturalhabitat the famous primates of Borobudur would be threatened. Habitat is all important for the establishment of new approaches to education.

8th October, 2020

All the airport lounges and kiosks serving food and drink were closed. The COVID-19 management application that I was asked to download only works on the latest version of an iPhone. At Adelaide Airport I was allowed to proceed to the efficient Qatar clerk who checked my luggage through to Cape Town and handed me a form to fill in for the South African Health Department that requested me to download the COVID-19 Safe Application. I tried but the application was only compatible with the latest version of iPhone. What a waste of government money that will be, I thought and I face problems when I arrive, but I am sure I will not be alone! People simply do not always have the latest iPhone and I doubt that South Africans will have one! Needless to say on my arrival in Cape Town I was not asked for this application and no one had downloaded it as it was indeed incompatible with most phones.

No vending machines were available for water. I waited for the international departures section to open and was issued a face shield before joining the queue and asked to wear it at all times. I put on the gear and waited, reading the news on my phone. Donald Trump had recovered from Covid 19 and was declaring his intention to attend rallies, or at the very least to host a rally from the White House Balcony.

The selfie I sent via WhatsApp to family and friends made me look like I was wearing a version of a bourka. We were called to board in zones. My seat was in zone 2 and I had a clear seat on either side. The flight was uneventful expect for the turbulence and the difficulty in eating food under the face shield. I had been told to expect no food at all and so had a box of bliss balls

(continued)

[16] In Indonesia, from Jan 3 to 6:51 pm CET, 13 January 2021, there have been 846,765 confirmed cases of COVID-19 with 24,645 deaths. Data last updated: 2021/1/13, 6:51 pm CET

[17] In India, from Jan 3 to 6:51 pm CET, 13 January 2021, there have been 10,495,147 confirmed cases of COVID-19 with 151,529 deaths. 2021/1/13, 6:51 pm CET

Box 1 (continued)
and an oat bar in my hand luggage which I decided to save for the time spent in Doha at the airport hotel.

Qatar is a state funded by oil, but striving to diversify and develop a knowledge based economy. One of the ways in which it thrives is through running one of the few active airlines.

In preparing to fly to Qatar I did a bit of reading on this part of the world and was told by a colleague that he had visited the wonderful Doha museum before the pandemic. I would have loved to explore the ancient art and culture in this this part of the world.

It is a place where Islamic culture prevails and women's place is still largely shaped by traditional patriarchal ideas. The hot desert outside the air-conditioned buildings can (almost) be forgotten by the athletes competing in air conditioned stadiums and the many tourists enjoying the elaborate lifestyle fuelled by Petro dollars.

The ancient Bedouin culture could be forgotten in the airport if the occasional white shrouded figures were not in the crowd.

Doha airport was deserted when we arrived and I found no difficulty locating the Oryx Hotel in the airport transit section. Unlike Adelaide Airport it was buzzing with activity. Every boutique, perfume and chocolate shop was plying its trade. Children were playing on elaborate play sculptures of giants made out of copper and in the thronging crowds were a few men dressed in desert whites and women in bourkas. Their face coverings seemed familiar whereas before they looked exotic. Covid-19 is teaching us many new ways of living and helping us to recognise that culture is indeed merely diverse responses to life.

12th October, 2020

The economy in South Africa has suffered and some of the bailout funding has been said to have been misappropriated. This morning on Cape Talk the discussion was not on corruption, but instead on the murder of a young farm manager allegedly by a known cattle thief which has been followed by protests at the Union buildings. Callers to talk back radio discussed the need for food security and for people to remain on the land as well as the roots of racial hatred and the lack of on the ground policing. Some stressed that whilst the plans to protect farmers are good, the lack of vehicles makes support in these areas negligible.

The next topic of conversation was the Township Economic Bill which specifies that South African residents should have priority. A caller stressed that given the economic hardship the bill made sense.

A representative for Lawyers for Human rights stressed that she was about to table a response to the bill because it could be seen as a recipe for a rise in xenophobia. She also stressed that the bill did not take into account the

(continued)

Box 1 (continued)
research which shows that the businesses that are set up by foreigners provide employment (on average) for at least two South Africans.

13th October, 2020

Today on Cape Talk the arrest of the deputy national police commissioner Bonang Mgwenya was accused of rigging a contract for the blue-light tender for police vehicles. Her 'kick back' was a BMW promotional vehicle. The news is filled with discussions on the breakdown of essential services. The post office workers have found that they do not have medical cover as payments for this security have been suspended due to government debt. The rail services have not been functioning well in the Gauteng area which has led to protests resulting in the Transport Minister Mbalula calling the Veteran's Association spokesperson 'a thug'[18] (Feketha and Mavuso, 2020) after he had been accused of allowing the rail system to collapse. An announcer on Cape Talk mentioned that traffic lights were not working and concerns about a reliable supply of energy have focused on Eskom and Koeberg. The former – because of the unsustainability of coal and – the latter because of the risks associated with nuclear power in the wake of an earthquake.

Whilst the Western Cape has received good winter rains, the Eastern Cape has a severe shortage of water, because the rain that did fall was not over the catchment dams. As a result the local residents in places such as Port Alfred are largely reliant on rain tanks for their supply.

All levels of government are impacted by accusations of mismanagement and fraud relating to assigning contracts inappropriately for a range of purposes ranging from funding for building social housing to funding for providing personal protective equipment for hospitals to water, waste and sewerage provision.

A 'Covid safe' Jerusalema dance was hosted by the Simons Town today, but I will be remaining in lock down for the 14 day incubation period for Covid. After that, all things being equal I will go walking on the beach.

This morning talk back radio focused on the extent of destitution facing so many South Africans including a range of performing artists who have had no income since the end of Feb 2020. This evening the topic on talk back radio was the futility of lock down in South Africa where the majority of people live in close confined housing and that COVID-19 has already affected about 60% of the population. Spikes of infection have been reported as a result of young people attending a night club in Cape Town.

The participants said that the focus ought to be on supporting the economy and not giving additional powers to a government that was not functioning

(continued)

[18] Feketha, S. and Mavuso, S. (2020). 'Mbalulu hits back at MKMVA, labels the Ace's thugs' Cape Times October 13th, page 1

Box 1 (continued)

well. By the 11th of December the situation had changed as a result of relaxing the restrictions on public movement and gatherings.

15th October

I went to purchase data for my modem and air time for my phone. The service was excellent but the young staff tend to wear their masks on their chins, barely covering their mouth or nose. I had to return the following day as the modem was not working and I wore a face shield as well as a mask and gloves. I explained that this was for our mutual protection.

The news today stressed that the rate of infections is rising and that precautions need to be maintained even though lock downs have eased and the number of people who recover is around 90%. An area of concern is racial classification. A young male teacher has been accused of fraud for classifying himself as African on one form for the education department and previously on another form as a so-called coloured. Callers were divided in their opinions. Some thought it important to put race behind us. Others said that self-classification should be consistent and that it was a way to ensure that transformation for the disadvantaged can be measured and benchmarked. A caller discussed the way in which changing classification in some cases could be seen as deception in some instances to gain advantage.

The level of disadvantage across African and coloured people can be similar, said another caller. An academic from Nelson Mandela university said that although race is a construct it does have particular meaning for people who experienced the realities of apartheid.

Clearly the issue of race remains a fraught issue in the context of rising unemployment and competition for scarce employment.

Another concern is the extent to which those who have employment have not been paid by public and private sector companies that have gone bankrupt.

Friends and family have handled lock down in different ways. One friend has limited her activity to on line connections and has hardly left her home, another has continued as many of her usual activities as possible. My elderly parents have remained in lockdown except for essential shopping and occasional visits by two service providers, until they were vaccinated in 2021.

The reconstruction and development plan was presented by President Ramaphosa who emphasised the need for local manufacturing and the improvement of both social and critical infrastructure including housing, rail and internet connectivity to support business development. It stressed that South Africa could rise from the fire of COVID-19 and could draw on the rich 'reservoir of resilience.' The potential for local production was emphasised.

(continued)

Box 1 (continued)

16th October, 2020

Senekal, Orange Free State is the area where the farm murder of Brendin Horner took place. Agri forum protesters and EFF protesters congregate in the area. The former to protest the lack of protection for farmers and the latter to protest and 'uphold the constitution'. A few weeks ago widespread unrest an damage to property followed the publication for an advert for hair care products seen as racist.

Comments made on talk back radio stressed the importance of maintaining a sense of balance. Ramaphosa stressed in social media that the murder was a tragedy and that all violence is unacceptable.

It is a truism that the majority of violent deaths are not suffered by white South Africans and the violent actions in Senekal that included the overturning of a police vehicle need to be seen in the context of concerns about the lack of security felt in farming communities. The singing of "kill the Boer" and tweets about target practice, do not help the situation. SABC news highlighted the extent of stock theft during COVID-19 and that this led to farmers laying off workers as they had less income. Two known stock thieves have been accused of the murder of farm manager in Senekal and needs to be seen within the context of unemployment, poverty and tension over land and the legacy of apartheid.

The opening of the borders to tourism is clearly for the benefit of the economy, but it does place South Africa at risk. Today there were 1770 new cases and whilst it may be tempting to say that the death rate when compared with the USA is comparatively low and that 90% of COVID-19 patients recover, the attitude towards lock down by many is that it is inappropriate for South Africa where the majority live in close proximity. The lock downs did not separate them from contact with others, it merely resulted in containing groups of people within restricted areas where it is estimated that possibly 60% of the population were infected by the virus.

Unemployment and food insecurity have resulted in Ramaphosa extending the R3500s COVID-19 -emergency payments. Whilst this policy measure strikes the right note it does not address the indigence that many face. For instance Vuso (2020)[19] highlighted the memorandum presented to the minister of Agriculture, Land Reform and Rural Development to mark International Rural Women's Day and World Food Day:

> Collette Solomon of the Women on Farms project said woman farm workers made a big contribution to the economy of the Western Cape but still did not have access to land and faced job insecurity, evictions and food insecurity.

(continued)

[19] Vuso, S. (2010). Women farmworkers march for access to land. Cape Times, Oct 16th pg3

Box 1 (continued)

Ironically, in Senekal in the Orange Free State black and white farmers have joined together to stress that stock theft is an issue faced by all farmers and that they need to stand side by side, rather than emphasising the racial aspects of the murder of a young white farmer, aged 21. One of the accused has not received bail. According to Maliba (2020: 6)[20]:

> The Senekal Magistrates Court in the Free State that was …a scene of confrontation between members of the EFF, farmers and right wing Afrikaner organisations armed with guns. The police stood as a buffer between the warring factions and prevented confrontation.

Maliba (2020: 6) stresses in the same article stresses that the president of the World Agricultural Organisation that Andre Pienaar was facing charges for inciting the violence and stresses that he is not a farmer and that many of the people involve were not farmers.

> The theft of livestock on farms is not race-based, so firstly we need not racialize this man's death. A crime has been committed and it is a crime against a farmer.

17th October

Today I went food shopping, the first step is taking a trolley wipe from a dispenser and wiping down the handles. The next is receiving a squirt of hand sanitiser and then shopping commences. I wore a mask and face shield and was glad of the protection as social distancing when queuing in not adhered to. The man weighing the fruit and vegetables advised an elderly lady not to lick her fingers:

> Mamma, please let me open the packets for you, please do not lick the fingers, because of Covid,
>
> "Quite right I said," to back him up, but the elderly lady did not listen.
>
> Don't worry about plastic packets for me. Just stick the label on the fruit and veg, we don't need plastic, do we?
>
> "Yes, mamma, too right," agreed the grocer.

Today the number of active cases rose to 700,000 which is possible an underestimation of morbidity, whilst the mortality rate is 18,309.

Chatted with D. who says that her four year old grandson is able to follow all the Carona Virus rules:

> He washes his hands and wears his mask without being told to do so…He even asked for hand sanitiser to carry with him.

(continued)

[20] Maliba, A. Mixed feelings about bail for suspect. *Weekend Argus*, 25th October

[21] Kos is Afrikaans for food

Box 1 (continued)
27th October
Spent morning doing chores and checking on requirements for my return flight. I assumed I would need to take a Covid Test 72 h prior to my Qatar flight, but was surprised to find that this was not a requirement for returning to Australia given the hotel quarantine procedures in place. This seemed problematic as it would be better to prevent the spread through prior testing!

29th October
As I worked on preparation of papers for a zoom conference on social enterprise organised by the Yunus network and an abstract for a health and wellbeing conference in Uganda, I heard radio news about the state of the economy and the extent to which corrupt government officials are able to 'get away with' misusing funds is the focus of discussion in the news. The South African government is in debt and some workers are not being paid, whilst the allocation of funding for education and health is under threat. The energy provider Eskom is bankrupt and good governance will need to be demonstrated if money is to be lent in the future.

It seems that the public sector is a battle ground of alliances and back stabbing. Some of the corruption – such as embezzling funds that were paid to clean up asbestos on luxury vehicles and instead of spending funds that ought to be allocated for basic infrastructure in hospitals – on office furniture for the managers, whilst the South African Defence Force was accused of buying Interferon from Cuba as a possible way to address the Covid 19 pandemic without following appropriate procedures. (Skitl, 2020). It was argued that a rapid response strategy underpinned the decision.

30th October

When I met a facilitator at Ocean View Organic Farm, it was explained to me that one of the ways they empower young people is to ask them to share their dreams and to tell stories. One of the founders said that she had been inspired by her own school nearby and by a Montessori course she had attended.

A spokesperson for the group, Sophie explained on an Cape Talk that they run a 'kos[21] system', rather than a 'dop[22] system' with the organic food 'gangsters' This is an inspirational success story that could help to change the lives of people living in Ocean View and it provides an example for the young people in the High School. All five of the women are graduates of the high school. As they have diversified they have also included an artist in residence. The participants hope that the co-operative would be able to generate more

(continued)

Box 1 (continued)

income for those who participate by setting up a restaurant, selling vegetables, boxes (ready planted with herbs and vegetables) and a range of self-watering vertical gardens using indigenous system of recycling plastic guttering and making holes for upturned halved plastic bottles that draw up the water. It makes it possible to maximise growing space and minimise watering. Herbalists teach traditional healing and other volunteers teach gardening and agriculture. The farm is also co-located close by the Ocean View High school, an informal Khoisan school teaching language and culture. Recycled shipping containers are used to house a bicycle shop where volunteers are taught how to mend and re-assemble bicycles. They make compost for their organic farms and they package vegetables for sale and they donate food (vegetables and bread) as part of a feeding scheme:

> We model and encourage good life styles and provide a good news story for people to look up to in Ocean View, rather than all the negative bad news stories.

The project is exemplary because it is owned by with and for the community. It has potential for expansion and further job creation.

> We also have a women's remembrance garden where we plant to commemorate women's lives...The art and music program is magical and it also serves as a way for women to express themselves as they gain increased belief in themselves...The project has developed quickly and it is beginning to have a life of its own as it is locally owned, they are able to provide good, healthy food cheaply to their own community ...

Structure and management

The group is run along the lines outlined in rules for co-operatives in South Africa that guide and discipline the participants.[23] The co-op comprises 5 actively involved women who follow a model constitution downloaded from the South African Government website and thy have a signed agreement and 3 partner growers. They hold an annual AGM which must have a quorum and they have a list of dos and don'ts to which they must all subscribe.

31st October

The stories of loss shared by my friend who works in social services were difficult to hear, because of the wind howling off the bay and the loud voices

(continued)

[22] Dop refers to the infamous practice of paying part of the wages of Cape farm workers with a tot of alcohol

[23] See https://www.sa.gov.au/topics/family-and-community/community-organisations/setting-up-a-community-organisation/register-a-co-operative

Box 1 (continued)
of forced, almost hysterical merriment of people at a nearby table after months of lock down. My friend confided that she was "never in lock down" as she needed to do outreach work in the community. Her pastoral work focuses on psychiatric and mental health issues, particularly with young people. She also delivers parcels for those out of work. Her own family have also been impacted and have lost part time jobs as a result of the pandemic.

28th November

Today in the supermarket three elderly women stood in a huddle, masks on their chins: "We are having a grandma moment in a huddle and not doing social distancing," they said to all who passed as they were ignoring the large signs requesting that we all wear masks.

I was glad I was wearing both a mask and visor!

Later in the day I met colleagues based in Indonesia, Australia and South Africa via Zoom. We explained to our Indonesian colleagues that the Camissa people are the indigenous people of the Cape, the descendants of the Khoi. Our colleague in Indonesia discussed the Sundanese Indigenous culture and their important role as custodians of the forest and how they wish to avoid being part of a tourism project. Thus they have not supported the tarmac roads that are proposed that will destroy the forest. They have lived the same way for more than 652 years and protect ancient strains of rice following the rituals of the rice goddess. Many rituals serve a pragmatic reason to remind people that there are appropriate processes to follow, such as procedures to ensure proper planting, management and sorting of seeds (Kelly, 2016). These rare rice strains are resilient and can be stored when harvested for decades if necessary. They do not wish to commodify the rice by selling it and if people visit they offer hospitality and ask for donations. They have an alternative economy and system of governance that has carefully integrated a few new sustainable development options to provide energy.

The government has not recognised their rights. They have a king and they want the right to live according to their traditions.

My Indonesian colleague stressed;

> I am Sundanese, I felt like I was travelling back in time. This is the Padjadjaran kingdom with an oral culture that supports forest management. Their rice is growing without pesticides according to the old rituals that respect the rice goddess. They would never introduce GMO. They want to fight back and protect the forest. They also want to encourage the young people to be educated and to be stay in the forest to protect it. Ida helped to set up a resilience village as part of the fight back during Covid-19. She took members of the village to visit Alam Endah and the Reggae farmers. They give rice to visitors; they do not sell it and they suggest that people can pay anything they would like to pay if they stay in the village they have micro hydro and some sustainable energy. They have their own school and they run their own radio and local TV. They have their own ministry of culture, information and water

(continued)

Box 1 (continued)

and artists teach children how to make their own videos. Anyone who visits needs to adopt their clothing styles and cover their heads.

P.R. continued by explaining: "when I wore a hat which they gave to me they said it was to ensure that people's heads do not get bigger! "He explained how he shared with them the idea of setting up Kampung IT and how organic products could be marketed and promoted. We explained that sustainable products are being promoted in line with the Indonesian policy of one village many enterprises.

We then explained that the focus of our research is to stress the need to protect food, energy and water and the need to prevent rapid rates of urbanisation, deforestation and the negative cascades caused by lack of resources. We explained that bamboo can be used for many purposes, and that as bamboo grows rapidly it reduces the need to cut down trees. We also explained that cassava could be used for bio plastics which would reduce the risks of pollution. This was topical given the spill of plastic nurdels floating in the ocean and contaminating the food chain or for that matter plastic clogging up the rivers and blocking drains.

A facilitator from the Organic Village Farm explained that it provides a window of opportunity in a gangster riven community, because the 'kos gangsters' provide an example of a new way of life into which they are invited as volunteers, partners or members. Before the Covid lock down they learned theory in school on how to take soil samples and the members of the Co-op graduated after completing a course in organic farming.

11th–25th December, 2020

A lesson on consumption was learned during my time in a medi-hotel in South Australia. Each meal was delivered in a brown paper bag that could be recycled and the disposable cutlery was wooden. Each item of food was packaged in disposable containers, some of plastic and some of polystyrene or tinfoil. Unfortunately the rule was to place items in plastic bags twice a day for collection. Over 14 days I generated a great deal of rubbish and I wondered to what extent it was recycled as there was no mechanism to pre- sort the rubbish.

During the lock down there was time to think, meditate and to do Thai chi. There was also time to listen to the Pope's message about Christmas not being about consumption. From my window the flow of people bustling to shops and to work continued during a reprieve from the spike in Covid cases a few weeks ago.

Covid-20 or the second wave of a more highly transmittable version of the virus is being discussed on the BBC news. Some countries in Europe have shut their borders to the UK. It is also being discussed as a threat in South

(continued)

Box 1 (continued)
Africa and it is wearing down the spirits of people who are already impacted by unemployment and social distancing. In Sydney the virus has not been flattened.

Aboriginal Knowing and Decentering the Colonial Mindset
24th January, 2021

In the days preceding Australia Day 2021 the ABC covered interviews with Indigenous families to ask them about their 'preferences' whether Australia Day should have a different date for celebrations of a united nation and whether the name should be Invasion Day. The issues are further underlined in a documentary called "In my blood it runs" which Maya Newell shows how a young aboriginal boy feels excluded from his own history. Clearly the issue needs to be addressed through an organisation that speaks by with and for Aboriginal Australians as suggested by the Uluru Statement of the Heart and that the need for an Aboriginal voice to parliament is overdue. In the meantime children are speaking up.[24]

On Australia Day, 26th January, 2021 Yvonne Weldon, Chair of the Sydney Metropolitan Land Council gave a moving welcome to country. She explained that the welcome is a spiritual event and she underlined the connections with her ancestors who had survived the colonisation of the land and that the welcome was to all Australians whom she invited to journey forward together:

> "regardless of what date or day you call it"…"We need to heal and we can only do this together."

Appleby (2019) in her article in *The Conversation*[25] explains that the Uluru Statement of the Heart based on representative dialogue across Australia makes a plea for a the creation of a representative organisation called "the First Nations Voice." She explains that the organisation:

> shall present its views to Parliament and the Executive on matters relating to Aboriginal and Torres Strait Islander peoples; and may perform such additional functions as the Parliament provides…

Muhammad Yunus (2020) stressed:

> Now that the economy has stopped, we can reorient it in a completely different direction. The pandemic has created an opportunity to abandon this suicidal path and

(continued)

---

[24] Vibeka Venema (2021). The 'smart and cheeky' Aboriginal boy teaching Australia a lesson - BBC News reports on the documentary by Maya Newell called https://vimeo.com/showcase/inmyblooditruns

[25] https://www.aph.gov.au/About_Parliament/Parliamentary_Departments/Parliamentary_Library/pubs/rp/rp1617/Quick_Guides)/UluruStatement

Box 1 (continued)

> create a new path which will lead us to a new world – a world of three zeros: zero net carbon emission, zero wealth concentration, zero unemployment...

Peter Turner (2020, personal communication) an elder from Alice Springs stressed to me over lunch in the garden:

> Mother nature is growling; it is time to change. It is a good thing that there have been changes to the National Anthem, but there is still a long way to go!

We agreed that being cynical was the easy route and the harder and more valuable route is to remain engaged and to keep trying to bring about change.

The issue is whether people can persuade governments to listen to their needs. The lead has been taken by young people who will need to live with the consequences of policy decisions that are outdated.

Entrepreneurship needs to be oriented towards job creation and all inventions need to be tailored to serve living systems Mohammad Yunus (2020) stresses that artificial intelligence is one such example where it should be used to create jobs, not to destroy them as stressed in this volume.

Similarly, we need to understand that what we do to others and the environment we do to ourselves. By filling the waterways with contaminants, we too become contaminated. By taking more than we need from others we rob ourselves and future generations of living systems.

A similar sentiment is expressed by Alexander McCall Smith (2019) in his popular series on Botswana in which he expresses this view through his fictional character Ma Ramotswe who explains that progress is not achieved by' planting concrete' instead of trees, because concrete suffocates and 'stops the earth from breathing'.

At the time of writing, COVID-19 cases were nearing the 2 million mark in Africa and probably higher if the cases were known. I watched a you tube video conversation with The Dalai Lama[26] (2020) who stressed the importance to change our thinking and behaviour with reference to climate change[27] and COVID-19 that all causes have consequences.

The nexus of the highest rate of unemployment in South Africa in 12 years along with the impact of poverty and lockdowns was expected to have a very high death rate in South Africa – given the high level of co-morbidities, such as TB which is endemic in populations with a high rate of HIV.

(continued)

[26] https://www.theguardian.com/world/2020/nov/11/buddha-would-be-green-dalai-lama-calls-for-urgent-climate-action

[27] https://www.youtube.com/watch?v=5inWZPOi7XE

Box 1 (continued)

Nevertheless by mid December 2020, the death rate in South Africa was still comparatively low with a 90% recovery rate[28] from COVID-19, attributed to high immunisation rates against TB and a rapid response to the pandemic through early lockdowns. But this does not address the high cost of gender based violence or the lack of access to basic health services for other illnesses or the economic costs impacting on family wellbeing, mental health and a range of chronic morbidities that will go untreated, such as cancer, heart disease and diabetes. Furthermore, many of the Covid-19 related deaths need to include mortalities associated with violence and lack of access to resources. The way people cope during the pandemic depends on their access to services, such as water, electricity and sanitation. The protests on the 12th November by residents of Khayelitsha underlined the way in which violent protests are the way in which residents react to being left out of service delivery.

The other core theme is corruption associated with on line crime and the design of applications such as the that simply do not work on technology unless it is the most up to date version that the majority of people are not using, namely the latest versions of the iPhone. It only works on the latest versions of iPhone[29], technology which South Africans are unlikely to be able to afford.

(continued)

[28] https://sacoronavirus.co.za/?gclid=Cj0KCQiAnb79BRDgARIsAOVbhRrkeqwuyLldwpwengHbN9yHu2bR_dsdwHLx-Hu22-V5go_sS_HR2CsaAiuREALw_wcB. 5,086,887 Tests Conducted, *746,945* Positive Cases Identified, *692,177* Recoveries, 20,153 Deaths.

[29] The web link cannot be downloaded, so it is a white elephant, not unlike the trains that could not be used because they did not fit the train lines, more waste age through lack of careful planning and research to ensure funds are used carefully: "Like many countries across the world, South Africa's National Department of Health has used Apple and Google's exposure notification framework to build our country's new Bluetooth contact-tracing app for COVID-19. The tech giants have, since April this year, worked together to develop a framework for governments, public health authorities and NGOs around the world to use, to custom-build local apps, and ultimately to empower ordinary citizens to work together to protect each other from COVID-19 infection. South Africa's Bluetooth contract-tracing app - COVID Alert SA – is now available for iOS and Android smartphones and can help tell you if you've been exposed to the COVID-19 virus. The app is user-friendly all the way. User privacy and security are core to the Apple-Google framework on which the COVID Alert SA app has been built. That means app users are anonymous and the app never asks for, or stores their identifying personal details, or location.

Box 1 (continued)
February 13th 2021 – 12th May 2021

It is worth reflecting on the unfolding issues surrounding the choice of vaccines. The need to respond rapidly to unfolding situations and to make the best use of resources is paramount. Sometimes the ideal, needs to be replaced with the best fit and making do.

Planning is fine, but resilience requires agility. I liked the analogy used by Ramaphosa[30] in his State of the Nation address about fynbos vegetation such as proteas requiring fire for regeneration. I also like the idea that re-generative pathways need to be explored with young people and I hope that some of the ideas expressed in this book could be of some use in enabling re-generative growth in harmony with the needs of living systems and the rights of all sentient beings.

18th April

The news of the fire sweeping across Table Mountain and University of Cape Town underline the risks posed by climate change, poverty and homelessness. A fire lit possibly to keep warm – resulted in far ranging consequences.

Critical Self Reflection

Whether or not a white South African should have a voice is not in question, but what authority it has, is in question. Being born white provided an unfair advantage in South Africa.

The notion that research can ever be undertaken without being mindful of the context is now centre stage in South Africa and the need for the academies to face up to the responsibility to de-centre, displace and re-establish or re-generate narratives that support a future that honours indigenous wisdom is paramount.

Thus this volume is a small step towards finding new non anthropocentric, non-Eurocentric ways and new stories to inspire and re-generate us. Nevertheless, I am a product of my environment and so I will need to rely on my colleagues to create pathways to a better way of living and learning from nature.

In some instances a pseudonym is preferable for an evidenced piece that protects the identity of those with whom one has engaged. Certainly in the context of civil war or a police state, this is vital in terms of ethics. Journalists should not reveal their sources or should try to 'leak' information anonymously. But these leaks by so-called whistle-blowers lead to severe

(continued)

[30] https://ewn.co.za/2021/02/11/full-speech-president-ramaphosa-s-state-of-the-nation-address

Box 1 (continued)
consequences as the Wikileaks sagas involving Julian Assange, Edward Snowdon[31] and Chelsea Manning[32] underline.

How one presents oneself in the struggle is always an issue, pitfalls abound and hubris or false humility are just two of challenges.

Presenting oneself as a shaper of events as in the case of Van der Post,[33] or suggesting that one is not at all political as in the case of Vera Buhrmann (Landman, 2019)[34] is in both instances inaccurate.

So to what extent can white South Africans have a voice? I am a South African not by birth but by registration as my parents are South Africans. I was born in Bulawayo, Zimbabwe where my father was working at the time. As a white South African I have lived a comparatively privileged life and I need to be circumspect when analysing as I see the world through the lenses of my socialisation. Unfolding the assumptions and values are vital when understanding and realising that all interpretations are filtered through our own experiences.[35]

Peter Ammann (2020) gave a Zoom lecture for the Jung society whilst I was in Cape Town which was followed by a lecture which prompted me to-read 'The lost world of the Kalahari' (Van der Post 1957) and the sequel 'The

(continued)

[31] https://www.theguardian.com/us-news/2020/oct/01/edward-snowden-book-permanent-record-court

[32] Michael Ray (2020). Chelsea Manning, Encyclopædia Britannica, December 13, URL https://www.britannica.com/biography/Chelsea-Manning. Access date, December 22, 2020

[33] Some of the lessons of the San may have been passed on through the history workshops, in the works by the now discredited Laurens Van der Post and Vera Buhrmann and in the lectures of Jungian analysts such as Peter Ammann. What can the San teach us about the protection of habitat? Their culture is more relevant than ever before as many people rely on food packages to survive. Biographical research reveals that Van Der Post was a fabricator of stories who abused the trust of a young woman and who tried to prevent Mandela from becoming South Africa's president. https://www.theguardian.com/world/2001/feb/04/uk.vanessathorpe

[34] According to the research undertaken by Landman (2019) Vera Buhrmann worked to support the Ossewabrandwacht, a far right organisation in South Africa and actively assisted the migration of children whom the Nazi party was trying to place after the second world war as part of a white immigration policy for South Africa, probably in the sincere belief that she was helping orphaned children and that she owed a debt to Germany as her family had survived the Boer War. Her Afrikaner credentials were important to her as she continued her work with the Far Right throughout her career, just as Van der Post tried to push the agenda of Buthelezi (a conservative) and to prevent Mandela's deal with FW de Klerk that led to a shared Nobel Peace Prize.

[35] The narratives woven by some figures such as Van der Post are largely fictitious. At one level Van der Post explores the stories of the hunter gatherers and their profound wisdom by analysing the lessons they have and their potential to span the boundaries between those who pursue profits at the expense of living systems. Their lessons are about elegance, sufficiency and not taking more than is needed. But Van der Post was also unfortunately not the sage old man of dreams. Clara his nurse did not exist and was a product of his imagination and he spent little time in the company of

Box 1 (continued)
Heart of the Hunter' (Van der Post, 1961) that provide insights into oral history of the San of the Kalahari. According to Van der Post (1958) they were the first people of South Africa and were probably driven South from Egypt and perhaps crossed land bridges from Asia when Gondwana land was one. They were driven out of the Western Cape by black and white pastoralists who believed in domesticating flocks and owning property. The themes in their stories have universal appeal and speak to the human spirit of the first peoples who saw themselves as part of the land and sky. The baboons in the Cape were known as 'the people who sit on their heels' (Van der Post, 1961: 195). The threat to the habitat of the baboons and their displacement by houses is a threat to the biodiversity of Table Mountain. The baboons are the gardeners who fertilise the fynbos. The distinction between human and animal taxonomy was blurred by the San.

Van der Post (1961) recounts the stories of hunters in the sky, the origin of fire stolen by their Mantis God (the trickster) from the Ostrich are just some of the stories that have layers of meaning that are unfolded by Jungian analysts such as Peter Amann in his lecture on early African stories.[36]

(continued)

the San, abused the trust of a 14 year old child with whom he fathered a child and whose conservative ideas led to his lack of support for Mandela as he mistakenly believed in the primacy of Zulu culture. Ironically although Van der Post explores the stories of the hunter gatherers and their profound wisdom by analysing the lessons which undercut the notion of wild and tame. We tame the environment and sacrifice ourselves. Although he comments on the terrible loss of identity caused by taming and controlling the San, he continues to use the language, of 'wild' and 'tame', even whilst campaigning for their rights. These cultural icons had been held up to me as a young person and the discovery that they had feet of clay was a cause of sadness, but also a vital turning point in realising how important it is to be open to critical reflection.

Another cultural icon who was created after completing my education in South Africa is Vera Buhrmann to whose work on indigenous healing I was made aware of her work when I was undertaking my MA research on indigenous healers from a socio-pollical stance on life chances of migrant workers at the University of Cape Town. I read her papers which were largely on the rituals and without any discussion of the social, cultural or political context. Recent research on her biography by a history scholar (Landman, 2019) at the University of the Western Cape confirms that her involvement was *indeed* political and that she had played an active role in the far right organisation, the Ossewabrandwacht with Nazi links and that she helped to select children (on behalf of the Nazis) to immigrate to South Africa. The post war migration policies in South Africa and Australia were an attempt to select healthy young white people to populate the colonies.

[36] The Mantis is central and plays a role as the re-generator of the slaughtered Hartebees who symbolises the San culture because he lives in small groups like the San. Mantis suffered for tricking the Ostrich (the largest bird) by stealing (some of the)fire from under her wing. As a result the ostrich lost the ability to fly. The Mantis also suffers and dies when he has fulfilled the creative act of procreating and his wife uses his body to provide food for her young. So there is a parallel with the crucifixion and resurrection story. The trickster gave the San fire, which is life, culture and safety and willingly sacrifices his life to through ensure the safety of his people. Mantis is s a god for his people. His little face looks oriental and he is always praying. Sadly the Dutch could not tell the difference between the Khoi (Hottentots) and the San (Bushman) so they mistakenly called

Box 1 (continued)

Throughout his life[37] Van der Post wanted to learn more about the San who had lived near his family farm, but who had been hunted and slaughtered. They were children whose families had been killed. Only a few remained who had been "tamed", to use the terrible language of the times, which he acknowledges as problematic, but nevertheless continues to use in his books, written in the 1950s which of course was a projection of the shadow from his colonial background. But to his credit (despite his limitations) he did try to address the culpability of those who had attempted to obliterate the San. Van der Post said that he realised the culpability of his own grandfather involved in the hunts and the loss, grief and guilt haunted him.[38]

Later after the Second World War, when Laurens survived the concentration camps in Japan, he made an effort to find the San in the Kalahari. Right at the very end of his first attempt to find them, he met a group at the Sip Wells whose dignity and musicality were profoundly moving to him. Despite the limitations of Van der Post's versions of the narratives they provide some profound lessons. The Kalahari San whom he met no longer had any artists left in their group. He wrote that they wept when he showed pictures of San art to them. Van der Post continued by explaining that for the San, hunting was only to meet the immediate needs of survival and never for sport.

Van der Post believed that the San had the ability to communicate over vast distances, through acute eyesight, hearing, ability to read the landscape and a sixth, intuitive sense. He was only able to scratch the surface of a culture that holds lessons for us today. Oral histories need to be kept alive as they have vital information for us today. They illustrate the dignity and aristocracy of living close to the land without any compromise.

(continued)

Mantis (Hotnotsgod, the Hottentots God). Vera Burmann (a psychiatrist) whose work I read when doing fieldwork with Xhosa Indigenous Healers discussed the Jungian symbolism in dreams that are shared during healing ceremonies.

[37] Laurens Van der Post was once offered a chair in Jungian psychology by an American who had founded a department of Jungian therapy, having (ironically) made a fortune in the oil industry. He had once heard a talk given by Vander Post that inspired him. Van der Post wisely declined, saying he did not know enough to take a chair, he was a story teller. The stories that had been shared by his San nurse, Clara who worked for his family, had planted a seed, a love of the bush and the magic evoked by the stories.

[38] Van der Post (1956) he stressed that he imagined that he heard the battles and cries in the bush echoing from the hills around the farm and believed that the white colonisers and the black armies who fought the colonisers were equally guilty in slaughtering the little hunter gatherers who had such wisdom and spirituality expressed in the art work he found in the caves. He also found little broken horns that had been used to carry paints. The farm workers believed the area was haunted. One of the very old farm workers confessed that as a child she had been told how people ate human flesh during the terrible wars between the black invaders who were competing across the tribal groups for power. The contest for land by the white and black pastoralists drove the San out further

Box 1 (continued)

Van der Post explains that those who disagreed with others, simply moved away. Their understanding of plant chemistry enabled them to make lethal poison and to create an antidote that they always carried with them. Humour, dance and laughter were used to entertain and educate and some of the dancing and singing helped to develop an altered state of consciousness that raised the energy, spirt or! Num that enabled healers to lay hands on people with illnesses.[39]

As I understood the work of Van der Post, the quest underpinning his journey (or perhaps his metaphors or fables) was to understand the divide between the hunter gatherer and the settled pastoralist who became a property owner. Once property (land, animals, food) are seen as belonging to the owner, the right to defend it produced rigid arguments couched in Calvinist terms, in the case of the Afrikaners and pragmatism in the case of the Khoi and other African competitors – who also fought amongst themselves and with the colonial invaders for survival. All pastoralist were against the raids by the bushman and justified hunting *them*, even though many had shared their knowledge to help those who they had found stranded in the bush. The terrible irony of their own behaviour towards Indigenous people remains a shadow.

It was serendipitous that whilst writing this chapter my colleague Prof Alexander Christakis sent to me some comments about a book written by Peter Turchin in 2016 that makes the case that *war* is one of the organisational forces responsible for making society co-operate. It is a very bleak thesis in my opinion, because co-operation and competition are equally important for human survival.

Frans de Waal (2009) in his book the 'Age of Empathy' makes a different argument. He said that we evolved because of our ability to develop trust within groups, which can of course support co-operation with in group members and conflict with outlaws with whom they compete for resources. The hunter gatherer mentality has been displaced by the soul of the accumulator and our group denial of our complicity. For Van der Post the quest underpinning his journey was to understand the divide between the hunter gatherer and the settled pastoralist/a property owner. Once property (land, animals, food) is seen as belonging to the owner, the right to defend it produced rigid arguments couched in Calvinist terms, in the case of the Afrikaners and

(continued)

and further into the desert of the Kalahari. According to Van der Post, they alone did not resort to cannibalism, but kept their dignity. They also kept their dignity when they were caught in ambush.

[39] According to Van der Post in some contexts when healing was not possible, members of a group were left to the desert.

Box 1 (continued)
pragmatism in the case of the Khoi and African competitors who also fought between themselves and with the 'outsider' for survival. The divide between the nomad and the pastoralist is the divide between *being* and *having*!

The close link between hunter gatherer and the land is lost when accumulation of profit, rather than sufficiency and survival provides 'justification' for displacement or extermination.

Multispecies Symbolism in Stories

Van der Post (1961: 152–163) re-tells the stories of origin of Dxui who created the world and who understood that they were part of a creation in which they participated. Dxui takes many forms as part of the living system. He becomes male and female, water, a lizard, is rescued by a bee and deposited upon a flower and become Mantis, the insect who bows and who has a face very like the face of the San. The bee is sacred and creates honey, the lizard is sacred as it re-generates when it loses a tail. A sense of the greatness in the small has been lost by the neo-liberal economy that "knows the price of everything" without understanding the value of wellbeing and the relationships on which it depends (adapted from Oscar Wilde's, 1893 definition of a cynic in Lady Windermere's Fan).

Bibliography

Aarhus Convention. (1998). On access to information. Public participation and access to justice in environmental matters, Denmark, 25 June. http://ec.europa.eu/environment/aarhus/

Ammann, P. (2020). https://speakingofjung.com/podcast/2020/3/20/episode-57-peter-ammann

Appleby, G. (2019). https://theconversation.com/the-uluru-statement-is-not-a-vague-idea-of-being-heard-but-deliberate-structural-reform-142820

Beck, U. (2010). Climate for change, or how to create a green modernity. *Theory, Culture and Society, 27*(2–3), 254–266.

Bollier, D., & Helfrich, S. (2012). *The commons strategies group*. Levellers Press.

Bostrom, N. (2011). *Existential risk prevention as the most important task for humanity*. Faculty of Philosophy & Oxford Martin School University of Oxford. http://www.existential-risk.org/

De Waal, F. (2009). *The age of empathy: Nature's lessons for a kinder society*. Harmony Books.

Dordley. (2020). https://www.capetownetc.com/news/nurdles-wash-up-on-cape-town-beaches/

Encountering the other: Jungian analysts and traditional healers in South Africa First published: 23 January 2020. https://doi.org/10.1111/1468-5922.12573

Feketha, S., & Mavuso, S. (2020). Mbalulu hits back at MKMVA, labels the Ace's thugs'. *Cape Times*. October 13th, page 1.

Field, J. (2019). *The case for a wellbeing budget to serve people* (p. 32). Sydney Morning Herald.

Florini, A. (2003). *The coming democracy*. Island Press.

Goldston, R. (2020). *Jah in the flesh: An examination of spirit, power and divine envesselment in Rastafari*. PhD thesis Published by ProQuest LLC.

Greenfield, S. (2015). *Mind change*. Random House.

Harper, P. (2020). Free state branches gun for Ace. *Mail and Guardian* 23–29. Page 3.

Higgins, P., Short, D., & South, N. (2013). Protecting the planet: A proposal for a law of ecocide. *Crime, Law and Social Change, 59*, 251–266. https://doi.org/10.1007/s10611-013-9413-6

IPCC. (2021). Summary for policymakers. In V. Masson Delmotte, P. Zhai, A. Pirani, S. L. Connors, C. Péan, S. Berger, N. Caud, Y. Chen, L. Goldfarb, M. I. Gomis, M. Huang, K. Leitzell, E. Lonnoy, J. B. R. Matthews, T. K. Maycock, T. Waterfield, O. Yelekçi, R. Yu, & B. Zhou (Eds.), *Climate change 2021: The physical science basis. Contribution of Working Group I to the Sixth Assessment Report of the Intergovernmental Panel on Climate Change*. Cambridge University Press. In Press.

IPPC Report. (2018). Global Warming of 1.5 degrees. http://ipcc.ch/pdf/special-reports/sr15/sr15_spm_final.pdf

Kelly, L. (2016). *The memory code*. Allen and Unwin.

Kotze, K., & Watkins, K. (2020). Prison Sewage leaks into estuary and river. *Echo News* 12th November. Fish Hoek: Report reveals alarming pollution along Cape Town's coast Ground Up 29 April 2020 Popular beaches are among 31 polluted sites, long-awaited water quality report see https://www.politicsweb.co.za/politics/beaches-at-risk-report-reveals-alarming-pollution

Landman, A. L. (2019). *From Volksmoeder to Igqira: Towards an intellectual biography of Dr Vera Bührmann (1910–1998)*. A thesis submitted in fulfilment of the requirement for the awarding of the degree Master of Arts in History, University of the Western Cape.

Maliba, A. (2020, October 25). Mixed feelings about bail for suspect. *Weekend Argus*.

Marella, M. R. (2017). The commons as a legal concept. *Law and Critique, 28*, 61–86. https://doi.org/10.1007/s10978-016-9193-0

McCall Smith, A. (2019). *To the land of long lost friends*. Anchor/Penguin.

McIntyre-Mills, J. (2006). *Systemic governance and accountability: Working and re-working the conceptual and spatial boundaries of international relations and governance*. Volume 3: 'C. West Churchman and related works series'. Series editor, John P. Van Gigch. Springer. 434 pages.

McIntyre-Mills, J. (2008). *User-centric design to meet complex needs*, New York: Nova Science. 423 page book. Appendix on software design by Denise de Vries. Output from Linkage LP0560406 Nova Science.

McIntyre-Mills, J. (2014). *Systemic ethics and non-anthropocentric stewardship* (p. 270). Springer. http://www.sense.nl/gfx_content/documents/20161115_Uitgevers.xlsx

McIntyre-Mills, J. (2017). *Planetary passport: Re-presentation, accountability and re-generation*. Springer International Publishing AG. eBook ISBN, 978-3-319-58011-1. https://doi.org/10.1007/978-3-319-58011-1

McIntyre-Mills, J. (2019). *The potential of eco-facturing to re-generate rural-urban balance through eco-villages and city hubs: Promoting social and environmental justice through vocational education and training hubs*. Proceedings of International Systems Sciences The 63nd Annual Meeting of ISSS, Oregon State University, Corvallis, Nature's Enduring Patterns: A Path to Systems Literacy, 2nd–5th July. http://journals.isss.org/index.php/proceedings63rd/article/view/3643

McIntyre-Mills, J. (2020). *The Covid-19 era: No longer business as usual*. Systems Research and Behavioral Science. https://doi.org/10.1002/sres.2745

McIntyre-Mills. (2021). *Communication and culture: A Multispecies endeavour within a shared habitat*. International Systems Sciences Year Book to appear in Systems Research and Behavioural Science (forthcoming) which develops this point.

McIntyre-Mills, J., Vries, D., & Binchai, N. (2014). *Transformation from Wall Street to wellbeing: Joining up the dots through participatory democracy and governance to mitigate the causes and adapt to the effects of climate change* (p. 253). Springer. Second and third authors provided the appendix on software user guide (pages 193–198).

Mellet, T. (2020). Camissa PeopleCape slavery & indigene heritage the research behind the founding of a proto-port of cape town before 1652 December 29, 2020 CAMISSA Camissa People (wordpress.com).

Nakamori, Y., Wierzbicki, A., & Zhu, Z. (2011). A theory of knowledge construction. *Systems Research and Behavioural Science, 28*, 15–39.

Nonaka, I., & Takeuchi, H. (1995). *The knowledge creating company: How Japanese companies create the dynamics of innovation*. Oxford University Press.
Ramaphosa. (2020). https://www.news24.com/news24/columnists/cyrilramaphosa/cyril-ramaphosa-gender-based-violence-it-is-time-to-turn-the-tide-on-the-scourge-20201123. https://www.news24.com/news24/southafrica/news/just-in-president-cyril-ramaphosa-to-face-motion-of-no-confidence-next-week-20201126
Ray, M. (2020). Chelsea Manning, *Encyclopædia Britannica* December 13. URL https://www.britannica.com/biography/Chelsea-Manning. Access Date December 22 2020. https://www.aquarium.co.za/blog/entry/the-great-nurdle-disaster-what-to-do-if-you-find-nurdles
Rayner, A. (2021). The natural inclusion of diversity in community: Swirls within swirls and variations around a central theme. In J. J. McIntyre-Mills & Y. Corcoran-Nantes (Eds.), *From polarization to multi-species relationships, contemporary systems thinking* (pp. 729–747). Springer.
Shiva, V. (2012). *Making peace with the earth*. Fernwood Publishing.
Shiva, V. (2020a). *Ecofeminism and the decolonization of women, nature and the future*. https://www.youtube.com/watch?v=hVbbov9Rfjg Mar 25, 2020 – Uploaded by San Telmo Museoa.
Shiva, V. (2020b). *Two paths to the future of food and farming – EcoFarm Keynote*. https://www.youtube.com/watch?v=9vNPGwqR4TQ
Skitl, S. (2020). SANDF hid R200 m spent on Covid drug. Mail and Guardian 23–29. Page 4.
Stiglitz, J., Sen, A., & Fitoussi, J. P. (2010). *Mis-measuring our lives: Why the GDP doesn't add up*. The New Press.
Synot, E. (2019). https://theconversation.com/listening-with-our-ears-and-our-eyes-ken-wyatts-big-promises-on-indigenous-affairs-120165
Thunberg, G. (2021). October 22 school strike week 166. Global climate strike in Stockholm #Uproot the system and #FridaysForFutyre#ClimateStrke.
United Nations. (2007). *Declaration of the rights of Indigenous peoples*. https://www.humanrights.gov.au/publications/un-declaration-rights-indigenous-peoples-1
United Nations Office for Disaster Risk Reduction (2015-2030). *Sendai framework*. http://www.preventionweb.net/drr-framework/sendai-framework/
United Nations Office for Disaster Risk Reduction. (2017). http://www.preventionweb.net/files/55465_globalplatform2017 edings.pdf
United Nations Paris Climate Change. (2015). Conference of the parties twenty-first session Paris, 30 November to 11 December 2015. https://unfccc.int/resource/docs/2015/cop21/eng/l09r01.pdf
Van der Post, L. (1958). *The lost world of the Kalahari*. Hogarth.
Van der Post, L. (1961). *Heart of the hunter*. Hogarth.
Vuso, S. (2020). Women farmworkers march for access to land. *Cape Times* Oct 16th, pg 3.
Venema, V. (2021). The 'smart and cheeky' Aboriginal boy teaching Australia a lesson – BBC News reports on the documentary by Maya Newell called https://vimeo.com/showcase/inmyblooditruns
Yunus, M. (2020). Pandemic has given us an opportunity to protect the world from its suicidal path – Muhammad Yunus. *The interview was published in Brazilian newspaper 'Estado de Sao Paulo'* on 18 December 2020.
Wilde, O. (1893). *Lady Windermere's fan* (1st ed.). The Bodley Head.

Web Links Cited

https://www.capetownetc.com/news/nurdles-wash-up-on-cape-town-beaches/
https://www.news24.com/news24/columnists/cyrilramaphosa/cyril-ramaphosa-gender-based-violence-it-is-time-to-turn-the-tide-on-the-scourge-20201123
https://www.news24.com/news24/southafrica/news/just-in-president-cyril-ramaphosa-to-face-motion-of-no-confidence-next-week-20201126
https://www.aquarium.co.za/blog/entry/the-great-nurdle-disaster-what-to-do-if-you-find-nurdles

The Power of Story: Ocean View Organic Farmers Cooperative, Cape Town, South Africa: Growing Food with Heart and Soul

Stefanie Swanepoel

Abstract The vignette gives an insight into an innovative project that provided inspiration for other projects mentioned in this volume. The website can be seen at https://www.facebook.com/kosgangsters/

Keywords Co-operative · Community engagement · Social and environmental justice

1 Introduction

Sometimes out of pain, poverty and violence, something new and stronger is forged, something innovative that talks to a better world, not yet born. It is so with the story of Sophia Grodes, Nikky Jacobs, Carrin Roberts, Laurica Petersen and Merna Booysen – the five women members of the Ocean View Organic Farmers Cooperative, locally known as the 'Kosgangsters' (Food gangsters) – who are collaboratively trying to change the narrative of their community – through food.

2 Background

The Kosgangsters live in Ocean View, a 'coloured' community on Cape Town's south peninsula. The apartheid government created the suburb in 1968 as a dumping ground for coloured people forcibly removed from the surrounding suburbs, which under the 1950 Group Areas Act were designated as whites only residential areas. The pain of its creation perhaps explains the melancholy that is sometimes palpable

---

S. Swanepoel (✉)
Ocean View Organic Farm, Fish Hoek, South Africa
e-mail: stef@africanearthrights.co.za

J. J. McIntyre-Mills, Y. Corcoran-Nantes (eds.), *Transformative Education for Regeneration and Wellbeing*, Contemporary Systems Thinking,
https://doi.org/10.1007/978-981-19-3258-8_4

in the people as they try to survive in an economy that has marginalised coloured people even further since democratic elections in 1994.

In the past few years, Ocean View has been overtaken by gang violence and is one of the least safe areas in the south peninsula. The cumulative effects of gangs, dangerous drugs, a culture of drinking and decades of social and economic marginalisation are tangible – but the end of Ocean View's story is not yet written.

There are many people working to change the narrative – through community-building activities, feeding schemes, art and dance therapy, storytelling and food. Sometimes all it requires is a dream.

3 The Origins

In 2018, Neighbourhood Farm (www.neighbourhoodfarm.org), a local non-profit organisation, established a set of market gardens in Cape Town's south peninsula with funding from the Western Cape Department of Agriculture. The founder of Neighbourhood Farm – Justin Bonello, a well-known filmmaker, author and celebrity chef – was a on a mission to "fix the broken food system in South Africa". He soon realised it was not enough to set up market gardens, but that to scale the impact of the initiative people would need to be trained as urban gardeners.

In 2019, the organisation opened an agri-training hub in Ocean View situated on the grounds of the Ocean View Secondary School – all of the cooperative members graduated from this school – some a few years ago and some more than 20 years ago. They took in 20 interns for a 3-month intensive training programme on establishing and running an urban farm. Ten interns were selected to complete the full 12-month programme. The five members of the cooperative were among these 10, each with their own motivation. Sophia was on a mission to uplift her community; Carrin wanted to be able to buy a house for herself, her mom and her two daughters; Laurica had a love of planting and wanted to develop it further; Merna wanted to build a better future for herself, her daughters and grandchildren; and Nikky wanted to be able to generate an income that would allow her to realise her ambitions. While none of them had any growing experience, they could all recall an instance from their youth where they had in some way been connected with soil and plants. For some it was spending time with grandparents helping them garden, for others it was a tree planted by their parents that they still sat under today.

They built the farm from the ground-up under the supervision of a dedicated trainer. In that first gruelling year, the 10 dug out the 'koffieklip' (coffee stone – iron Ferricrete) with their bare hands from the soil, installing poles and shade cloth to make windbreaks, setting out the beds and eventually planting out this 6000-square meter market garden. They went through a year of preparation, planting, production, harvesting and marketing to prepare them as urban market gardeners. It was not easy. Many struggled with the discipline required, the theoretical and practical learning and the sheer hard work that is farming. But the five women never gave up – their dreams were within their reach and the sacrifices needed would be worth it.

In January 2020, on completion of their training, the farm was handed to the qualifying interns to take forward as agri-entrepreneurs. The journey of the Ocean View Organic Farmers Cooperative started to take shape (see Box 1).

Box 1: Overview, Vision and Mission

The Ocean View Organic Farmers is an urban farming cooperative located in the food insecure suburb of Ocean View on Cape Town's south peninsula. There are five women members that completed an intensive 12-month accredited urban farming training programme in early 2020. They produce vegetables and herbs on a 6000-square metre urban farm, which is fully equipped with irrigation, cold rooms, offices and a small on-site bakery, café and nursery.

They have a written Constitution and elected Chairperson, Secretary and Treasurer to provide leadership, governance and financial oversight. The cooperative is currently in the process of setting up key performance indicators across the various revenue streams. Rent is paid to the school and the farm has a separate electricity meter, borehole and pumping system.

The cooperative's vision is to develop and maintain a prosperous and sustainable urban farm that supports good livelihoods, builds community unity and boosts local food and nutrition security, with a particular focus on youth. When they took over management of the market garden, they set out three primary goals and objectives. These are:

- We are a well-capacitated organisation
  - We can manage and coordinate activities by end-2021
  - We have undergone the training necessary to complete our roles and responsibilities by end-2021
  - We are a legally and financially compliant organisation by end-2021.
- We are regarded as a trusted organisation to work with
  - We have secured external partnerships to support our work, including with government and donor agencies by end-2021
  - We have made strong, long-term partnerships within the community with key stakeholders by end-2021
- We can fund ongoing and planned operations
  - We have identified additional revenue streams to supplement food production by end-February and ongoing in 2021
  - We have secured long-term production contracts with at least three procurers by end-2021

To a signifant extent, despite the constraints imposed by COVID-19 lockdowns, the cooperative has attained its objectives.

4 Ocean View Organic Farmers Cooperative

Managing an enterprise is not something one is taught at school; it was also not taught during the training. It requires a structure, a strategy, a plan and resources. None of these were in place. Many, many meetings were held among the group about the best way forward. One of the men wanted to return to his work as a mosaic artist and relinquished his claim on a farming block for a container in which he could base his studio at the farm. Three of the other men were not keen on being held to a working schedule and wanted to move forward with their 'share' of the garden as they saw fit – and subsequently signed partnership agreements with the cooperative. One man relinquished his farming block for a cash payout. The others were keen to work together to deepen the impact of their work by collaboratively creating a brand – thus building something bigger, a legacy business that would benefit them more in the long run than working on their own. The idea of a cooperative was born – equal shares, equal responsibilities and equal benefits. The five women and one man joined, but he resigned a few months later, so the story of the Ocean View Organic Farmers Cooperative is about five women on a mission to change their worlds and that of their community.

Their journey is one of hope and despair, of love and hatred, of tolerance and prejudice, of creativity and crassness – but more than the extremes of emotions and experiences undergone in their first year of business, their journey is characterised by endurance, creativity and resilience.

Within two weeks of the decision made to form a cooperative, South Africa went into a hard lockdown because of COVID-19. On 26 March 2020, residents were told to stay in their homes, businesses were shut, markets were closed, and strict health protocols imposed. The cooperative had not yet established a market, it had no savings to provide stipends to its members, it had no resources to call on except for the crops in the ground and the women's own pragmatic creativity.

This meant finding a way to sell their produce into a marketplace that didn't know who they were, and to do it in such a way that also combatted the escalating food insecurity in their community (see Box 2).

Box 2: Food Security

Ocean View is classified as a food-insecure suburb. A recent study found that more than 80% of households experience hunger on a regular basis. Almost 40% of households rely on state social grants. While the average cost of feeding one person a month in South Africa was a minimum of R614.58 in 2019, the average spent in Ocean View was R250.00. This was prior to the COVID-19 lockdowns in 2020 that saw many lose their work and livelihood opportunities. Soup kitchens and feeding schemes in the community cannot keep up with the demand. A more alarming trend is growing nutritional insecurity, as diets are focused on processed food with little nutrition.

(continued)

Box 2 (continued)

The Kosgangsters response has been three-fold to the food and nutrition crisis. First, the provision of organic, nutrient-dense vegetables into feeding schemes and soup kitchens. Second, a Perpetual Veggie Box scheme (the box, organic compost and seedlings, planted out and cared for by the women on an ongoing basis) in which boxes are donated and/or sold at a subsidised rate into key spaces in Ocean View to support easy access to nutritous food. These boxes have been placed at old age homes, shelters for abused women, the police station and Mosque – all identified as being places that cook for large groups of people. They were also sold to the general public with a third of the sale price then donated directly to a feeding scheme. Third, allotment space has been opened up on the farm to enable feeding schemes and soup kitchens to themselves grow organic produce. This will also help them to be less reliant on donor funding. Future plans include offering a Pay-as-you-can day at the Hoenderhok cafe, in which everyone is welcome to eat and can pay what they can afford.

5 Innovation in Response to COVID-19

Connections were quickly made with local feeding schemes and food security groups in the peninsula working on food distribution systems to ensure that nobody went hungry. The women were able to position themselves as central to ongoing food provision and because they gave freely of produce at the start of the lockdown, they started receiving donations to continue providing organic vegetables to local soup kitchens. In addition, the cooperative was able to bake up to 300 loaves of bread a day in a donated wood-fired oven. One and a half years into lockdown, donations for the provision of organic vegetables and bread to local feeding schemes have been formalised into contractual agreements, providing the cooperative with a third of its monthly income.

The 'story' of the Kosgangsters who were donating food and bread quickly spread through the larger community attracting radio, newspaper and television coverage. A story of five women from a poor community on a mission to feed and to do it in such a way that did not harm the health of the planet inspired many.

This exposure drew people to the farm to come and meet the women and to volunteer, opening the opportunity for a weekly veggie box scheme to generate revenue. Demand grew for the boxes, particularly because more people were conscious of the need for nutritious food. The number of people visiting the farm to collect their boxes increased and so the women opened an on-site café – The 'Hoenderhok' (chicken coop) in a spacious, unused brick and wood chicken coop selling fresh vegetable juices; pizzas, scones and lasagnes cooked in the wood-fired oven; and salads. This kept their baker employed full-time and allowed the cooperative to provide waitressing opportunities to three young women in the community. The

Hoenderhok was Ocean View's first sitdown restaurant – it remains the only one to date. As a venue, this farm-style café became popular as a great place to meet interesting people and the cross-cultural conversations, which often sparked other initiatives, are a hallmark of its charm.

By September 2020, the market garden had become a local attraction with some tour groups coming through, a regular stream of visitors to the garden and the café, a starting-to-thrive veggie box scheme and continued donations for food and bread to feeding schemes. A monthly Talent Exchange market was established, regular events scheduled (training workshops, public holiday festivals), children's after-school activities hosted and free yoga and Nia dance sessions offered on a Friday. It seemed that the only way was up.

6 Partnerships

Since inception, the cooperative has had strong relationships with local Community Action Network (CAN) groups. Various initiatives have been supported by the Clovelly and Kommetjie CANs – these are based in two suburbs in the south peninsula more privileged than Ocean View. As trust was built, these partnerships have been formalised into monthly financial commitments from the CANs towards providing seedlings to local grower groups, baking bread for local feeding schemes, and hosting garden competitions in Ocean View. These garden competitions aim to elevate gardening as a hobby in the community with prizes awarded to those growing food, encouraging biodiversity and cultivating indigenous plants.

A long-term partnership with Rokpa Seeds (of Nourishment), an initiative of the Kagyu Buddhist Centre, has enabled the cooperative to provide dedicated produce to the Open Door Centre in Ocean View that works with vulnerable women and children.

There is a close relationship with the school, and the cooperative sits on the school committee focused on developing the property over the next 50 years. There is an additional 12,000-square metres of land available on school grounds, and the cooperative plans to extend production to some of that land, once funds are raised for the additional irrigation and security measures.

7 Current Projects

In December 2020, the market garden started to attract those wanting to work with indigenous food and medicine plants. A plant medicine woman Karen Parkin dedicated herself to creating a space for these plants, and it is here that local Khoisan chiefs and the 'bush' doctors (African medicine men) started gathering to exchange their traditional knowledge. Within the year, this space will host seedlings for wild harvesters to replenish the Slangkop Mountain that is a landmark backdrop to Ocean

View. Several workshops have been hosted around plant propagation, how to grow and use indigenous herbs, and how to plant indigenous food plants and prepare them for eating and medicine.

In early 2021, chairlady of the cooperative, Sophia Grodes, was selected to participate in an Ecological Organic Agriculture Pollinator Programme that trained 20 people around South Africa to establish participatory guarantee systems (PGS). The cooperative is a proud founding member of the Slangkop PGS, which comprises four other market gardens in the south peninsula, three household growers and various consumers, retailers and food system experts. PGS is a collaborative system that provides assurance of organic production, enabling the use of a seal to that effect. In addition, the collaborative nature makes it easier to engage with local government about providing support to urban farming – these discussions started in May 2021. The manager of the cooperative, who has acted as a mentor for the group since they entered training, is also part of the management team of the GIZ-funded Knowledge Hub for Organic Agriculture in Southern Africa. It is one of five knowledge hubs in Africa. This relationship has enabled the cooperative to act as a site for International Federation of Organic Movements (IFOAM)'s PGS training sessions and inspections, with the cooperative's video and other communication outputs being shared on regional training platforms.

In June 2021, space was opened at the farm to enable feeding schemes and soup kitchens to grow organic produce. As donor funds for feeding schemes are starting to dwindle in the economic slump brought on by the lockdowns, it is critical that these organisations (often micro-organisations) look to become self-sufficient. The allotments will allow them to grow food at no cost, except for their time and effort. It will also encourage greater cross-community collaboration between Ocean View and the black African informal settlement of Masipumelele about five kilometres down the road.

8 Behind the Scenes

While the public story of the Kosgangsters is one of success and great impact, behind the scenes there have been tensions between the five women over leadership, management and financial issues. Subsequent lockdowns in 2021 have affected the revenue streams of the cooperative. Lack of in-depth understanding of production planning and succession planting have made continuous production difficult. Restrictions on gatherings have slowed attendance at events and the café. Increasing incidences of theft and vandalism have resulted in loss of produce and necessary infrastructure such as windbreaks. Carrying the weekly wages of the women, the full-time baker and the necessary two security guards, as well as rent and electricity put further strain on an undercapitalised business. Insufficient training and knowledge of these areas and the lack of capital to invest in such training and tools such as computers, smartphones and electronic payment gateways resulted in a lack of trust, which has hindered progress.

These tensions and economic hardships have not, however, deterred the women in steadfastly pursuing their dreams. It was back to the drawing board time and time again, shifting responsibilities for different areas of work to other members, cultivating relationships with outside experts to gain their insights, supporting micro-growers in the community to produce organic food that could be sourced at a lower cost than from commercial organic growers to supplement the veggie boxes, and long discussions about what it means to be a leader – in the cooperative and in the community. Beyond all of their troubles and travails, what always remains is that they have a vision – for the farm and for their future.

9 Why Story Matters

It is not the relative success that they have had at times, or the recognition they have received from the surrounding communities and far further afield in South Africa – and even at international events such as the Oxford Real Farming Conference – that has kept the cooperative going. It is the power of story. From inception, the cooperative has used the languge of love, of creativity, of ceremony to describe its interaction with the Earth to produce food. The women have crafted – through their personal interactions, interviews with the media, their social media channels and on-site activities – a story that speaks to Arundhati Roy's note that "Another world is not only possible, she is on her way. On a quiet day, I can hear her breathing."

Each new idea – whether it worked or failed – was incorporated into a larger vision of radical beauty and kindness – to themselves, to the community and to the Earth. Such a vision tolerates failure, allows for the learning from mistakes and encourages the alignment of day-to-day actions with one's higher self.

Organic production is not just about farming without chemicals, it is about farming with values – care, health, ecology and fairness. These principles apply to both social and ecological systems. An errant porcupine that sporadically raids the crops – beetroot is his favourite – was not summarily dispatched. Rather an animal whisperer was called in, a meditation session undertaken and finally a porpcupine restaurant garden planted out. The porcupine appears in the cooperative logo.

It is the vision of themselves as five elderly ladies in decades to come, owners of a prosperous business that has not only brought food and nutritional security to their community and permanent jobs and livelihoods, but that has also delivered hope. And delivered it to a community often steeped in hopelessness.

The women speak the language of emancipation – of freedom from an exploitative and destructive food system and of a future in which all are treated equally and in which their children and grandchildren thrive.

This language of emanicpation is not war songs, but rather the calling of the wild – of 'free' plants and spaces that beckon with beauty, always appealing to the expression of the higher self. Solstice and equinox ceremonies, open to the public and supporters, are held in which the women rededicate themselves to the land and their vision. Sometimes the singing and prayer at these occassions seems to resonate so strongly with the Earth that it is as if she is singing along.

There is power in story – both to create visions of alternative realities/futures but also to hold people to those visions when obstacles emerge, when the going gets tough, when the women lose their way. What they have created is a story bigger than themselves, one wound so deeply into their soul's aspirations that its tendrils can gently hold and soothe the sometimes harsh reality they face.

10 Looking to the Future

Given the COVID reality and a new normal that has not quite landed, it is not clear how the cooperative will adapt and thrive. What has been made abundantly clear in this period is that the global, national, regional and local food systems need to change. They need to deliver both ecological and socioeconomic justice, particularly in the face of climate change and the expected impacts on agricultural production.

A long-term goal is that the cooperative is a food hub in the south peninsula that provides training in urban agriculture, hosts a seed bank of open-pollinated organic and indigenous plant seeds, provides a space for community collaboration and has processing facilities, which would enable the generation of livelihoods in the community.

The cooperative is an emerging social enterprise in that it aims to not only uplift the community, but also to support decent work and livelihood opportunities. It is an example of community-level innovation in that the space and the context that the women found themselves in has led to creative conceptualisations of what it means to farm and to feed, different revenue models and forms of cooperative governance. This iterative process of experimentation in a safe-to-fail space and futuring is what is needed to bring that new world – the one we can hear breathing just over the mountain – into being.

Collective Action for Regeneration of the Web of Life in the Face of Disruptive Injustice

Francis Akena Adyanga and Norma Romm

Abstract In this chapter, we explore the ways in which collective awareness of possibilities to act in the face of social and environmental justice became activated in a community in Uganda. We concentrate on community participants' joint reflections around the operations of a foreign-owned factory that was set up in Koch Goma Subcounty, Nwoya District in northern Uganda. The factory was set up for the agroprocessing of fresh cassava to produce denatured alcohol that could be used for cooking and lighting purposes. In principle, the (constructive) idea as mooted to the locals was that the company would contribute to the local economy in various ways, for example, by employing workers in the processing factory and by paying local farmers to produce the needed cassava. However, our research – based on the facilitation of four focus groups in the area – concluded that the modus operandi of the factory turned out in the main to be a source of anguish to the locals. Their experiences, as reflected upon in their community meetings and further discussed in our focus group fora, have demonstrated that the unregulated operation of foreign owned investments can become a major source of poverty, economic disempowerment, and public health concern. The chapter focuses on some of the ways in which community participants developed a sense of collective agency to draw the attention of the district leadership and national environmental protection agencies to the disruptive effects of this factory as well as a foreign-owned road construction company and how this panned out to date.

Keywords Community-oriented knowing processes · Building collective agency · Environmental protection · Social and environmental justice · Ubuntu

---

F. Akena Adyanga
Faculty of Education, Kabale University, Kabale, Uganda

Research Associate, Department of Adult, Community, and Continuing Education, University of South Africa, Pretoria, South Africa

N. Romm (✉)
Department of Adult, Community, and Continuing Education, University of South Africa, Pretoria, South Africa

J. J. McIntyre-Mills, Y. Corcoran-Nantes (eds.), *Transformative Education for Regeneration and Wellbeing*, Contemporary Systems Thinking,
https://doi.org/10.1007/978-981-19-3258-8_5

1 Introduction

This chapter addresses the question of how informal educational processes towards collective decision-making and action in a community, can serve transformative purposes for regenerating the cycle of life in response to foreign investment initiatives that fail to respect this cycle. While foreign-owned investments are argued by certain economists to bolster economic growth (Durham, 2004; Carkovic & Levine, 2005; Li & Liu, 2005), an interdisciplinary ecological economics (Martinez-Alier, 2009; Spash, 2009; McIntyre-Mills & Romm, 2019) and indigenous cultural wisdom (Chilisa, 2017; Harris & Wasilewski, 2004; Osuji, 2018) suggests that an ecosystemic perspective has to be borne in mind for "investment" to become regenerative rather than destructive of social and ecological wellbeing. In the case discussed in this chapter, we first heard about protests in a community in Nyowa district of Uganda against the operations of certain foreign enterprises, through Francis, the first author of this chapter. Francis became aware of this in early 2020 while travelling through one of the villages and speaking to some of the villagers. Based upon what he heard, we organized several focus group (FG) sessions with participants from the various villages. This, we hoped, would enable participants to reflect further on the issues at stake, including about their increased awareness of the possibility of working together towards realising their power to initiate transformative thinking and action regarding the problems they were experiencing (This intent was in keeping with Indigenous and transformative paradigms of social research – cf. Cram et al., 2013; Ndimande, 2018; Romm, 2015a, 2020).

As far as social positioning is concerned, most of the participants in the study community were subsistence farmers who had been asked by a foreign-owned agro-processing factory to grow cassava. There was also a small segment of participants – mainly youth – who had become employees of the factory. This group also belonged to the peasant/subsistence farmer category before the factory came into their community. With the coming of the factory, they got jobs mainly as manual laborers. These were the people who complained of long working hours with little and sometimes no pay, sexual exploitation, and generally poor working conditions, as also seen from the FG discussions as presented below.

The FG discussions were held with participants across several villages which had been negatively affected in various ways, socially and ecologically, by the operations of this factory. The factory is seen as the main culprit but additionally, they were also being affected by a road construction company operating in the area at that time. The FG sessions were held with 40 participants who were chosen to participate in four FG discussions (10 in each group) respectively. The sessions were conducted in three villages surrounding the factory in Nwoya district. The participants were selected in terms of gender considerations (equal number of men and women). Working collaboratively with a trusted village Chief, the names of potential participants were generated, and they were approached by the researchers who requested that they participate freely in the study, having been briefed about it by the researchers. What emanated from the FG discussions was that the members of the community had managed to mobilize to ensure that the factory was closed pending its management taking steps to organize cleaner modes of production and proper waste disposal in terms of national environmental regulations. The road construction company was instructed to build a borehole to furnish the community with clean water prior to its moving on to new areas of operation.

Regarding the actual facilitation of the FG sessions, two sessions were conducted by research assistants who were trained by Francis in the use of a guide written by Francis and Norma to aid them in their facilitation function. Each of these FG sessions lasted about three quarters of an hour. Two of the FGs were conducted by Francis and lasted about one and a half to two hours each, as he prompted the participants to reflect deeper on the issues that they identified (The questions that were posed based on the interview guide were the same across all the FGs). When comparing the FG material from the four sessions, Francis noticed the FGs he had facilitated repeated the issues that had been raised in the sessions facilitated by the research assistants; however, he had used his research skills to *probe participant responses more deeply and also to engage the participants in more in-depth reflection and discussion during the FG sessions*. Therefore, the results as we discuss them in this chapter, and the quotations that we supply to substantiate our narrative, are all derived from the FG sessions facilitated by Francis.

During the FG sessions, we learned from the participants that at first community members did not become involved in collective deliberations around their plight and around what could possibly be done to rectify the experienced injustices, but with some prompting by women in the community, meetings were set up to consider what might be done collectively to gain the attention of certain district leaders and national government environmental services (National Environment Management Authority, NEMA). Interestingly, the various FG members, both men and women, recognized the role of women in urging community members to realize the need for collective discussion in the community, including the discussion of options for action in the face of the damaging consequences of the factory's operation. We consider the role of women in this process in the light of academic controversies concerning the contribution of women to political processes as part of African cultural heritages (cf. Adyanga, 2019; Buswell & Corcoran-Nantes, 2018; Chilisa & Ntseane, 2010). We indicate how this example points to a respect displayed by men for women's insisting that men put the community-related issues on the agenda for discussion in their elders' meetings; and for indicating the need for action in this regard, while also taking part in the mobilization effort.

While we do not suggest that women in African cultural heritages always had and have political power equal to men in striving for social and ecological justice, we do suggest that Western-oriented readings of their general lack of power in ancient Africa can be re-interpreted as detailed by Adyanga (2019). Adyanga (2019, p. 38) avers that in some states/communities in pre-colonial Africa, men and women were regarded as equally able to participate fruitfully in political life whether they themselves occupied leadership positions and/or were respected for their advice regarding issues being confronted. He suggests that this points to a different reading of the role and status of women than the one often accorded in conventional and sometimes blanket-styled accounts of African patriarchy which do not look at the specific dynamics at play in different communities.

Adyanga further points out that the deficit reading in Western-oriented history of pre-colonial Africa as generally being undemocratic, is a misreading of the way in which consensually oriented decision-making was often valued and implemented (2019, p. 38). For example, he points out that a Village Council was often responsible for "equitable uses of land" and he argues that council members needed to be trusted by the community otherwise they could be removed from the Council or any other leadership position. He remarks that post-colonial African leaders, insofar as this suits them, are often supportive of foreign involvement in political processes which is buttressed by foreign-led economic interests. This may not be in the interest of the country/region in terms of the overall wellbeing of the people (Adyanga, 2019, p. 41).[1] Adyanga argues that the selfish interests of many current leaders in Africa with support from countries outside of Africa can be contrasted with ancient democratic traditions in Africa, where there were various mechanisms in place to generate a collective style of decision-making which could serve the common good and which would expect chosen leaders to solicit community participation in decision-making. Adyanga's statements regarding ancient governance in Africa are consistent with Bernal's book entitled *Black Athena* (1987), which shows that contrary to historical readings which suggest that White people in Athens generated democratic thinking, many Black people from Africa as well as people from Asia had been involved in Athenian society, including contributing ideas regarding modes of governance. The conventional belief of democracy as a Western tradition is thus ill-conceived. Harris and Wasilewski make a similar point when they speak of a tradition of dialogical decision-making in indigenous cultural heritages (2004, p. 498).

Our account in this chapter shows the manner in which the FG participants related to one another during the FG discussions as expressing their participative and dialogical spirit, as they collectively reflected on the FG questions posed to them. In this way, they generated ideas together and enriched the overall discussion; also demonstrating their appreciation of the fact that people can introduce ideas into the discussion for further input towards collectively-generated insights. People did not expect everyone to agree immediately with initial statements; some built on or even disputed previous statements – all with the aim of arriving at a more "collective" voice. The FG sessions offered us a glimpse of how, as part of their community mobilization, they had been moving from a stance of "individually based"

[1] In the case of Uganda, further to the 2021 elections, President Museveni and his main opponent (Hon. Robert Sentamu Kyagulanyi) disagree on the role of foreign governments in domestic affairs. While in the 1990s and early 2000s there were understandings between President Museveni and the USA, who appreciated, *inter alia*, his market-friendly policies, after 2021 the relations have shifted; the USA is threating sanctions due to the way the 2021 elections were conducted. Hon. Kyagulanyi has suggested to Aljazeera that "General Museveni is showing the world that one can be a dictator but still be accepted by the world's leading democracy [the USA]." Kyagulanyi is requesting the USA to follow through its threats (https://www.aljazeera.com/news/2021/3/4/us-threatens-sanctions-against-ugandan-leader).

complaints to community-based thinking and acting – which they expressed in various ways during the sessions. We give more detail in this regard below.

Before we proceed with the next section, we also wish to draw attention to the way in which various FG participants referred to indigenous understandings of the interconnectedness of the web of life, and to the need for humans to connect to, and feel connected to, all that exists (Chilisa, 2020, p. 24). In their statements in this regard, they were resuscitating traditions and knowledge systems that recognise that the destruction of interdependencies in the web of life is anathema to both social and ecological wellbeing. And they alluded to "the commons", which should be treated as indeed for the common good, namely, water and land as the source of life, was no longer being treated as such. To understand their views concerning this, it is necessary to offer a brief account of the traditional land tenure system, and how the dismantling thereof provided leeway for foreign investors to ignore "the commons".

2 Traditional Land Tenure System in Pre-colonial Times: Some Consequences of Its Dismantling

In pre-colonial times, in Uganda as elsewhere in Africa, a communally-owned land tenure system was usually in place and traditional leaders were accorded the role of settling land conflicts over the use of land, based on seeking consent-based solutions (Arko-Achemfuor, 2019; Adyanga, 2019). In the aftermath of colonialism, however, these leaders often became stripped of their political powers by the national governments, and what became left of their power as only ceremonial authority, mainly cultural and social. Traditional ways of equitable sharing of "the commons" have thus become eroded in Africa. McIntyre-Mills and Wirawan (2020) indicate that this can be regarded as a global problem. With reference to Stiglitz et al.'s (2010) conception of "wellbeing stocks" they note:

> The aim of the "wellbeing stocks" concept is to enable people to re-evaluate economics and to become more aware of the way in which we neglect social and environmental aspects of life. The pursuit of profit at the expense of people and the environment is a central problem for democracy and governance. (McIntyre-Mills & Wirawan, 2020, Slide 17)

They suggest furthermore that:

> Alternative forms of organization are possible to support "wellbeing stocks" but this needs to be achieved through creating a change in values through reaching out and supporting one another. Participation by those who are to be at the receiving end needs to inform policy with practice. (McIntyre-Mills & Wirawan, 2020, as cited in McIntyre-Mills & Wirawan, 2020, slide 17)

While McIntyre-Mills and Wirawan refer to the need for a change of values, given the dominance across the globe of economic and political systems that fail to support wellbeing stocks, including a concern with the wellbeing of nature, this chapter shows the consequences of the erosion via colonization of indigenous systems that might have better supported these. In the case of Uganda, the disruption of the customary land tenure system has resulted in an increased wave of insecurity in the region, also following the war between the Lord Resistance Army (LRA) rebels and the Ugandan army (Uganda People's Defence Forces – UPDF) that lasted over two decades from 1986 to 2008. The insecurity has led to destruction of property, disruption of settlement as well as the disruption of agricultural productivity (Rugadya et al., 2008; Annan & Brier, 2010; Mabikke, 2011; Mugizi & Matsumoto, 2020a). What was hitherto communally-owned land such as hunting grounds, rivers, graze land, forests and national parks has been encroached upon for individual gains (Akena, 2017; Mugizi & Matsumoto, 2020b). In the process, violent confrontations often erupted over competition for land accumulation.

With the rise in globalization, the government has been accused of trying to take land from the locals for multinational companies (Refugee Law Project, 2014) in the name of attracting foreign investments. Negotiation of benefits from multinational companies are often done without the involvement of locals. The absence of their voices at the bargaining table means they are vulnerable to exploitation by foreign investors.

In 2016, the Bukona Agro Processors Ltd. factory was established in the northern Uganda district of Nwoya. According to the Indian-owned Managing directory, as stated on its website, the factory was set up at a cost of six million dollars ($6 million USD). At the construction site ground-breaking in early 2016, Christina Melmberg, World Bank Country Director, praised the project as a breakthrough for locals to have increased bargaining power for their cassava crops. On its website, the factory states its purpose as being to generate ethanol from fresh cassava so as to produce denatured alcohol for cooking and lighting. Specifically, the factory processes low-cost fuel from fresh cassava (https://www.udc.go.ug/gallery/bukona-agro-processors).

The setting up of Bukona Agro Processors Ltd. drew excitement from many of the locals and government because of the potential for job creation for hundreds of people and boosting revenue collection. For its part, the government sent the Prime Minister (PM) flanked by local leaders to inspect the physical infrastructure in 2016 after the completion of construction. The inspection trip was an enormous success as local leaders commended the government for bringing investors to bolster economic growth in the region. The Prime Minister pledged government support to the factory. A few months later, the proprietors of Bukona Agro Processors Ltd. met with Uganda's president at State House in Entebbe. The meeting was embraced by key members of the ecclesiastical community and local leaders from Northern Uganda. Impressed with the development, President Yuweri Museveni also pledged government support to the project which he echoed will significantly reduce deforestation activities in the region. Specifically, the President appreciated the use of denatured ethanol fuel stoves by the factory as supporting the government's efforts of reducing the challenges faced by people in northern Uganda due to deforestation. In March 2020, Bukona Agro Processors Ltd. was formally inaugurated by the President. In his inauguration address, the President again reiterated his praise to the factory proprietors and promised tax exemption. The factory had started production in 2018, 2 years prior to its formal inauguration.

Nevertheless, despite this positive support at national level, when Bukona Agro Processors Ltd. started production in 2018, the local community bore the brunt of the fact that the factory disposed its chemical liquid waste in a seasonal stream nearby. In brief, when the Bukona factory started production, they built a long pipe that runs underground and ends in the riverbed. Using this pipe, they disposed of industrial waste into the river, drawing complaints and condemnation from the villagers after their livestock stock started dying from drinking the contaminated water and the fish dying in large numbers too. A team from NEMA's regional office confirmed that the factory was indeed undertaking activities that were harmful to the river, impacting significantly on fish and other aquatic life (NEMA, 2019). When the villagers' complaints intensified, and because they were able to pressure the district leadership (local councillors III & V, District Chairperson, and the Resident District Commissioner), Bukona dug a giant soak pit about 30–40 m further and used this to dispose of their industrial waste. Francis saw the pit and took pictures thereof. But this pit, like stagnant water, emits a terrible stench causing, *inter alia,* air pollution. Francis bore witness to this first-hand. The villagers again complained to their local leaders and made phone calls to radio stations about the dreadful stench. Due to the villagers' complaints, the District Environment Officer proceeded to invite experts from NEMA. Some of the participants surmized during the FG sessions that perhaps the councillors eventually came on board and called in NEMA because, as one female participant stated, "the radio station was talking about our ordeals … and they could have feared that the central government would hear and ask them what they did to resolve the problem". Some of the men surmised that perhaps the councillors were concerned that the villagers would "attack the Indians" – referring to the fact that Bukona was owned by Indians – as they were realizing how unhappy the villagers were and had heard rumours about ensuing protest action. After NEMA was contacted, they paid a physical visit to the factory to assess and verify the complaints. They indeed verified the dreadful stench and the open waste disposal and ordered the closure of the factory until a proper waste management mechanism could be put in place.

To add some more detail to this brief narration, the villagers' protest leading to the intervention of the national NEMA via the district, took place at the gate of Zhongmei Engineering Group Ltd., a Chinese-owned road construction company. The Chinese multinational construction company had earlier won a multimillion-dollar road construction project from the government of Uganda. They were building a highway through the same community in 2019. Like Bukona Agro Processors, Zhongmei Engineering Group Ltd. discarded chemical waste into the same seasonal river used by the locals for domestic consumption and source of water for their animals. There was once again an outcry from the local community about this; but their outcry was initially ignored by the company and by the district leaders. It was only after their animals started dying from drinking the contaminated water that they decided to take matters into their own hands. One morning, the villagers protested by blocking the gate to the Chinese firm so that no one would enter or leave the compound. The villagers who were determined to prevent the deliberate water contamination, had succeeded in organizing themselves into a kind of social movement to peacefully confront the engineering company at their gate. In the discussion below, we indicate their decision to handle the matter peacefully. The protestors insisted that Zhongmei Engineering Group Ltd. drill a borehole for clean water for them before they would let go the besieged property. They also insisted that chemical waste disposal into the water stop forthwith. Probably for fear of this story making it to the press had they not complied, the proprietors of Zhongmei yielded and drilled the borehole immediately. They also ceased dumping their hazardous chemical waste into the seasonal river. However, what remains unclear is the new mode of waste disposal that the engineering company adopted. At the time of the FG sessions, they had left the area, moved their offices out and continued with road construction. Road construction companies keep moving offices and staff settlements as the road construction progresses.

We concentrate in this chapter on how the members of the community expressed during the FG discussions their view of these foreign-owned enterprises. We also concentrate on how they mobilized at community level to confront the disruptive injustice with a view to regeneration of the "web of life".

As one male participant during one FG session, speaking about Bukona factory specifically, put it:

> *(Man speaking): I am an old man and was at pain to see these newcomers to our community turn us into their slaves. .... Our culture cannot be treated with contempt as if there are no young men to defend it. You see everything you find in our society living or non-living depends on each other to be able to regenerate. So, when the Indians pour waste into the river, they are disrupting the web of life.*

Francis probed this statement and asked: *"What do you mean by culture treated with contempt?"*, to which the man replied:

> *(Man speaking): Oh yes, you see in our culture everything is treated with respect, you treat the Rivers with respect because that is the source of life to humans, animals, plants, trees and future generation, you treat women with respect and you don't use young ladies or take advantage of them because of jobs [this was a reference to sexual exploitation reported by some of the females]– that is an insult to culture and we the custodians of culture cannot keep quiet.*

The above participant interpreted the disruption of the web of life by the factory as antithetical to indigenous cultural traditions grounded in treating everything with respect. His reactive submission was in a way premised on conceptualization of the web of life as inclusive of living and non-living things. He expressed his pain that the foreign owners felt that no-one would be "up to the task" of defending these holistically-oriented traditions. The disruption, according to the participant, is the antithesis of African culture that considers young people as the carrier of culture into the next generation – hence the assertion of culture being treated with contempt "as if there are no young men to defend it". The submission by the male participant concerning the need to treat everything with respect, including humans, animals, plants, etc., is supported by a female voice from another FG session, who likewise referred to the importance of foreign-owned companies respecting all life:

> *(Woman speaking): Let other people be careful next time they want to receive investors from outside our country. Now I know that if care is not exercised and investors are allowed in the community without proper laws guiding them, they end up abusing the community generosity. They should be accountable to the people and their waters, trees, land, soil, religion, and culture but should not behave like Bukona factory that has no regards for all this.*

The participants' reflections and expressions during the FG sessions were clearly based on their indigenous recognition that the mode of operating of foreign businesses to the exclusion of social and environmental justice is anathema to indigenous appreciations of the web of life. They did not express that all foreign-owned enterprise must be relinquished; but they expressed that operating out of maximization of profit in terms of a non-ecological and socially irresponsible business model, could not be "just" and they reflected, in tune with traditional dialogical modes of knowing together, how as a community they could mobilize their collective power to act.

The fact that the factory had been closed and remained closed at the time of the FG discussions, for them was crucial. They had become increasingly aware that any promise of "employment" that goes hand in hand with a profit maximization incentive on the part of the business involved, and which leads to exploitation of people and of the environment, is not the development path that should be followed. It cannot be called "development" if this is measured in terms of wellbeing of people and the planet (See also McIntyre-Mills, 2019, p. 440).

Evidently, the local community as stakeholders in "development" here had effectively been ignored until they put pressure on the factory through the district environmental offices who then invited NEMA. In this regard, they were able to draw on environmental protection laws to, as McIntyre-Mills puts it, albeit in another context, "protect the commons". She observes that "this can only be achieved by active agency" (2019, p. 440). McIntyre-Mills refers to the history of colonization across the globe which of course cannot be overlooked in discussions about trajectories of "development" – the history of colonization accompanied by capitalist encroachment and extractive use of people and resources has issued in the dominance of a narrative of "growth" as being the only viable economic path to follow, irrespective its social and ecological effects.

Concerning the question of active participation of stakeholders in processes that influence their lives, the indigenous communitarian principle of people participating in defining trajectories in terms of a conception of the common good, became ignored – until they insisted on participating. Or, as Adyanga mentions (2019), the ancient democratic traditions which he sees as domain to Africa, and that of many indigenous communities worldwide, had been shifted in favor of the power of capital and its lobbying influence on the state, to the detriment of genuine, democratic discussion around social and environmental justice. Chilisa (2012, 2020) also indicates how the African philosophy of *Ubuntu* as a way of relationally oriented knowing and being resonates with other indigenous traditions which question the self-interest model of conventional economics. In this regard, see also Murove, 2005, for a historical exposition of how this model came to dominate social discourses across the globe, and Arko-Achemfuor & Dzansi, 2015, for an example of an alternative.

In the light hereof, in the next section of the chapter, we provide some detail on the FG participant views concerning foreign-owned enterprises (offered in response to our first four questions that we posed to them in the FG sessions). And in a further section, we point to their expressions of how they began to forge community-based protest rooted in community dialogue and aimed at regenerating the "web of life". We have not tried to cover all the questions as posed in the FG sessions as some of these are covered in other forums.

2.1 Foreign-Owned Enterprises: Participant Dialogues

2.1.1 Development of the Conversation Around This in FG One

In the first FG session's flow of conversation as presented hereunder, different views were expressed, all leading to the idea that insofar as the foreign-owned companies (they were familiar with Bukona and the road construction company) operated fairly, their operations can be accepted. But insofar as they engage in forms of exploitation such as paying less to local workers than to others; expecting long hours of work; polluting the land and water, and repatriating most of the profits, it becomes unacceptable. It is also worth noting that the first speaker, in response to the first question posed by the facilitator, invokes the *group,* thus pointing to an indigenous conception of "we" – a communal outlook – where he suggests that he is making his contribution to "our thoughts as a group". He expresses that insofar as the companies are from countries with "financial stability", in principle they can make the lives of locals better. But the next speaker qualifies this statement and the following speaker again qualifies it (as we can see from the box below). Further to their engaging with Question 1, with some different opinions expressed as they added to one another's thoughts, they mentioned in response to Question 2, the companies with which they were familiar. Then, in response to Question 3, they expressed some concerns about the way the land had been bought; thus alluding to the fact that the communal land tenure system was no longer in place. We can see that in response to Question 4, there was some disagreement about the relationship with the foreign companies, where the first female speaker indicated that some employment had been created, while the next one indicated why the relationship was "not so good" and the third speaker, also female, supporting the latter idea. The first man to contribute to the discussion offers two opinions – that the borehole built by the Chinese company was helpful, but that workers "from outside [the country] are paid better". Then two men raised further issues regarding firstly, the problem that the foreign companies mostly repatriate profits, which means the country itself receives little benefit, and secondly, that the working hours expected are very long and not conducive to the workers' travelling home after work. Below we provide the exact course of the discussion, so as not to try to split the material expressed from the context in which it was said during the discussion, as people built up the joint story:

1. There are many foreign investors in the country these days. What is your impression of foreign investors?

   (Man speaking): Our thoughts as a group, we think these people are from an area which has a better financial stability they come to improve our lives or make our lives better,
   (Man speaking): We welcome them as long as they are prepared to follow the laws of Uganda.
   (Man speaking): We welcome these visitors with open hands as long if they can cooperate with us. What we hate is suppressing our youth and polluting the river.

2. Do you know of any facility (e.g., factory, big businesses, company, etc.) set up by foreign investors in your community?

   (Man speaking): We see one company down the road called Bukona Sugar factory.
   (Man speaking): There is one even though I don't know the name of the owner dealing in farming in Barkawera, village.
   (Woman speaking): The Chinese road construction company. Zhongmei Construction company limited.

3. If yes, how did the investor/s acquire land on which they set up the facility?

   (Man speaking): Bukona processing company limited bought the land from the landowners,
   (Man speaking: Farming company bought theirs from the landowners.
   (Man speaking): There is UNRA [Uganda National Roads Authority] road company also who have compensated people who are near the road.

4. What relationship exists between your community and the facility?

   (Woman speaking): There is good relationship between us because they are helping us by offering employment to our children. This has helped in reducing the level of stealing/theft.
   (Woman speaking): The relationship is not so good because of the mode of payment/ salary to our children after employment. Sometimes people work for one to two months and they are paid for only one month. We were also deceived into planting cassava as their raw materials, then suddenly they refused to buy the cassava and no reason was given for not buying our casava (They said if you can't produce a certain number of tonnages (5 ton onwards) they will not buy), they should have collected us and informed us on the reason why. They have been bringing workers from other areas and they have polluted the area with their

(continued)

waste in to Kulu Ayago [River Ayago] and we only realized with the stench/smell. We realized with the death of aquatic animals.
(Woman speaking): They poured their chemicals and started causing problems, they were stopped from pouring the waste in the stream and they instead dug a hole in which they now pour their waste, but the stench is too much when the wind blows in our direction.
(Man speaking): I have also seen they have come out with the plan of drilling the bore hole which serves both the community and their workers as well. What I have seen is workers from outside are paid better than the local workers from here and sometimes the salary takes long to be paid.
(Man speaking): Most investors repatriate the profits, so any profit which we would get like paying school fees has not been realized.
(Man speaking): Working time is not favorable to the time for going back home, as some people work from 8 am to 9 pm

2.1.2 Development of the Conversation in FG Two

In FG two that was held in a different village, the first response to Q1 makes the point that foreign investors can be welcomed insofar as they "*bring us development*". The group then answers the probing question of Francis and the next person (male speaker) speaks about the potential of foreign-owned enterprises, but suggests by allusion that this is not the case thus far. This person's views on the potential offers a non-simplified response, which points to the conditions under which the companies might be welcomed, for example, if they could create a "good relationship" then there would be a connection which is "good for development". The participant invokes the indigenous view of relationality, where "good connections' are possible and where this leads to "development". His understanding of development here is clearly not defined as profit at the expense of people and the environment! He also expresses hope that the voices of the FG participants could "reach the government; some citizens like youths within the community where there are good factories should be allowed to get better jobs, or those without knowledge if the government can allow then it should bring some training opportunities". Here he points to the current lack of stakeholder involvement, but clearly points to this potential as an ideal. In response to Q2., the participants who answer this question refer to companies with which they are familiar (*Bukona* and *Zhongmei).* In response to Q3, the man speaking raises concerns about the buying of land and raises suspicion. He indicates that it would be better if the motives of the company were more transparent and if the villagers were more aware of these motives. The other speakers do not add to this, implying that they agree. In response to Q4, the two men and the women point to problems that are of concern to them. The first man suggests that "*nothing really big, they have done for people of this area*". The next speakers elaborate, offering detail that is consistent with the sentiments being expressed in the FG. At the end, the female speaker suggests that they had been deceived by

Bukona in that Bukona had initially suggested that the cassava that the villagers planted would be bought from them, "*we were deceived to plant cassava, there was no markets others were sending [to the factory] their cassava on credit, and no payment has so far been made*".[2]

The conversation proceeded as follows:

1. There are many foreign investors in the country these days. What is your impression of foreign investors?

   (Man speaking): I thank you [from] university for coming to take our ideas: we welcome the investors, we see as a good thing because they bring us development to help the youth because right now there are no work for the youth even if you are educated, they also help us with different things like sanitizer from Bukona
   (Francis): Is the sanitizer given for free to the local community?
   (Group response): No, it is not, we buy the small bottle at shs 2000, they also promised to bring us sugar factory, but it has not yet been brought. That is what I see as the advantages of investors from out.
   (Man speaking): I see these companies that come from outside when coming they should not have brought people from other districts; I see their coming into our land they are not doing well because they let foreigners ... get better benefits than those of us from here. When it comes payment, it's those from the other side [Foreigners] who are paid huge sums of money compared to us who on any kind of work are placed in sweeping the rubbish pits and other things which are not important. So, they are seeing the neighbours where they built their factory in a good way. If good relationship was there, you see the person you started connection with is good for the development to be better. So, if our voices can reach the government some citizens like youths within the community where there are good factories should be allowed to get better jobs, or those without knowledge if the government can allow then it should bring some training opportunities to help them gain the knowledge so that the youth can gain skills because education for some of the youths got spoilt. So, if the government doesn't see it, then it won't be good.

(continued)

[2] From the discussion with the villagers it transpired that originally the factory had agreed verbally with the villagers that an acre of cassava would be bought at the price of 8 million shillings (about 2200 USD) and the factory would be responsible for uprooting and transporting the fresh cassava to their facility for processing: every farmer would have his/her cassava bought straight from the garden. However, when the cassava was ready, Bukona told the villagers that they would only buy **dry** cassava (the original verbal agreement was that cassava would be bought fresh from the gardens) and they set the prices for each kilogram of dry cassava @ 400 shs (the original verbal agreement was that an acre of cassava would be bought @ 8 million shillings). They also said for each subsistence farmer to qualify, s/he must bring a minimum of one ton (1000 kg) of dry cassava – (this was not part of the verbal agreement). When the villagers failed to meet their minimum requirement, Bukona started buying dry cassava from other districts.

2. Do you know of any facility (e.g., factory, big businesses, company, etc.) set up by foreign investors in your community?

   (Man speaking): From here we know one factory called Bukona near Ayago River, it is just there down the road, but no, big market isn't there. There is no farmer who has hired land for farming.
   (Woman speaking): The Chinese were also here with the road construction company called Zhongmei. They built that road which comes from Gulu into our district.

3. If yes, how did the investor/s acquire the land on which they set up the facility?

   (Man speaking): Bukona got land when they came from around Ayago and they met up with one elderly man from the other side of the river and he sold them a small piece of land and later connected with a person whose land is connected to the previous seller. They (Bukona) bought the land later. So, they have since that time been expanding by buying other people land. Others are being threatened to sell off their land. So now we all fear that we may lose our land to Bukona. If we knew this was their intention, we would not have welcomed them here.

4. What relationship exists between your community and the facility?

   (Man speaking): There is no good relationship because there is nothing really big they have done for people of this area.
   (Man speaking): The smell from the factory waste disposed in the river have killed the snakes in the water, cattle.
   (Woman speaking): Poor disposal of the factory waste, issue of getting work for the children in our homes, their late payment, they return late from work from 8 am to 9 pm. The time keeps changing as when the boss decides and the payment if you start work from 8 am to 5:30 pm you are paid 7000. From 5:30 onwards is paid as overtime paid 750 shillings per hour worked. They said over time would be paid every after two weeks but later it turned into two months of which you are paid one month like that so that you don't leave the work and for us the farmers, we were deceived to plant cassava, there was no markets others were sending their cassava on credit, and no payment has so far been made.

2.2 Routes Towards Community Mobilization: Participant Dialogues

In this section, we have again chosen to concentrate on the FG questions particularly pertinent to the topic of this chapter (These were Questions 8–10 in the interview guide). For this section, we have taken an example of the flow of conversation only from Focus Group One. What we are concentrating on here is the way in which the women were acknowledged by both women and men for their contributions in moving the protest from individual complaints which were not made public but were expressed "in our homes", to collective action on the part of the community (In both FG sessions the participants made mention of the role of women in the community mobilization efforts and also referred to their contribution in suggesting that a peaceful strategy should be employed. This suggestion was supported by many of the men, and it was adopted as the community's mode of action).

2.2.1 Development of the Conversation Around This in FG One

As can be seen in the box below (from FG One), in response to Question 8, the first statement is made by a female participant in FG one. She states that "we" mobilized, thus noting the role of women in the mobilization. She also points out that some of the youth were more inclined to fight, but this was not supported within the community. We can see from the later statements where the conversation moves to the participants considering which mechanisms had prompted action on the part of some of the district leaders, that the elders, who were respected in the community, also did not endorse a strategy of violence. This was the case even though initially, the officially elected leaders, namely, the district councillors were not making headway in advancing any changes. In both groups, the participants expressed that initially when some of the councillors went to speak to the factory owners, no action was taken, and the participants suggested that some money probably changed hands such that the councillors did not put pressure at first. It was only later, when community members contacted the radio and started protesting at the gate of Zhongmei Construction Ltd. to show their strength, that the district environment officer invited NEMA to make an inspection of the factory. It was only at this point that the Bukona factory was closed. In the extract from the actual conversation below, it is noteworthy how the women are suggesting that in some sense they took the lead in prompting action and in defining strategy and the men are not disputing their claims either, with some of them pointing to the support of the elders too.

In response to Question 9 below, a female participant starts by expressing the women's active role in the protest movement. The man who responds to Question 9 points to his appreciation that Francis chose to initiate the FG session, because as he states, the further discussion also helps to consider ways forward, especially regarding the stench from the factory pit. As an aside, we wish to highlight that the FG discussions became one mechanism for the community members to examine their responses to the injustices and to recognize their strength as a community. They also asked Francis to speak to the Resident District Commissioner (RDC) and if possible to a member of parliament, which indeed he did. Francis also discussed the integrated research report which included all the FG sessions with the RDC and with a diplomat running for parliament at the time. The report was also discussed in a dissemination workshop held at the home of a village chief about a month after the FG sessions and further to this Francis again contacted the diplomat as well as an MP in the study area[3] to share with them the final write-up of the dissemination workshop.

In response to Question 10, a male participant started by indicating that at first the community was not functioning as a community – people were simply expressing individual complaints in the confines of their homes. It was only later that they started to realize that this strategy would not be fruitful. He mentioned that the men started to discuss this at meetings "in the marketplace". A female participant teased him for implying that it was the men who initiated the idea of putting the issues on the agenda for discussion at meetings, whereas the women should be given credit for this. The next male speaker does not dispute this and stated that it was clearly important that women become involved in planning in this as in other instances.

In responses to further questions, the participants made mention that at first the district leaders who should have acted did not, until the community put increased non-violent pressure.

The flow of the conversation is provided below:

Q8. Please consider in particular how you think the community created spaces for discussing the experienced injustices and for considering together ways of getting them attended to by the multinational corporations.

(Woman speaking): We mobilized meetings and talked about the problem and plot the next course of action. Some of our young men ... wanted to fight but we said no to use of force. We can talk to our leaders.
(Man speaking): I agree there were a few people who made a phone call to Mega [local radio station in Gulu city]) and reported and when the complaints became much, our leaders went to the factory and talked to them.

(continued)

[3]When Francis presented the written report to the Resident District Commissioner (RDC) and discussed it with her, she acknowledged that some of the issues raised by the villagers during the FGDs were true. Specifically, she mentioned sexual exploitation of the young ladies, environmental pollution, and low pay for local staff which she said had reached her desk. She expressed her view that the CEO of the factory was a "nice and responsible" person who may not even know that the exploitation was happening. She then said, "with this well-written report from you, I will now arrange to meet with the CEO and present the report to him". Francis was unable to arrange that she would invite him too to the meeting, but he did arrange to speak to a local diplomat running for parliament at the time, who was very interested in the case. Further to the dissemination workshop held with participants about 2 months later, he learned that the diplomat with whom he had spoken earlier was from a neighboring constituency. So, he arranged and met with the MP elect of the study context. He also called the diplomat (now MP elect for the neighboring constituency) when he was with the workshop participants. This diplomat assured him that he would work closely with the MP from the study district to ensure that concerns of the locals were addressed.

(Woman speaking): I think when we threatened action and actually protested at the gate of Zhongmei, then later the Indian also knew that things were going to get out of hand. That was when the government had to send some people to see for themselves the level of reckless waste disposal. This led to the closure of the factory. My sister works there, and she said the government environment inspectors who came were very harsh on the factory bosses, so they had to close. They were quarrelling and wondered how a responsible person could dump waste in River Ayago and also in the open where people can smell.

(Man speaking): The elders encouraged us, and this was a good space to talk also. You see when you know that your elders support you then you can overcome fear. The first obstacle to our taking action was fear which had to be overcome. I think the elders saw it in us and felt they had to do something, or their society is gone.

Q9. How did the community continue to think of ways forward when initial raising of the issues was not addressed?

(Woman speaking): We protested their horrific act – we mobilized quietly in the evening and agreed that we have to stop this injustice. So, we sent words to all the youth, adults, and elders to come out and we meet at the gate of Zhongmei. The informants took words throughout the village and people were aware that there was a call for collective action. The next morning many people were there, and the gate was besieged. We could not let anyone in or out of the gate. After listening to our grievances, the company agreed to stop contaminating our water. The Chinese people stopped, and we didn't know how they were disposing the waste. When the [Bukona] factory was set up, they also started dumping their waste into Ayago River.

(Man speaking): Talking to you is another way of continuing to think of the way forward because we are not yet happy when the stench from the pit is making life miserable for us in our homes.

Q10. How did the community members share ideas together to develop a way forward?

(Man speaking): First it was not easy to get together as people complains individually in their homes. But later we realized that this strategy was not going to help so we changed plan. We then started speaking in men's gatherings when we meet at the marketplace.

(Woman speaking): Now listen to him taking credit for the men. If we the ladies had not forced you, you were too afraid to even talk

(continued)

about the issue of the river pollution fearing that the white man will put you in jail --- laughter and clapping in the background.
(Man speaking): Of course, that is why men bring women home not just to cook food but also to help men with plan in situation such as this but in this case we men are the ones who plan and took action with everyone one – scattered laughter in the background.

3 Our Reflections on Participant Understandings of Lessons Learned: A Movement to Resuscitate a Relational Way of Life as Part of "Development"

We contend that the participants' meetings and discussions as they reflected on the issues arising during their protest movement and in the continued dialogues that we facilitated, and again continued in the workshop, provide a valuable entry into discourses in academia and outside of it on social movements. Langdon suggests that when considering the emergence of social movements, we should recognize the contexts within which they are developing. Referring to a specific "learning-in-action" project undertaken in Ghana (against a foreign-owned mining company), he suggests that in this case,

> a clear sense of a critique of the global and national economy is intertwined with the issue-based critique of the rights of rural communities to land access and decision-making rights to their land and natural resources. When this is coupled with strong spiritual and ancestral connections to these resources and land, the implications not only on everyday livelihood issues but also on cultural reproduction become clear. (Langdon, 2009, p. 95)

He states that what emerged in the Ghanaian case was a

> critique that combines the erosion of cultural ways of life as well as livelihoods into a strategy that uses both human rights discourse as well as a pointed and well-articulated critique of capital exploitation to defend their ways of life. (Langdon, 2009, p. 95)

The protest movement, as we can call it, that arose in this community in Uganda was spurred by the growing awareness on the part of the local community that the capitalist-type business approach used by foreign investors was anathema to their own valued heritage and that the erosion of the commons had dramatic implications for their way of life. Although they did not use language such as capitalism to express their concerns, clearly, they had been building a sense that the foreign investors in this case were "up to no good". They did not, however, wish to return to the past in the sense of denying the value of "development" – but they insisted that development which involves some "foreign" involvement must take place on neighbourly terms – such that the investors do not offer a snake and expect that the community will accept this. The insights that they offer, and the recommendations for the government to support *investors who act in terms of respect for the community and for the commons rather than in terms of an attitude of exploitation, reflect an indigenous understanding that these forms of relationship are possible and that business does not need to operate in terms of selfish interests.* People – including business investors – can be more "*neighbourly*". Insofar as governments welcome these businesses, the community in turn could welcome them as aiding the development process. However, development must be redefined not in terms of provision of "employment" at all costs – but in terms of decent employment and respect for the people and what should be "the commons".

In all their dialogues, what came to the fore very strongly (as we see it) is the need to establish "good relations" with foreign investment companies – which they defined in some detail during their discussions. They insisted that the "common good", defined in terms of maintaining good relations between humans and with nature, needs to be pursued. They appealed to various players including their local elected leaders and also national government officials to recognize the importance of inviting into the country companies only after doing "background checks" on their ways of operating and also that the government should not allow land to be bought up for use by the factories which threatens the way of life of the local people. Furthermore, one of their suggestions was that in consultation with the local farmers, the government should come up with price stabilization for agricultural products so that the local farmers are not exploited by big companies such as Bukona.

Langdon identifies that in the Ghanaian "learning-in-action" project which he discusses, the group discerned that dialogue-based movements are most likely to emerge around communally held resources whose protection *implicates a community* as a whole (Langdon, 2009, p. 95, our italics).

Likewise, in the Ugandan case, the community increasingly recognized that they were being implicated *as a community* – and that it was not a matter of complaining at household level. This was a community-based issue, namely, the erosion of their way of life and their sense of the importance of restoring the "web of life", which includes human and non-human life. The questions that we posed in the FG sessions enabled the participants in the sessions to consider together and to share with us their emergent deliberations and sense that action needed to be taken to restore some balance in the web of life.

Before we close our reflections on this, we wish to point readers to our admittedly non-neutral role in facilitating this research.[4] We regard our setting up of the FG research as linked to our own commitment to deflect the dominance of the "modern" state-business nexus and to explore with participants the potential of community-oriented forms of life (including nature seen as part of the community) that can be said to be more or less "present" in various forms in indigenous communities. Thus, as we noted above, we entered the arena with our understanding that governments themselves can come to act in terms of selfish interests when they join forces with capitalist economic interests geared to extractive profit making, thus advancing what can be termed an imperialist agenda. Therefore, our commitment is to look for and explore ways of agitating this in theory and in practice. We were interested in setting up the FG discussions to explore with participants how they managed to some extent to create such a space – as we believe also that lessons can be learned for wider audiences in Africa and indeed around the globe.

[4] See also Romm (2015b), for an account of conducting FGs in terms of a conception of indigenous ways of knowing, and Romm (2018, 2020), for an exposition of the justification for value-committed research.

The Ugandan case (similarly to the Ghanaian case) draws implicitly on traditional notions of "the commons" and an implied critique of current trends towards de-commonizing of the commons. This is relevant for any movement aimed at desisting mainstream conceptions of "development" and the interests tied to this, and offering alternative world visions as part of its praxis. In this regard, we concur with White (2018), in his introduction to his interview with Escobar entitled "Farewell to Development" that:

> As inequality and environmental degradation worsen, the search is on not only for alternative development models but also for alternatives to development itself [insofar as it is not experienced as "development"] (https://greattransition.org/publication/farewell-to-development)

During the interview (2018), Escobar pleads for us as humanity to create models which enable us to appreciate and enact "ourselves as communal and collective beings". We suggest that the insights advanced by this Ugandan community as expressed during the FG sessions and in the workshop, offer the basis for models – as applicable to different contexts – for national and international relations built on a conception of "neighbourliness". This was expressed succinctly by some of the participants, whose expressions resonated with the tenor of the discussions that were generated. The tenor of the discussions built up during the dialogues shows the participants' reflections on how to recognize genuine "development" as tied to an inclusive wellbeing and possibilities of creating movements to activate this quest.

4 Conclusion

The aim of this study was to examine how the seemingly "voiceless" peasants could use their collective agency effectively to resist social and environmental exploitation by foreign-owned and multinational corporations. Specifically, it examined, using different FG discussions held in 2020 amongst a peasant community in Uganda, how the oppressed organized themselves in a social movement style at no cost in order to push back against diverse forms of social and environmental injustices perpetuated by big businesses. Categorized as an indigenous people's social movement because the movement was locally generated by the study community, we suggest that Indigenous Knowledge (IK) offers powerful tools and pathways to assert the intellectual agency of African people and validate their lived experiences. This heritage empowers them to become active agents of change instead of passive recipients awaiting foreign liberation as demonstrated in the case of the Ugandan community. Armed with raw courage and the will to achieve justice, the "voiceless" peasants did not yield to what they considered to be the slavery-style treatment by foreign investors to supress their sense of values and defuse their cultural resilience. The actions taken by the community against their oppressors ripped the veils from the face of the foreign investors and exposed how heartless certain groups in society can be when given powers over others. The behavior of the two foreign-owned companies, namely Zhongmei and Bukona factory defies belief as to how callous some people can be, to cause so much senseless social, economic, and emotional anguish to others, whom they perceive to be powerless.

However, leveraging the power of social agency, the oppressed peasants unleashed their potential by pushing back. Elsewhere, in a different context, Fanon (1963) identifies the peasants as the real revolutionary class but not the urban proletariats. The proletariats, according to Fanon, are pampered by the colonial regime, accumulated wealth, status, and developed complacency in colonial rule. This group cannot launch any meaningful resistance because they have everything to lose. For the study community in Uganda, the peasants, are the true *wretched of the earth*. They can be relied upon to resist oppression and exploitation using any means necessary including violence because in the words of a male participant "*we certainly had to act because we had nothing to lose but everything to gain*". They had everything to gain from their resistance against the foreign-owned and exploitative businesses. Although some sections of the community, especially the youth, preferred the use of physical violence, as endorsed by Fanon (1963), the elders, drawing on their long years of experience, discouraged this. The elders knew that physical violence as an instrument of regaining control does not only affect the oppressors but also the oppressed peasants on different scales. This is because the oppressed peasants who would survive the physical violence directed at the oppressors would have to live with the post-violence complications such as post traumatic stress disorders for the rest of their lives. The women too referred to their role in steering the movement in a peaceful direction. Hence one of the women, speaking on behalf of women, stated that "*we said no to use of force*".

Finally, we believe that the case of the peasants' exploitation articulated in this chapter, demonstrates how detached most of the officials in government, both local and national, are from the locals. This is amplified by a submission from a male participant who, in talking about a lack of elected (government) leaders paying attention to their plights, asserted "*we elected some of them into office and why aren't they coming to talk to us like you people from the University [researchers] have done – what are they fearing? Why did they seek our votes if they are not willing to be our voices?*". This position is reinforced by the government's positive appraisal of Bukona factory in the national media on aspects like job creation for the locals, boosting economic growth, and contributing to environmental protection among others – the reality on the ground sharply differs from government claims. On the contrary, the entry of the foreign-owned businesses ushered in economic disempowerment of the locals and increased environmental contaminations hence disrupting the web of life. It is, therefore, the responsibility of the government to reconceptualize its approach to attracting foreign investors as suggested by participants. Participants called for increased background scrutiny of all investors coming into the country and enacting/enforcing appropriate legislation that concerns the protection of both the citizens and foreign investors from all forms of exploitation.

Acknowledgments The production of this manuscript was made possible through the financial support provided by the University of South Africa (UNISA) and Kabale University (KAB).

References

Adyanga, F. A. (2019). Ancient governance in Africa. In N. Wane (Ed.), *Gender, democracy and institutional development in Africa* (pp. 37–65). Palgrave Macmillan.

Akena, F. A. (2017). Customary land tenure and ecological sustainability in Acholiland, Northern Uganda. In McIntyre-Mills & Romm (Eds.), *Balancing individualism and collectivism* (pp. 221–237). Springer.

Annan, J., & Brier, M. (2010). The risk of return: Intimate partner violence in Northern Uganda's armed conflict. *Social Science & Medicine, 70*(1), 152–159.

Arko-Achemfuor, A. (2019). Putting communal land into productive use through collaborations, networking and partnerships in rural South Africa. In J. J. McIntyre-Mills & N. R. A. Romm (Eds.), *Mixed methods and cross-disciplinary research towards cultivating eco-systemic living* (pp. 251–266). Springer.

Arko-Achemfuor, A., & Dzansi, D. Y. (2015). Business doing well by doing good in the community. *The Journal of Commerce, 7*(2), 53–68.

Bernal, M. (1987). *Black Athena: The Afroasiatic roots of classical civilization volume I: The fabrication of ancient Greece 1785–1985*. Free Association Books.

Buswell, C., & Corcoran-Nantes, Y. (2018). Precarious liaisons: Gender, moral authority and marriage in colonial Kenya. In J. J. McIntyre-Mills, N. R. A. Romm, & Y. Corcoran-Nantes (Eds.), *Balancing individualism and collectivism* (pp. 295–414). Springer.

Carkovic, M., & Levine, R. (2005). Does foreign direct investment accelerate economic growth? In *Proceedings of the conference organized by the Institute of International Economics* (Washington, 2004): Does foreign direct investment promote *development?* (pp. 195–220).

Chilisa, B. (2012). *Indigenous research methodologies*. Sage.

Chilisa, B. (2017). Decolonizing transdisciplinary research approaches: An African perspective for enhancing knowledge integration in sustainability science. *Sustainability Science, 12*(5), 813–827.

Chilisa, B. (2020). *Indigenous research methodologies* (2nd ed.). Sage.

Chilisa, B., & Ntseane, G. (2010). Resisting dominant discourses: Implications of Indigenous, African feminist theory and methods for gender and education research. *Gender and Education, 22*(6), 617–632.

Cram, F., Chilisa, B., & Mertens, D. M. (2013). The journey begins. In D. M. Mertens, F. Cram, & B. Chilisa (Eds.), *Indigenous pathways into social research: Voices of a new generation* (pp. 11–40). Left Coast Press.

Durham, J. B. (2004). Absorptive capacity and the effects of foreign direct investment and equity foreign portfolio investment on economic growth. *European Economic Review, 48*(2), 285–306.

Fanon, F. (1963). *The wretched of the earth*. Grove Press.

Harris, L. D., & Wasilewski, J. (2004). Indigeneity, an alternative worldview: Four Rs (relationship, *responsibility*, reciprocity, redistribution) vs. two Ps (power and profit). Sharing the journey toward conscious evolution. *Systems Research and Behavioral Science, 21*(5), 489–503.

Langdon, J. (2009). Learning to sleep without perching: Reflections by activist-educators on learning in social action in Ghanaian social movements. *McGill Journal of Education/Revue des sciences de l'éducation de McGill, 44*(1), 79–105.

Li, X., & Liu, X. (2005). Foreign direct investment and economic growth: An increasingly endogenous relationship. *World Development, 33*(3), 393–407.

Mabikke, S. B. (2011). Escalating land grabbing in post-conflict regions of Northern Uganda: A need for strengthening good land governance in Acholi Region. In *International conference on global land Grabbing* (pp. 6–8).

Martinez-Alier, J. (2009). Ecological economics: "Taking nature into account". In C. L. Spash (Ed.), *Ecological economics* (pp. 39–61). Routledge.

McIntyre-Mills, J. J. (2019). Fostering ecological citizenship through recognizing non-anthropocentric right to habitat. In McIntyre-Mills, J. J. & Romm, N. R. A. (Eds.), *Mixed methods and cross-disciplinary research: Towards cultivating ecosystemic living* (pp. 429–451). Cham: Springer.

McIntyre-Mills, J. J., & Romm, N. R. A. (Eds.). (2019). *Mixed methods and cross-disciplinary research: Towards cultivating ecosystemic living*. Springer.

McIntyre-Mills, J. J., & Wirawan, R. (2020). *No longer top predator and no longer business as usual: Implications of Covid 19 for multispecies relationships and wellbeing*. Presented online at the 7th annual conference of the Society for the Advancement of Science in Africa (SASA), on Health Research, Policy, and Care Across Africa in the Age of the COVID-19 Pandemic.

Mugizi, F. M., & Matsumoto, T. (2020a). Population pressure and soil quality in Sub-Saharan Africa: Panel evidence from Kenya. *Land Use Policy, 94*, 104499.

Mugizi, F. M., & Matsumoto, T. (2020b). From conflict to conflicts: War-induced displacement, land conflicts, and agricultural productivity in post-war Northern Uganda. *Land Use Policy, 94*, 105149.

Murove, M. F. (2005). *The theory of self-interest in modern economic discourse: A critical study in the light of African humanism and process philosophical anthropology* [Doctoral thesis]. University of South Africa.

Ndimande, B. S. (2018). Unraveling the neocolonial epistemologies: Decolonizing research toward transformative literacy. *Journal of Literacy Research, 50*(3), 383–390.

NEMA. (2019). *NEMA suspends operations of Bukona Agro Processors in Nwoya*. https://www.nema.go.ug/media/nema-suspends-operations-bukona-agro-processors-nwoya

Osuji, P. (2018). Laudato Si' and traditional African environmental ethics. In G. Magill & J. Potter (Eds.), *Integral ecology: Protecting our common home* (pp. 184–208). Cambridge Scholars Publishing.

Refugee Law Project. (2014). *Compendium of conflicts in Uganda. Findings of the National Reconciliation and Transitional Justice Audit*. School of Law Makerere University.

Romm, N. R. A. (2015a). Reviewing the transformative paradigm: A critical systemic and relational (Indigenous) lens. *Systemic Practice and Action Research, 28*(5), 411–427.

Romm, N. R. A. (2015b). Conducting focus groups in terms of an appreciation of Indigenous ways of knowing. *Forum: Qualitative Social Research, 16*(1) (Art. 2).

Romm, N. R. A. (2018). *Responsible research practice: Revisiting transformative paradigm in social research*. Springer.

Romm, N. R. A. (2020). Justifying research as conscious intervention in social and educational life: Activating transformative potential. *Educational Research for Social Change, 9*(2), 1–15.

Rugadya, M. A., Nsamba-Gayiiya, E., & Kamusiime, H. (2008). *Northern Uganda land study analysis of post-conflict land policy and land administration: A survey of IDP return and resettlement issues and lesson: Acholi and Lango regions*. World Bank.

Spash, C. L. (2009). The development of environmental thinking in economics. In C. L. Spash (Ed.), *Ecological economics* (pp. 62–83). Routledge.

Stiglitz, J., Sen, A., & Fitoussi, J. P. (2010). *Mis-measuring our lives: Why the GDP doesn't add up*. The New Press.

White, A. (2018, February). Farewell to development. Interview by Allen White. *Great Transition Initiative*. https://greattransition.org/publication/farewell-to-development

Part II
Extending the 'Frontiers of Justice'

Learning from Nature's Classroom: Reframing Economics, Accounting and Accountability

Janet J. McIntyre-Mills, R. Wirawan, and I. Widianingsih

Abstract This learning community approach (adapting Senge, The fifth discipline: the art and practice of the learning organization. Random House, 2006) is facilitated through Yunus Social Enterprise and UnPad (Widianingsih) to support social inclusion and green, re-generative entrepreneurship. The paper proposes transforming educational systems to support an alternative cyclical economy based on work in progress projects that support rural-urban balance through participatory action research and learning. This relational approach to education is linked with social enterprise using local products, such as bamboo and coffee. The approach uses a gender and species sensitive lens with a view to mainstreaming opportunities.

Keywords Vocational training · On line engagement · Circular economy · Regeneration

---

J. J. McIntyre-Mills role is collaborative designer, facilitator, thesis supervisor and contributing to needs assessments.

J. J. McIntyre-Mills (✉)
College of Education, University of South Africa, Pretoria Central, Gauteng, South Africa

College of Professions, Adelaide University, Adelaide, Australia
e-mail: janet.mcintyre@adelaide.edu.au

R. Wirawan · I. Widianingsih
University of Adelaide, Adelaide, Australia

Universitas Padjadjaran, Sumedang, West Java, Indonesia
e-mail: rudolf.wirawan@adelaide.edu.au; ida.widianingsih@unpad.ac.id

J. J. McIntyre-Mills, Y. Corcoran-Nantes (eds.), *Transformative Education for Regeneration and Wellbeing*, Contemporary Systems Thinking,
https://doi.org/10.1007/978-981-19-3258-8_6

1 Introduction: Aims, Praxis and Process

The twofold aim of the chapter is to

1. Build on and extend several small pilots which strive to be transformative (Mertens, 2017) by testing out a collaborative approach to support action learning to enable people to earn while they learn and grow a future together.
2. Address alternative ways of supporting a community of practice through local hands on intervention by facilitators plus digital engagement and listening (Dobson, 2012).

2 Research Approach

The exploratory case studies use mixed methods to explore what social engagement processes work, why and how to support the development of a circular economy. A circular economy works with, rather than against nature, whilst a re-generative approach helps to grow social, economic opportunities through protecting the environment. The project addresses social, economic and environmental indicators of wellbeing to mitigate and adapt to climate change through operationalising the concept of wellbeing stocks (Stiglitz et al., 2010 and extending the work of Elkington, 1999 on triple bottom line). The approach uses a gender and species sensitive lens to mainstreaming opportunities.

The ongoing work in progress strives to learn from an existing community empowerment projects – by enabling multi-site collaboration via an on line engagement portal. This chapter focuses on Indonesian case studies. The project is supported by Universitas Padjadjaran, the Indonesian Diaspora and Adelaide University and is linked with the Yunus Centre.

Social enterprises have the support of Indonesian government (including in kind support and law enforcement) to protect social enterprise which is in line with the Presidential Decree to support One village, one Enterprise. The project complies with the ethics requirements of Adelaide University and has been approved by the University of Padjadjaran.

The project builds on current connections with Alam Endah, Cibodas, Tarumajaya (West Java), East Kediri and Universitas Padjadjaran to support setting up an eco-village enterprise with the support of the Indonesian Diaspora and the Yunus Centre. Wirawan has set up a computer training social enterprise on which this project builds. McIntyre and Wirawan are research fellows at Unpad and build on research collaborations that have developed since 2010 with Widianingsih who heads the Center for Decentralization and Participatory Development Research at the faculty of social and political sciences, Unpad and who co-hosted research in Alam Endah and Cibodas on which this research builds.[1] Widianingsih explains that villages that were once isolated are being increasingly connected through road construction in West Java and with better internet connectivity villagers will be able to market their goods.

The cases include Alam Endah, Cibodas, Tarumajaya, Kasepuhan Ciptagelar and East Kediri to explore the activities of community co-operatives that operate as social enterprises:

- Alam Endah, an hour's drive from Bandung has a strong women's group engaged in a range of social enterprises.
- Cibodas in a neighbouring area has a strong group of young people who have formed the Reggae group to re-generate interest in farming and in Tarumajaya the women's group and local youth groups are digitally connected and well organised. Tarumajaya is one of the most significant villages in Bandung District, West Java because the area is part of upper Citarum watershed. Under Jokowi's leadership, the Indonesian government started to introduce a new approach towards revitalising the Citarum river using a more collaborative model to clean up the river (Widianingsih et al., 2019; Abdoellah et al., 2019). The use of the military to undertake the clean up of the upper river system has resulted in some improvements and some tensions around fish farming and land usage by the military which need to be carefully managed to ensure environmental and social concerns are addressed.
- Kasepuhan Ciptagelar village community comprising about 30,000 people has practiced Sundanese culture to protect the forests and their traditional rice seeds. The engagement with Sudanese participants in Kasepuhan Ciptagelar around the border area of Banten and West Java Provinces who have been living for more than 600 years in the forests of West Java will be a central focus of this research to demonstrate the way in which rice production based on protecting seeds and the surrounding forests has enabled the community to maintain high yields even during droughts. The team engaged in similar projects in Venda, South Africa (Lethole et al., chapter "Aa!!Venda Women and Social Enterprise: Stepwise Process to Regenerative and Sustainable Living" in this volume) builds on these case studies in Indonesia and the inspiration from the Ocean View Project in South Africa. Indigenous wisdom and custodianship is central to the approach.
- East Kediri has a group who are specifically interested in the potential of IT as Wirawan has already set up a digital engagement project in Kediri on which this project builds.

---

[1] McIntyre-Mills, J., Corcoran-Nantes, Y, Widianingsih, I. Wirawan, R. 2019. Chapter 7 Alam Endah: Rural Camelot in West Java—A Case Study of Empowerment and Integrated Rural Development In McIntyre-Mills, J, Romm, Corcoran Nantes, Y. 2019, *Democracy and Governance for Resourcing the Commons: Theory and Practice on Rural-Urban Balance.* Springer. Cham. pp.179–192. DOI https://doi.org/10.1007/978-3-030-04891-4 Print ISBN: 978–3–030-04890-7 Human Research Ethics LIVING VIRTUOUSLY AND WELL 7213.

These case studies will explore what social engagement processes work why and how to support a circular economy. The community of practice is supported through Yunus Social Enterprise, Universitas Padjadjaran as well as local NGOs and village level authorities. Scoping has already been undertaken in the above areas and pilots on local enterprise and development underpin this next phase of the research (see McIntyre-Mills et al., 2019a, b, 2021).[2]

We advocate that education needs to be transformed to support local *hands on* intervention by facilitators who help rural engagement processes that are co-operative and that aim to support social and environmental justice and try to not only sustain but re-generate the commons.[3]

World hunger is increasing at an alarming rate according to the Report by the Secretary- General and this will accelerate as a result of climate change.[4] Food security could become more of an issue in regional areas of West Java if landlessness

---

[2] McIntyre-Mills, J., Corcoran-Nantes, Y, Widianingsih, I. Wirawan, R. 2019. Chapter 7 Alam Endah: Rural Camelot in West Java—A Case Study of Empowerment and Integrated Rural Development In McIntyre-Mills, J, Romm, Corcoran Nantes, Y. 2019, *Democracy and Governance for Resourcing the Commons: Theory and Practice on Rural-Urban Balance.* Springer. Cham. pp. 179–192. DOI https://doi.org/10.1007/978-3-030-04891-4 Print ISBN: 978–3–030-04890-7 Human Research Ethics Living Virtuously and Well 7213.

McIntyre-Mills J., Corcoran-Nantes Y., Wirawan R., Widianingsih I., Saptenno I., Sari N.H. 2019. Chapter 8. Freeing the Luwaks and Escaping the Iron Cage: Vignettes on Surviving in the Anthropocene by Joining Up the Social, Economic and Environmental Policy Dots—An Application of Critical Systemic Thinking to Areas of Concern. In: McIntyre-Mills J., Romm N., Corcoran-Nantes Y. (eds) Democracy and Governance for Resourcing the Commons. Contemporary Systems Thinking. Springer, Cham. pp. 195–232 https://doi.org/10.1007/978-3-030-04891-4 Print ISBN: 978-3-030-04890- Human Research Ethics LIVING VIRTUOUSLY AND WELL 7213.

McIntyre-Mills, J. with Wirawan, R and Widiansingh, I and Riswanda (2021) Social engagement to protect multispecies habitat: implications for regeneration and food security From Polarisation to Multispecies Relationships in the Age of Mass Extinctions, Springer Nature Singapore.

[3] As stressed in 'Democracy and Governance for Resourcing the Commons' (McIntyre-Mills et al., 2019b) the rate at which world hunger is increasing has been highlighted in the Report by the Secretary- Generaland urbanization has increased beyond expectations (See UN report on urbanisation, 2014). By 2050 the majority of the world's population will be in Asia and Africa. Unemployment in cities and social exclusion (as highlighted by the Sendai Risk Platform) will pose a human security risk. As stressed elsewhere (McIntyre-Mills & Romm 2019; McIntyre-Mills et al., 2019b, 2021a) the Asian and African demographic growth provides both a dividend and a liability. The potential for investing in social and environmental justice through *job creation* is the focus for this chapter.

[4] https://sustainabledevelopment.un.org/sdg2 Report of the Secretary-General, Special edition: progress towards the Sustainable Development Goals

> An estimated 821 million people – approximately 1 in 9 people in the world – were undernourished in 2017, up from 784 million in 2015. This represents a worrying rise in world hunger for a third consecutive year after a prolonged decline. Africa remains the continent with the highest prevalence of undernourishment, affecting one fifth of its population (more than 256 million people). Consistent with the continued growth in undernourishment, 770 million people faced severe food insecurity in 2017…

associated with the privatisation of tea plantations continues which results in the displacement of workers who are landless. Those with access to land are being encouraged to engage in agricultural regeneration in a 'reggae' movement by young people who are making a success of farming through forming co-operatives and motivating young people to remain farmers as it can provide a good living. They make farming appear 'cool' by wearing uniforms and supporting one another in successful farming ventures. Another successful example is the way that Alam Endah has worked together to support successful enterprises in line with Jokowi's notion of 'One Village, One Enterprise Approach' (2014) decreed by the President of Indonesia, Jokowi. Another example is the intergenerational transfer of knowledge on bee farming in Cibeber from father to son spanning more than four generations. Through creating a community of practice network at a post-national level, we have considered the application of the "one village one enterprise" notion in the South African context to help share local wisdom to support local capacity building and its potential to support democratic engagement (Habermas, 2001). The UN 2030 Agenda[5] is:

> the new global framework to help eradicate poverty and achieve sustainable development by 2030. It includes an ambitious set of 17 Sustainable Development Goals.... The 2030 Agenda for Sustainable Development sets out the global framework to eradicate poverty and achieve sustainable development by 2030.

But in order to have a hope of achieving the UN Sustainable Development Goals (2030) we propose the need for a re-generative approach based on every day strategies to earn, learn and grow a future in eco-villages.

3 Transformative Approach

The pandemic has resulted in a game changing environment in which meetings and transactions need to occur in a digital environment. Local engagement can be supported by a digital community of practice. The research explores plausible pathways to enhance wellbeing and regeneration. Clearly, this has implications for education. Covid-19 has resulted in normalising digital engagement to support a digital community of practice (Wenger et al., 2009) to foster reciprocal relationships to support institutional capacity building and learning networks.

Public education needs to take the opportunity to address biopolitics, a key area of concern raised by Foucault. Taxonomies are linked with power and knowledge (Foucault & Gordon, 1980). What we decide to include or exclude, depends on our values. The Indigenous, African/first nation philosophies that stress interdependency could also provide a way forward as suggested in chapters "Aa!!Venda Women and Social Enterprise: Stepwise Process to Regenerative and Sustainable Living" and "Women, Education and Socialisation: Transforming Gender and

[5] http://europa.eu/rapid/press-release_MEMO-15-5709_en.htm

Power in the MST, São Paulo, Brasil" on spirituality and totemic links in Ghana and Venda, discussed in this volume. Public education needs to focus on:

- Protecting habitat and food security for diverse species and redressing the impact of rural -urban imbalance
- Food energy water security through social engagement projects provide a way to explore the conceptualisation of food, what ought we to eat or to avoid and why? What constitutes ethical decision making in relation to food? What forms of engagement support democracy and governance to support the commons.

Top down versus bottom up options have been discussed as commentators have compared the way in which governments have reacted to the challenges posed by the virus – but little attention has been paid to the environmental implications of unethical food production that are central to the problem of pandemics.

Scott (1998) in 'Seeing like a State' extends the approach at which Foucault (2008, Foucault & Gordon, 1980) hints by giving detailed examples of the way in which the state has crushed diversity – from designing monocultures in forestry to ironing out diverse political opinions. If we accept the concept of the banality of evil (see Arendt, 1963) – the notion that everyday decisions by many can collectively result in a normalization and acceptance of evil – then we have to re-think many of the aspects of governance, democracy, education and economics that we take for granted today. A way forward is protecting oceans, rivers, wild, rural and urban habitat through more public education and more monitoring by local residents. The environment needs to be protected at a post national regional level in order to reduce the likelihood of pandemics and mass extinctions. Australia, in particular needs to re-examine its policies on climate change, according to the IPPC (2018: 4):

> ...Australia's greenhouse gas pollution levels have continued to rise over the past three years, and we have no credible policy settings to reverse this trend.

The 26th UN Climate Summit in Glasgow (31st October–12th November 202) emphasised the regional implications of climate change on rising sea levels, rising temperatures which will require transformative praxis within the region. Public education has a central role to play by creating partnerships with rural and regional communities to assist them to create employment opportunities that are clean and green. By scaling up projects such as these real steps towards lowering emissions and demonstrating alternatives that protect against existential risk (Bostrom, 2011; Higgins, 2018).

4 Theory and Key Concepts: Relationality and Nature's Classroom

Social engagement to promote social enterprise based on social, economic and environmental accountability is explored by comparing and contrasting what works why and how across areas comprising a strong women's network, a network for young farmers, a network to support the potential of IT built upon indigenous wisdom on protecting biodiversity. Indigenous wisdom and custodianship is central to the approach. The notion of extending solidarity to organic and inorganic life is explained by Lethole who is a custodian of a sacred forest and the elephants with whom she shares totemic kinship and for whom she is responsible. This philosophy also resonates with Australian Aboriginal culture. Olive Veverbrandts (see McIntyre-Mills, 2003) explains that women give totemic names to their children when the foetus moves, the totems could be birds, a lizard or a rock formation). The case is made for giving more rights to sentient beings, simply because they are sentient and that this requires liveable habitats for them (see McIntyre-Mills et al., 2021a; McIntyre-Mills, 2021a, b, c, d, 2022) which refers to Goodall (2020) who explains that protecting habitat is central as species that develop immunity within their own habitats should not be thrust into contact through trafficking, transporting and marketing.[6] Goodall (2020) has explained that protecting habitat is central as species that develop immunity within their own habitats should not be thrust into contact through trafficking and marketing has resulted in this pandemic (along with many previous pandemics).

4.1 A Community of Practice to Support an Ecological Mindset

In the case study areas the community of practice is a step towards empowering the marginalised and thus an intersectional, gendered and multispecies mainstreaming approach is used to support diversity and biodiversity.

Social inclusion and Indigenous knowledge are key concepts that build on Freire's legacy (1970). Kabeer's (1999, 2015) gender mainstreaming approach will be applied and extended by means of an intersectional approach (adapted from Dhamoon, 2011) to address age, ability, cultural and other life chance factors that affect the participants. The notion that profit can be extracted by commodifying people, animals, plants /nature or treating it as an object is challenged by drawing on the potential of alternative approaches suggested by Bateson (1972), Pauli (2010) and Rayner (2010, 2017a, b) respectively who make a case for thinking systemically to develop 'an ecological mindset', working with nature and 'natural

[6]A move towards vegetarianism and veganism could perhaps lessen the potential for pandemics whilst also helping to reduce the emissions linked with climate change. It could also help to boost immune systems through encouraging reliance on unprocessed, fresh vegetables, fruit and exploring the potential for alternative forms of protein, for example from pulses, insects and algae which could replace reliance on meat consumption as opposed to transitioning to alternative forms of protein (van der Weele, 2019). The responsibility to protect animals subjected to transportation as part of the so-called live meat trade lies with voters who need to redraw the boundaries of what is acceptable decent treatment of animals and what constitutes unacceptable commodification of sentient beings.

inclusion', to achieve regenerative cascading benefits, rather than extractive degenerative profits.

Indicators of social, economic and environmental wellbeing support a circular economy. A co-operative in this context means a group who work together for the common good in a transparent and accountable manner. For the purpose of this research social enterprise means activities aimed at supporting social and the environmental wellbeing as well as providing a living for the participants.

The case is made for giving more rights to sentient beings, simply because they are sentient and that this requires liveable habitats for them. A move towards vegetarianism and veganism could perhaps lessen the potential for pandemics whilst also helping to reduce the emissions linked with climate change. It could also help to boost immune systems through encouraging reliance on unprocessed, fresh vegetables, fruit and exploring the potential for alternative forms of protein, for example from pulses, insects and algae which could replace reliance on meat consumption as opposed to transitioning to alternative forms of protein (van der Weele et al., 2019). The responsibility to protect animals subjected to transportation as part of the so-called live meat trade lies with voters who need to redraw the boundaries of what is acceptable decent treatment of animals and what constitutes unacceptable commodification of sentient beings.

4.2 A Systemic Approach: Climate Change, Pandemics and Unemployment: Implications for Public Education

Future scenarios need to consider the implications of increased floods, fires and the displacement of people and animals resulting in increased levels of food insecurity. Top down versus bottom up options to governance have been discussed as commentators compare the way in which governments have reacted to the convergent social and economic challenges posed by the virus and that the environmental implications of unethical food production are central to the problem.

An alternative way of living needs to be explored through balancing individual and collective needs (McIntyre-Mills & Romm, 2018; McIntyre-Mills et al., 2019a, b, 2021) through developing eco-villages and enabling people to think critically and systemically about the consequences[7] of their choices in order to build on their lived experiences to bring about a new transformations (Polanyi, 1966, 1968).

[7] This requires being guided by *a priori* norms for praxis and considering the consequences of one's choices through careful consideration aided by means of critical heuristics, also known as thinking in terms of 'if then' scenarios. This process of ethical consideration is known as 'expanding pragmatism', in order to consider the consequences for self and other living systems. (McIntyre-Mills, 2014).

5 The Potential of Participatory Learning Networks to Protect Habitat and Food Security

The Nobel Peace Prize in 2021 was awarded most appropriately to the Food program. The paper proposes an alternative architecture based on work in progress projects in Indonesia and South Africa. Our reason for working in these contexts is both pragmatic (we have access to these case studies) and because the rate of urbanisation in Africa and Asia will result in the majority of the world's population living in urban areas in these regions.

The small case studies provide some indicators of a way forward that could enable balancing individual and collective needs to protect food, energy and water. Each one of the projects makes a small contribution to understanding the potential of the eco-village concept.

The development of eco-facturing using local products such as cassava for bioplastics, bamboo for biochar, fair trade coffee, mushroom production (using the coffee grains) and honey are discussed as examples of products in Indonesia that could help to foster a green economy and reduce reliance on animal husbandry that adds to the size of the carbon footprint through heavy reliance on water and adding to methane emissions. This is a challenge for Australia, piloting alternative ways to reduce emissions will become increasing relevant for mitigating and adapting to climate change within the region. The 'one village, one enterprise approach' decreed by Jokowi has been adapted as 'one village many enterprises'.

Core design principles need to underpin the decision-making process, namely salience, trust and engagement to protect living systems and the people who are affected. The community of practice approach outlined in this chapter follows these principles in order to ensure that small groups of people are able to work together in co-operatives to manage the co-operative using block chain to ensure that the transactions cannot be altered.

These principles are discussed together with the importance of 'being the change' through expanding pragmatism to consider the social, economic and environmental implications of choices. Ethical decisions need to honour freedom and diversity to the extent that freedom and diversity are not undermined by power imbalances. A case is made for liveable habitat for human beings, other animals and plants that support organic and inorganic life. The axial theme is that systemic ethical decisions need to be based on both the priori rights of all sentient beings and the a posteriori monitoring of social, economic and environmental indicators of wellbeing.

In West Java agricultural regeneration is a focus amongst young people who are making a success of farming through forming co-operatives and giving farming a good name as a vocation that can provide a good living. Through creating a community of practice network at a post-national level, we have considered the application of the 'one village one enterprise' notion in the South African context to help share local wisdom to support local capacity building. Research on ways to support action and learning based on earning, whilst learning and growing a future together in a circular economy, The purpose of the research is to reconceptualise education for re-generative development and to inspire hope based on systemic ethics and local wisdom.

5.1 *Vocational Education and Training*

'Earn while we learn and grow a future together' is the slogan for working together with rural communities supported by vocational education and training in collaboration with a network of providers.

The challenge is to enable vocational education to support water, food and energy security. Ecofacturing relies on natural resources, the sun, wind and water for power, the earth to support seed germination and the protection of pollinators and organic workers of the soil spanning many species (see Pauli, 2010).

In Cibodas, for example, an entrepreneurial cattle farmer has diversified the cattle co-op to support biogas and adding value to cattle farming by producing a range of milk products, vegetable growing and improving the diet of the cattle.

The potential to move into coffee production will be discussed as a way to diversify and to develop more sustainable farming by reducing reliance on cattle which may not be a very popular idea unless alternative options can be suggested. The leader of the co-operative was nominated for an award by President Jokowi for his successful leadership.

The Sundanese forest community called 'Kasepuhan Ciptagelar' which for centuries has been isolated from mainstream society is collaborating with Commonroom Network foundation and Universitas Padjadjaran, in order to preserve their sustainable way of life and to avoid the encroachment of developers. They have their own sustainable infrastructure and their culture protects the multiple species within their forest community (Abdul Haris et al., 2019). Their rituals to protect a rare and ancient rice crop can be held up as a success story[8] in protecting healthy seeds and crops as advocated by Shiva (1989, 2002, 2012a, b).

[8] According to Widianingsih (drawing on Abdul Haris et al., 2019), internationally social forestry is a concept that has been implemented since the 1970s to provide access to land for the people, to enhance welfare of the people, and to sustain environmental carrying capacity. The concept has also been implemented in Indonesia since 1995. The indigenous community of Kasepuhan Ciptagelar has been applying a form of integrated forest management for 652 years within the territory of Kasepuhan Ciptagelar to balance conservation and the development of social, economic, and environmental system within the region. By considering the lessons learned from integrated territorial management by Kasepuhan Ciptagelar, there are several factors that hinder achievement of implementation of social forestry and, in general, the comprehensive management of the region by Kasepuhan Ciptagelar, as well as other indigenous communities in Indonesia with a similar characteristics. Those factors are the absence of a comprehensive legal and institutional framework along with a lack of geospatial data and information, on local knowledge. To achieve balance between conservation and development of social, economic, and environmental system by means of social forestry and acknowledgement of community's territory, the above mentioned problems need to be addressed through participatory engagement with and acknowledgement of the territory of indigenous community of Kasepuhan Ciptagelar through a pilot project.

Wirawan has a long standing relationship with members of a Kediri Training group that he set up. Kediri is a regency located in East Java which acts as a trading centre for sugar. It also has the largest cigarette industry in Indonesia, two of the most problematic products for health. Indonesia has limited control on cigarette advertising and no public education on the harmful effects. Not far from Kediri is Malang where cigarette manufacturing is also a prime activity. Village co-operatives produce tobacco and children start smoking at a very young age because advertising on television makes cigarette smoking appear glamorous. Bill boards are everywhere. The companies also sponsor education and sporting events.[9] Two hundred thousand die of tobacco related deaths per year and two thirds of Indonesian males smoke according to this program. The young age at which children are targeted makes this a health pandemic. Child labour results in their being affected by secondary smoke, but also through handling the green leaves which exude toxin. International companies target developing countries because of the lack of regulation. According to Kumboyono et al. (2020) cigarette smoking is subject to more government controls, but the widespread advertising continues. Alternative industries to tobacco are needed to replace job opportunities and to end the pandemic of smoking related illnesses.

5.2 *Policy Context*

The project is in line with Indonesian Gender Mainstreaming Law and the 'one village, one product/enterprise' (OVAP, Morihiko Hiramatsu – Governor of Oita prefecture, 1979, Yogyakarta, 2014) approach that has been applied by President Jakowi in 2008–2009.

This research is underpinned by a previous project, titled 'Living virtuously and well' for which McIntyre received ethics clearance and completed at Flinders University and now covered by two ethics approvals for studies in South Africa (University of South Africa) and Indonesia (Universitas Padjadjaran and Adelaide University). We follow the Australian Government guidelines[10] outlined in the National Statement on Ethical Conduct in Human Research.

[9] Indonesia's Tobacco Children | Unreported World, https://www.youtube.com/watch?v=BsUAAw2qLB8

[10] By meeting the Human Ethics Requirements the team will provide an audit trail for the data stored on the Adelaide University network for 5 years. Links will be shared across the named Investigators. Hard copies of research materials will be stored in a locked filing cabinet to which only the named investigators have access. Once the data has been collected and recorded the original de-identified transcripts will be stored and accessible to reviewers and policy makers. Care will be taken at all stages to ensure that de-identified data are presented in ways that ensure anonymity in publications through careful management of the data and team. The secure website managed via Adelaide University will ensure that the system is safe for users (https://www.nhmrc.gov.au/about-us/publications/national-statement-ethical-conduct-human-research-2007-updated-2018).

The team link social enterprise to learning about the importance of ecosystems in West Java region. The project team have set up a community of practice to support the group in line with Gunther Pauli's (2010) approach to developing a production, distribution and consumption cycle that supports regenerative social enterprise and reduces the risks of trafficking highlighted by a women's group (McIntyre-Mills et al., 2019).

The data collection focuses on narratives of what works why and how. These stories from self-selected participants will be used to explore the specific socio-cultural, economic and environmental contexts of social enterprise and how best to support social engagement within a circular economy which is the focus of the research.

Participants in the case study areas will be engaged within and across multiple sites in a COVID-19 safe manner and will use an on line approach. All the participants will self-nominate at the invitation of the administrative office of the Universitas Padjadjaran. The approach enables participants who self-nominate to participate and no risks are envisaged in helping to support social enterprise.

The mixed methods research combines qualitative narrative and quantitative data mapping to explore what works why and how to support a circular economy. By using several case studies it will be possible to compare and contrast the social engagement process by working with the co-operatives, particularly women and young people to add value to their products. Data collection will be enabled through the on line data collection process.

The circular economy could be supported by institutional governance to better manage the global commons. Elinor Ostrom won the Nobel Peace Prize in 2009 for her empirical research on the rational choices made by groups of people who worked collaboratively to protect the common good. Her work showed that in some specific cases conditions for co-operation could be achieved and she argued that Hardin's (1968) thesis which he called "Tragedy of the Commons" is not inevitable and that a few freeloaders will not inevitably drain the common resources. Ostrom's research (1992, 2008, 2014) demonstrates that people can and do share resources under certain circumstances, these are summarised as 8 principles which relate to setting clear boundaries and clear rules, appropriate decision making processes and monitoring linked with appropriately applied sanctions and processes. She also stresses that the local community rights need some recognition by wider government and that the local systems need to be nested within layers of governance. The challenge which we face is *what* motivates people to change their thinking and practice and how can learning organisations and communities motivate people to move from business as usual towards supporting wellbeing stocks and enthusiasm for managing the commons?

6 Transformative Action Learning

The social enterprise mission has potential if decoupled from the potential of misuse by neoliberal microcredit systems that requires borrowing and pressure to repay. The first step is to work with those who have local lived experience (see Romm, 2015; Romm & McIntyre-Mills, 2020) through facilitation and learning from one another, whilst being mindful that the political and economic context in which the projects operate may have positive or negative cascading effects.

The Yunus Social enterprise approach provides inspiration – it focuses on working with the public, private, volunteer and civil society sectors. Those who are currently excluded from mainstream business as usual need to be prioritized and the emphasis needs to avoid lending micro-credit if it is linked with punitive repayment conditions or using community pressure (DeLong-Bas, 2013). Social enterprise needs to rely on creating hope and opportunities not creating fear and debt. The formative, process and summative evaluations need to involve the recipients to ensure that the so-called 'neoliberal' agenda of individual responsibility and profit at the expense of the community and the environment does not occur. Social enterprise (in the context of this paper) is a little different from social entrepreneurship[11] as it focuses on balancing individual and collective concerns and is non-anthropocentric. The development of eco-villages supporting the 'one village many enterprises' concept currently applied in Indonesia relies on responsive design informed by the principle of subsidiarity and Ashby's rule, namely that policy decisions need to be made at the lowest level possible and that the complexity of design decisions need to match the complexity of the local residents who act as caretakers for local living systems. The design of eco-villages (as we conceptualise it) is also informed by Pikketty (2015) who stressed the importance of data trails to ensure fairness and reciprocity in his 13th Nelson Mandela Address. Although transparency is vital for public trust a further step is required, namely the need to protect the environment through everyday decisions, as stressed previously by Wangaari Maathai in the 3rd Nelson Mandela Lecture.

[11] Huybrechts and Nicholls (2012: 3) discuss the diverse definitions of social entrepreneurship: "All the definitions of social entrepreneurship agree on a central focus on social or environmental outcomes that has primacy over profit maximization or other strategic considerations. A second defining feature is innovation. Innovation can be pursued through new organizational models and processes, through new products and services, or through new thinking about, and framing of, societal challenges."

6.1 Vignette on the Potential of Integrated Development and Action Learning

In January a landslide in Java[12] resulting in the loss of life of 11 villagers underlines the importance of halting de-forestation and promoting the planting of bamboo and terracing to restored ground cover and the regeneration of the environment.

The first step is to work with or develop co-operatives and to work with the women, young people and those who are marginalised in ways that facilitate gender mainstreaming and opportunities for those who are usually excluded.

The construction of bamboo forests could be a cost effective first step which could also generate income through marketing a range of products through the on line eco village website. Websites to support the projects have been developed so far for Venda and Taramajaya:

- https://www.wirasoftfoundation.org/en_GB/web/ecovillage/venda1
- https://www.wirasoftfoundation.org/en_GB/web/ecovillage/tarumajaya1

The second step is to develop training to understand the nature of a sustainable development goals and the need for a re-generative economy by looking at the following links to explain a circular economy:

Participants could also learn that what appears to be a static map is actually a construct of a dynamic process. The implications of recognising patterns in nature could provide valuable insights for learning about the constructive or destructive consequences of everyday decisions made at community level.

Furthermore, the opportunity to apply participatory action learning and research *as a component of core subjects* could create the context for a new approach to re-generative education that achieve systemic integration.

The third step is to engage in the on line training program supported by a community of practice that links those who are able to inspire and learn from one another.

The third step is to report what works why and how by means of a survey which will gather de-identified data to enable the eco-village team members (co-authors of this chapter) to improve the engagement process.

The fourth step is to share de-identified stories or requests for further training or assistance.

Here is a hypothetical example of how it could be used to help young people:

A young woman is living in a small village outside Bandung in West Java. She wants to develop a small business selling vegetables, mushrooms and coffee beans, but Covid-19 has limited her access to markets, furthermore, her small village is facing water insecurity and the temperatures have been soaring. The clearance of trees has resulted in the top soil being blown away and her business is not progressing. Added to this, there was a landslide behind her house caused by the clearance

[12] https://www.abc.net.au/news/2021-01-11/indonesia-landslides-kill-11-as-search-goes-on-for-survivors/13047258

of the trees. The potential to plant bamboo to support the hillsides is being explored with her through discussions hosted by Eco Villages and it is suggested that re-vegetating the hillsides with the original diversity of plants is a step forward to restore the micro-climate. Shelters of bamboo can protect the vegetables as well. She is asked to explore what she has, what she needs, what she is willing to add or discard in terms of adding training and discarding the wrong practice of clearing forests for vegetable growing on hillsides that have become barren. She is asked to explore the barriers and the turning points for the better and worse and to participate with those in her community who want to develop green social enterprises.

6.2 De-coupling Circular Economics from Neo-Liberalism: The Potential of Action Learning and Social Enterprise

We assume that economics and accounting need reframing to support relational approaches that protect living systems through circular economies that value well-being, not merely the economic bottom line (Stiglitz et al., 2010; McIntyre-Mills & Wirawan, 2020). The purpose of the research is to:

- Explore how education could be improved by learning from nature's classroom through practical community engagement processes from primary, secondary, tertiary an VET levels.
- Address the issue of *a priori norms* and *a posteriori indicators* for transformation towards re-generative living and preventing displacement of people, plants, animals, transformative research
- Contribute to understanding the commons as a process and a sense of connection to living systems, rather than as a resource 'held in common', to cite Bollier (2011, Bollier & Helfrich, 2012).

Stiglitz et al. (2010: 15) use a multidimensional measure of systemic wellbeing that can be used to enable participants to re-evaluate economics and to become more aware of the way in which business as usual tends to neglect social and environmental aspects of life. The pursuit of profit at the expense of people and the environment is a central problem for democracy and governance (Keane, 2009). Alternative forms of organisation are possible (see Aarhus Convention, 1998; Florini, 2003; McIntyre-Mills, 2017a, b) to support 'wellbeing stocks' but this needs to be achieved through creating a change in values through reaching out and supporting one another.

Participation by those who are to be at the receiving end needs to inform policy (McIntyre-Mills, 2008). In order to manage the commons, mutual agreements need to be negotiated and records need to be kept, in order to protect the interests of stakeholders (see Planetary Passport, McIntyre-Mills, 2017a, b, Systemic Ethics, McIntyre-Mills, 2014; Wirawan & McIntyre-Mills, 2019). Covid-19 provides an opportunity to re-think our place within nature and the need to apply more relational approach to other species.

6.3 Circular Economy and Non-linear Relational Approach

A circular economy is an economic system aimed at ensuring that there is no waste of human or natural resources as it supports the continual use, re-use, sharing and re-purposing of resources to create a closed-loop system, minimising the use of resource inputs and the creation of unnecessary waste by excluding people, polluting the environment, shifting the costs of production to future generations, eliminating and moving away from a carbon based economy. A circular economy works with, rather than against nature. Kirchherr et al. (2017) stress that the way the concept circular economy is used is diverse. For the purpose of this research the concept of circular economy is based on a recognition that we rely on multiple species in the web of life (Capra, 1996). Our systemic approach retains the central tenets of reducing, re-using and recycling and recognises that all living systems are part of one fabric of life and that this requires a shift in values. The notion that a circular economy works with nature rather than against it. Case studies could provide some indicators of a way forward that could enable balancing individual and collective needs to protect food, energy and water.

The development of eco-facturing using local products in Indonesia include cassava for bioplastics, a range of local vegetables, herbs, bee keeping, coffee or tea production. By carefully ensuring that co-planting is undertaken within the forest environment, the dangers of soil loss, loss of ground water, micro climatic changes could be avoided. These are lessons that have been learned in the wake of the so-called Green Revolution in West Java that has resulted in many of the farmers experiencing the impact and flow on effects of clearing forests and losing species diversity.

The 'one village, one enterprise approach' decreed by Jokowi has been adapted as 'one village many enterprises' and applied (in part) in Kediri (East Java) by Wirawan and Alam Endah (West Java) by Widianingsih.

By emulating the way in which living systems work within nature, a biomimicry approach strives to make the best use of the local resources, such as bamboo for biochar to support energy, building, furniture, clothing within a niche habitat. Gunther Pauli calls this approach 'ecofacturing', because it means working with nature.

New forms of enterprise are needed to enable a better balance between urban and regional areas to prevent risk and to enhance resilience by recognising the need to protect habitat for all living systems, both plants and animals.

Gunther Pauli (2010) explains that nature has no waste and that if we work with nature we can re-generate the land and opportunities for all. Vocational Education in Rural Universities and Training colleges could help to promote the value of agriculture to learners who could be tasked with the vital issue of food and water security. In order to enable lifelong engagement by active citizens requires action learning to address areas of perceived policy concern.

The patterns will be used to create the influence map (Wirawan & McIntyre-Mills, 2019) and in turn this map can be used to inform policy decisions based on social, economic and environmental indicators of wellbeing.

Data from the user stories can be categorised as follows: (a) Re-generative Living, which means working with nature to support wellbeing stocks, (b) Small Changes to move away from business-as-usual towards wellbeing and (c) Business as Usual which focuses on profit at the expense of the environment and people. These data categories can be used to create maps and the influencer maps can be used to guide social, economic and environmental regeneration to ensure that they meet and extend the United Nations Sustainable Development Goals.

7 Conclusion

Online communities of practice could play a role to support re-generative social enterprises linked with opportunities for learning linked with learning organisations. See link The Web site http://wirasoftfoundation.org/en_GB/web/bic-sig for further details on the potential to protect organic food production that is not undermined by attempts to produce cheaper food using chemicals or adding cheaper additives to bulk out products. The value chain and the food chain could be protected by means of tamper proof ledgers (Al-Saqaf & Seidler, 2017).

To sum up, vocational learning needs to include face-to-face engagement supported by capacity building through on line training to enable the development of re-generative social, economic and environmental systems Pilots could be scaled up to inspire hope and creativity. A non-anthropocentric approach is needed if curricula are to have a hope of achieving the rapid transformation required to address the concerns raised by the IPPC report (2021).

Bibliography

Aarhus Convention. (1998, June 25). *On access to information*. Public participation and access to justice in environmental matters, Denmark. http://ec.europa.eu/environment/aarhus/

Abdoellah, O., et al. (2019, April). Pemetaan Sosial dalam Perencanaan Program Pengembangan Ekowisata Berkelanjutan di Citarum Hulu, Kumawula. *Jurnal Pengabdian Kepada Masyarakat 2*(1), 53–66. Universitas Padjadjaran. http://jurnal.unpad.ac.id/kumawula/article/view/24461

Abdul Haris, R., Kusdiwanggo, S., Nurlinda, I., Dwiartama, A., Fikry Pratama, M., Harriman, Iskandar, G., Widianingsih, I., & Rakasiwi, S. (2019, Desember). Perhutanan Sosial, 560 Pengakuan Wilayah Masyarakat Hukum Adat, dan Pembelajaran dari Pengelolaan Hutan oleh 561 Masyarakat Hukum Adat Kasepuhan Cipta Gelar. *Jurnal Analisis Sosial, 23*(2), 48–71. Yayasan 562 Akatiga, Bandung. https://www.researchgate.net/publication/344365649_perhutanan_sosial_ 563 pengakuan_wilayah_masyarakat_hukum_adat_dan_pembelajaran_dari_pengelolaan_hutan_ 564 oleh_masyarakat_hukum_adat_kasepuhan_ciptagelar. Accessed 5 Sept 202.

Al-Saqaf, W., & Seidler, N. (2017). Blockchain technology for social impact: Opportunities and challenges ahead. *Journal of Cyber Policy, 2*(3), 338–354. https://doi.org/10.1080/23738871.2017.1400084

Arendt, H. (1963). *Eichmann in Jerusalem: A report on the banality of evil*. Viking Press.

Bateson, G. (1972). *Steps to an ecology of mind*. Ballantine.

Bollier (2011). 15 July 2011. http://www.bollier.org/commons-short-and-sweet

Bollier, D., & Helfrich, S. (2012). *The commons strategies group*. Levellers Press.

Bostrom, N. (2011). *Existential risk prevention as the most important task for humanity*. www.existential-risk.org

Capra, F. (1996). *The web of life: A new synthesis of mind and matter*. Harper Collins.

De Vries, D. (2008). Appendix on pathways to wellbeing. In J. J. McIntyre-Mills (Ed.), *User centric policy design to address complex needs*. Nova Science.

DeLong-Bas, N. J. (2013). *The Oxford encyclopedia of Islam and women*. Edited by Oxford Reference Economics and Finance.

Dhamoon, R. K. (2011). Considerations on mainstreaming intersectionality. *Political Research Quarterly, 64*(1), 230–243. https://doi.org/10.1177/1065912910379227

Dobson, A. (2012). Listening: 'The new democratic deficit'. *Political Studies 60*, 843–859. https://doi.org/10.1111/j1467-9248.2012.00944.X see http://onlinelibrary.wiley.com/doi/10.1111/j.1467-9248.2012.00944.x/abstract

Donaldson, S., & Kymlicka, W. (2011). *Zoopolis: A political theory of animal rights*. Oxford University Press/Capstone.

Elinor Ostrom Nobel Prize in Economics Lecture. https://www.youtube.com/watch?v=T6OgRki5SgM gives an update on the 2009 lecture.

Elkington, J. (1999). *Cannibals with forks: The triple bottom line of 21st Century business*. Capstone.

Florini, A. (2003). *The coming democracy*. Island Press.

Forsyth, T., & Johnson, C. (2014). Elinor Ostrom's legacy: Governing the commons and the rational choice controversy. *Development and Change, 45*, 1093–1110. https://doi.org/10.1111/dech.12110

Foucault, M. (2008). *The birth of biopolitics: Lectures at the College de France 1978–79*. Palgrave Macmillan.

Foucault, M., & Gordon, C. (Eds.). (1980). *Power/Knowledge*. Harvester/Brighton.

Freire, P. (1970). *Pedagogy of the oppressed*. Penguin Books.

Goodall. (2020). https://www.cnn.com/videos/us/2020/03/19/coronavirus-jane-goodall-acfc-full-episode-vpx.cnn

Habermas, J. (2001). *The postnational constellation. Political essays*. Polity.

Hardin, G. (1968). The tragedy of the commons. *Science, 162*(3859), 1243–1248.

Higgins, P. (2018). *The 2018 Hague Freedom lecture by Polly Higgins*. https://www.youtube.com/watch?v=kQn8oA6e9To. http://www.ipcc.ch/pdf/session48/pr_181008_P48_spm_en.pdf

Human Rights Council Advisory Committee (on the advancement of the rights of peasants and other people working in rural areas), UN doc. A/HRC/19/75, 24 February 2012.

Huybrechts, B., & Nicholls, A. (2012). Social entrepreneurship: Definitions, drivers and challenges. In C. Volkmann, K. Tokarski, & K. Ernst (Eds.), *Social entrepreneurship and social business*. Gabler Verlag.

IPCC. (2021). Summary for policymakers. In V. MassonDelmotte, P. Zhai, A. Pirani, S. L. Connors, C. Péan, S. Berger, N. Caud, Y. Chen, L. Goldfarb, M. I. Gomis, M. Huang, K. Leitzell, E. Lonnoy, J. B. R. Matthews, T. K. Maycock, T. Waterfield, O. Yelekçi, R. Yu, & B. Zhou (Eds.), *Climate change 2021: The physical science basis. Contribution of working group I to the sixth assessment report of the intergovernmental panel on climate change*. Cambridge University Press. (in press).

IPPC. (2021). *Summary for policy makers*. https://www.ipcc.ch/report/ar6/wg1/downloads/report/IPCC_AR6_WGI_SPM.pdf

IPPC report. (2018). *Global warming of 1.5 degrees*. http://ipcc.ch/pdf/special-reports/sr15/sr15_spm_final.pdf

Kabeer, N. (1999). Social relations approach. In C. March, I. Smyth, & M. Mukhopadhyay (Eds.), *A guide to gender-analysis frameworks* (pp. 102–119). Oxfam.

Kabeer, N. (2015). Gender, poverty, and inequality. *Gender and Development, 23*(2), 189–205.

Keane, J. (2009). *The life and death of democracy*. Simon and Schuster.

Kirchherr, J., Reike, D., & Hekkert, M. (2017). Conceptualizing the circular economy: An analysis of 114 definitions. *Resources, Conservation and Recycling, 127*, 221–232. https://doi.org/10.1016/j.resconrec.2017.09.005

Kshetri, N. (2017). Will blockchain emerge as a tool to break the poverty chain in the Global South? *Third World Quarterly, 38*(8), 1710–1732. https://doi.org/10.1080/01436597.2017.1298438

Kumboyono, K., Hamid, A. Y. S., Sahar, J., & Bardosono, S. (2020). Community response to the initiation of smoking in Indonesian early adolescents: A qualitative study. *International Journal of Adolescence and Youth, 25*(1), 210–220. https://doi.org/10.1080/02673843.2019.1608273. To link to this article: https://doi.org/10.1080/02673843.2019.1608273

Lethole, P., McIntyre-Mills, J. J., Wirawan, R., & Makaulule, M. (2022). *AA!!Venda women and social enterprise: Stepwise process to regenerative and sustainable living*. Springer.

Makaulule, M. (2013). https://www.culturalsurvival.org/publications/cultural-survival-quarterly/there-no-longer-time-mphatheleni-makaulule-agency

Makaulule, M. (2021). https://www.youtube.com/watch?v=IszP5Uq-X50

McIntyre-Mills, J. (2003). *Critical systems praxis: Participatory governance for social and environmental justice* (The contemporary systems series). Kluwer/Springer.

McIntyre-Mills, J. J. (2008). *User centric policy design to address complex needs*. Nova Science.

McIntyre-Mills, J. (2010). Wellbeing, mindfulness and the global commons. *Journal of Consciousness Studies, 17*(7–8), 44.

McIntyre-Mills, J. (2014). *Systemic ethics and non-anthropocentric stewardship*. Springer.

McIntyre-Mills, J. (2017a). *Planetary Passport for representation, accountability and re-generation* (Contemporary systems series). Springer.

McIntyre-Mills, J. (2017b). Recognising our hybridity and interconnectedness: Implications for social and environmental justice. *Current Sociology*. https://uk.sagepub.com/en-gb/journals- permissions

McIntyre-Mills, J. (2018). Policy design for non-anthropocentric pathways to protect biodiversity and regenerate the land. *International Journal of Educational Development in Africa*. https://doi.org/10.25159/2312-3540/2865

McIntyre-Mills, J. J. (2019a). Summary and key themes: We are the land and the waters In McIntyre-Mills, J, Romm, 2019 Mixed Methods and Cross Disciplinary Research: Towards cultivating eco-systemic living. Springer. Cham. pp. 1–13

McIntyre-Mills, J. J. (2019b). Dynamic weaving together strands of experience: Multiple mixed methods approaches to resilience and re-generation based on intra-, inter- and cross-disciplinary approaches In McIntyre-Mills, J, Romm, 2019 Mixed Methods and Cross Disciplinary Research: Towards cultivating eco-systemic living. Springer, Cham, Switzerland. pp. 15–58

McIntyre-Mills, J. J. (2021a). *From polarisation to multispecies relationships in the age of mass extinctions*. Springer.

McIntyre-Mills, J. in conversation with Rudolf Wirawan, Dennis Finlayson, & Tom Flanagan. (2021b). *Architectures to protect the commons: A community of practice to inspire agency and hope, International Systems Sciences*. Paper delivered 65th annual meeting of the international society for the systems sciences, "The Art and Science of the Impossible: The Human Experience". https://www.isss.org/online-2021/

McIntyre-Mills, J. (2021c). Chapter 9. Social engagement to redress the banality of evil and the frontiers of justice: Limitations of the social contract to protect habitat and why an international law to prevent the crime of ecocide matters. In *From polarisation to multispecies relationships* (pp. 141–172). Springer.

McIntyre-Mills, J. (2021d). Chapter 26. Biopolitics and food security to protect social and environmental justice. In *From polarisation to multispecies relationships* (pp. 575–602). Springer.

McIntyre-Mills, J. (2021e). Chapter 28. Systemic praxis: Steps towards re-generation. In *From polarisation to multispecies relationships* (pp. 575–602). Springer.

McIntyre-Mills, J. J. (2021f). From old to new taxonomies of rights, relationships and responsibilities to protect habitat. In J. J. McIntyre-Mills & Y. Corcoran Nantes (Eds.), *From polarisation to multispecies relationships in the age of mass extinctions* (pp. 31–55). Springer International.

McIntyre-Mills, J. J. (2022). The importance of relationality: A note on co-determinism, multispecies relationships and implications for COVID-19. *Systems Research and Behavioral Science, 39*(2), 339–353. https://doi.org/10.1002/sres.2817

McIntyre-Mills, J. J., & Romm, N. R. A. (2018). Research note on social ecology: Critical systemic reflections. *Systems Research and Behavioral Science* (in preparation).

McIntyre-Mills, J. J., & Romm, N. R. A. (Eds.). (2019). Mixed methods and cross disciplinary research: Towards cultivating ecosystemic living. Springer. https://doi.org/10.1007/978-3-319-58014-2

McIntyre-Mills, J., & Wirawan, R. (2019). Chapter 14. Cascading risks of climate change political and policy dynamics of water crisis: 'Consequences of Modernity' and implications for transformative praxis. In J. McIntyre-Mills, N. Romm, & Y. Corcoran-Nantes (Eds.), *Democracy and governance for resourcing the commons. Contemporary systems thinking* (pp. 415–446). Springer.

McIntyre-Mills, J. with Wirawan, R. (2020). *De-coupling social enterprise from neo-liberalism: The potential of ecovillages* [9th social Business Academia conference 2020, November 5th as part of the program from 4–6, 2020]. SBAC 2020.

McIntyre-Mills, J. with Wirawan, R. (2021a). Chapter 17: The potential of eco-facturing: Towards social and environmental justice through vocational education and training. In J. McIntyre-Mills & Corcoran-Nantes (Eds.), *From polarisation to multispecies relationships* (pp. 391–424). Springer.

McIntyre-Mills, J., & Wirawan, R. (2021b). Chapter 18: Creating a post national learning community. Systemic intervention. In J. McIntyre-Mills & Corcoran-Nantes (Eds.), *From polarisation to multispecies relationships* (pp. 391–424). Springer.

McIntyre-Mills, J., Corcoran-Nantes, Y., Wirawan, R., Widianingsih, I., Saptenno, I., & Sari, N. H. (2019a). Chapter 8. Freeing the Luwaks and escaping the iron cage: Vignettes on surviving in the anthropocene by joining up the social, economic and environmental policy dots—An application of critical systemic thinking to areas of concern. In J. McIntyre-Mills, N. Romm, & Y. Corcoran-Nantes (Eds.), *Democracy and governance for resourcing the commons. Contemporary systems thinking* (pp. 195–232). Springer. https://doi.org/10.1007/978-3-030-04891-4. Print ISBN: 978-3-030-04890- Human Research Ethics Living virtuously and well 7213

McIntyre-Mills, J., Romm, N. R. A., & Corcoran Nantes, Y. (Eds.). (2019b). *Democracy and governance for resourcing the commons: Theory and practice on rural-urban balance*. Springer. https://doi.org/10.1007/978-3-319-58014-2

McIntyre-Mills, J., Corcoran-Nantes, Y., Wirawan, R., & Widianingsih, I. (2021a). The potential of ecofacturing: Towards social and environmental justice through vocational education and training. In J. J. J. McIntyre-Mills & Y. Corcoran-Nantes (Eds.), *From polarisation to multi species relationships: Re-generation of the commons in the era of mass extinctions* (pp. 391–424). Springer.

McIntyre-Mills, J. J., Lethole, P. V., Romm, N. R. A., Makaulule, M., & Wirawan, R. (2021b). Re-designing education for regeneration and wellbeing: Exploring the potential of digital engagement. In A. L. Fymat (Ed.), *The COVID context society for the advancement of Science in Africa, proceedings from 2020*, 7th annual conference: COVID-19: Perspectives across Africa.

McIntyre-Mills, J. J., Wirawan, R., Widianingsih, I., & Riswanda, R. (2021c). Chapter 15 social engagement to protect multispecies habitat: Implications for re-generation and food security. In J. J. McIntyre-Mills & Y. Corcoran Nantes (Eds.), *From polarisation to multispecies relationships in the age of mass extinctions* (pp. 309–368). Springer International.

Meadows, D., & Randers, J. (1992). *Beyond the limits: Global collapse of a sustainable future*. Earthscan Publications.

Mertens, D. M. (2017). Transformative research: Personal and societal. *International Journal for Transformative Research, 4*(1), 18–24. https://doi.org/10.1515/ijtr-2017-0001

Murove, M. F. (1999). *The Shona concept of ukama and the process philosophical concept of relatedness, with special reference to the ethical implications of the contemporary neo-liberal economic practices* [MA thesis]. University of Natal.

Murove, M. F. (2005). *The theory of self-interest in modern economic discourse: A critical study in the light of African humanism and process philosophical anthropology* [Doctoral thesis]. University of South Africa.

Nussbaum, M. (2006). *Frontiers of justice: Disability, nationality, species membership*. Belknap Press.

Nussbaum, M. (2011). *Creating capabilities: The human development approach*. The Belknap Press.

Oekan, S., Abdoellah, O. S., Sunardi, S., Widianingsih, I., & Cahyandito, M. F. (2019, April). Pemetaan Sosial dalam Perencanaan Program Pengembangan Ekowisata Berjelanjutan di Citarum Hulu, Kumawula. *Jurnal Pengabdian Kepada Masyarakat* 2(1), 53–66. Universitas Padjadjaran. http://jurnal.unpad.ac.id/kumawula/article/view/24461

Ostrom, E. (1992). Governing the commons: The Evolution of Institutions for Collective Action, 32. *Natural Resources Journal, 415*. Available at: https://digitalrepository.unm.edu/nrj/vol32/iss2/6Ostrom

Ostrom, E. (2008). *Design principles of robust property-rights institutions: What have we learned?* [Elinor Ostrom workshop in political theory and policy analysis]. Indiana University Center for the Study of Institutional Diversity Arizona State University. http://www.indiana.edu/~workshop

Ostrom, E. Published on 25 June (2014). *Elinor Ostrom Nobel Prize in economics lecture*. https://www.youtube.com/watch?v=T6OgRki5SgM

Pauli, G. (2010). *The blue economy: Report to the Club of Rome*. Paradigm Publications.

Pikketty, T. (2015). *13th Nelson Mandela annual lecture*. https://www.nelsonmandela.org/news/entry/transcript-of-nelson-mandela-annual-lecture-2015

Polanyi, M. (1966). *The tacit dimension*. Chicago University Press. Foreword by Amartya Sen, 2009.

Polanyi, M. (1968). *The Great transformation: The political and economic origins of our time*. Renehart.

Quan-Baffour, K. P., McIntyre-Mills, J., & Romm, N. R. A. (2019). Ubuntu: Dialogue on connectedness, environmental protection and education. In J. McIntyre-Mills & N. R. A. Romm (Eds.), *Mixed methods and cross disciplinary research: Towards cultivating eco-systemic living* (pp. 221–250). Springer.

Rayner, A. (2010). Inclusionality and sustainability – Attuning with the currency of natural energy flow and how this contrasts with abstract economic rationality. *Environmental Economics, 1*(1), 98–108.

Rayner, A. (2017a). Natural inclusion. In J. J. McIntyre-Mills, N. Romm, Y. Corcoran-Nantes, J. J. McIntyre-Mills, N. R. A. Romm, & Y. Corcoran-Nantes (Eds.), *Balancing individualism and collectivism: Social and environmental justice* (pp. 461–470). Springer.

Rayner, A. (2017b). https://en.wikiversity.org/wiki/Natural_Inclusion

Renaldi, E. (2020). *Indonesia's latest natural disasters are a 'wake-up call', environmentalists* ***say.*** https://www.abc.net.au/news/2021-01-22/indonesia-hit-by-series-of-disasters-in-the-first-weeks-of-2021/13075930

Romm, N. R. A. (2015). Reviewing the transformative paradigm: A critical systemic and relational (Indigenous) lens. *Systemic Practice and Action Research, 28*(5), 411–427. https://doi.org/10.1007/s11213-015-9344-5

Romm, N. R. A., & McIntyre-Mills, J. J. (2020). Systemic thinking for re-generative development. In G. Midgley (Ed.), *Handbook of systems thinking*. Routledge.

Scott, J. (1998). *Seeing like a state*. Yale University Press.

Senge, P. M. (2006). *The fifth discipline: The art and practice of the learning organization*. Random House.

Shiva, V. (1989). *Staying Alive: Women, ecology and survival in India*. Zed Books.
Shiva, V. (2002). *Water wars: Privatization, pollution and profit*. Pluto Press.
Shiva, V. (2011). *Earth democracy*. Portland University. www.youtube.com/watch?v=UOfM7QD7-kk/
Shiva, V. (2012a). *Making peace with the earth*. Fernwood Publishing.
Shiva, V. (2012b). *Monocultures of the mind*. Third World Network.
Stiglitz, J., Sen, A., & Fitoussi, J. P. (2010). *Mis-measuring our lives: Why the GDP doesn't add up*. The New Press.
The Nobel Peace Prize. (2020). NobelPrize.org. *Nobel Media AB 2020. Thu. 29 October 2020*. https://www.nobelprize.org/prizes/peace/2020/press-release/
The UN Declaration of the Rights of Indigenous Peoples. (2008). http://www.ohchr.org/
Ulrich, W., & Reynolds, M. (2010). Critical systems heuristics. In M. Reynolds & S. Holwell (Eds.), *Systems approaches to managing change: A practical guide* (pp. 242–292). Springer: London.
United Nations. (2014). *World urbanisation prospects: The 2014 revision*. https://esa.un.org/unpd/wup/publications/files/wup2014-highlights.Pdf
United Nations Declaration on the Rights of Indigenous Peoples. (2007). https://humanrights.gov.au/our-work/un-declaration-rights-indigenous-peoples-1
United Nations Paris Climate Change. (2015). *Conference of the Parties Twenty-first session Paris, 30 November to 11 December 2015*. https://unfccc.int/resource/docs/2015/cop21/eng/l09r01.pdf
van der Weele, C., Feindt, P., Jan van der Goot, A., & van Mierlo, B. (2019). Meat alternatives: An integrative comparison. *Trends in Food Science & Technology, 88*, 505–512.
Von Foerster, H. (1995). *Cybernetics of cybernetics* (2nd ed.). Future Systems, Inc.
Wadsworth, Y. (2010). *Building in research and evaluation. Human inquiry for living systems*. Allen and Unwin.
Wenger, E., et al. (2009). *Digital habitats: Stewarding technology for communities*. CPsquare.
Widianingsih, I., Riswanda, & Paskarina. (2019). Rural-urban linkage and local government capacity in coping with water crisis: A brief note from Indonesia'. In *Proceedings of the International Conference of Democratisation in Southeast Asia (ICDeSA 2019)* (pp. 350–354). https://www.atlantis-press.com/proceedings/icdesa-19/125923308
Wirawan, R., & McIntyre-Mills, J. (2019). Innovation for social and environmental justice; a way forward? In J. McIntyre-Mills, N. Romm, & Y. Corcoran Nantes (Eds.), *Democracy and governance for resourcing the commons: Theory and practice on rural-urban balance*. Springer.

Transformative Education: Employing the Balanced Scorecard for Regenerative Development

S. Algraini and Janet J. McIntyre-Mills

Abstract Education plays a multitude of roles in securing society's wellbeing and ensuring the development of a sense of social and environmental awareness and responsibility. Leveraging on this premise, the objectives of this chapter are twofold. First, it aims to identify the norms that could transform education to support regenerative development. Second, it attempts to determine how these norms can be achieved in terms of strategic planning and measurement. The paper starts by theorising the capability approach as an a priori norm that can be used to guide education. Further discussion focuses on the importance of various integrated measures that appreciate the broader schooling perspectives to ensure that education expands students' learning and life capabilities. For this purpose, the chapter introduces the balanced scorecard as a strategic tool that highlights the steps required to address the values that underpin education policies and the measurement of performance. The balanced scorecard shows that a successful transformation starts with rethinking educational institutions' mission and vision. Furthermore, indicators of performance are developed to evaluate our progress towards an incredible future.

Keywords Education · Regenerative development · Capability approach · Balanced scorecard

S. Algraini (✉)
Graduate of the School of Social and Policy Studies, Flinders University, Adelaide, SA, Australia

J. J. McIntyre-Mills (✉)
College of Education, University of South Africa, Pretoria Central, Gauteng, South Africa

College of Professions, Adelaide University, Adelaide, Australia
e-mail: janet.mcintyre@adelaide.edu.au

J. J. McIntyre-Mills, Y. Corcoran-Nantes (eds.), *Transformative Education for Regeneration and Wellbeing*, Contemporary Systems Thinking,
https://doi.org/10.1007/978-981-19-3258-8_7

1 Introduction

Education plays a multitude of roles in shaping an individual's personality, enabling one to develop a sense of social and environmental rights and responsibilities in the short, medium and long terms. However, the economic view to education dominates most of the world's education systems (Walker, 2005; Giroux, 2011; Algraini, 2021). The challenge for educational policies is to address the wider role of education and expand students' development opportunities, and not merely focus on economic preparation. The broader perspective to education needs to address what has been taught and how to decide what is worthwhile to promote current and future wellbeing socially, economically and environmentally. To transform education into a factor that challenges the traditional view of development, we need to rethink the a priori norms that guide education and educational institutions' measurements and apply the regenerative development praxis to their everyday activities.

This chapter will contribute to regenerative development in two ways. First, it provides a better understanding of what it takes to transform education to achieve regenerative development, not just in terms of academic achievements but in creating well-rounded human beings, which require looking at the whole system and the environment that it serves (the social environment and the ecosystem). This chapter addresses the norms of education for regenerative development by arguing for an education that fosters sustainable development and organic multispecies relationships (McIntyre-Mills, 2020). This argument is based on the capability approach. The capability approach provides a framework for contemplating a priori norms for education, which have the potential to address development issues in social, economic and environmental terms. One implication is to encourage schools to consider some of the broader social and environmental challenges for students to be able to achieve their learning and life goals. By focusing on capabilities, we can achieve an education system that values lifelong learning and well-rounded human beings, which are vital in securing society's wellbeing.

The second contribution is determining how to achieve education's norms in terms of measurement. Thus, the second part of this chapter provides more details about the importance of measurement in investigating progress. It aims to ask whether the system's purpose is 'reflected in our metrics' (Stiglitz, 2010, p. 15). When making measurement suggestions, a set of measures should be included to measure educational outcomes, as well as a posteriori measures that incorporate awareness of the United Nations Sustainable Development Goals (UN SDGs). Hence, performance measures need to be much broader to consider core skills as well as core wisdom that reflects environmental awareness to support living systems. This challenge of measuring education comprehensively has been addressed by Nussbaum's (2011) 10 central capabilities. The list represents a normative approach for measuring education's success. The list includes basic learning and life capabilities.

The third section will discuss how to achieve regenerative development in terms of strategic planning and measurement. Using the balanced scorecard, this strategic tool could supplement the current traditional planning approach. To increase education's powerful effect, the balanced scorecard emphasises rethinking the purpose of organisations and measurements. Measurements are indicators developed to measure the extent by which organisations are achieving their strategic goals. The expanded form of assessment could be used to encourage students to address wider areas of concern, such as building awareness on social and environmental rights and responsibilities. Thus, schools will have more potential for achievements that matter most to graduating future leaders and society members.

2 A Priori Norms to Guide Education: A Capability Perspective

Building a new vision for education and a brighter mission for schools require prior norms that transform schools to support regenerative development. In this respect, one potential element comes from the capability approach (Sen, 1999; Nussbaum, 2011), which is normative and based on a priori norms to guide education. The capability approach focuses on shifting the development perspective from economic outcomes to wellbeing, that is, to what extent each person is able to do and to be. In education, the fundamental shift that the capability approach provides is in the role of education in human development and agency, and not only for the economy. Therefore, the capability approach provides a more extensive view of education that helps address the extent by which schools contribute to regenerative development. Thinking of education this way means education is an endeavour that involves adapting our attitude to the environment and thinking about alternative ways of living rather than merely focusing on knowledge achievement alone. In this way, education could go beyond the economic paradigm perspective. A broader understanding of education extends its role in dealing with global issues, such as sustainability and environmental problems, that affect our capability to live a better life.

In integrating education's various roles, focusing on capabilities can help people achieve multiple outcomes. The capability approach to education aims to enable students to fulfil their learning goals and maximise their potential. Through this human development lens, education's role is to fully develop children's personalities, as well as their capacity to benefit themselves and the community (Alkire, 2013, p. 14). This potential comes from the notions of functioning and capability, which emphasise freedom and achievements (Sen, 1987, 1999).

Sen and Nussbaum's work cited many examples of the potential of the capability approach in education. For instance, Sen (1997) pointed out that a full understanding of education's roles in development needs to consider three interrelated roles: 'the wellbeing and freedom of people, their indirect role through influencing economic production and their indirect role through influencing social change' (Sen, 1997, p. 1960). Similarly, Nussbaum (1997, pp. 10–11) argued that education should cultivate three fundamental capacities: a critical examination of one reality and life, world citizenship that includes a common concern of global issues and a narrative imagination of others' perspectives. These potential implications encourage students to think differently and to move beyond the traditional and unsustainable approach. This change in education could include thinking about a new, post-carbon future and ways to protect current and future generations' wellbeing.

From the human capability perspective (Sen, 1999; Nussbaum, 2011), it is necessary to rethink the process of learning, that is, how learning can bring about change in students' ability to think, reason and develop lifelong learning skills. The approach urges schools to cultivate critical thinking and reasoning. According to Nussbaum (2011), each human being should be 'able to use the senses, to imagine, think, and reason—and to do these things in a "truly human" way, a way informed and cultivated by an adequate education' (p. 33). This capability highlights the importance of thinking capacity for students. One implication is encouraging them to be environmentally conscious. Education that promotes thinking is vital in tackling global issues, such as the wellbeing of the planet and the rapid pace of climate change. Thinking also plays a crucial role in moving our society towards more sustainable ways of living. Many people are trapped in limited thinking. They do not think about how their social, economic and environmental choices impact the environment of which they are a part (McIntyre-Mills, 2013). This has implications for the ecological stewards beyond the nation state's limits (McIntyre-Mills, 2010, 2020). In this regard, the curricula need to enable students to formulate ways to protect the environment, rethink our rights and responsibilities and encourage cognition about the UN SDGs. Those efforts are essential to preserving wellbeing stocks[1] for current and future generations.

The human capability perspective also prioritises environmental stewardship. Nussbaum (2011) believes that the capability to connect to the environment is a central human capability. This capability that she called 'other species' means 'being able to live with concern for and in relation to animals, plants and the world of nature' (p. 34). This capability represents the importance of ecological diversity to achieve an environment that is ecologically sound and protected. Environmental concern gives schools more recognition of the right of non-human creations, as well as awareness of climate change and the increased level of pollution globally. All of these issues put more pressure on education systems to adapt to change and be part of the solution, not the problem.

The role of education in expediting sustainable development is now highly recognised by the UN. The 17 SDGs put 'quality education' as the fourth goal that should achieve sustainable development by 2030, that is, to 'ensure inclusive and equitable quality education and promote lifelong learning opportunities for all' (2021). Accordingly, when we deliberate about education, we should think about

[1] Wellbeing stocks refers to a concept developed by Stiglitz et al. (2010, p. 15) indicating a multidimensional measure of wellbeing spanning the following aspects: (1) material living standards (income, consumption and wealth); (2) health; (3) education; (4) personal activities including work; (5) political voice and governance; (6) social connections and relationships; (7) environment (present and future conditions); and (8) insecurity that is of an economic or physical nature.

equipping young people with the necessary skillset to enable them to enhance their overall sustainability. In this respect, schools need to teach students how to connect with the natural environment and, in particular, to understand the systemic importance of protecting and regenerating environmental conditions that maintain food and water security. This is an essential lifelong learning skill that is vital for human sustainability. A fundamental shift in education also needs to happen in valuing people and the environment as ends in themselves and not as means to an end. Thus, intrinsic values need to be appreciated rather than emphasising instrumental benefits that are not sustainable.

The wider view of the capability approach also echoes the essential characteristics of education for sustainable development developed by the UN (2007). According to the report Education for Sustainable Development (ESD), the characteristics of education for sustainable development are:

- based on the principles and values that underlie sustainable development;
- deals with the wellbeing of all three realms of sustainability, environment, society and economy;
- promotes lifelong learning;
- is locally relevant and culturally appropriate;
- is based on local needs, perceptions and conditions, but acknowledges that fulfilling local needs often has international effects and consequences;
- engages formal, non-formal and informal education;
- accommodates the evolving nature of the concept of sustainability;
- addresses content, taking into account context, global issues and local priorities;
- builds civil capacity for community-based decision-making, social tolerance, environmental stewardship, adaptable workforce and quality of life;
- is interdisciplinary: no one discipline can claim ESD as its own, but all disciplines can contribute to ESD;
- uses a variety of pedagogical techniques that promote participatory learning and higher-order thinking skills(p. 6).

These characteristics are consistent with our argument that education aimed only at the economy leaves significant questions unaddressed. Therefore, to encompass more aspects, it is important to utilise the human capability perspective. By integrating the different roles that education can play, focusing on capabilities can advance students' lives and their environment. The capability approach to education involves students being able to achieve their learning goals as well as their full potential. This potential comes from the notions of capabilities and functioning, which challenge the limitation of education for economic growth (Walker, 2005, p. 104). The capability perspective is a significant contribution that enables schools to work beyond the old paradigm, for example, teaching the students practical tips about recycling and saving water and energy sources.

Overall, the previous theoretical and practical applications presented show how the capability approach can be utilised to advance education's potential for sustainable development and organic multispecies relationships to support living systems. The assumptions and values underlying education policy discourses constitute the starting point for an education that supports sustainable development, and not fundamental economics. According to this view, the role of education capability is to expand people's thinking and learning capabilities to create a better future.

3 Measurement

The previous section highlights the potential of the capability approach to inform policy goals and strategies for education, particularly in developing education policies that have more potential to people and the environment. Nevertheless, the capability approach has been criticised for focusing more on opportunities rather than achievements (Fleurbaey, 2002; Crocker, 2008), for instance, creating conditions and opportunities or removing obstacles for people to pursue their goals to be healthy, learn new skills and be free from harm. Sen (1985, pp. 204–5, 1987, p. 23, 1999, p. 74) argued that we should look at people's capabilities and their ability for achievement. Nussbaum (2011) acknowledged this criticism of focusing on capability in terms of opportunity and freedom rather than achievement. Based on her perspective, people's choices are essential and cannot be compromised by forcing them to adapt to specific achievements (p. 26). However, there is an inconsistency in this argument. Sen and Nussbaum adopted different positions when capabilities are related to children. Sen (as cited in Saito, 2003) argued that children's future capabilities need to be considered and their current freedom may be comprised for the sake of preparing them well for the future:

> If the child does not want to be inoculated and you nevertheless think it is a good idea for him/her to be inoculated, then the argument may be connected with the freedom that this person will have in the future by having the measles shot now. The child, when it grows up, must have more freedom. So, when you are considering a child, you have to consider not only the child's freedom now, but also the child's freedom in the future. (p. 25)

Sen also emphasised the importance of measuring students' achievement in his work with Drèze and Sen (2013). They argued that 'if a large proportion of children learn virtually nothing for years on end in a particular school, it is important to know it, well before they are sent for slaughter in the Board Examinations' (p. 138). Similarly, Nussbaum (2011) pointed out the importance of paternalistic policies to promote children's functioning, such as basic education (p. 26). Therefore, it is vital to promote and simultaneously measure students' functioning, especially children, in basic education.

Considering the limitations of the capability approach, the question that arises is how to achieve its a priori norms. To address the limitations of the capability approach, we need to work on both the values that underpin policy choices and the governance measures to achieve it. In other words, the education policy needs to address a priori norms for good governance to guide performance and a posteriori measures to guide assessment (McIntyre-Mills, 2014, p. 87). As McIntyre-Mills (2014) pointed out, governance needs to look at a priori norms. Thereafter, it has to look at ways to ensure that the process is appropriately implemented in terms of outputs and outcomes. Giving students the freedom to achieve human development is essential. At the same time, we need to measure their progress and achievements. Both the values underpinning education policies and the measurement of performance are equally important—a balance between giving students freedom and making sure that students are achieving specific performance standards.

To address regenerative development, the performance measures in education need to be much wider than just narrow academic scores. They need to include indicators of how education addresses development issues in social, economic and environmental terms. The expanded form of assessment would encourage schools to address more expansive areas of concern, such as building awareness of social and environmental rights and responsibilities. Thus, the measurement needs to include how students are doing in their learning and in their ability to contribute to the world. In this view, performance measures are wider in scope than just narrow curricula and narrow academic performance standards, as they ensure broader participation in the framing of education policies and education outcomes.

In building a broader assessment, we first need to discuss what type of capabilities should be measured. This kind of question is important to maximise student potentials, developing their capacity to think critically and acquire lifelong learning capabilities. One way to encourage a more comprehensive measurement to education is to apply Nussbaum's (2011) 10 central capabilities, namely: (1) life; (2) bodily health; (3) bodily integrity; (4) senses, imagination and thought; (5) emotions; (6) practical reason; (7) affiliation; (8) other species; (9) play; and (10) control over one's environment. The 10 central human capabilities cover wide aspects of life that are essential to all human beings and are relevant to other sentient beings (Nussbaum, 2011, pp. 33–34). The list could be used to measure learning goals that give students the ability to be creative and critical. At the same time, Nussbaum's approach provides a holistic approach to support essential capabilities that facilitate regenerative development, such as strengthening connections to nature and promoting positive emotions and values towards our planet.

How can the previously mentioned capabilities be achieved in terms of measurement? This can be done by asking whether the system's purpose is 'reflected in our metrics' (Stiglitz, 2010, p. 15) because 'what you measure is what you get' (Kaplan & Norton, 1992, p. 71). It requires a posteriori measures of performance (McIntyre-Mills, 2014) for the governance process to address the policy goals. Thus, the challenge for education systems worldwide is to integrate the ideals of the capability approach in shaping policy norms. At the same time, performance measures should be considered so that the consequences of teaching and learning choices can be assessed.

This is undertaken by using the balanced scorecard (Kaplan & Norton, 1992) to enable strategic planning and management system. The focus includes planning for a more rounded approach to education through new vision and mission for schools. The balanced scorecard also provides better performance measures to balance giving students freedom and making sure that they are achieving a specific standard of performance. A posteriori measures would help schools incorporate awareness of students' rights and responsibilities as ecological citizens. Performance measures also need to be much broader to consider core skills as well as core wisdoms that reflect environmental awareness. Expanded performance measures would balance academic achievement with other important factors, such as students' understanding of their social and environmental rights and responsibilities.

4 The Balanced Scorecard

After discussing the prior norms and the importance of measurement, this section outlines how to achieve these norms in terms of strategic planning and measurement. One way to achieve this purpose is through the balanced scorecard. The balanced scorecard is a strategic tool that helps organisations build and implement their strategies. The Balanced Scorecard emphasises the importance of setting prior norms through vision, mission and core values. At the same time, it urges organisations to develop initiatives and measurement systems to realise plans. The mission, vision and performance measures are essential elements in articulating and implementing a successful strategy.

The balanced scorecard provides an overall approach to develop a strategy and monitor its progress. Kaplan and Norton introduced this strategic tool in the article "The Balanced Scorecard—Measures that Drive Performance" published in *Harvard Business Review* in 1992. Their idea of a balanced scorecard was aimed at strategy formulation and performance management. According to Kaplan and Norton (1992):

> Many companies today have a corporate mission that focuses on the customer. 'To be number one in delivering value to customers' is a typical mission statement. How a company is performing from its customers' perspective has become, therefore, a priority for top management. The balanced scorecard demands that managers translate their general mission statement on customer service into specific measures that reflect the factors that really matter to customers. (p. 73)

The quote above highlights the translation of the mission statement into specific measures. In other words, the balanced scorecard interprets the strategy into quantifiable targets. Those targets balance the performance of the organisation to be perceived from different perspectives. The perspectives that usually drive performance are financial or stewardship, customer and stakeholder, internal process and organisational capacity (Rohm et al., 2016, p. 58). These four perspectives support the idea of measuring achievement using a much wider approach. Thus, for deeper and broader insights into an organisation's performance, the four perspectives function as guides for strategy and performance management. It is worth mentioning that the four perspectives are not fixed. The type of perspectives and their order vary depending on the organisation type (Rohm et al., 2016, p. 45). In mission- driven organisations such as schools, the steward or stakeholder is the first to assess the performance.

Rohm et al. (2016, p. 58) explain the main components of the balanced scorecard as comprising key values comprising operational, strategic, service excellent and culture change at an organisational level involving stewardship, enhancement of engagement to ensure that the ideas are tested out by participants who feel a sense of co-creating a new organisational culture based on interactive communication to foster the key values.

To measure performance systemically, based on the four perspectives, the organisation needs to develop its strategy and set its directions. The first step in this process is the mission and vision statements. The above figure shows that the mission and vision are placed at the top of the strategy framework. Developing the mission and vision requires a participatory approach wherein the staff, students, parents and the wider community are involved. Students' capabilities need to be considered and guided by the normative aspects of Nussbaum's (2011) capability approach when the mission and vision are developed. Another key aspect of the balanced scorecard is measurement. The balanced scorecard highlights the importance of measuring how the activities address the mission and accomplish the vision. The way we measure things is shaped by our values as reflected in the mission, vision and strategic goals. In this regard, a set of indicators is required to evaluate progress and address the different strategy perspectives. As Kaplan and Norton (1992) pointed out, 'no single measure can provide a clear performance target or focus attention on the critical areas of the business' (p. 71). In this respect, the indicators could revolve around income and customer satisfaction levels or relate to the internal process. In education, we need to approach measurement from different perspectives to achieve good education performance. Based on the normative concepts of the capability approach, the indicators can include Nussbaum's (2011) 10 central capabilities, as they would supplement the current traditional grading system to achieve well-rounded students.

5 Application of the Balanced Scorecard in Schools

The balanced scorecard provides a practical way to involve schools in regenerative development. This potential comes from its systemic planning and implementation tools that involve setting the vision and mission and developing a measurement system to ensure that efforts are coordinated and evaluated. In this regard, schools could build a vision and mission that fosters regenerative development, specifically the students' ability to develop capabilities to address development issues in social, economic and environmental terms. The strategic objectives, measures and initiatives ought to appreciate the broader role of education. In the following paragraphs, we will discuss these potentials in more detail.

The first step in building a strategy is developing a vision and mission. We will start with vision. Vision is about the future. For schools, the vision for children's education needs to be ambitious and clear. The vision should be around equipping young people and positioning a learning system to address their needs. Based on our normative approach, the vision could revolve around what kind of capabilities the schools are aiming for their students to have when they graduate. Notably, students should become future citizens and leaders that support the UN SDGs. In doing so, the vision would have more potential to transform students' lives beyond preparing them for work. As Sen (1999) put it:

> We must go beyond the notion of human capital after acknowledging its relevance and reach. The broadening that is needed is additional and inclusive, rather than, in any sense, an alternative to the "human capital" perspective. (p. 296)

While vision is about the future, the school's mission focuses on achieving the desired end. Participants need to conduct a qualitative mission exercise to describe how the school operates to ensure the students' future. This exercise entails working with all key stakeholders to explore their hopes for social, economic and environmental wellbeing. Following this exercise, the mission statement should describe education's purpose at the primary, secondary or tertiary level. It needs to describe in a few words what the education institute aims to achieve within the learning community in which it is located. Thus, the school's mission should move from graduating students to promoting business as usual based on old economic growth indicators. Since the pandemic, it has become clear that business as usual is no longer relevant to ensuring a viable future in which young people thrive as they mitigate the impact of the old carbon economy and find ways to adapt to climate change by fostering social, economic and environmental capital. Hence, schools need to provide high-quality education that would facilitate the students' success in different aspects and prepare them as future leaders and good members of the society. The new mission represents the difference between a viable economy or one that hurtles towards 'existential risks' (Bostrom, 2011) and a dire future. Thus, the new mission should transform learning in schools. Learning needs to emphasise practical reasoning in terms of considering wellbeing and alternative ways of protecting the environment, and not only employment success. This new or updated mission will help us be more critical about the process and the content of learning and the extent by which students build environmental citizenship and connect with their wider environment.

Measurements need to be shaped in a way that supports the new mission and vision for schools. Therefore, schools that rely only on tests based on limited materials will fail to achieve this purpose. To avoid an assessment trap that celebrates memorisation and rote learning, measuring students' success should go beyond academic achievement. Based on a priori norms, schools' measurements of success need to include the 10 human capability indicators. Nussbaum's (2011) 10 central capabilities represent a normative approach for measuring educational success, as they consist of basic learning and life capabilities. Thus, the approach provides a framework through which policymakers and administrators could expand the role of education and at the same time provide a good life for students to achieve an education that supports the development of well-rounded human beings who have the capacity to benefit themselves and the community (Alkire, 2013, p. 14).

The way the 10 central capabilities are applied in education can have two important applications (Algraini, 2017; Algraini & McIntyre-Mills, 2019). The first is the potential to measure the process of teaching and learning. Two central capabilities from the list, namely, 'senses, imagination and thought' and 'practical reasoning' require learning that enables students to communicate their ideas critically, reflect on their learning and build lifelong skills. From this perspective, schools need to develop the ability to be creative, reasonable, imaginative and critical.

According to the first implication, the measurement needs to evaluate the extent by which schools foster critical thinking, problem solving and finding ways to answer challenging questions. Thus, students' potential will not be restricted by school reports that are narrowly directed to the standard curriculum without an adequate evaluation of critical abilities. Advancing students' learning across different subjects and skills enables students to achieve their learning capabilities. In this respect, the challenge is related to the current evaluation's nature and the kind of narrowly defined tests that cover limited aspects of students' learning and evaluation. Learning assessments in schools should formulate ongoing evaluations using workbooks that foster 'going beyond and diving deeper' to use the current pedagogical terminology for teaching that inspires lifelong learning that can make a difference. Therefore, the grading systems should address the wider understanding of education's role in development through more formal and informal learning opportunities, for example, to think imaginatively about the market and how it can be constructed to support the whole system. How could the students' careers be designed to transform the market as stewards of new ways of living sustainably?

The lack of recognition of many learning capabilities in learning evaluations is a dilemma faced by many education systems worldwide. The first author's work on Saudi education found that the education policy focuses on economic benefits and gives little consideration to the intrinsic value of education for whole-person development and lifelong learning (Algraini, 2021). This finding is also consistent with other studies. Biggeri and Santi (2012) found that creativity and critical thinking are usually neglected in academically focused education systems. Lozano et al. (2012) analysed European higher education and found limitations in achieving only a few competencies. They suggested complementary measurements using the capability approach as it provides a wider understanding and implication of education's role in development.

The second potential of the capability approach in measurement goes beyond learning and academic achievement. It addresses the full spectrum of capabilities, namely, life, bodily health, bodily integrity, emotions, affiliation, other species, play and control over one's environment. Meeting the second potential requires indicators to advance students' wellbeing. The general view is used to investigate the quality of life and other important capabilities that are essential for all human beings in an education setting, for example, health, safety and empowerment. This view is related to human capabilities in a general perspective 'by virtue of their humanity' (Nussbaum, 2011, p. 62). Thus, the role of strategic planning in promoting health and providing a decent life for students is a fundamental requirement. Therefore, according to the second perspective, the measurement should open more possibilities for children to address their interdependence on living systems and their potential role as stewards to support the SDGs, rather than being trained to perpetuate 'business as usual'.

6 Conclusion

This chapter investigates *a priori norms* and *a posteriori measures* that could inform education to achieve regenerative development. The capability approach is utilised to promote an educational policy that appreciates current and future wellbeing socially, economically, and environmentally. The theoretical contribution of the capability approach is to encourage a more extensive role of education. The study draws on Nussbaum's capability approach to highlight students' complex needs pertaining to the 10 central capabilities. The 10 central capabilities emphasise that students should achieve their learning and life goals while simultaneously urging schools to consider some of the broader social and environmental challenges. This chapter also discusses how to achieve regenerative development in terms of strategic planning and measurement. Using the balanced scorecard, the emphasis is on rethinking schools' mission and vision. An expanded form of assessment is used to encourage students to address wider areas of concern, such as building awareness on social and environmental rights and responsibilities. Thus, students have more potential for achievements that matter most to themselves and to our environment.

References

Algraini, S. (2017). *A critical systemic approach to human development in education: A case study of the Ministry of Education in Saudi Arabia* [Doctoral dissertation]. Flinders University.

Algraini, S. (2021). Education for human development: a capability perspective in Saudi public education. Compare. *A Journal of Comparative and International Education, 51*(3), 416–432.

Algraini, S., & McIntyre-Mills, J. (2019). Education policy and governance of secondary schools in Saudi Arabia: A critical review informed by Nussbaum's capabilities approach. In *Democracy and governance for resourcing the commons: Theory and practice on rural-urban balance* (pp. 295–369). Springer.

Alkire, S. (2013). Well-being, happiness and public policy. *OPHI*. Research in Progress (1a).

Biggeri, M., & Santi, M. (2012). The missing dimensions of children's well-being and well-becoming in education systems: Capabilities and philosophy for children. *Journal of Human Development and Capabilities, 13*(3), 373–395.

Bostrom, N. (2011). *Existential risk prevention as the most important task for humanity*. Faculty of Philosophy & Oxford Martin School University of Oxford. http://www.existential- risk.org/

Crocker, D. A. (2008). *Ethics of global development: Agency, capability and deliberative democracy*. Cambridge University Press.

Drèze, J., & Sen, A. (2013). *An uncertain glory: India and its contradictions*. Penguin Books.

Fleurbaey, M. (2002). Development, capabilities and freedom. *Studies in Comparative International Development, 37*, 71–77.

Giroux, H. (2011). *On critical pedagogy*. Continuum International Publishing Group.

Kaplan, R., & Norton, D. (1992, January–February). The balanced scorecard: Measures that drive performance. *Harvard Business Review*, 70–79. https://hbr.org/1992/01/the-balanced-scorecard-measures-that-drive-performance-2

Lozano, J. F., Boni, A., Peris, J., & Hueso, A. (2012). Competencies in higher education: A critical analysis from the capabilities approach. *Journal of Philosophy of Education, 46*(1), 132–147.

McIntyre-Mills, J. (2010). Wellbeing, mindfulness and the global commons. *Journal of Consciousness Studies, 17*(7–8), 44–72.

McIntyre-Mills, J. (2013). Anthropocentrism and wellbeing: A way out of the lobster pot? *Systems Research and Behavioral Science, 30*(2), 136–155.

McIntyre-Mills, J. with De Vries, & Binchai, N. (2014). *Transformation from wall street to well-being: Joining up the dots through participatory democracy and governance to mitigate the causes and adapt to the effects of climate change*. Springer. Second and third authors provided the appendix on software user guide (pp. 193–198).

McIntyre-Mills, J. (2020). The COVID-19 era: No longer business as usual. *Systems Research and Behavioral Science, 37*(5), 827–838.

Nussbaum, M. (1997). *Cultivating humanity: A classical defense of reform in liberal education*. Harvard University Press.

Nussbaum, M. C. (2011). *Creating capabilities: The human development approach*. Belknap Press of Harvard University Press.

Rohm, H., Wilsey, D., Perry, G., & Montgomery, D. (2016). *The institute way: Simplify strategic planning and management with the balanced scorecard*. The Institute Press.

Saito, M. (2003). Amartya Sen's capability approach to education: A critical exploration. *Journal of Philosophy of Education, 37*(1), 17–33.

Sen, A. (1985). *Commodities and capabilities*. North-Holland.

Sen, A. (1987). *The standard of living: Tanner lectures*. Cambridge University Press.

Sen, A. (1997). Editorial: Human capital and human capability. *World Development, 25*(12), 1959–1961.

Sen, A. (1999). *Development as freedom*. Knopf.

Stiglitz, J. (2010). From measuring production to measuring wellbeing. Australian productivity. *Commission Review of Government Service Provision*. Retrieved from http://www.pc.gov.au/news-media/lectures/stiglitz. Accessed on 25 February 2021.

Stiglitz, J., Sen, A., & Fitoussi, J. (2010). *Mis-measuring our lives: Why GDP doesn't add up*. The New Press.

United Nations. (2007). *The UN Decade of Education for Sustainable Development (DESD 2005–2014): The first two years*. UNESCO.

United Nations. (2021). *Sustainable development goals: Knowledge platform*. United Nations. Retrieved from https://sustainabledevelopment.un.org/topics/education. Accessed on 20 Mar 2021.

Walker, M. (2005). Amartya Sen's capability approach and education. *Educational Action Research, 13*(1), 103–110.

Education and Gender Mainstreaming in a Patriarchal Setting: What Really Matters

Shajeda Aktar and SM. Mokhlasur Rahman

Abstract This paper explores the views and perceptions of women from various education and employment backgrounds in the context of Bangladesh. It collects data from fifteen highly-educated, fifteen illiterate and fifteen microcredit borrowers, and offers an analysis using a modified framework for gender mainstreaming. It is widely suggested that education is a pre-requisite and an essential driver to combat poverty and inequality, in order to foster gender justice in a classic patriarchy such as Bangladesh. Behind this hypothesis is the assumption that education would boost productive employment for women which in turn could enhance their self-esteem, capability, agency, socio- political network and authority in decision-making. Women's mobility outside home, participation in household decision-making, economic freedom and control over productive assets and activities could result in a significant rise as the education of women rises. However, there are also views raising concerns that without necessary safeguards, education of women alone may not produce the results that are expected. In particular, education without sufficient job skills, job creation, entrepreneurship, innovation, and industry-aligned to skill may fail to empower women. It can even backfire as education itself is a costly investment in terms of monetary resources, productive energy and invested-time. If investment in education is not supported by development that mainstreams opportunities for women it may fail to increase women's socio-economic power.

Keywords Gender mainstreaming · Education · Employment · Agency · Mobility · Bangladesh

---

S. Aktar (✉)
Department of Public Administration, University of Rajshahi, Rajshahi, Bangladesh
e-mail: aktar96@ru.ac.bd

SM. M. Rahman
Department of Political Science, University of Rajshahi, Rajshahi, Bangladesh

J. J. McIntyre-Mills, Y. Corcoran-Nantes (eds.), *Transformative Education for Regeneration and Wellbeing*, Contemporary Systems Thinking,
https://doi.org/10.1007/978-981-19-3258-8_8

1 Introduction

The conventional wisdom suggests that education is an important driver to boost women's confidence and capability not only participating in economic activities but also to participate in decision-making contexts (Moser, 1993). The importance of education is also critical for social change and political transformation including gender justice and women's empowerment (Betts, 2003; Freire, 1973). The societies in the developing countries are often clouded with century-old norms and practices restricting women's access to certain places and activities by defining and restricting their roles. Education can help mitigate prejudices and help societies overcome age-old stigma including those against women's value and position in the society (Betts, 2003). Education is a critical component for liberation of the underprivileged groups including women (Rogers, 2001). "Education focused on social justice and equity attempts to change institutions and the distribution of power, promoting new behaviours, relations and ways of viewing the world" (Stacki & Monkman, 2003: 173). Education is undoubtedly a means to end inequality, injustice and hence a catalyst to empowerment of the marginalized community including women. Yet, the relationship between education and gender empowerment is not straightforward and can be seen from different ways (Freire, 1973; Leeve, 2001). The mainstream argument that education facilitates economic achievement in terms of employment, efficiency and productivity combined with health, nutrition and fertility rates (King & Hill, 1993). It is also argued that education can help marginalized communities to overcome structural barriers, eliminate oppression and enable creating a more just and equitable societies (Leeve, 2001; Maruatona & Cervero, 2004).

Historically, women were valued and regarded higher places in primitive ages for their reproductive role and their contribution as nurturers of humanity (Goel, 2004; Smith, 2008). As a result, the ancient mythologies describe women as mother and saviour. For example, the Greek mythology used to name several Goddesses for the well-being and safeguard of humankind while the Indian scripts termed women as Devi & Maata (Noreen & Khalid, 2012). However, the situation changed once primitive society evolved as a hunter-gatherer society in which masculinity started to dominate through hunting animals and gathering food from the forests. Men's central role in the hunter-gatherer society possibly laid the groundwork of patriarchal dominance in the primitive society (Arnot, 2002; Noreen & Khalid, 2012). The domestic role of women in this society is assigned perhaps for convenience rather than dominance (Arnot, 2002). Men in the hunting society provided food supply and protection. This increasing valuation of men ultimately resulted in male dominance and the patriarchal control of society (Aikman & Unterhalter, 2005; Arnot, 2002). Since then, the world, specially the developing world sees male dominance and control in every spheres of life. In the traditional Asian and African societies males were exclusively the family heads and women are tied to the domestic chore with various customs and norms (Ahmed, 1992). It is only after the European Renaissance and after the two world wars that the notion of women's rights, equal citizenship, agency in household decision-making and participation in development activities got attention and importance (Goel, 2004).

The Sustainable Development Goals (SDGs), sets by the United Nations, therefore, categorically argues for championing equality in education as well as achieving gender equality in order to promote sustainable development across the world. The SDG goal #4 explains the goals and targets for the member countries in bringing equality in education, while goal #5 sets the goals and targets to empower women and promote gender equality, which is summarized in the Box 1 below:

Box 1: Sustainable Development Goals

SDG #4: Ensure inclusive and equitable quality and promote lifelong learning opportunities for all.

4.1: By 2030, ensure that all girls and boys complete free, equitable and quality primary and secondary leading to relevant and effective learning outcomes.
4.3: By 2030, ensure equal access for all and men to affordable and quality technical, vocational and tertiary , including university.
4.5: By 2030, eliminate gender disparities in and ensure equal access to all levels of and .

SDG #5: Achieve Gender Equality and Empower all and Girls

5.1: End all forms of discrimination against all and girls everywhere.
5.2: Eliminate all forms of violence against all and girls in the public and private spheres, including trafficking and sexual and other types of exploitation.
5.3.Eliminate all harmful practices, such as child, early and forced marriage and female genital mutilation.
5.4: Recognize and unpaid care and domestic work.

UNESCO (2020), in this context, reveals mixed evidence in terms of addressing inequality in education in the past three decades. The report (UNESCO, 2020) mentions that though girls' educational attainment has been improving in most of the countries across the world in the past three decades, the inequality is increasingly seen in terms of digital literacy. The enrolment in primary and secondary education sees parity between boys and girls in 2018—marking a 10% elimination of disparity in 1995. But despite such improvement, the disparity still exists in terms of girls' studying in technical subjects including engineering and ICT. For example, globally less than 25% of the students in technical subjects are girls. Even in OECD countries, about 14% of the high achieving girls intend to study science and engineering compared to 26% of high achieving boys (UNESCO, 2020). Researchers and experts, therefore, argue that policy makers need to pay attention to the rights of women in order to address improvements in post-secondary education (Lewis & Lockheed, 2006; Miske et al., 2007). They argue resources need to be applied to improving enrolment in schools but also emphasize more on access to affordable and quality education. This study, in this context, investigates the relationship between education and women's empowerment in the context of Bangladesh. It explores the dynamic role that education can play in enabling women's capability, agency and empowerment in this patriarchal society while uncovering the challenges and underlying conditions in such pursuit.

2 Background: Gender Inequality in Bangladesh

The women in Bangladesh lag behind men in terms of education, employment and income. The female literacy rate stands at 54% compared to that of male at 73%. The labour force participation is also lower (34% compared to 79% for male) (BBS, 2020). The cultural stereotypes prevailing in Bangladesh including that women's are responsible to take care of family and children; that women are not suitable for leadership position and decision-making skill; the domestic chore belongs to women while the works outside home goes with the men; and women can work outside if necessary but that doesn't excuse her from domestic role (CAPWIP, 2000; Kabeer, 2001). The country is often touted as a symbol of classic patriarchy where the husband is the master of the household (Kabeer, 2011; Mahtab, 2007). In the case of a husband's death the adult son or in-law becomes the decision-maker as head of the household. Throughout her life a women's subordinate position and identity is controlled by men including fathers, husbands and sons. Women's dependency indicates their economic and social vulnerability which scholars conceptualise as "patriarchal risk". Women in Bangladesh have rights on parental properties by law but the cultural norms often force them to waiver these rights in favour of their brothers (Kabeer, 1999, 2011; Panday, 2008). As a result of losing rights over parental properties renders women hugely dependent on their husbands and forces them to accept subordinate positions (Anderson & Eswaran, 2009). The pressure for dowry goes up in the society as husbands perceive that the wives cannot access parental properties later—causing brutal pressure on women including physical violence in many instances (Bates et al., 2004; Naved & Persson, 2005). In the marriage market, good girls are highly valued who do not challenge gender norms, practice purdah[1] and are good-looking with fair-skin (Chowdhury, 2009). Women are not allowed to talk to outside men. The older is the girl, the harder it becomes to find a groom for marriage. Families often decide strategically to focus on a girl's marriage when she is young, rather than a girl's education (Chowdhury, 2009). Even with rising education, the investments and preferences of the families continue to favour boys as investments rather than investment in girls who are supposed to go to their husbands' home after marriage (World Bank, 2008). In this patriarchal society, girls do not have the right to select their life partners. The father is the ultimate one who decides the daughter's groom, neither the bride nor her mother has any say in this regard. The social norm is that a husband is next to God for a decent, modest and pious wives (Panday, 2008). Bangladesh is a Muslim majority country (with 90% population) where religion dictates cultural norms and legal provisions including women's role and position in the society. The dominant religion, *Islam*, is conservative in allowing women's freedom compared to that of men. "Muslim women generally face tremendous opposition and are often criticized by religious fundamentalists on the ground that there is no provision for women to be state leaders under Islam" (Halder, 2002: 49). The Muslim Family Law sets out the provision of parental property in which a daughter inherits half of what a son inherits. This leads to bias in preferring the birth of boys (Anderson & Eswaran, 2009; Halder, 2004). The birth of a son is marked by a call for prayer (Azaan) which is not allowed for a girl's birth (Huque & Akhter, 1987; Panday, 2008). Women are supposed to practice purdah[1] to conform social norms (Kabeer, 2011; Mahmud et al., 2012; White, 2010). Women's mobility outside home is severely restricted. They need to be accompanied by male guardians while going outside (Kabeer, 2011; White, 2010). Women are blamed for giving birth to girl children and are forced to give birth until a son is born (Kabir, 2002; Nath, 1981).

The incidence of fatwa was also very common in which religious clerics (who are all men) u to punish women for accusations including extra-marital affairs (Aktar, 2014; Kabir, 2002). There is hardly any incidence of fatwas against men. These fatwa include harsh punishment including lashes, stoning to death, boycotting from society, driving out from village, etc. (Aktar, 2014; Rashiduzzaman, 1997). In case of divorce, or even words of divorce, women are forced to follow Hilla marriage[2] if the couple wishes to continue their marital relationship (Hamid, 1996; Kabir, 2002). Even if there is an incidence of rape, the religious clerics and influential people in the community often label the victim as a 'bad character' or 'bad girl'. The fatwa in these contexts goes against the victim with brutal physical punishment while the perpetrators pay fines for the monetary benefit of social leaders (ASK, 2019; Rashiduzzaman, 1997). The incidence of fatwa has remained high with as many 944 cases in the past two decades, causing 24 fatwa-related deaths (ASK, 2019).

In Bangladesh, the political landscape is highly restricted for women, though the country has been led by a female prime ministers for the past three decades. The two major parties are both headed by female descendants of their iconic past leaders, yet these parties accommodate only 20% women in their ranks and file membership(Panday, 2008). Politics tends to be dominated by money and muscle where women can seldom compete (Halder, 2002). As a result, only a handful of women candidates contest elections, but nevertheless, the 2018 national election resulted in the highest number of female candidates in parliament ever recorded, namely 22 of 300 seats (Prothom Alo, 2014, 2019).

---

[1] Purdah is Islamic tradition in which women need to cover their face and head while going outside, often referred as Hijab/Niqab.

[2] Hilla marriage is an interim marriage in Islamic Sharia. This is applicable if a husband divorces his wife, even in verbal words, but wishes to continue their marital relationship. The religious verdict in such is that the wife needs to marry a third person, live with him for at least 40 days, then get a divorce from that newly-wed man, and only after this she will be eligible to remarry her ex-husband. This is a harsh punishment considering the fact that the women in Islam has no right to divorce, but in case of a divorce by the husband it is the wife who has to go through such medieval ordeal of marrying and sleep forty nights with a third person, whom she doesn't love and doesn't consider husband, before she can claim back her old partner!

3 Gender Mainstreaming: Theories and Dimensions

This study uses the concepts and notions of gender mainstreaming and women's empowerment described by Sen (1999), Nussbaum (1999, 2011), Rowlands (1997) and Stromquist (1995), all emphasizing the discourse of gender and development (GAD). The GAD approach argues that development initiatives must focus on change of structural conditions that create and sustain inequality including the division of labour based on gender (Friedmann, 1992; Kabeer, 1999; Rathgeber, 1990). Following the GAD hypothesis of gender equality, this study examines the role of education in addressing gender inequalities both in the socio-cultural and economic dimensions in Bangladesh.

In Sen's (1999) Capability Approach, development is defined as the improvement of people's capability of doing things of choice including education, work, shelter, politically active and healthy life. To Sen, development should prioritize people's capabilities—their potential for *doing* and *being* rather than their income or consumption: "the ability to satisfy certain crucially important functionings up to certain minimally adequate levels" (Sen, 1999: 41). Accordingly, resources are enabling factors only which may not guarantee one's well-being as there might be socio-political or cultural constraints. The per capita income, for example, cannot tell us the law and order situation or the quality of health services available in case required. The Capability Approach, in this context, emphasizes people's *agency*: the capability of people doing and pursuing own goals with interests (Sen, 1999). It focuses on people's ability and choice at the centre: how individual act or disagree to act; and how she accepts or refuses. It considers individuals as *active agents* instead of passive actors of doing something whether it comes to income, employment or non-economic decisions.

Nussbaum (1999, 2011) links capabilities with certain human rights. Accordingly, certain human capabilities are aligned with universal value. For women, this includes capability of having a dignified life as human being with truly human functioning. The conventional measures of development often overlooks equality and justice by focusing on economic growth and ignores the real value of life (Nussbaum, 1999, 2011). Accordingly, GDP growth cannot answer our social norms, political freedom or environmental quality that very much shapes our living and being as a citizen. Therefore, a meaningful development strategy must focus beyond GDP and prioritize certain areas of human capabilities necessary to command a dignified life (Nussbaum, 1999, 2011). She mentions 10 capabilities central for a dignified life: enjoying life; bodily integrity; bodily health; control over own environment; sense and emotion; affiliation, etc. Accordingly, these ten capabilities are a pre-requisite for equality, social justice and empowerment of the disempowered; and failure to ensure these capabilities means a society is far away from being a just society whatever its economic prosperity might be. Malhotra et al. (2002) also opined that these human capabilities are universal in nature and are essential to foster meaningful development and empowerment from the roots.

Rowlands's (1997) model describes power as personal, relational and collective attributes which can be either destructive or productive. Power can be used by individuals or groups to retain a certain order of relationship or to change it. Rowlands (1997: 15) describes personal empowerment as the development of "self and individual confidence and capacity, and undoing the effects of internalised oppression". Accordingly, empowerment is the development of one's "ability to negotiate and influence the nature of a relationship and decisions made within it" (Rowlands, 1997: 15). Such abilities and negotiations can take shape in the social settings as well as decision-making within households. Collective empowerment, on the other hand, is the ability of the people joining together, "to achieve a more extensive impact than each could have had alone" (Rowlands, 1997: 15). The model, however, warns that development in one area may not necessarily lead to development and empowerment in other areas and dimensions for which careful strategies and collective actions are required.

Stromquist's (1995) model frames empowerment into four specific dimensions. Accordingly, cognitive empowerment of women indicates understanding of the causes of women's weaker positions and achieving knowledge and skills necessary to overturn existing gender-power relations. The psychological empowerment refers to developing self-esteem and confidence that women can do certain acts for the improvement of their living and being. The economic dimension of empowerment, as Stromquist (1995) argued, indicates engagement of women in productive activities necessary to support their living and exercise autonomy. Finally, the political dimension of empowerment refers to the ability to participate in activities and organizations for representation and change. This model captures the diverse notion of empowerment necessary to change the male dominance in the society and enabling women's meaningful participation and agency in the decision-making process in every spheres of life.

4 Research Approach

Uncovering the nexus between education and gender mainstreaming is a complex task considering the complex nature and duality of the relationship. Education, in one hand, is a pre- requisite to eliminate social stigma and help women overcome patriarchal barriers, on the other hand, mother's pioneer role is a significant catalyst in order to expand education. The motto of this paper, however, is to explore the significance of education in mainstreaming gender in the society of Bangladesh. To investigate the issue, we collect data from thirty respondents in three categories: (i) fifteen highly educated women (with graduate/post-graduate degree); (ii) fifteen illiterate women, and (iii) fifteen women microcredit borrowers with primary to secondary level education. About half of the highly educated women respondents are working in paid jobs and half are housewives; the microcredit borrower women are working in various income generating activities; and the illiterate women respondents are mostly housewives with no regular income.

The data were collected from the respondents using a semi-structured questionnaire with both closed and open-ended questions, followed by in-depth interview and participant observation. Data for the highly-educated women are collected from the divisional city of Rajshahi, while data for the illiterate and the microcredit borrowers are collected from five different villages of Puthiya Upazilla of Rajshahi district. For each of these categories, respondents were randomly drawn from a list of one hundred populations. The list of microcredit borrowers in Puthiya Upazilla were collected from several NGOs including Grameen Bank, BRAC & TMSS operating in the area. The list of illiterate women was collected from the office of the Union Parishad, while the list of the highly educated women was collected from Councillor's office of the Rajshahi City Corporation. The data were collected by visiting respondents at their suitable time with full consent and keeping anonymity. The questionnaire was prepared in local language and female only interviewers were employed to make sure respondents feel free and to conform to the social norms in the survey area.

The collected data show that most of the women respondents fall within the age category of 24–48. Most of them are permanent residents of the survey area, although some highly-educated respondents migrated here in the past 5–10 years. All of the respondents except one are married and one of them is widow. About 80% respondents are Muslim and the rest are Hindu.

5 Results and Discussion

Employment and Income Education can help employability and productivity of women considerably that in turn can help increasing agency and empowerment (Kabeer, 2011; Noreen & Khalid, 2012). The respondents of this study mentioned that educated women can find suitable jobs and earn higher income than the uneducated housewives. They further mentioned that the uneducated women can find laborious jobs but can earn only a small amount. In contrast education can help women to find formal jobs with better salary and career opportunity. As one of our respondent stated,

> *I had a dream that I will be independent. I get admitted to the university dreaming that one day I will qualify for public service and will be a government officer. I was chasing my dream from the beginning of my university day. After completing my graduation I sat for several job tests and finally selected for first class officer position. Today I am District Social Welfare Officer and am supervising the entire district office under my jurisdiction. I strongly believe that education can help women fulfil their dream. Without education I couldn't be what I am today.*

Another woman mentioned that despite having same backgrounds and sharing same opportunities, education can make big difference between people's employment and career. As she argued:

We are four sisters. Two sisters got married after completing their secondary education and become housewives. But I and my youngest sister continued to pursue higher study at the university. I completed my Honours and Master's degree in Business and my younger sister earned her bachelor degree in engineering. Today I am a banker and my younger sister is an engineer in a reputed organization. It is obvious that education surely has contributed to establish our career.

Respondents mentioned that without education women cannot break out of the vicious cycle of unemployment, poverty and deprivation. Even the illiterate respondents mentioned that they are trying their best to educate their children so that they get better employment and earn decent income in their life. This finding is consistent with the empowerment framework of Moser (1993), Sen (1999) and Stromquist (1995). Moser (1993) and Sen (1999) consider empowerment as women's capability and confidence in participating in economic and non-economic activities of choice. Stromquist (1995) on the one hand considers economic empowerment of women as engagement of women in certain productive activities to support their living. On the other hand, increased knowledge, awareness and confidence are prerequisites for empowerment (Arnot, 2002; Moser, 1995; Nussbaum, 1999).

5.1 Enhancing Confidence and Self-Esteem

Education is also critical to help boost women's self-esteem and confidence.

Education can open the horizon of knowledge; eliminate stigma and aware women of their rights (Maslak & Singhal, 2008). The respondents of this study suggested that education is important in developing confidence and self-respect. As one educated respondent suggested,

Education gives me confidence who I am. It opened my eyes to see what is there in this world. Today when I talk to my friends, my relatives, neighbours—they all listen to me carefully as I have wisdom and knowledge. I can go to any office if necessary, do any work like a man can do—all because I got the education. But for my beloved illiterate womenfolk, they cannot do what I do as they lack education.

The illiterate women agreed with it. They mentioned that since they are illiterate they do not understand many things of the outside world but being dependent on others. As one illiterate woman stated that lacking education she has no confidence of doing anything without advice from husband or in-laws:

Thus, the empirical findings of this study document role of education in enhancing self-confidence and self-esteem of women in the survey area. Since women are undervalued and are subject to manifold restrictions in this patriarchal society, the role of education appears critical in eliminating social stigma and help boost women's self-confidence of doing and achieving the outcomes they deserve. The findings of the study are consistent with Rowlands's (1997) *personal empowerment* which indicates women's self-confidence and capability to pursue their own goals and undo the internalized oppression. The empirical findings are also in line with the argument of Stromquist (1995) in which the *cognitive* and *psychological* dimension of empowerment refers to women's self-esteem and confidence of doing certain things to improve their living.

5.2 *Increasing Status in Family and Society*

Education is imperative to improve women's position within household and in the society (Mitra & Singh, 2007). Without education, women cannot exercise agency in the household decision-making. But education enables their voices to be heard and honoured. As one respondent mentioned,

> *I am the youngest of four daughters-in-laws in my husband's home. My father-in- law is a school teacher and a social worker. He is a wise man. But sometimes he asks me this and that including some complex issues that he might encounter. To him, I am a university graduate, and I must have understanding and knowledge of the government policies, outside world etc. Other daughters-in-laws of the house have education up to secondary school level. I see that because of my higher education my father-in-law sees me in a higher position than others.*

Another respondent mentioned that education also enables social respect for women. To her:

> *My father-in-law was a chairman of a Union Parishad (local government body). He passed away last year. When the election schedule of the Union Parishad was declared, our relatives, family members and long-time supporters sit together. They all decided that I am a university graduate and if I contest this election, it would be easier for them to convince the voters and bring back victory. I was youngest in the family and was nervous. But they all argued that this Union Parishad has never seen a highly-educated candidate, all we have seen so far are illiterate or candidates with school level education. With such continuous pressure from family and supporters, I finally contested in the election. During election campaign I noticed enthusiasms among the voters for me, and even I saw respect from my opponents. I won the election convincingly, probably because voters preferred to elect an educated candidate as their representative.*

The respondents agreed that an educated woman is typically valued more than an uneducated woman both within the family and in the community. They mentioned that when the educated women talk, they talk with information, logic and scientific point that enable their voices to be heard easily than the illiterate women who are ignorant of everything. This finding is consistent with the findings of Betts (2003) and Stacki and Monkman (2003) that mentioned role of education in enhancing women's value and position in the society.

5.3 *Raising Awareness About Women's Rights and Reducing Dependency*

Respondents also mentioned implication of education in raising awareness amongst women about their rights. While the illiterate respondents are found mostly ignorant of their legal rights, the educated respondents are aware of their rights and safeguards. As one respondent mentioned her misery being illiterate:

> *My husband married second time without informing me. I am an illiterate person; I earn nothing and have three children. I had no way but to accept his marriage. But after few months, he started torturing me and at one stage drove me out of his house. I came back to parental home and started living at the mercy of my brother and his wife. My friends and neighbours asked me to go to the court and seek legal measures against my husband. But I don't know anything of my rights, and the legal processes. So I didn't go to the court.*

The opposite is seen in case of educated women who are well aware of their rights. As one woman shared her experience:

> *I got married two years back. In my in-laws house, I noticed soon after my marriage, that they were looking for dowry. I am an educated woman and I know this is against law, and against a woman's dignity. I understood their intention and protested by cleverly saying that, 'my parents were looking for a groom who might be poor, but they were not looking for someone who is a wealthy beggar'. I understand my rights and safeguards and conveyed them. Eventually they were forced to stop pressuring on me and my parents.*

The empirical findings, thus, document positive role of education in raising awareness and women's rights—essential for women's empowerment in this conservative society. Previous literature (e.g., Leeve, 2001; Betts, 2003; Maruatona & Cervero, 2004) also document education as catalyst to raising awareness, creating a more equitable relations and just societies.

Women respondents also highlighted role of education in reducing women's dependency on men in various aspects of their life. As one respondent mentioned,

> *I am lucky that I am educated. I don't need to depend on anyone even on my husband. I earn my living by my own, do my shopping, go to the bank when necessary and go to hospital to visit doctor. For these I don't need to depend on my husband or in-laws. But for other women in my neighbourhood, I see they are always dependent on their husbands as they are not educated. My husband is a businessman and he often goes to distant towns for business purpose. During this time I oversee his business, do banking activities for him, take children to school and hospital comfortably. My husband relies on me, especially in running his business and banking tasks, whenever he is out of the town.*

The illiterate women on the other hand found heavily dependent on their husbands. As one woman respondent mentioned,

> *I am an illiterate woman. I don't know anything about government office, police station or anything. I depend on my husband for any outside activities. I even don't know which school is good for my children, whether should they study or not, which fertilizer or pesticide suits our crops, etc. My husband decides on these. Sometimes he shares and takes advices from his friends but not from me. Frankly speaking, lacking education I am unable to make a serious decision but depend on my husband.*

The respondents overwhelmingly agreed that the illiteracy is the major cause why women are dependent on their male family members. It is also the key factor why they are often ignored and undervalued in the family and in the community. Findings of this study is consistent with the arguments of Nussbaum (1999, 2011) that seeks women's control over own environment. Stacki and Monkman (2003) also evidenced that education plays important role in changing institutions and the distribution of power.

5.4 Increasing Mobility Outside Home

In Bangladesh, women's mobility outside home is severely restricted due to patriarchal culture. The role of education is critical in such a situation in increasing women's mobility outside home. As one respondent mentioned,

> *I am married and living in this village with my in-laws for past 16 years. In all these years, I have never been to the Upazilla or District head-quarter. I have never been to these places for shopping or for any other reason. This is because I am illiterate and I am afraid of going outside. However, this year I join a microcredit group. I visited their office in the Upazilla head-quarter and attended a training session there. This is the first time I have ever met outside people and officials in my life. It opened my eyes that to engage in productive activities and to prosper in life, woman must go and explore the outside world. But lacking education, women remain ignorant of this opportunity.*

Another woman echoed the same highlighting role of education in her life:

> *I was born in a conservative family. My parents are very pious. My father is an Imam of the local mosque. My mother never visited a mosque, nor did she visit any village except her parents in her entire life. I was a very good student from childhood. When I did brilliant result in my secondary school examination, my father—who is very religious and conservative Imam, enrolled me in a girls' college in district head-quarter. My education and potential to do better in future convinced my conservative parents to let me go outside and pursue higher study. I moved to the college hostel and achieved highest grade in higher secondary examination. At this stage my parents wanted me to get married. But I requested them to let me pursue my dream to be a doctor and promised them to remain religious and wear purdah. They were convinced. After that I got chance to study in a government medical college—and my parents are very proud that I am a doctor now! Today I can visit any place, village or town, home or abroad, day or night without any restriction. This is obviously thanks to my education and position. I also argue parents to let their girls pursue higher study for their own betterment. At the same time, I argue girls not to confront with parents but to convince them with humble behaviour, modesty and thoughtful presentation of your case.*

The educated women all agreed that they can visit any public places including government offices, hospitals, shopping centre, medicine shops, etc., like men without anyone's permission— which is in contrary to the views of the illiterate women. The illiterate women respondents in contrast mentioned not to visit outside villages without husbands' accompany and permission. This finding is consistent with the empowerment frameworks provided by Sen (1999), Nussbaum (1999, 2011), Rowlands (1997) and Stromquist (1995)—all of which emphasize on increasing capabilities and acquiring necessary skills and confidence necessary for women's empowerment.

5.5 Reducing Pressure for Dowry

Scholars have argued for long that education can eliminate social curses and malpractices such dowry, bride price and domestic violence (Aktar, 2014; Chowdhury, 2009; Kabeer, 2011). In this study we also asked respondents about the role of education in combating social stigmas in Bangladesh. Most of the respondents agreed that combating social curses like dowry required raising awareness through education. Respondents agreed that dowry is common among the illiterate families but are less common among the highly educated families. As one respondent mentioned,

> *Dowry is common among the illiterate and lowly-educated families. Since uneducated people are unaware of the laws and regulations of dowry, and since they are not taught the consequences of dowry in the society, they take it as a convention and a short-cut way of getting some cash. But one cannot ask dowry when the bride is well-educated. Even if someone asks for dowry, a higher- educated girl and her parents will protest it.*

Another educated respondent mentioned her experience:

> *My in-laws took dowry from the wedding of their first two sons. I was married to their third son. Before marriage they didn't demand any dowry. But after I came to their home, I realized that they very often refer to what the other two daughters-in-laws brought to their house. It is an indirect pressure on me for dowry. I was very upset that being a university graduate if I need to pay dowry then what will happen with ordinary girls of this society. I challenged them one day that what I brought to this house is my university degree, my knowledge and wisdom that are no match to the lump-sum dowry they got from* their previous two lowly-*educated sons' wedding. I told them that dowry is forbidden and demanding such is a punishable crime. Since then they never uttered it anymore.*

An illiterate microcredit borrower shared how training can help raising awareness and eliminate social curses like dowry and child marriage among people who are not educated. She mentioned:

> *I knew that dowry is very oppressive for helpless parents. Being an uneducated person, I didn't know that there was a law strictly forbidding dowry declaring it a punishable crime. When I took loan from the NGO, I attend a training session where I first learnt about it. The training also informed that there is a hotline number (333) to seek legal and police help to stop child marriage and dowry. The training also mentioned that the national hotline number (999) can also be dialled using any mobile phone without charging fee for the same purpose. Now as NGO borrower we know it and seek police help whenever come to know any incidence of dowry. As a NGO member, we are also promised not to demand dowry in our son's marriage and not to pay dowry in our daughter's marriage.*

Thus, our findings suggest that girls' education can significantly reduce pressure for dowry and dowry-related violence and oppression. This is critical in this patriarchal society where there are numbers of dowry related violence are reported in the newspapers every year. This finding conform the hypothesis of Nussbaum (1999, 2011) arguing empowerment as capabilities of enjoying dignified life, bodily integrity and bodily health. The findings are also consistent with.

Maruatona and Cervero (2004) and Leeve (2001) that education can effectively eliminate oppression and enable creating an equitable and just society.

5.6 *Education and Upbringing of Children*

The respondents also mentioned importance of education in educating and upbringing of children. They mentioned that education enhances mothers' knowledge about child health, nutrition and vaccination which are important for the upbringing of children in right direction. As one respondent mentioned:

Throwing away the foremilk was a tradition in our village. But as an educated lady I learnt the importance of the first milk for immunity and nutrition of the children. I didn't follow the tradition and also motivated other lactating mothers to feed the first milk to their children. My education also helped me to plan for my children's study, leisure, physical activity and socialization.

The illiterate mothers on the other hand were found unprepared to upbringing their children well. As one illiterate woman said,

I cannot help my children in reading or writing that an educated mother can do. I don't understand why my children are not doing well in their school examination. Sometimes I feel that it is my children's misfortune that they born in a family where parents cannot understand their need or problem. Being illiterate is a curse indeed, especially for the children.

The role of education, thus, appears critical for the personal and cognitive development of women (Rowlands, 1997; Stromquist, 1995).

5.7 Health and Family Planning

The respondents of this study also mentioned that education is an enabling factor in increasing family planning and contraceptive prevalence. As one university-graduate woman respondent mentioned:

I was married at the age of 25. Many of my illiterate and lowly educated friends by this age got three to four children. Being an educated woman I have a plan about my career and family. After my marriage, I told my husband that I will pursue for a good job first and only after that I will take a baby and I convinced him. With my passion and persistence I got a job in two years and then our first child born. I am currently on contraceptives as I have no wish to have a second baby very soon.

The illiterate women on the other hand found to have no such plan about children or contraceptive. As one microcredit borrower woman mentioned,

I gave birth to four daughters with a hope and pressure for a boy child. I was so ashamed every time I was giving birth to a daughter. My in-laws were all blaming.

me for not having a son. I was also thinking to blame myself. Sometimes I thought to commit suicide for this. But I was afraid of doing so thinking of my children. When I join this microcredit program, I then learnt that it is not my fault at all. The trainee sister also taught us that with proper education a girl child is equally worthy like a boy child. Now I am sending all my daughters to school and hope their life with education will be a different one than that of mine.

The role of education is, thus, evidenced in promoting health and family planning in the survey area. The results suggest that policy makers aiming at increasing health and contraceptive prevalence, must also focus on educating girls.

5.8 Access to Finance and Microcredit

The respondents also mentioned education's enabling role in accessing finance and banking services. As one educated respondent mentioned,

> *Education is important in accessing financial services of the banks. I am running a boutique house and at one stage I needed bank loan to expand my business operation. I went to several banks and found various loan products in offer. I collected their brochures, compared their terms and conditions and finally selected one. Without education and critical understanding one cannot chose the right financial product or cannot find the best deal.*

Another woman respondent mentioned how her university degree helped her finding a loan without collateral:

> *I run an online shop and supply various indigenous products countrywide. My sources of product supply are indigenous craftsmen. They are very poor and often need to pay them in advance. I always need cash for this. When we need finance I see that these craftsmen are unable to get loan by themselves lacking collateral assets. One day I came across a circular in the daily newspaper that the government is providing a collateral-free loan for the young-educated entrepreneurs through Karmosangsthan Bank. I went to the Bank manager next day and applied for a loan. They asked my original academic certificates to support my loan application and I provided them. Soon I get the loan without any asset collateral, thanks to my academic degrees.*

The respondents mentioned that economic emancipation is a must in order to women's freedom, empowerment and self-esteem. Without economic freedom women are dependent on men and as a result of that men's dominant position goes unchallenged in the family and in the society. The findings are consistent with (King & Hill, 1993; Moser, 1993; Stromquist, 1995).

This study also identifies several challenges that need to be addressed in order to harness the benefit the potential of education in mainstreaming gender in a developing country like Bangladesh. The study evidences various challenges summarized below:

5.9 Shallow Job Market

The job market in Bangladesh is very shallow. Every year almost half a million graduates are joining the job market but only half of them find a job (Prothom Alo, 2014). The situation is worse for the college graduates who are 68% of the total graduates. Among the college graduates, about 66% male and as much as 77% female remain unemployed (Daily Inqilab, 2019). As a result, the educated-unemployment becomes a major headache for the country. The salary in the private sector is often very low that the educated women prefer their housewife role instead of working. As one of the respondent mentioned,

> *The problem with educated women is that they cannot do what the illiterate can do for living. The educated women are not supposed to do low-paid laborious jobs. On the other hand, formal jobs are very few. If an educated woman does not get a job, her education hardly comes to help her in building self-confidence, voice and household bargaining. In the family she is just an added burden as she is even unable to do what an illiterate woman can do.*

Another woman respondent mentioned that lacking employment, educated women become worthless in the marriage market, causing a social problem. She mentioned:

> *When a girl completes her graduation, she is already late in the marriage market. When she is trying further for a job for two-to-four years, she becomes too old. She loses glamour and beauty. Then the only way she is demanded by the matchmakers is to come with a job. When the girl does not find a job, it becomes very difficult to get married. Many guardians suffered from this difficulty and many others in the society feel reluctant to educate their daughters.*

The creation of employment, thus, appears critical to mainstream gender in this society. The findings is consistent with the previous studies by Moser (1993), King and Hill (1993) and Stromquist (1995) that argue role of education in economic empowerment of women.

5.10 Corrupt Political System

The political system is a corrupt and dysfunctional one in Bangladesh (Panday, 2008). Voter frauds in election, use of power, money and muscle are so rampant that the Economist Intelligence Unit put the country as a "hybrid regime"—with substantial irregularities in election being free and fair. In this regime, "government pressure on opposition parties and candidates may be common. Serious weaknesses are more prevalent than in flawed democracies—in political culture, functioning of government and political participation" (Daily Star, 2021). All these put the prospect of political empowerment of educated women into a dash. As one respondent mentioned,

> *I contested the municipality election in 2020. I didn't get ruling party nomination as I couldn't buy it paying huge sum the leaders were demanding. My supporters wanted me to contest as independent candidate. I had a huge support base being an university graduate and being involved with various socio-cultural organizations since my school life. But during election, the ruling party candidate and her supporters continuously attacked my supporters, my rally, ripped off my posters and banners. On the election day,* they even drove out my agents from many polling booths. I lost the election as a result, to a candidate who has just completed her secondary school.

During interview, most of the respondents mentioned about election irregularities, dominant use of money and muscle as major hurdles that discourage them to pursue a political career. To Stromquist (1995), such a situation will impede women's political empowerment in the society.

5.11 Strong Patriarchal Culture

The society in Bangladesh is characterized with strong patriarchal culture (Kabeer, 2011; Mahtab, 2007). The respondents mentioned that they also suffered from patriarchal malpractices in many occasions. As one respondent mentioned:

> *My brother used to go to coaching centre without reservation. But when it comes to me, my parents were reluctant for security concern—that I cannot come back before the sun set. Sometimes I also hear that being a girl I will marry and go to husband's home—so they shouldn't worry that much for my study. Yet I performed better than my brother in the secondary and higher secondary examinations. But then my parents arranged my marriage blocking my desire for higher study; whereas they sent my brother to Dhaka to study further as he was a son.*

The respondents mentioned that social mobilization and social awareness are essential in order to eliminate gender bias from the society.

5.12 Poverty and Inequality

Bangladesh is a developing country with per capita income of US$ 2064. The poverty rate was 21.5% in 2019 but rose to 29.5% in 2020 due to Covid-19 (Dhaka Tribune, 2020). The low level of income and poverty is a critical hurdle in mainstreaming gender in the country. Poverty put a limit on the achieving desired level of education for many women surveyed in this study. As one woman mentioned:

> *I couldn't complete my school as my parents were poor. I was a very good student. My teachers of the school liked me very much. But due to poverty, I have to leave school without completing secondary education. I knew that without education the world will close its door for me but I was helpless. Now I want my children to get proper education. But God knows how far I will be able to support them.*

5.13 Strong Religious Norms and Tradition

The society of Bangladesh is dictated by religion in many cases (Chowdhury, 2009; Panday, 2008). The religious doctrines are parts and parcels of this conservative society which often put impediments to mainstream gender (White, 2010). As one woman respondent mentioned:

My parents sent me to a Madrasa for Islamic education though I wanted to go to a general school. But the tradition in my village is that every parent must send at least one child to Madrasa—in order to demonstrate to be religious and good people. So my parents sent me to Madrasa no matter what I think of it. I wanted to go to the school, study science and dreamt to become a doctor. But my hopes are dashed by going to Madrasa.

The respondents mentioned that religious faith and culture often put restrictions for girl children in not going to co-education schools, not going to co-education coaching centres, not even going to private tutor alone, not staying outside after sun set, and so on. Girls are not allowed to take physical education, music, dance and art classes once they turn puberty for safety concern. All these put girl students in a disadvantaged position in achieving better education and their attainment in extra-curricular activities.

6 Conclusion

The role of education in eliminating poverty and inequality is established in the literature. This paper investigated the dynamic role of education in mainstreaming gender in a patriarchal society like Bangladesh and in such explores its role in achieving the Sustainable Development Goal #4—bringing equality in education & Goal #5—empowering women and promoting gender equality. We consider women's empowerment from the theoretical rigours of Sen (1999), Nussbaum (1999, 2011), Rowlands (1997) and Stromquist (1995) models in which empowerment refers to choice, agency and capability of women for dignified living and being that is able to fulfil the fundamental and strategic gender needs. The study documents the efficacy of education and the underlying challenges in mainstreaming gender by summarizing the survey results of a sample of women in three categories: the highly educated, the illiterate and the microcredit borrower women in Bangladesh. The findings of the study show importance of education to improve employability and job salary of the women which is critical for women's emancipation from oppression and deprivation in this patriarchal social setting. The respondents, across all strata of education, further emphasized on educating their children for better job and higher income.

The empirical findings also show that education is important in raising awareness, building self-esteem and confidence among the women. The respondents mentioned the role of education in enhancing knowledge, eliminating social stigma, and raising awareness about women's rights. Respondents further mentioned the role of education in increasing acceptance and status within household and in the society. They mentioned that education enhances their position in the society as educated women raise their voice with words of wisdom, scientific knowledge and information which is difficult to ignore. On the other hand, illiterate women are ignored and seldom heard as they are ignorant of the outside world. The empirical findings also document positive role of education in combating social malpractices like dowry and child marriage in this patriarchal society. The role of education was also mentioned by respondents in better upbringing children, their health and nutrition. Education was also found important in enhancing family planning and contraceptive use. It transmits knowledge about modern contraceptive tools and eliminates age-old stigma against contraception. The findings also document role of education in accessing finance and markets which crucial for women's economic emancipation.

The study also documents certain challenges in mainstreaming gender in this society. It identifies those challenges in order to harness the maximum benefit of education in empowering women in this conservative society. The findings of the study document shallow job market for educated women as the major hindrance in empowering women in the survey area. This is consistent with the national data where it is found that as much 77% of the female college graduates are unemployed in the country. The findings, thus, suggests appropriate policy action to streamline graduate skill with job market and creation of job opportunities for empowering women. The findings also show that a dysfunctional political system combined with election irregularities, use of money and muscles by the hooligans as the reasons why educated women are reluctant to participate in politics and achieve political empowerment. The findings also document poverty as the major cause why families often marry their girl child rather than educating them. They mentioned that many households are unable to bear the cost of education and finding helpless position they see marriage of their daughters as the easy escape. The findings also show that the dominant patriarchal culture and religious faith hinder female students' mobility outside home, their participation in co-education, and in extra-curricular activities including physical education, martial art, music, dance and art classes due to concern from safety and security. All these hinder girl student's curricular and extra-curricular attainment in this traditional society. The empirical findings, therefore, argue for necessary safety and security measures in order to facilitate girls' education in the country.

References

Ahmed, L. (1992). *Women and gender in Islam*. Yale University.

Aikman, S., & Unterhalter, E. (2005). The education MDGs: Achieving gender equality through curriculum and pedagogy change. *Gender and Development, 13*(1), 44–55.

Aktar, S. (2014). *Empowering women through direct election in reserved seats: A comparative study of rural and urban local government institutions in Bangladesh*. Flinders University.

Anderson, S., & Eswaran, M. (2009). What determines female autonomy? Evidence from Bangladesh. *Journal of Development Economics, 90*, 179–191.

Arnot, M. (2002). *Reproducing gender?: Essays on educational theory and feminist*. Rutledge.

Ayeen-O-Shalish Kendra (ASK). (2019). Fatowabaji berei choleche (in Bengali), in *Daily Prothom Alo*, November 24, 2019, Dhaka.

Bates, L., Schuler, S., Islam, F., & Islam, K. (2004). Socioeconomic factors and processes associated with domestic violence in rural Bangladesh. *International Family Planning Perspectives, 30*(4, Gender-based violence and reproductive health), 190–199.

BBS. (2020). *Statistical year book*. Bangladesh Bureau of Statistics.

Betts, J. (2003). Literacies and livelihood strategies: Experience from Usulután, El Salvador. *International Journal of Educational Development, 23*(3), 291–298.

CAPWIP. (2000). *Women in Legislative Chambers: Asia-Pacific Region, Centre for Asia-Pacific Women in Politics*. http://www.capwip.org/participation/womenchambers.html.

Chowdhury, F. D. (2009). Problems of women's participation in Bangladesh politics, The Round Table: The Commonwealth. *Journal of International Affairs, 98*(404), 555–567.

Daily Inqilab. (2019). https://www.dailyinqilab.com/article/197549/শশিতিবেকারে·াড়ছে. *Daily Inqilab*, 4 April 2019, Dhaka.
Daily Star. (2021). https://www.thedailystar.net/frontpage/news/democracy-index-bangladesh-moves-four-notches-2038753. *The daily Star*, 4 February, 2021, Dhaka.
Dhaka Tribune. (2020). https://www.dhakatribune.com/health/coronavirus/2020/08/12/covid-19-impact-national-poverty-rate-rises-to-29-5-as-of-june. *Dhaka Tribune*, 20 August 2020.
Freire, P. (1973). *Pedagogy of the Oppressed*. The Seabury Press.
Friedmann, J. (1992). *Empowerment: The politics of alternative development*. Black-well.
Goel, A. (2004). *Education and socio-economic perspectives of women development and empowerment*. The Flammer.
Halder, N. (2002). *Effective representation of women legislators in parliaments: A case study of New Zealand in theoretical and global perspectives*. Canterbury University.
Halder, N. (2004). Female representation in parliament: A case study from Bangladesh. *New Zealand Journal of Asian Studies, 6*, 27–63.
Hamid, S. (1996). *Why women count: Essays on women in development in Bangladesh*. University Press Limited.
Huque, A. S., & Akhter, M. Y. (1987). The ubiquity of Islam: Religion and society in Bangladesh. *Pacific Affairs, 60*(2), 200–225.
Kabeer, N. (1999). *Reversed realities: Gender hierarchies in development thought*. Verso.
Kabeer, N. (2001). *Reflections on the measurement of Women's empowerment, in discussing Women's empowerment-theory and practice* (SIDA studies no. 3). Novum Grafiska.
Kabeer, N. (2011). Between affiliation and autonomy: Navigating pathways of women's empowerment and gender justice in rural Bangladesh. *Development and Change, 42*(2), 499–528.
Kabir, A. H. M. (2002). Outlawing unauthorized fatwa through judicial activism. *Interventions, 4*(2), 224–227.
King, E. M., & Hill, A. (Eds.). (1993). *Women's education in developing countries: Barriers, benefits, and policies*. University Press Books (for the World Bank).
Leve, L. G. (2001). Between Jesse Helms and Ram Bahadur: Participation and empowerment in women's literacy programming in Nepal, *PoLAR. Political and Legal Anthropology Review, 24*(1), 108–128.
Lewis, M. A., & Lockheed, M. E. (2006). *Inexcusable absence: Why 60 million girls still aren't in school and what to do about it*. Center for Global Development.
Mahmud, S., Shah, N., & Becker, S. (2012). Measurement of women's empowerment in rural Bangladesh. *World Development, 40*(3), 610–619.
Mahtab, N. (2007). *Women in Bangladesh: From inequality to empowerment*. AH Development Publishing House.
Malhotra, A., Schuler, S., & Boender, C. (2002). *Measuring women's empowerment as a variable in international development*. Background Paper, World Bank Workshop on Poverty and Gender.
Maruatona, T., & Cervero, R. M. (2004). Adult literacy education in Botswana: Planning between reproduction and resistance. *Studies in the Education of Adults, 36*(2), 235–251.
Maslak, M. A., & Singhal, G. (2008). The identity of educated women in India: Confluence or divergence. *Gender and Education, 20*(5), 481–493.
Miske, W., Monkman, K., Miske, S., & Hydrie, Y. Y. (2007). *Transition to post-primary education with a special focus on girls: Medium-term strategies for developing post-primary education in Eastern and Southern Africa*. UNICEF Eastern & Southern Africa Region Office, Education Section.
Mitra, A., & Singh, P. (2007). Human capital attainment and gender empowerment: The Kerala paradox. *Social Science Quarterly, 88*(5), 1227–1242.
Moser, C. (1993). *Gender planning and development: Theory, practice, and training*. Routledge.
Moser, C. (1995). *Gender, planning and development: Theory, practice and training*. Routledge.
Nath, J. (1981). Beliefs and customs observed by Muslim rural women during their life cycle, in. In T. S. Epstein & R. A. Watts (Eds.), *The endless day* (pp. 13–28). Pergamon.

Naved, R., & Persson, L. (2005). Factors associated with spousal physical violence against women in Bangladesh. *Studies in Family Planning, 36*(4), 289–300.

Noreen, G., & Khalid, H. (2012). Gender Empowerment through women's higher education: Opportunities and possibilities. *Journal of Research and Reflections in Education, 6*(1), 50–60. http://www.ue.edu.pk/journal.asp

Nussbaum, M. (1999). Women and equality: The capabilities approach. *International Labour Review, 138*(3), 227–245.

Nussbaum, M. (2011). *Creating capabilities: The human development approach*. Chicago University Press.

Panday, P. (2008). *Women's political participation in Bangladesh: Institutional reforms and actors*. City University of Hong Kong.

Prothom Alo. (2014). https://www.prothomalo.com/bangladesh/এে◌ারই-সেছেছে-বৌ-নারীর-জে, 2 March 2014, *Daily Prothom Alo*, Dhaka.

Prothom Alo. (2019). https://www.prothomalo.com/bangladesh/উষ্চশিশিতেছর-প্রাে-অছধেকই-বেকার, 1 January 2019, *Daily Prothom Alo*, Dhaka.

Rashiduzzaman, M. (1997). The dichotomy of Islam and development: NGOs, women's development and fatwa in Bangladesh. *Contemporary South Asia, 6*(3), 239–246.

Rathgeber, E. M. (1990). WID, WAD, GAD: Trends in research and practice. *Journal of Developing Areas, 24*, 489–502.

Rogers, A. (2001). Afterword: Problematising literacy and development. In B. Street (Ed.), *Literacy and development: Ethnographic perspectives* (pp. 205–222). Routledge.

Rowlands, J. (1997). *Questioning empowerment: Working with women in Honduras*. Oxfam.

Sen, A. K. (1999). *Development As Freedom*. Oxford University Press.

Smith, B. G. (2008). *Oxford encyclopaedia of women in world history*. Oxford University Press.

Stacki, S. L., & Monkman, K. (2003). Change through empowerment processes: Women's stories from South Asia and Latin America. *Compare, 33*(2), 173–189.

Stromquist, N. (1995). The theoretical and practical bases for empowerment. In C. Medel-Añonuevo (Ed.), *Women, education, and empowerment: Pathways towards autonomy* (pp. 13–22). Hamburg.

UNESCO. (2020). *Global education monitoring report 2020: A new generation: 25 Years of efforts for gender equality in education*. UNESCO.

White, S. (2010). Domains of contestation: Women's empowerment and Islam in Bangladesh, *Women's Studies International Forum, 33*, 334–344.

World Bank. (2008). *Whispers to voices: Gender and social transformation in Bangladesh, South Asia sustainable development department*. The World Bank.

Gender Roles in Vietnam: A Metalogue on the Traditional and the New and Suggestions for Transformation

Huong Nguyen, Janet J. McIntyre-Mills, and Yvonne Corcoran-Nantes

Abstract The paper is organized as a virtual dialogue and iterative commentary using email and phone calls during lock down. The authors address the potential for public education to redress the gendered inequality towards women. Historically women have shown their capabilities to take on essential roles in times of danger, in order to help the country gain independence and to recover from war, to help with reconstruction and development but unequal gender norms persist that need to be addressed. The tensions across traditional folklore and Buddhism which emphasizes harmony across the genders and some of the patriarchal notions in Confucianism remain a challenge for gender identity and roles. The paper concludes by suggesting that public education needs to continue to challenge the more conservative approach of Confucianism towards the role of women and shift the traditional gendered social norms towards more equality and more opportunities for the advancement of human development goals and the prevention of harassment. Nevertheless, the first author stresses that since the end of the Vietnam war considerable progress has been made.

The methodology applied is qualitative using a semi-structured question interview and draws on secondary data and because of the limitations caused by Covid, most of the interviews were conducted on line. The information cited in this paper is a part of the first author's primary data collection for her PhD research project on "sexual harassment of working women". As part of the research, she aimed to understand how people perceive gender relations and gender roles of women and how traditionally gendered social norms are associated with gender inequality and sexual harassment.

H. Nguyen (✉) · Y. Corcoran-Nantes
Flinders University, Adelaide, Australia
e-mail: yvonne.corcoran-nantes@flinders.edu.au

J. J. McIntyre-Mills
College of Education, University of South Africa, Pretoria Central, Gauteng, South Africa

College of Professions, Adelaide University, Adelaide, Australia

J. J. McIntyre-Mills, Y. Corcoran-Nantes (eds.), *Transformative Education for Regeneration and Wellbeing*, Contemporary Systems Thinking,
https://doi.org/10.1007/978-981-19-3258-8_9

Keywords Regeneration · Respect · Values · Gendered social norms · Gender equality public education · Subordination

1 Introduction to Metalogue

Janet: I like the idea of our using some auto-ethnography (based on your lived experiences) as so-called 'warm data' in a metalogue – dialogue with a purpose – to draw out some of the themes in your research data. This could provide 'local, lived experience which is 'contextual, relational data', not static de-contextualized data. Here is what Nora Bateson says about her concept 'warm data':[1]

> The International Bateson Institute [IBI] exists to generate and give access to information that offers a wider vision. The focus of inquiry is on the inter-relational processes between and among systems. It can involve recognizing how patterns repeat and reflect each other among multiple contexts and across multiple systems–many of these systems' maintenance and renewal is critical in the coming decades…. We call this information "Warm Data".

This is a quote from chapter "Regenerative Education and Safe Habitats for Diverse Species: Caterpillar Dreaming Butterfly Being" in this volume:

> Nora Bateson created enthusiasm and energy when she discussed the potential for thinking differently by learning to see differently through developing an 'ecology of mind' [1] and *through side by side learning* in different contexts. … Her concept of 'warm data' [2] created in context resonated as did her notion of learning in transcontextual ways. The notion that telling stories can change the way people see their own stories resonated as did her notion of the importance of making sense of people's experiences in context …

Janet: From my reading of some draft chapters for your thesis and a reading of a published thesis on critical issues on gender mainstreaming by Binh Thi Le (2012, 120), although considerable progress has occurred, there is still room for improvement through education to support gender mainstreaming, the thesis concludes:

> The research revealed only limited and superficial communication within and between the organizations about the implementation of gender mainstreaming. This finding was found in the INGO as one of the challenges facing the process. In the organization, some people were active while others did not listen to and accept others' ideas for improvement. Questioning could be interpreted as a way of accusing, challenging, and looking down on someone. Therefore, people tended to be passive in these sharing sessions. The concept of 'losing face' might be another barrier preventing people from sharing more deeply with each other. With partnerships, the idea of getting all partners together to reach proper understanding of notions of gender equality did not occur.

---

[1] https://batesoninstitute.org/warm-data/
https://www.youtube.com/watch?v=jJ4Rdkm1iBA
https://norabateson.wordpress.com/2017/05/28/warm-data/
Nora Bateson Plenary, July 12th International Systems Sciences https://zoom.us/rec/play/_oe07ngyK1jHXqZxIaT5nBr_c95ReZ2KWOwXCrYw7u1K6rKfJRUPtUH_urEVNtlI9m1WDZ-bqZFxJ8k7i.po_IqnXsYX9swRJr?continueMode=true&_x_zm_rtaid=u3XXrLmPQCaoiZM3sR9LHA.1626085969999.920d3aa8d2c700ad3a46aaaca293852c&_x_zm_rhtaid=301

It seems that whilst progress has been made with gender mainstreaming, tensions exist across how gender roles are perceived. Whilst traditional ideas of strong women exist in agricultural folk lore and harmonious relations are stressed in Buddhism, some traditional patriarchal ideas remain, despite modern communist ideas about women's rights. Please would you elaborate on this? It seems that even in NGOs, INGOs and GOs there are still strong beliefs rooted in Confucianist notions that women ought to be carers, feminine and focused on their domestic roles and subservient to those who are above them in the patriarchal hierarchy?

These are some of the questions that emerged from reading your chapters and from the metalogue on some of the data:

- Please tell us about your in depth interviews, the design, the purposive sample and how these data inform your interpretations and grounded theory? I understand that you are researching the extent of harassment that still continues in the workplace of women working in offices and factories?
- How have the women's roles changed in the last few decades?
- Please draw from your personal experiences and testimonies, these can serve as auto ethnographic insights on the topic of gender roles and the need for changes in the attitudes towards women. This could help us understand the way women have taken responsibility and helped to ensure the survival of families from the war years onwards.
- Another question that springs to mind is: How could public education help to redress the imbalance in gender rights and roles? To what extent do you think the NGOs, INGOs and GO's can play a role?

Huong: Yes, I can draw on my experience and fieldwork.[2]

Janet: Yvonne, we would be glad if you would join the metalogue by drawing on your research in Vietnam?

Yvonne: My fieldwork would certainly complement some of the material here albeit that my research looks at the past and women's role in the resistance, the recalibration of gender relations throughout both the French and American wars in Vietnam and how far that impacted on women's lives in the post war period.

Huong: Yes, let us explore the material from my fieldwork and I can draw on my own experiences. The primary data were collected from in-depth interviews on both women and men aged 20–55 as a part of a PhD thesis on gender issues. So far I have interviewed thirty five participants, including five key informants of government agencies and NGOs representatives in 5 factories in six provinces in the North, South and Middle of Vietnam(Hanoi, Vinh Phuc, Hung Yen, Hai Phong, and Dong Nai, and Thanh Hoa). These research participants are working in a different industries including garment, electronic and footwear. Ten female officers are based in state agencies.

[2]At the time this paper was written the first author had completed 72 interviews, including 45 workers, 17 female officers, ten key informants and 350 responded to the online survey.

In my research interviews with the young men and women workers in factories, their response provides information about the change of women roles, for example:

> We are now different from our mothers' generation who had less education opportunity and mostly work at home as farmers and looked after the family. We can go to work in a factory and have more autonomy in our life. I think people also do not strictly ask young people to follow tradition custom like a daughter in law needing to stay in the parent-in-laws house to take care of them (28-year-old female worker).

> In my family, I am equal with my husband. He helps me do housework, gives me his salary to cover the family's cost, and makes decisions for the family. Of course, there is some traditional social conduct that I follow. For example, I am in charge of all the inside duties and responsibilities (kinship activities, greetings, visiting or taking care of family relatives and look after children) and my husband deals with the outside activities (32-year-old female worker).

Public attitudes to women's role and contributions have changed positively due to the improvement of gender equality policy and education and the proven contribution of women and their participation in all spheres of life (Institute for Social Development Studies, 2020).

Fifty to sixty years ago, the view of women as devalued, incapable, and submissive was prevalent in my grandmother and mother's generation influenced by feudal Confucian gender ideology. I heard that my father was born after much sacrifice and became the principal hope of my grandparent after my grandmother gave birth to three daughters. He saved my grandfather's marriage because my great grandparents wanted to have a grandson to maintain their father's kinship.

My parents have three children, including me and my two younger brothers, so my mom did not come under pressure trying for a son. In fact, my father always favoured me as I am the only daughter in the family and as such, he did not follow Confucianist values. Instead, my brothers and I were encouraged to go to school and share the housework at home. Nonetheless, we were all influenced by the idea of what constitutes a 'good woman' and how to be a 'true man'. For example, girls/women should be gentle and charming and more emotional, while boys should be strong, brave, and not cry like girls.

Some of my peers dropped out of school when they were in their final high school year and soon married. The response from their parents was "it was not necessary to have higher education for women; who will marry, so it's better to do it earlier".

When my daughter was born, many relatives, colleagues and friends visited me to congratulate me on the new born baby; some innocently asked me saying: "Good to have the first child as a daughter, but … try to have a son". Sometimes on my business trips or visiting my hometown, people would ask, "when do you plan to have a boy?" when they heard that I have two daughters. This illustration describes how traditional gendered social norms and expectations are significant in Vietnamese people's mindsets and life.

Yvonne: This reminds me of one of my research trips to Vietnam. I was in Hue and invited to my friend's village celebration of 50 years and I got to meet the rest of the family. She was one of two girls; her parents had no

> boys. In the ancestral family house there was a family tree mapped out on the wall her father's line stopped with him – after the birth of his two daughters. While his brother's family lines continued through their children because they had boys. Ironically, my friend has two boys so her husband's family line will continue through her children.

The popularity of a son remains in Vietnamese society, both in the urban and rural areas (Khuat, 2004; Bélanger & Khuat, 2018; Institute for Social Development Studies, 2020; Phạm et al., 2021). Men were believed to be strong and decisive in handling heavy and complex tasks, capable of important positions, maintaining a kinship system and taking care of their ancestors. In my village, most of the women in my grandparents and mother's generations worked at home and on farms, and those who had a job were men, either in a factory, a state agency or they served in the Vietnamese military. Consequently, older people who have retired are mostly male and their wives depend on the husband's income. Even in contemporary society, women in lower-paid industries with unequal pay for equal work and lower positions are compared to men because of gender discrimination toward women. Most of the management positions in all workplaces at all levels are occupied by men.

> In my factory, most of the managers are men. I don't know, but in other factories, managers are also mostly men. Maybe men are more knowledgeable, stronger and have more time spending on work than women because we have to take care family (36-year-old female worker – research interview).

Traditional Confucian social norms continue to influence public attitudes toward women as a subject of compliance and blamed for any mistakes (Khuat, 2004, 2016). Both men and women hold that perception (Institute for Social Development Studies, 2020). When a woman has a family problem (domestic violence or marriage's dispute), the common advice from family or friends could be "Being a woman, better to give in to keep warm in a family" or "don't behave like a woman" is a warning to a boy or a man if they behave differently with other men. The success of women is evaluated by their performance at home and their family good management regardless their social success positions (Hoang, 2020; Khuat, 2004).

My interview with workers at different ages also confirm this attitude toward women:

> Because we are women, sometimes I feel like my opinion is not worth hearing. – 26-year-old female worker
>
> They would not believe it if I said I was harassed at work by a male manager because they think that women are talkative and troublemakers (35-year-old female officer).

A study of contemporary Vietnamese youth through discussions in the press and on social networks conducted in 2020 points to that the norms of women characteristics as "gentle," soft", "tender", "sympathetic", "calm", "persistent", "subtle", "sensitive" as strong points of women that contribute to their success (Phạm et al., 2021). These norms were drawn from t traditional Confucianism. Many interviews with successful women in the press emphasize the gender identity of women that has made them successful. Women's characteristics influence the "work assignment" for them, although for the purposes of "equality", but it turns to inequality in employment opportunities and promotion (Phạm et al., 2021, 46).

> Women are sensitive and soft, so when … [I] feel …. harassed, I often find the way to reject men politely to not cause the relationship broken but avoid unwanted touch. We have our own weapon as a woman to be gentle and smartly deal such situations (47 year old female officer).

> When my peer was promoted to work in the factory office instead of in the production line, people may say: she is beautiful, charming and gentle, so she fit with that job (30 year old female worker).

> If a woman is in leadership position, there will be gossip that she clam up this chair because of her beauty. A former female minister made significant contribution to the industry she led, but she was often adjusted by her appearance and wearing (a local NGO director).

The changing social, cultural and economic development with a significant increase in women's education, employment and economic growth have resulted in some changes in the public attitude toward women and there is now more acceptance of women's rights and equality (Institute for Social Development Studies, 2020; Goverment, 2015).

In contemporary Vietnamese society, women continue to believe in the importance of building a happy family and their economic roles (Institute for Social Development Studies, 2020; Khuat, 2016). Women make up a high percentage of workers in agriculture and in export industries like textile, garment manufacturing, and food processing, As a result of their economic empowerment, women are more confident at home and work. However, Vietnamese women are at the crossroads of balancing traditional ideas and modern ideas about their position and roles, for example, according to one interviewee:

> My husband and I are all working, so we need to share responsibility at home. I think women now improve their knowledge and understanding about themselves and become more independent and equal with all members in the family [34-year-old female worker].

> Of course, my husband share the domestic duties with me. He takes my kids to school and pick them up. I will do cooking for my family. The most challenging for women (like me) when going for business trips is family matter. My husband cannot take care children well like me (42 year old female officer)

Janet: So it is worth noting that the variety of workplaces and geographical contexts were selected to provide a sense of the extent to which context (being more rural or regionally based versus urban based) could play a role in shaping perceptions on gender roles?

Huong: Vietnam's natural and ecological conditions are considered the basis for the two characteristics of traditional Vietnamese culture and living life: rivers and plants. The culture of community organization (collective life and personal life) and the culture of harmonized behaviour with the natural environment are core elements that construct the gender roles and perception of equality in Vietnam (Trần, 2001).

So by inviting women and men in the research, I wanted to understand the different perceptions of men and women in viewing women gender roles. They are at their prime working age, but they could carry over the traditional socio-cultural values on gender and the value of the environment and agriculture. Now they have opportunities to change as Vietnam responds to the challenge to develop in ways that re-generate and sustain a balanced way of life. My interviews were conducted via phone due to travel restrictions during the Covid pandemic. To recruit participants, I invited a research assistant who works in a local NGO and has a strong network with workers to send invitations to workers.

Janet: Please tell us a bit more about your approach?

Huong: I also did a review of the literature that demonstrates the change in women's roles in different periods in Vietnam history. A woman's role is traditionally as primary family caregivers and domestic duties. The family becomes a central value of women's life and thus responsibility for maintaining family happiness is vested in mothers and wives. Women roles are mostly domestically oriented, serving their husbands, bearing and rearing children, taking care of the housework and looking after parents-in-law in their old age (Khuat, 2004, 2016; Institute for Social Development Studies, 2020; Hoang, 2020). These gender roles of women still firmly remain in the mindset of Vietnamese across all social strata. Notably, women deeply internalize the gendered social value and are willing to compromise their wellbeing, advancement and interests (Institute for Social Development Studies, 2015)

Women played a critical role in the Vietnam Liberation Revolution against the French in 1945 to gain independence and liberate women from feudal Confucianism. In the American war, the role of Vietnamese women emphasized warfare during the national defence. The social expectation for women included both the traditional responsibility of a mother, collective agriculture, and warfare (Hoang, 2020; Le & Mai, 1978; Le, 2002).

My literature review demonstrates the change of women roles in different periods in Vietnam history. In the American war, the role of Vietnamese women emphasized warfare during the national defence. The social expectation for women's role included both the traditional responsibility of a mother, collective agriculture, and warfare. Such as, the roles of women displayed beyond the traditional norms of domestic roles inside the family to be significant for the nation's interest of independence and reunion.

Yvonne: During the war all norms were essentially suspended in favour of a more egalitarian interdependent set of gender relations which underpinned life in the resistance. After the war it was business as usual as far as gender relations were concerned for those in the resistance and for those who maintained their lives on farms, in factories and in the public service. The Confucianist norm of the 'Three criteria' which underpins gender relations and gender norms up to the present time. Even the

Vietnam Women's Union underpins its socio-political agenda and commitment to change with a continued adherence to the three criteria which reinforce the traditional domestic role of women and re-criteria women with exceed roles and pressure at home and at work (Schuler et al., 2006)

Huong: The perception of a 'good woman' with key characters mentioned above demonstrates clearly and in new expression form of self-sacrifice for national defence. Women participated in the country resistance was the evidence to strengthen matriarchal values that originated from indigenous customs. When women participated in the American war, their leadership role was equal to that of men and it proved that women are as strong.

This related to the matriarchal value of how women are determined and strong to lead community development. There was a time when Vietnamese women proudly implemented Vietnam's Women Union campaigns which encouraged women to develop their capabilities optimally, taking men's roles at home, at work, and on the frontline when most Vietnamese men were in the American war. At the time, Vietnamese female heroes and patriotism were significantly portrayed in the country history. Women played a critical role as a backup force of the Vietnam military for the national defence. The image of Vietnamese women was portrayed as the nation's pride. Women excelled themselves successfully undertaking the multiple role of a mother, a farmer to produce food, a fighter to protect the village and youth volunteers in the frontier battles.

In Vietnam's Doi Moi period, the women's role shifted back to building a happy family and contributing to the country's development in a new manner of women's household economic contributors to reduce poverty and childhood malnutrition. Although this shift may reinforce the traditional role of women assigned with family and domestic work and the notion of womanhood, it also recognizes women's capacity and contribution to the country in the intensive rebuilding after the American war. For example, in 1989, the Vietnamese General Confederation of Labour launched the movement of "being good at public work, doing well at home housework". It meant that the movement officially reinforced the domestic and family roles of women but still required women to be accomplished in the workplace. Which for women meant a double or triple burden.

When the war ended, and men no longer had to go to the battlefield, thus the wartime gender complementarity and nationalist work ethic was no longer appropriate (Hoang, 2020).

In industrialization, modernization, and contemporary society, women's roles have significantly changed to be more equal and advanced. The main roles of family caregivers and domestic work shift to more educated people who meet the demand for economic integration. They are creators of a new modern family while maintaining their traditional culture and holding an important economic role in the public service and in many industries. Moreover, women are increasingly participating in the labour market, in politics and they assume leadership roles in business.

Vietnam recently had the first-ever female acting as President of Vietnam and the Chairwoman of the National Assembly of Vietnam in 2016–2021. Vietnam has one of the highest female labour-force participation rates. As I read the ILO gender report, 79% of women aged 15–64 are in the labour force, compared with 86% of men in 2019 (Barcucci et al., 2021). The National Assembly election 2021 recognized the remarkable increase in women politics, which result in 3026 female delegates. Vietnam has the second-highest rate of senior female leaders in Asia, with about 36%.

Janet: So, in fact the gap between gender roles is narrowing.

Huong: Yes, that is so, but the gender gaps remains at the leadership level, because women often take a lower position than men in senior roles and they often take the role of deputy instead of director.

Yvonne: Women irrespective of their role are expected to take a secondary role in leadership. Discrimination against women continues and they are expected to conform to socio-cultural norms irrespective of how powerful they may be or whether they have an important leadership role in the workplace or community.

Also, Vietnamese women carry a disproportionate double burden, and they face multiple and persistent inequalities at the work regarding retirement age, equal pay for equal work. The pressure of "being good at public work, doing well at home housework" makes modern women more responsible than before. It cannot be denied that Confucianism powerfully manipulates gendered social norms of women's role conservatively in the past and still has influence today. In my interviews with many respondents they said that life has changed, and women are now more equal to men and that "it is our traditional culture". For example, a man of 31 said that "of course, I am a man, a husband, so I need to make important decisions". Or "Because they are men and are managers, we have to listen to them". Even some women said that they are equal with their husband, they still think that "my main responsibility at home is to take care family and children. I am in charge of sorting out internal duties like looking after family relative relationships". A simple and common answer from women I interview were "because we are women". This means that when defining as women, they are influenced by normative ideas and role models as to what constitutes femininity.

Under the influence of Confucianism gendered social norms and expectations underline the importance of women's role as primary family caregivers.

From my family story, my parents sometimes shared their life stories with us and with their grandchildren in my family. I understand it was the most difficult time in their life like other Vietnamese at period of time. Because Vietnam just came out of the war and there was considerable damage and much pain.

For instance, my grandmother worked hard to take care of her five children when my grandfather was a way for his business in a hospital. She usually stayed up late until midnight to finish all chores and woke up at 4.30 am to prepare for an early working day. My mother, like other women took care of the children and parents-in-law when their husband was in the army. My mother shared how she was very poor and never had enough food for serving the family. Still, none of them complained about the extremely difficult living conditions and being away from their husbands to manage all family independently.

It seemed that it was a voluntary sacrifice for a family which is typical of Vietnamese womanhood. However, when I grew up, I realized that it was also the way women comply with social principle of morals that meets with the belief that obedience to men was central to being a good woman. Anyone who behaved against such social conduct and norms would be reprimanded strictly by family and community, particularly during wartime. "Sacrifice" can be a good word to summarise and describe Vietnamese women's key characters and what society expects from them. This is because the way they work hard and compromise their interest and personal desire can be understood as their 'sacrifice".

Yes, I also grew up with that notion of being a good girl and good woman to sacrifice myself for my beloved family voluntarily. When I was small, the meaning of sacrifice was not clear for me, but when I got married and had children, I realized that I could compromise my own interest for my children if I passed on traditional gendered roles to the next generation.

Women like my mother also worked hard as a worker in the factory or farmer in the rice paddy and took care family with shortage food. However, they were very brave and strong to overcome all the difficulties and the way they accepted it with calm minds and great effort.

Janet:

- How have the attitudes towards women changed after the Vietnam war? Have you thought about that? Strong, bold women are evident in Vietnam's history in re-generating village life and agriculture in Vietnam?
- Can you recall stories told by your family? How have attitudes to women changed since the war? From reading your field notes on your current research, it appears that women are only referring to their current roles. Have they forgotten the roles they played in history and during the Vietnam war? You mentioned the way traditionally women were seen as strong farmers and are respected in traditional folklore and Buddhist philosophy?

Huong: My father was a soldier of 30 years serving the Vietnamese military. He also talked about his brave female peers whom he met on the way to battles. Their stories were like a reminder or memory of the past of which they proud. These stories tell of how people and women overcame challenges together

In my conversations with eight workers for my thesis, they mainly focused on their current life and work, so we did not mention Vietnamese women's roles in history and during the Vietnam war. However, it did not appear that they had forgotten the role of women in the past. In fact, they also referred to the traditional role of their mothers or mother's in law who experienced more difficulties than their generation. They were successful in managing their family and raised kids to have a better future.

When I think of their responses it also raises the question as to how social and cultural education programs could help to inspire the younger generation to draw on the good traditional culture of women that respect women as capable contributors taking difficult duties as equal as men, but also to challenge the harmful cultural norms toward gender in equality.

The role of women in Vietnam has been subject to many changes throughout Vietnam's history (Le & Mai, 1978; Turley, 1972; Khuat, 2004; Rydström, 1998). They have taken on varying roles in society, including mothers and wives, warriors, farmers and workers in different industries and businesses and as activists. And thus, public attitude toward women was not only seeing women in the "inside the kitchen" but taking many roles in national and public spheres.

The traditional respect for women highlighted by Buddhism and harnessed by the Communist gender regime to liberate women helped to empower women. This spirit is vital for regenerative education and for developing opportunities for women in the `new economy that encourage the balance and equal of men and women as well as the harmony with nature acknowledges women's monetary values as equal as men.

In my interview with a key informant, a director of a local NGO, it was fascinating when she talked about the need for more social and environmental harmonization. She argued that things became worse because of imbalances in status of men and women and imbalances with nature.

Janet: So this is interesting, could public education stress the need for rethinking the need for social, economic and environmental harmony and how people *work with* nature?

Huong: Yes, it is critical given the harmful treatment of nature and the severe consequences. Disasters now threaten our current and future way of life. The director of a local NGO stressed:

> ….The way we deal with nature is also not harmonized. We launched the movement like clearing wild forests, digging mountains and filling the sea. We considered that we [could win over] the nature. However, we are receiving the consequence from these initiatives. Why don't we live in harmony with nature? Thus we can sustain our lives for generations? ….[we need] harmony and not dominance …..

Janet: Please tell me more about what you mean by the traditional role of women, Confucianism and Buddhism?

Huong: A good woman was supposed to be hardworking, devoted to family, personal sacrifice and endurance. This creates the perception that "family is the priority of women life". In addition, women were traditionally also farmers supporting an agricultural economy in an underdeveloped country with limited employment options due to endless war. Women continue to play an essential role in agriculture- a key economic industry of Vietnam and count for 70% of the labour force in this area currently and in other industry, but their monetary values is not equally recognized as men's (Institute for Social Development Studies, 2020).

Let me refer you to Box 1 on the traditional role of women, Confucianism and Buddhism to inform our discussion. Buddhism entered to Vietnam around the 2nd century AD and quickly integrated with local culture to become a profound influence religion of the Vietnamese spiritual life The process of Buddhism localization to Vietnam occurred in a good match of Vietnamese agriculture cultivation characters of flexibility, soft behaviour, tolerance and harmony with Buddhism philosophy of nonviolence teachings and respect for women. The Buddhism interconnected with the matriarchal family the traditional values of the woman is considered important to uphold the role of women in both society and family.

Women play an important role in the Vietnamese economy, particularly in the main economic sector of agriculture and the emerging industry of Vietnam. The concept of harmony with not only social relations but also with nature should be emphasized in the contemporary cultural education. The current climate change impact in Vietnam provides profound evidence of overused natural resources and exploitative threats on nature resulting in an unbalanced ecosystem and nature disasters. It is important for women and men taking the role as "the first teachers in a family" to educate a new generation to be in harmony with nature and shift their practice to be environmentally friendly.

Yvonne: It is worthy of note that pathways to empowerment have also been sought within agricultural regions the obstacles however are illustrated well in Donna Donae and Phillipe Doney's article on the Greater Mekong Sub region (2015)

Huong: My personal experience could explain how Confucianism and mixing Buddhism shapes the traditional role of women. At the same time, it is important to acknowledge that it was a critical change in myself to go beyond the traditional boundary of gender roles to define my own way of education, life, and employment.

Such change is essentially a result of education. I mean both family education and other variety learning channels. From the school early education, family education, and culture practice in a rural area, I was growing up with the notion that: family is the most important for every woman as the 'nature function of childhood, womanhood and motherhood". This is not an exception for anyone in my generation, previous generations, and later generations. Because as a girl, I was educated that I need to keep my dignity by learning the "four virtues' of Cong, Dung, Ngon, Hanh. These virtues mean women are compelled to comply a rigid fastidious propriety in behaviours, gesture, language and facial expression and meet the particular qualities of frankness, modesty, politeness and fidelity (Le & Mai, 1978, 43). As a girl, I was told that I must learn to be charming, gentle, tolerant, sympathetic, and harmonious. I also learned how to take care of family and to do domestic work. Listening to father and grandfather, respecting older people at all levels of kinship relations were a requirement at high moral standards as influenced by Confucianism (See Bonnin & Turner, 2014).

Joining the community activity as a culture of collective community spirit was typically robust in rural areas. I remember that the pathway of growing, getting married and having a family was considered a natural destination and desire of any girl like me. Even though family importance is still stable in my mindset and any decision I make daily, having a family and caring for family members are central to my happiness.

Yes, I learned women's roles from school education and, most importantly, family education and cultural observation. At the same time, the changing of gender roles and education has happened within family. My extended family, including my grandparents and parents, always encourage every child regardless of whether they are boys or girls, go to school, and further education for a better future, this reflects the changes away from traditional values. Although my grandparents and my mothers were assigned in traditional roles of domestic duties, they had an open-mind about women's advancement and about the need for gender roles to change.

As a Vietnamese woman who was born during peacetime, I have benefited very much from Vietnam's significant achievement of gender equality and women's advancement. Living in the countryside, studying and working in the capital of Vietnam, having opportunities to study abroad in some developed countries and returning to work in one of the oldest women's organizations, Vietnam's Women Union. Later in an International Organization, my pathway can be served as a small piece in a big picture reflecting how women's role has changed in the last few decades.

My work at Vietnam's Women Union allowed me to understand better-changing women's roles from the domestic sphere to the international sphere, from farmers to workers, engineers, doctors, teachers, and politicians and businesspeople. The changing role of women reflects the significant change of Vietnam in economic and gender equality coupled with an increase in public understanding about women. I met with different female heroes in each position they were and the situation they faced and overcame. For example, they were female drivers of military trucks in the Vietnamese forest. This was a fierce war road to supply and deliver food and weapons to the frontlines. Women were great farmers who took the initiative and were responsible for improving rice production and food, supporting the military for Vietnam's Victory and Reunion 1975.

Janet: So, once again you are showing how the gap between genders is narrowing for some women.

Huong: Confucian patriarchy significantly continues to influence Vietnamese gendered social norms. The way Confucianism unequally structures men and women relationships creates a power imbalance that leads to violence.

In Vietnam domestic violence is considered a family matter and fear of violence and adjustment is often strong enough to prevent women from challenging their traditional roles or leaving their subordinated position. Patriarchal hierarchy and gender inequality are interconnected in defining gender roles. A woman's identity was entwined with that of her family, and her benchmarked social standing is first evaluated by her performance in the domestic sphere (Hoang, 2020, 289). This is a part of the long-lasting influence of Confucianism on Vietnamese traditional culture and gender norms toward both women and men.

Janet: So this emphasis on home and domestic duties is surely counterproductive for women who want to have a more public role?

Huong: As mentioned in more detail in my thesis, some social campaigns organized by Vietnamese Women's Union (VWU), Trade Union and others contribute to traditionalized women's roles by highlighting and emphasizing women's family duties. But Vietnam's gender policies, promote gender equality as changing attitudes, which is very important as patriarchal power norms are still pervasive and can lead to the overuse of violence at home and the exposure of violence of young boy may lead to violent behaviour in later life.

Violence against women persists in both the domestic and public spheres and sexual harassment arises in the workplace as a consequence of gender inequality. Violence may be used to maintain the social order in a family and reinforce men's power in society. However, honouring women as a value of Buddhism together with the traditional matriarchal roles can be emphasized in transformative education programs to help to balance gender roles and to redress gender based violence.

Janet: So you are saying that the role of women as primarily care givers is emphasized even public organisations that are ostensibly trying to promote equality?

Huong: Yes, this is still a challenge.

Janet: You mentioned the other day that women are playing a strong leadership role now during the pandemic? Clearly care giving would fit this norm, but it also provides an opportunity to stress the role women are showing as strong leaders?

Huong: Yes, indeed. National media have covered the pictures of women as contributing significantly to the success of Vietnam Covid management policy by practicing strict prevention measure in various forms of lockdowns and social distancing (targeted lockdowns, travel bans, business closures, mass quarantines, and widespread testing). Women as the majority workers in various industries are willing to stay and work at the factories to contribute to the government dual targets of economic development and Covid prevention.

Female doctors and nurses count for high percentage of health workers are working as the frontier force in the fight to Covid, which is elevating the status and respect of women. Mass media has covered many stories of doctors who leave their little kids at home for months to stay in hospitals to carry their duties. Local women cook good food and send to hospital for doctors and health workers to keep their good health for long shifts under pressure.

Women participate and contribute in-kind support by donating money, food, labour to the people in lockdown areas. The cultural expression is Vietnamese women are like trees: "the good leaves protect the worn-out leaves".

Janet: To what extent can leadership in other spheres be more strongly emphasized in public education? In many parts of the world the service workers (teachers, nurses, doctors, shop keepers, food producers and cleaners, for example) have played a vital role.

Huong: Yes, in this grey picture of Covid, the Vietnamese people, particularly women clearly demonstrate their leadership roles in overcoming tremendous difficulties. It is also a time when traditional cultural values of community collective sprit, solidarity, social conformation and women's kindness is on display and the mass media has raised the profile of women.

Vietnam, has experienced 4 waves.[3] COVID- 19 continues to affect life and livelihoods on a global scale. In Vietnam by the end of 2020, 30.8 million out of 51.8 million employed workers were impacted by epidemic, of which 2.4 million workers lost their jobs, 17.6 million suffered income loss, which amount for 57.3% of total affected people (Nguyen et al., 2020). It also doubled the demand for shelters for victims of domestic violence.[4] The regressive effect of Covid during the current 4th waves of Covid and has not yet been documented. Vietnamese national media currently reports the exodus of millions of migrant workers from southern regions back to northern and middle regions of Vietnam to find a safe place in their hometown. Many of them have family of wife or husband and little kids, even new-borns returning by any means of personal transportation, mostly motorbikes, bicycles or on foot for the journeys of up to 2000 km.

Many women are community volunteers to work in various duties to help migrant workers on the way they return home. Current media have reported the emotional stories of two women who used their own savings (of 100 million Vietnam Dong each) who stood for long days in the national highway to give money and food to workers on the way they return to home. They did it for a simple happiness to see that many workers would be safe and have some support on the way to go home at a very difficult situation. Although women are more vulnerable they still show responsibility to care for the community.

Yvonne: There are many examples of women's leadership and duty of care activities within their communities. In various urban centres women have formed community kitchens putting out hundreds of meals per day to feed the vulnerable and those who are unable to work during the global pandemic. They do all of this through contributions from local businesses and market farmers – it is incredibly impressive.

Huong: These days, it is very common images of people help each other in several places across Vietnam during the pandemic. The "Zero Vietnamese currency markets – Siêu Thị 0 đồng) and rice ATMs[5] have been established in many locations to support poor people and people who lose incomes. This affirms that traditional culture is great social capital when

---

[3] Until early May 2021, Vietnam was among the countries with lowest number of reporting cases and deaths and Vietnam's management of the COVID-19 pandemic was highly effective. However, the 4th wave of Covid with new type of Dental has vanished the achievement of Vietnam in managing the pandemic.

[4] https://www.spotlightinitiative.org/fr/node/22788

[5] https://tuoitre.vn/phong-bi-500-000-dong-va-cua-hang-0-dong-tiep-suc-nguoi-tu-nam-ve-que-20210731154202726.htm

it is practiced without stigma or bias to gender roles, and women are always great supporters to the country and community.

Janet: How are women viewed if they emphasize their strong historical role as food producers and community leaders?

Huong: Throughout history, women have been proven to do all the work that men do. In fact, they have been in the significant role of food producers before, during and after the war. That roles are undeniable. As I said previously, women count for 70% of the agriculture labour force and a high proportion in Vietnam exported industry. Having emphasized women's roles as food producers and activists illustrates that women are the key factor of economic development and agents of change. In fact, they have brought the changes for whole society, including family building, country defence and reconstruction and development. Without women contribution, Vietnam's history would have been written differently than it is today.

Women roles have been praised by society as mothers and wives and their contribution to the country defence, reconstruction and development, however, women are not fully recognized for their monetary value even they earn equal or better than men. This is because their value is more often assessed in family welfares than their outsider performance and the notion that men are a breadwinner.

Janet: By reflecting on the role of women in the regeneration of society after the war and the devastation, what can you tell me about this aspect of women's identity and history?

Huong: From my side, I would say, the country's history would have been written in a different way without the success and peace that we have today. Coming out from the war, the Vietnamese proved their role as not only farmers, workers and soldiers, but also in their new role in reconstructing the country. Women's liberation was an important part of the Vietnam 1945 revolution and the country victory and reunion 1975. They have mobilized for the economic construction of the nation as an important half of the country's labour force. The role of women in the family remains significant as the Vietnamese Government also identified the critical goal of building family as an important cell of society. This is illustrated through the Vietnamese Law of Family and marriage 1986, 2001 and 2014 which continue to emphasis the importance of building happiness family (Hoang, 2020).

Janet: So public education for transformation seems to be making a difference as you have mentioned many examples and cited references to the fact that such traditional patriarchal norms are not so pervasive and there are elements of more equality and some women feeling that they are treated more or less equally. What role can trade unions, public sector and other organizations play in educating gender equality through challenging gendered social norms to help women achieve gender equality?

Huong: Trade Unions, VWU, and public sectors play essential roles in gender education to challenge social norms and shift the focus of traditional Confucianism toward more equality and women advancement. Moreover, with their political, social resources and authority, these organizations have tremendous power to influence and shape public understanding and gender ideology to be equal and equitable through policy advocacy and public discourses.

Janet: To summarise how can changes be brought about to re-balance the gender roles so that they are more harmonious?

Huong: Education policy needs to advocate for a focus on challenging gendered social norms that assign women as the primary family caregivers. It is essential to change public understanding and attitudes and strategies need to enable women to hold key leadership positions, to have equal participation in employment, education, and health while reducing gender-based violence Advocacy needs to be supported by implementation to understand and acknowledge women's economic roles and contributions. Vietnam needs to have critical changes in family policy and programs that do not emphasise and re-rationalize women's double burden. The education program, policy and intervention could target men and challenge masculinity to influence men to be more sensitive to women's rights.

Anti-discrimination legal frameworks are necessary to limit institutional and individual gender discrimination,[6] through gender mainstreaming and as suggested from research scholars (Nguyen & Simkin, 2017, 609).

Companies and workplace policies should be more gender-sensitive and aware of women's rights.

---

[6] There are some critical laws on gender equality:

Gender law to legalize the importance of gender equality mainstream in Vietnamese Law and Policy development (National Assembly of Vietnam, 2006).

Law on promulgation of legal documents 2015 amended in 2020 requests Regulatory Gender Impact (National Assembly of Vietnam, 2020). Gender impact assessment is compulsory of any new law formulation. This requirement helps to boost gender mainstreaming in the policy.

Gender impact assessment brought significant gender mainstreaming in the Labor Code revision 2019 (MOLISA and UN Women, 2019). It aimed to:

Narrow and gradually eliminate the retirement age gap between male and female employees.

Ensure and promote gender equality and non-discrimination based on gender regarding opportunities, conditions, and capacity to exercise rights and enjoy benefits at work in performing maternity and child-rearing functions in accordance with gender characteristics.

Improve effective and reasonable responsibility-sharing mechanism between the State and employers in organising day care facilities and kindergartens and in assisting employees with sending their children to day care facilities and kindergartens.[3]

Improving the legal mechanism to prevent and address sexual harassment in the workplace guarantees a safe and sound working environment (MOLISA and UN Women, 2019). See The Vietnamese Labour Code 2019 which enacted provisions applicable to female employee and assurance of gender equality (National Assembly of Vietnam, 2019) and The Vietnamese Law on Domestic Violence Control and Prevention 2007 (National Assembly of Vietnam, 2007; Actionaid International Vietnam 2014; UNFPA 2016).

Janet: Thank you for telling us your story and reflecting on your work in progress towards your thesis.

Yvonne: Yes it has been a very interesting exercise.

Box 1: Gender Construction: Historical Context
There is clear evidence of matriarchal life reflected in Vietnamese history that portrayed women as veritable giants and more capable than men in domestic and public spheres (Le & Mai, 1978, 12). Vietnamese notions of femininity were early illustrated through the legendary Vietnamese myth of uniqueness as Mother Au Co-Mẹ Âu Cơ (Chiricosta, 2010). Au Co is the most popular of the folk tales. It is a fairy story that has been handed down for the generations about Vietnamese origins and women's leadership roles.

References

Actionaid International Vietnam. (2014). *Policy brief: For women and girlds: Can dreams come true*. Hanoi.

Barcucci, V., Cole, W., & Rosina, G. (2021). Research brief: Gender and labor market in Vietnam. *ILO Vietnam*. https://doi.org/10.1080/10556699.1995.10603080

Bélanger, D., & Khuat, T. H. (2018). Too late to marry: Failure, fate or fortune? Female singlehood in rural North Vietnam. In J. Werner & D. Bélange (Eds.), *Gender, Household, State Doi Moi in Viet Nam* (pp. 89–110). Cornell. https://doi.org/10.7591/9781501719455

Binh, T. L. (2012). *Critical insights into gender mainstreaming in non-governmental organizations based in Vietnam*. A Thesis Submitted to the Victoria University of Wellington in Partial Fulfilment of the Requirements for the Degree of Masters in Development Studies, Victoria University of Wellington.

Bonnin, C., & Turner, S. (2014). *A good wife stays home: Gendered negotiations over state agricultural programmes, upland Vietnam, Gender, Place & Culture, 21*(10), 1302–1320.

Chiricosta, A. (2010). Following the trail of the fairy-bird: The search for a uniquely Vietnamese women's movement. In M. Roces & L. Edwards (Eds.), *Movements in Asia: Feminisms and transnational activism* (pp. 124–143). Routledge.

Goverment, Vietnam. (2015). *CEDAW Report*. https://tbinternet.ohchr.org/_layouts/15/treatybodyexternal/Download.aspx?symbolno=CEDAW%2FC%2FVNM%2F7-8&Lang=en

Hoang, L. A. (2020). The Vietnam Women's Union and the Contradictions of a Socialist Gender Regime. *Asian Studies Review, 44*(2), 297–314. https://doi.org/10.1080/10357823.2019.1699023

Institute for Social Development Studies. (2015). *Social determinants of gender inequality in Vietnam* (pp. 1–161). http://vietnam.embassy.gov.au/files/hnoi/ISDS_Report_Binhdanggioi_EN_PDF-2.pdf

Institute for Social Development Studies. (2020). *Men and masculinities in a globalising Viet Nam*. Hanoi.

Khuat, T. H. (2004). In L. Drummond & H. Rydstrom (Eds.), *Gender practices in contemporary Vietnam*. Singapore University Express.

Khuat, T. H. (2016). Women and development in Vietnam: Caught between social tradition and economic globalization. *Regions and Cohesion, 6*(2), 110–119. https://doi.org/10.3167/reco.2016.060207

Le, T. N. T. (Ed.). (2002). *Images of the Vietnamese Woman in the New Millennium* (Thi Nham Tuyet Le, Ed.). Gioi.

Le, T. N. T., & Mai, T. T. (1978). *Women in Vietnam*. Foreign Language Publising House.

MOLISA, and UN Women. (2019). *Policy brief: Gender equality at work*. Hanoi.

National Assembly of Vietnam. (2006). *The Vietnamese Gender Law*. National Assembly of Vietnam. http://www.chinhphu.vn/portal/page/portal/chinhphu/hethongvanban?class_id=1&mode=detail&document_id=28975

National Assembly of Vietnam. (2007). *4. The Vietnamese law on domestic violence control and prevention*. National Assembly of Vietnam. https://thuvienphapluat.vn/van-ban/Van-hoa-Xa-hoi/Luat-phong-chong-bao-luc-gia-dinh-2007-02-2007-QH12-59647.aspx

National Assembly of Vietnam. (2019). *The Vietnamese labour code*. National Assembly of Vietnam.

National Assembly of Vietnam. (2020). *Law on promulgation of legal documents*. National Assembly of Vietnam. https://thuvienphapluat.vn/van-ban/Lao-dong-Tien-luong/Bo-Luat-lao-dong-2019-333670.aspx

Nguyen, T. Q. T., & Simkin, K. (2017). Gender discrimination in Vietnam: The role of personal face. *Journal of Gender Studies, 26*(6), 609–617.

Nguyen, H. T. T., Nguyen, T. T., Dam, V. A. T., Nguyen, L. H., & Vu, G. T. (2020). COVID-19 employment crisis in Vietnam: Global issue, national solutions. *Frontiers in Public Health, 8*(December), 19–22. https://doi.org/10.3389/fpubh.2020.590074

Phạm, Q. H., Le, Q. B., Nguyễn, M., & Huye. (2021). *Khuôn Mẫu Giới và Vấn Đề Việc Làm: Một Nghiên Cứ về Giới Trẻ Việt Nam Đương Đại Qua Các Thảo Luận Trên Báo Chí và Mạng Xã Hội*. Ha Noi.

Rydström, H. (1998). *Embodying morality: Girl's socialization in a North Vietnamese Commune*. Linköping University. Publication in Lund University Research Portal.

Trần, N. T. (2001). *Tìm Hiểu về Bản Sắc Văn Hóa Việt Nam* (3rd ed). Hồ Chí Minh: Nhà Xuất Bản Thành phố Hồ Chí Minh.

Turley, W. S. (1972). Women in the Communist Revolution in Vietnam. *Asian Survey – University of California Press, 12*(9), 793–805.

UNFPA. (2016). *Policy brief ending gender-based violence at home in Viet Nam*. https://tuoitre.vn/phong-bi-500-000-dong-va-cua-hang-0-dong-tiep-suc-nguoi-tu-nam-ve-que-20210731154202726.htm, https://www.spotlightinitiative.org/fr/node/22788

Dis-ability, Differently Abled and Sentient Beings: Transformative Regeneration: The Need for Public Education on Vulnerability and Interbeing

Janet J. McIntyre-Mills

Abstract Public policy is framed in terms of the social contract based on extending rights to people of voting age. Dis-ability policy (at best) is framed in terms of the veil of ignorance, an awareness of intersectional disadvantage and its impact on welfare and being voiceless. However this paper makes the case that dis-Ability needs to be seen as a form of solidarity with differently abled sentient beings and that the boundaries between human beings and other living systems needs to be re-thought. A case is made for a recognition of concepts such as 'natural inclusion' and our interbeing and that focusing on ways to re-generate living systems ought to be mainstreamed in educational policies.

Judith Butler's plea for a recognition of our shared vulnerability is highlighted as a departure point for policy. She stresses that recognition of shared vulnerability is as important as independence and individualism. Both aspects are relevant to a re-generative approach to social education, circular economics and sustainable environments. Interbeing is a concept increasingly recognised across the social sciences and natural sciences and holds in common some of the key concepts of Buddhism and many other religions. The paper proposes that solidarity with those who *cur-*

This chapter builds and extends some of the ideas that have been developed over the years and expressed in a series of Springer volumes as detailed in the acknowledgements on page XX of this volume. In Systemic Ethics (McIntyre-Mills, 2014) I explore extending rights to other species by building on the work of Martha Nussbaum. My understanding of the continuity of living systems or what Rayner calls 'natural inclusion', or what Buddhists call 'interbeing' was fostered by mentors such as Adelaide Dlamini, an indigenous healer who taught me about the notion of connectivity across many species and Olive Veverbrandts in Alice Springs who taught me about totemic kinship with organic and inorganic life. These ideas led to "From Polarisation to Multispecies Relationships (McIntyre-Mills and Corcoran Nantes, 2021).

J. J. McIntyre-Mills (✉)
College of Education, University of South Africa, Pretoria Central, Gauteng, South Africa

College of Professions, Adelaide University, Adelaide, Australia
e-mail: janet.mcintyre@adelaide.edu.au

J. J. McIntyre-Mills, Y. Corcoran-Nantes (eds.), *Transformative Education for Regeneration and Wellbeing*, Contemporary Systems Thinking,
https://doi.org/10.1007/978-981-19-3258-8_10

rently fall outside the mantle of protection by virtue of their age, gender, level of ability or species needs to be addressed.

Keywords Transformative · Regeneration · Interbeing · Vulnerability · Extending social and environmental justice · Rights · Sentient beings

1 Introduction

The framing of people as *disabled* is the starting point for *disempowerment.* Judith Butler (who holds the Hannah Arendt chair) and Sunaura Taylor (Taylor, 2009)[1] discuss the importance of realising that a *recognition of our shared vulnerability* is a starting point for transformation. The message of resilience needs to be balanced by a message of the right to care. Judith Butler, whose work extends the thinking of Hannah Arendt asks: what would the world be like if we did not recognise one another's vulnerability? The next step is to state explicitly that *by not recognising* our shared vulnerability as human beings (and importantly our connection with other species) we are denying the everyday injustice towards the differently abled. Thus policy can be said to deny the 'banality of evil' inherent in everyday life (Arendt, 1963; see McIntyre-Mills, 2019a, b).

'*Dis-Ability*' needs to be seen as a form of solidarity with *differently abled* sentient beings who have the right to be respected and the right to live lives that are worth living, simply because they *are sentient* (Nussbaum, 2006). Drawing on the literature a case is made that we need to think in terms of social and environmental justice for all sentient beings.

'Natural inclusion' (Rayner, 2017) goes beyond interconnectedness as it recognises the ongoing energy flow across many species. 'Interbeing', a concept applied by a Vietnamese Buddhist monk Tiep Hien during the Vietnam War has been taught more widely by Thich Nhat Hanh.[2]

The importance of responding in ways to address the wounds to body, mind and spirit during the Vietnam War prompted the movement to support interbeing. It is also worth noting here that the destruction of the forests using the exfoliant Agent Orange also resulted in the movement to support the Ecocide Law starting with the work of a group of academics at Harvard for a national law supported by only some nations and extended by Higgins et al. (2013) as a plea for an international law to which all nations ought to comply to protect the conditions of a peaceful life for 'all the inhabitants' of a region.

All these (very different concepts) avoid the notion of categories by avoiding speciesism and taking as a starting point the notion that human beings and other species are part of a 'web of life' (Capra, 1996) and that we need to understand the

[1] https://www.google.com/search?q=butler+Judith+and+examined+life&oq=butler+Judith+and+examined+life&aqs=chrome..69i57j0i22i30i457.9443j1j7&sourceid=chrome&ie=UTF-8

[2] https://uplift.love/thich-nhat-hanhs-14-principles-of-engaged-buddhism/also

implications of dualism or us/them thinking. Leading plant and animal researchers stress that *communication and decision making span multiple species* and that communication can occur as signalling through sound, movement or chemicals (Meijer, 2016a, b) that convey information when used consistently. Recognition of the rights of sentient animals is overdue.

Plant researchers (Gaglioni et al., 2018) also make a case that plants communicate and make decisions and thus respond to information.

Indigenous Knowledge Systems in many parts of the world acknowledge kinship not only with organic life, but also inorganic life. The elements of the earth, atmosphere and waters nurture life. When living systems die they return to the elements and the cycle continues. This notion is confirmed by Addae (2022a, *pers. comm* and see Addae and McIntyre-Mills, 2022b):

> From the Ghanaian traditional perspective, human beings are part of the wider cosmology where our purpose is to protect the different species to help sustain the ecosystem. So, in many communities in Ghana, there are days set aside as *no farming days* which offers Asaase Yaa (the earth) the opportunity to rest. We believe that Asaase Yaa is the mother of all being offering us food and shelter from the land.

Kinship with features of the landscape, plants and animals can be recognised in the form of totems as explained by my mentor in Alice Springs, Olive Veverbrandts (2003), an Arrerente elder. Similarly in South Africa, Pat Lethole describes her totemic kinship relationships (see Romm & Lethole, 2021; Lethole et al., chapter "Aa!!Venda Women and Social Enterprise: Stepwise Process to Regenerative and Sustainable Living" in this volume).

The work of Joanna Macy et al. (2014) whose deep understanding of systems thinking and a career spent doing hands on development and cross cultural engagement (linked with a range of faith based NGOs) is characterised by an understanding that '*coming back to life*' requires an understanding of intergenerational responsibility for multiple species.

2 Education Needs to Focus on Our Interconnectedness

If we unpack the full implications of Butler's conversation with Sunaura Taylor (Taylor, 2009) about the need to recognise vulnerability as a starting point, then the need for re-framing policy becomes apparent. Sunaura explains to Judith Butler that participation in policy making, based on lived experience should be a right. Someone who uses a wheel chair and who has difficulty shopping, for example should explain what is required to make their engagement in every day activities easier. Judith Butler's strident call during the Occupy Movement[3] on 24th October 2011 is about demanding change and that if change is impossible then human beings *need to demand the impossible*.

[3] https://www.versobooks.com/blogs/765-if-hope-is-an-impossible-demand-then-we-demand-the-impossible-judith-butler-at-occupy-wall-street-video

The era of climate change and pandemics has underlined that human beings are not 'top predators' and that business as usual has undermined living systems (see McIntyre-Mills, 2020, 2021a, b) as stressed by Urry (2010) "Consuming the Planet to excess" and the most recent IPPC report (2021) on the implications of a carbon economy. The next step in the convergent social, economic and environmental crisis is the displacement of many species and the destruction of our interdependent habitat.

2.1 Shaping the Policy Narrative

Ferwerda (2016) underlines the importance of narrative in critical feminist theory, in order to reshape boundaries. Public policy is shaped according to assumptions and values that currently give little support to the differently abled. In general it is framed in terms of the social contract based on extending rights to people of *voting age* with little consideration of the needs of those who are (as yet) without representation.

The "veil of ignorance", means making a policy decision based on the notion that one should be willing to be at the receiving end of such a policy decision as it is based on universally applied criteria. Rawls notion of rights – based on the citizenship accorded to all people (who meet the requirements within the boundaries of the nation state— no longer goes far enough. Rawls (1999) himself raised the point that changing contexts perhaps need to be taken into account. Those who are differently abled and or voiceless need to be represented so that their interests are served. Disability policy needs to be framed in terms of the *veil of ignorance*, the right to a decent quality of life for all sentient beings and not merely welfare for those who are deemed to be citizens or children of citizens.

Nussbaum's capabilities for sentient beings (2011) are presented as a human development approach and are thus anthropocentric, but to her credit, she takes a step towards emphasising the rights of all sentient beings.[4]

Disability rights for *the voiceless and frail* have led to Judith Butler stressing the need for a recognition of human interdependency and care (see Butler & Taylor, 2009; Butler & Athenasiou, 2013). But we need to stop, pause and re-consider what this actually means.

---

[4] In Frontiers of Justice, Nussbaum (2006) stresses the need for the right to a life worth living to be extended to all sentient beings. Nussbaum's (2011) ten capabilities are directly concerned with wellbeing, a life worth living and the extent to which these can be addressed. These capabilities, include: (1) living a life that is not cut short prematurely (2) bodily health, (3) bodily integrity, (4) sense, imagination, and thoughts,(5) emotions, (6) practical reason, (7) Affiliation, (8) links to other species, (9) Play, (10) Control over one's environment. In Nussbaum's version, the Capabilities approach focus on "the protection of areas of freedom so central that their removal makes a life not worthy of human dignity" (p31). Nussbaum, however, extends these capabilities *to all sentient beings* and thus a small step towards non-Anthropocentrism in terms of recognition of rights, but her focus is still on human development, not on the rights of animals and habitats.

2.2 *Interbeing and the Protection of Shared Habitats for a New Manifesto*

In her conversation with Butler, Taylor[5] explains how she was labelled and referred to in a derogatory way because of the way she moved as a result of her physical makeup. She explained how she re-framed and responded by saying that monkeys are her favourite animals. The next step is to stress that all sentient beings have rights simply because they are sentient. Gibson et al. (2015) develop a *Manifesto for living in the Anthropocene* that stresses a new approach to relating to multiple species. Many different intelligences can be employed to make sense of our world. Howard Gardner (2008) stresses the need to draw on diverse forms of human intelligence including: 'bodily, linguistic, musical, mathematical or logical, naturalistic, spatial, interpersonal and intrapersonal". But this does not go far enough. The kinds of intelligence of different species is only now beginning to be realised ranging from plant to animal intelligence.

A sociology and philosophy of education applied to education policy needs to go one step further and recognise the vulnerability of sentient beings with whom we share our genetic and neural make up. The contributions of a range of thinkers in the social sciences, humanities have contributed to setting aside the Cartesian notion of dualistic thinking that has been used as a basis for defining human thought: "I think therefore I am" has been used as a basis for distinguishing human beings from other beings.

- If we accept that all living beings have consciousness and that what distinguishes some sentients from others is having different forms of intelligence rather than merely higher order intelligence, this is a first step. Instead species specific intelligence and sub species intelligence related to being able to respond to a local habitat needs to be a starting point.
- If culture is about responding to one's local, changing environment, then we need to *expand our notion of which species have a form of culture* which is the means by which living systems respond to their environment. It is thus a form of teleology. Animal behaviourists and plant researchers have attributed decision making ability to a range of living systems (Meijer, 2016a, b; Gaglioni et al., 2018) whilst De Waal (2009) has demonstrated in his research that animals also have a sense of fairness and reciprocity, whilst Ackerman (2016) explains that birds learn and pass on knowledge and thus they too apply a form of cultural learning. Greenfield (2000) stresses that consciousness is a continuum and which raises implications for social justice for all sentient beings.
- If we unpack the full implications of Butler's conversation with Taylor about the need to recognise vulnerability as a starting point, then the need for re-framing policy becomes apparent as we recognise that we are all first and foremost *inter beings,* because we cannot survive on our own or in isolation from other species.

[5] Butler (2010) stressed in a videoed conversation with Sunaura "that in this world we need each other, in order to meet our basic needs" Nominalist Way 6 videos (Examined Life—Judith Butler and Sunaura Taylor 720p.avi)."

2.3 Extending Butler's Plea to Demanding Regeneration and Multispecies Relationships to Foster Interbeing

In this section I make the case that Butler's plea for a recognition of our shared vulnerability is vital in moving towards a re-generative approach to social education, circular economics and sustainable environments. In order to extend Butler's work I draw on Anna Tsing's (2017) approach to drawing together strands of narrative in " the mushroom at the end of the world". Her work is important for a number of reasons, one of which is the wide ranging way in which she narrates the re-generative potential of nature and the need to recognise the power and potential of transformation.

By telling a different narrative, from a starting point of rights and respect for diverse forms of living systems and diverse intelligence, education could be greatly enhanced. She demonstrates multispecies relationships and how the love of matsutake mushrooms that are no longer readily grown in Japan are foraged in a range of contexts and how despite despoiled environments, the mushroom has survived.

3 Regenerative Education Policy

Policy is about who gets what when, why and to what effect and the need to reframe re-generative policy is overdue to protect vulnerable human beings by recognising our interdependency and the varying intelligences with in and across species. Interbeing and vulnerability are as important as resilience and independence, we need a better balance and a better understanding that resilience flows from a recognition of our place in the web of life.

3.1 Recognising Our Shared Vulnerability

During the pandemic reliance on IT has soared, the ability of people with autism to find ways to address IT design is now being recognised for pragmatic reasons because they are able to bring an alternative vision that benefits organisations employing people in artificial intelligence (Carson, 2020).The International Day of Persons with Disability is marked on third of Dec, 2020. In South Africa the rights of disabled need a great more attention as the tragic Esidemini report[6] underlined the vulnerability of the elderly, mentally frail who were sent to private care homes where they suffered degrading deaths as a result of a lack of care and malnutrition.

The vulnerability of people to domestic violence during lockdowns was highlighted in South Africa on talk back radio and Cyril Ramaphosa, President of South Africa, highlighted the cost to society. The high levels of unemployment and the

[6] https://www.scribd.com/document/338095858/Esidimeni-Report#from_embed

poverty facing families has resulted in higher levels of domestic violence and child abuse. The cost of domestic violence to the South African economy is more than 40 million rands per year according to a 2017 report cited by Ramaphosa.[7] On Cape Talk callers highlighted the extent to which women and children are at risk and the terrible instances of rape, in particular the case of a baby raped by his 17 year old father on 26th November.

Whilst physical disability and resilience to overcome the odds is the usual trope, it is time to also embrace the implications of recognising our shared vulnerability. Whilst the young man in Indonesia – who overcame his disability and "calls the shots" by becoming a talented photographer who frames his shots from a different angle is an internet success which demonstrate his skills and the support of friends who helped him to build his own vehicle is inspiring— it is not the whole story. He had to work very hard to feel worthwhile because of the way he felt rejected by his family and society for many years.

Mental health is a topic that needs more attention as the COVID-19 pandemic demonstrates. Social distancing comes with a cost for many who have lost their jobs and members of their family and friends. The number of people who have relied on companion animals has increased in Australia (Roy, 2020; Selinger-Morris, 2020) and elsewhere (Morgan et al., 2020). Recently there have been reports of people wanting to 'dump' animals when lockdowns have ended.[8]

To date the record of care in public health is dire for those who are voiceless. According to the executive summary of the Esidimeni-Report,46 the Ombud established that:

> A total of ninety-one (94+) and not thirty-six (36) mentally ill patients (as initially and commonly reported publicly in the media) died between the 23rd March 2016 and 19th December 2016 in Gauteng Province.

Capri et al. (2018) stress that the deaths linked with Esideni (ironically meaning dignity) are merely indictive of a much wider ranging problem and the treatment of the disabled needs to be addressed through re-considering improved educational systems.

The first step is the policy will to make a difference and recognising the rights of the differently abled and our shared vulnerability to address the everyday *"banality of evil"* towards the differently abled. The next step is realising that all sentient beings have rights, simply because they exist. This is the high road to morality. The low road is to realise pragmatically that what we do to others we do to our selves – there is no us versus them and no economic externalities in one living system.

Donna Haraway's recent work (2015, 2016a, b, 2018) stresses the need to extend a sense of solidarity and care to other species. We need to move away from the sense of boundaries across-species and re-think our values and our relationships. This requires a new perspective on our place within living systems.

---

[7] https://www.news24.com/news24/columnists/cyrilramaphosa/cyril-ramaphosa-gender-based-violence-it-is-time-to-turn-the-tide-on-the-scourge-20201123

[8] Cornish, R (2021) *New pet owners urged to ask for help before surrendering animals post-lockdown.* https://www.abc.net.au/news/2021-08-11/covid-dog-adoptions-returns/100365042

3.2 Education Policy Needs to Be Re-framed

The problem with the social contract as it is currently structured is that it gives voice and agency only to those of voting age, but excludes young people, the frail, the disabled, refugees, asylum seekers and non-human animals. Human beings are one strand within the web of life with a range of diverse species intelligences that we are only on the brink of discovering. Furthermore, some different intelligences are vital for our shared survival. Compassion and a recognition of rights could help us to reverse the current trajectory towards disaster as a result of increased competition for resources and greater gaps between haves and have nots. Social and environmental justice remain a priority for those who are least powerful and unable to protest. This applies to young people, the elderly, disabled, undocumented migrants, the frail and animals.

Education policy needs to be re-framed to take into account the notion that human beings are part of a living system and that rights need to be addressed and modelled through educational praxis that gives a voice to those who are currently excluded from the social contract.

3.3 Mainstreaming Intersectionality and the Capabilities Approach in Order to Extend the Frontiers of Justice

Mainstreaming intersectionality means that irrespective of age, gender, level of ability that individual capabilities for all sentient beings[9] will be recognised to support a life that is worth living. This entails mental and physical health and safety and the ability to express oneself fully through engaging in meaningful activities that give voice to needs and enables a sense of fulfilment and a political voice. The rights of the differently abled need to be expressed through their own voice, alternative forms of communication or through trained advocates who recognise the rights of all sentients to a life that is worth living and their right to express their capabilities in the fullest sense.

[9] We need to recognise that consciousness is a continuum. Elsewhere the case is made (see McIntyre-Mills, 2021a, b:153) that participation and the way we relate to others shapes the world we live: "Both Bohm and the Dalai Lama believe that perception, research design and measurement shape the way we see the world. Bohm designed an experiment to study waves and particles and explained that whether we 'see' a wave or a particle, depends on the process of measurement. Bohm's interest in the relationship between the observer and the observed led to conversations with spiritual leaders including the Dalai Lama as they both considered that thinking matters quite literally. The Dalai Lama said that Bohm's thinking and Buddhism have much in common, namely thought experiments and meditation on 'what if scenarios'. The focus of their conversations was on relationships and the implications for knowledge and its potential for making a difference to the way we choose to live our lives. By thinking about our thinking we can try to address self-deception and what West Churchman (1979) called 'the enemies within'. These are our assumptions and values that can cloud our thinking."

3.4 Interbeing and the Transformative Approach

Education needs to take place not only in the classroom, but by engaging in and with nature. Learning can take place through engaging with nature through walking, exploring and working to restore the environment and deepen our engagement with diverse species. One of the ways is through:

- supporting better rural urban balance
- re-generating rural-urban balance and reversing the plantationocene (Mitman in conversation with Mitman, 2019 and see Tsing, 2011, 2013, 2015).
- protecting appropriate habitat and to reverse current trends.
- creating ecovillages (Shiva, 2020a, b) and green enterprises drawing on Gunter Pauli's (2010) concept that in nature there is no waste and that sources of abundance can be found by working with nature and through designs that mimic nature. The notion that the mantle of citizenship should only be given to those of voting age and with the right to cast a ballot is problematic. The environment on which we depend is also entirely controlled by the voting citizens of nation states. In 'Frontiers of Justice', Nussbaum (2006) develops an argument for extending the social contract to those sentient beings who are not protected. Her starting point is to stress the need for individual capabilities to be protected, in order to be able to live a life worth living. Her argument includes being able to live in an environment that *supports a life in which capabilities can be achieved.* Current debates hinge on whether cosmopolitan universal rights can be given to sentient beings as a whole or whether rights for human sentients and animal sentients should differ.

One of the most insightful educational documentaries "My octopus teacher"[10] stresses the healing potential of learning every day from the environment as detailed in chapter "Regenerative Education and Safe Habitats for Diverse Species: Caterpillar Dreaming Butterfly Being." The octopus reaches out a tentacle like a finger to communicate with Craig and so begins a lesson in the profound way in which communication can span species. As trust develops she leaps into his hand to greet him. Craig learns more about the octopus, her capabilities and the way she lives in her kelp forest home and the way the ecosystem functions as an interdependent system.

Craig realises that the octopus is not only sentient, curious and intelligent[11] but capable of trust. She defends herself against predators by disguising herself with shells, by changing shape and colour. She has a very special kind of intelligence.

---

[10] https://archiwum2020.wkinach.mdag.pl/17/en/warszawa/movie/Osmiornica%252C-nauczycielka

[11] https://www.youtube.com/watch?v=dKWssIQplw8 The documentary, My octopus teacher released on Netflix was filled off the cost of South Africa. The documentary shows Craig Foster's underwater journey to learn more about a kelp forest which he visits every day. He befriends an octopus who teaches him about transformation and resilience.

She manages to outwit predators on a number of occasions. Then she survives a shark attack and Craig manages to find her after he loses contact with her for many days. At first he feared she had died as she has been severely mauled. But after enclosing herself in a rock cavity she recovers and over time re-generate a lost tentacle.

The documentary ends with the octopus producing offspring and so she achieves her destiny to create a new generation of octopi. Craig needed time to re-connect with himself and his family after a bout of depression. The film shows his own recovery and his deepening relationship with his son whom he taught about the kelp forest.

The documentary demonstrates the importance of learning from and in nature through creating ongoing relationships. Craig's daily dives to the same area of kelp forest, helped him to understand the many relationships across different species. He started to learn about his own vulnerability and his relationships with people and with nature changed.

Another important lesson is provided in an interview of Jane Goodall about the Covid pandemic[12] and her vision for the future (Watts, 2021). Goodall stressed that 75% of diseases are caused by the destruction of animal habitat which creates disease vectors that would not occur if animals were left in their natural habitat and not abused by confining them in factory farms, wet markets or trafficking them.

She also stressed that one of the ways in which habitat can be protected is by educating women so that they are able to earn a living and so that they have personal agency. Educated women help to bring about a demographic transition, in other words they have fewer children and they are able to make better decisions about living in more sustainable ways.

COVID-19 has underlined that human beings are no longer top predators (McIntyre-Mills, 2020) and that unless we protect our habitats we face existential risk (Bostrom, 2011). This requires 'extending the frontiers of justice' (Nussbaum, 2006) and ensuring that this extension is reflected in transformative education policy that is rooted in mainstreaming intersectionality to support a capabilities approach. This requires extending solidarity to the differently abled, based on a recognition that our fate is linked with multiple species comprising a dynamic living system (see Rayner 2010; McIntyre-Mills, 2021a, b; Romm & Lethole, 2021).

The destruction of forest habitat will impact on the ability of the planet to manage carbon emissions and habitat that is required for diverse species, in order to prevent pandemics (McKie, 2020).

Sociology needs to go one step further and recognise the vulnerability of sentient beings with whom we share our genetic and neural make up. If a laboratory rat shares, according to Couglan (2002) share 97.5%[13] of our genes and thus cannot be thought to be an automaton, nevertheless in 2004 a report on the sequencing of the rat genome by Washington University claims:

---

[12] https://www.cnn.com/videos/us/2020/03/19/coronavirus-jane-goodall-acfc-full-episode-vpx.cnn

[13] Coughlan, A. 2002 Just 2.5% of DNA turns mice into men New Scientist https://www.newscientist.com/article/dn2352-just-2-5-of-dna-turns-mice-into-men/

> The laboratory rat is an indispensable tool in experimental medicine and drug development and has made inestimable contributions to human health. The new data expand and consolidate its role as a research resource.

The rat is sentient and is not a tool. The Cartesian notion of categorizing human and other is based on the mistaken notion that homo sapiens sapiens is different from other species in her ability to think. Descartes declared, the most basic assumption in philosophy is "I think, therefore I am."

The notion of the ability to think as an indication of being a categorical member of the human race was used as a justification for drawing lines of exclusion between fully functional human beings (who play a useful role in society as citizens) and those who are labelled frail, voiceless, disabled, criminals, non-citizens or members of other species. But many other sentient species are conscious and are able to make decisions and to communicate. Communication per se does not require human voices. It can rely on text, signaling and symbols. Sounds, movements, chemicals can convey information and the whole realm of communication is opened up by Meijer (2016a, b) and Gagliano et al. (2018) on animal and plant communications, respectively. We need to focus on a non-anthropocentric future (see McIntyre-Mills, 2021a, b).

Fortunately, contributions by a range of thinkers in the social sciences and humanities have contributed to setting aside the Cartesian notion of dualistic thinking that has been used as a basis for defining human thought: 'I think therefore I am' has been used as a basis for distinguishing human beings from other beings.

This paper builds on the case developed in "Communication and Culture: a Multispecies endeavour: recognising kinship with multiple species (McIntyre-Mills, 2021a, b).

4 Critical Heuristics and Re-generative Design

A design of inquiring systems approach (West Churchman, 1979) and re-generative design (Wahl, 2016) is adapted and informed by Donna Haraway's work on re-framing boundaries and applied to education policy and praxis. The notion of what is sacred and profane (Douglas, 1978) needs to be re-thought now that it is evident that Cartesian 'us/them' conceptualisations are flawed.

We stand and fall together as living systems sharing one planet. In the words of Kenneth Boulding (1966) we need to understand we are part of one space ship and not in competing life boats. This applies as much to reconceptualising rights and responsibilities towards those with differing abilities as it does to recognising the diverse intelligences within and beyond our species. Current educational curricula need to develop a greater understanding through fostering an ecological mindset (Bateson, 1972) and bolstering the ability of post national regional areas to protect the environment through an ecocide law (Higgins, 2012) that will enable individual companies, organisations and nations to be held to account.

4.1 Rethinking Policy, Praxis and Curriculum

All policy needs to be framed in terms of context. The social, economic and environmental challenges are convergent and we face the challenge of increasing levels of urbanisation (United Nations, 2014) which will place humanity at risk (Sendai Risk platform, 2014–30).

Thus the case is made that education needs to embrace environmental learning and an ecological mindset (Bateson, 1972) to address the existential risks that we are currently facing. The learned and lived experiences of the differently abled need to be included in policy scenarios and policy plans and the pedagogical environment needs to address the life chances of the oppressed (Freire, 1970) and extended to address capabilities of sentient beings (Nussbaum, 2006).

If we are able to undertake a boundary critique and realise that we need to move away from 'polarisation to recognising multispecies relationships' (McIntyre-Mills, 2017c, 2021a, b).

Wicked problems by definition are complex. They comprise many, interrelated variables that are perceived differently by different stakeholders and must be explored contextually (see Flood & Carson, 1993; Rittel & Webber, 1984; West Churchman, 1971, 1979, 1982). However, the collaboration across stakeholders needs to be guided by the axiom that: 'We can be free and diverse to the extent that our freedom and diversity does not undermine the common good of both current and future generation of life' (McIntyre-Mills, 2014). This axiom has been explored in depth in previous work (see McIntyre-Mills 2006, 2014, 2017a, b, c). It explores the notion that our fate is determined by a realization of our interdependence. It aims to increase an understanding of life chances and dynamics of vulnerable population groups.

A more fluid and less categorical understanding of the world is vital. Rayner (2020)[14] stressed that we have evolved with the binocular vision of a predator, rather that more peripheral vision of a buck who needs panoramic vision.

The notion of *us and them, self and others* need to be replaced with a way in which we can bridge the divides by realising that we are the boundaries and our values matter quite literally. By finding ways to bridge the divided middle ground using dialogue, sharing knowledge, mapping it and using it to guide further decisions could help to ensure that people are able to hold in mind more variables and thus make better decisions. Much has been written about the fact that human beings cannot manage more than about 7 variables and that this is why we make such bad political decisions than jeopardize our futures.

Intersectionality (Crenshaw, 1991; Dhamoon, 2011) is suggested as a way forward, but only if we are also able to extend our solidarity to recognise solidarity with other species and our approach to science so that we also realise the vital importance of interbeing – rather than compartmentalised ways of knowing and being.

---

[14] Daniel Wahl with you. Recorded last month. https://www.youtube.com/watch?v=IKBlGtzEJoo

Daniel Wahl (2016: 5) stresses:

> Creating regenerative systems is not simply a technical, economic, ecological or social shift: it has to go hand-in-hand with an underlying shift in the way we think about ourselves, our relationships with each other and with life as a whole.

5 Analysis

By focusing on re-generative education in a range of contexts to support opportunities for the marginalised. The focus on indigenous people, young people, the disabled and displaced peoples is important. Creating employment opportunities in the area of food, energy and water security is a theme underpinning the work and builds on pathways to wellbeing to address complex health, housing and social inclusion as well as mitigation and adaptation to climate change. User centric design where possible supported by stewardship needs to address these policy concerns.

The displacement of people needs to be linked with displacement of plants and animals and how we need one another to ensure a sustainable way of life that works with nature – instead, of against nature.

In chapter "Sacred Groves of the Tolon District of the Northern Region, Ghana: Where Spirituality Meets Education for Sustainable Development" (this volume) we discuss the importance of African taboos and customs to protect sacred sites and their diversity from destruction. We make the case for the potential for public education to drawing on this wisdom in order to help prevent deforestation. It is through initiatives to protect habitat that we can stand together as Forsetto (2020)[15] Nobel Peace Prize laureate has pleaded to protect the planet from deforestation. This will require recognising our shared vulnerability and the need for a ethic of care for multiple species, in order to address climate change and to prevent pandemics, because the issues are interlinked (Goodall, 2020; Gorbalenya et al., 2020).Without habitat species that normally do not come into contact interact and become vulnerable to infection and without forest habitats we destroy the 'lungs of the planet.' (Centre for International Forestry Research, 2015).[16]

Diversity in nature is vital and the social contract (as it is currently defined) does not go far enough to protect living systems of which we are a strand.

It is vital that the United Nations Rights of Indigenous People be more broadly applied as a policy underpinning to protect habitats that have been protected for centuries and are now under threat as urbanisation increases (UN, 2014) rapidly in Asia, Africa and Southern America.

Transformative praxis could include protecting people and habitat through the implementation of an ecocide law to address 'Existential risk' (Bostrom, 2011) which continues to escalate. The crime of 'ecocide' (Higgins, 2016) is not yet recognised as part of international law even though it poses a new form of 'genocide'.

---

[15] https://news.mongabay.com/2020/02/chief-raoni-nobel-peace-prize-nominee-presides-historic-meeting-with-over-600-indigenous-leaders-in-brazil/

[16] https://www.youtube.com/watch?v=KAuTxIPoV_k

6 Conclusion

Politically fragmentation and populism have become the new order driven by capitalism, anthropocentrism, speciesism, nationalism and racism. The case is made that liberalism has progressed too far in undermining collective (cosmopolitan) responsibility. The paper makes a case for educational policy to address institutional capacity and powers to implement UN 2030 Sustainable Development Agenda by drawing on the differently abled, local wisdom and species specific intelligences.

This could be supported by extending Vocational Education and Training using a gender mainstreaming approach that foregrounds opportunities for the marginalised and disabled in remote, areas with few employment opportunities and high levels of poverty in order to try to address the high rates of urbanisation and encourage better rural urban balance. Creating opportunities with young people through setting up communities of practice that support rural-urban balance through eco-villages is a priority (McIntyre-Mills, 2019a, b). This can include creating opportunities with differently abled who have aptitudes in arts, sciences or technology such as computing that maximises the skills of young people through re-building, re-purposing and fixing technology so that wastage is reduced and the notion of externalities is re-thought by learning that 'there is no waste in nature' (Pauli, 2010).

Bibliography

Abram, D. (2020). Foreword. In S. Kaza (Ed.), *A wild love for the world: Joanna Macy and the work of our time* (pp. 1x–xv11). Boulder.

Ackerman, J. (2016). *The genius of birds*. Scribe.

Addae, D., & McIntyre-Mills, J. J. (2022a, this volume). Chapter 16: Sacred groves of the Tolon district of the Northern Region. *Transformative Education for Regeneration and Wellbeing*. Ghana: Where Spirituality meets education for Sustainable Development.

Addae, D., & McIntyre-Mills, J. J. (2022b). Balancing individualism and collectivism: A conversation on Afrocentricity in Olivia. In T. A. Kwapong, D. Addae, & J. K. Boateng (Eds.), *Development education in Africa*. Springer Nature.

Arendt, H. (1963). *Eichmann in Jerusalem: A report on the banality of evil*. Viking Press.

Bateson, G. (1972). *Steps to an ecology of mind*. Ballantine.

Bohm, D. (2002). *Wholeness and the implicate order*. Routledge, Routledge Classics.

Bohm, D., & Dalai Lama, H. H. (2018). *From Fragmentation to Wholeness: David Bohm & Dalai Lama Life and mind enfolded in everything*. https://www.youtube.com/watch?v=V2He4oU32sg

Bostrom, N. (2011). *Existential risk prevention as the most important task for humanity*. Faculty of Philosophy & Oxford Martin School University of Oxford. http://www.existentialrisk.org/

Boulding, K. (1966). Space ship earth the economics of the coming spaceship earth Kenneth E. Boulding. In H. Jarrett (Ed.), *Environmental Quality in a Growing Economy* (pp. 3–14). Resources for the Future/Johns Hopkins University Press.

Butler J. (2011). *Precarious life: The obligations of proximity*. The Neale Wheeler Watson Lecture, Nobel Museum, Svenska, 28 May. MIT Press. Available at: www.youtube.com/watch?v=KJT69AQtDtg

Butler, J., & Athanasiou, A. (2013). *Dispossession: The performative in the political*. Polity.

Butler, J., Taylor, S., Interdependence in Taylor, A. (2009). *Examined life: Excursions with contemporary thinkers* (pp. 185–214). New York Press.

Butler, J., & Taylor, S. (2010). Taylor, A. *Examined Life. Excursions with contemporary thinkers*. The New Press. https://cjgl.cdrs.columbia.edu/article/performing-interdependence-judith-butler-and-sunaura-taylor-in-the-examined-life-2/

Capra, F. (1996). *The web of life: A new synthesis of mind and matter*. Harper Collins.
Capri, C., Watermeyer, B., Mckenzie, J., & Coetzee, O. (2018). Intellectual disability in the Esidimeni tragedy: Silent deaths. *SAMJ: South African Medical Journal, 108*(3), 153–154. https://doi.org/10.7196/samj.2018.v108i3.13029
Carrington, D. (2020). *Pandemics result from destruction of nature, say UN and WHO. Experts call for legislation and trade deals worldwide to encourage green recovery.* https://www.theguardian.com/world/2020/jun/17/pandemics-destruction-nature-un-who-legislation-trade-green-recovery
Carson, E. (2020). *Autism as an advantage in AI jobs: This organization is making an argument for neurodiversity for Daivergent, matching tech workers with autism to jobs isn't an act of charity.* https://www.cnet.com/news/autism-as-an-advantage-in-ai-jobs-this-organization-is-making-an-argument-for-neurodiversity/
Churchman, C. W. (1971). *The design of inquiring systems: Basic concepts of systems and organisations*. Basic Books.
Churchman, C. W. (1979). *The systems approach and its enemies*. Basic Books.
Churchman, C. W. (1982). *Though and wisdom*. Intersystems.
Coughlan, A. (2002). Just 2.5% of DNA turns mice into men new scientist. https://www.newscientist.com/article/dn2352-just-2-5-of-dna-turns-mice-into-men/
Clack, B. (1999). René Descartes 1596–1650. In B. Clack (Ed.), *Misogyny in the Western philosophical tradition*. Palgrave Macmillan. https://doi.org/10.1057/9780230212800_8
Crenshaw, K. (1991). Mapping the margins: Intersectionality, identity politics and violence against women of color. *Stanford Law Review, 43*(6), 1241–1299.
De Waal, F. (2009). *The age of empathy: Nature's lessons for a kinder society*. Harmony Books.
Descartes, R, 1596–1650. (1986). *Discourse on method*. Macmillan/Collier Macmillan.
Descartes, R. (2017). *Meditations on first philosophy* (J. Veitch, Trans.). Toront d from http://bradleymurray.ca
Dhamoon, R. K. (2011). Considerations on mainstreaming intersectionality. *Political Research Quarterly, 64*(1), 230–243. https://doi.org/10.1177/1065912910379227
Douglas, M. (1978). *Purity and danger: An analysis of the concepts of pollution and taboo*. Routledge and Kegan Paul.
Ferwerda, S. (2016). *Narrating the Anthropocene. Threading contemporary feminist theory and fiction in a human-dominated world.* MA thesis. http://dspace.library.uu.nl/handle/1874/330543
Flood, R., & Carson, E. (1993). *Dealing with complexity: An introduction to the theory and application of systems science* (2nd ed.). Plenum.
Forestry Research. (2015). https://news.mongabay.com/2020/02/chief-raoni-nobel-peaceprize-nominee-presides-historicmeeting-with-over-600-indigenous-leaders-in-brazil/ https://www.youtube.com/watch?v=KAuTxIPoV_k
Forsetto, R. (2020). *Chief Raoni, Nobel peace prize nominee, presides over historic meeting with over 600 indigenous leaders in Brazil*, on 5 February, 2020.
Freire, P. (1970). *Pedagogy of the oppressed*. Penguin Books.
Gagliano, M., Abramson, C. I., & Depczynski, M. (2018). Plants learn and remember: let's get used to it. *Oecologia, 186*, 29–31. https://doi.org/10.1007/s00442-017-4029-7
Gardner, H. (2008). *Multiple intelligences: New horizons in theory and practice*. Basic Books.
Gibson, D., Rose, B., & Fincher, R. (Eds.). (2015). *Manifesto for living in the Anthropocene* (pp. 17–22). Punctum books.
Goodall, J. (2020). https://www.cnn.com/videos/us/2020/03/19/coronavirus-jane-goodall-acfcfull-episode-vpx.cnn
Gorbalenya, A. E., Baker, S. C., Baric, R. S., et al. (2020). The species severe acute respiratory syndrome-related coronavirus: Classifying 2019-nCoV and naming it SARS-CoV-2. *Nature Microbiology, 5*, 536–544. https://doi.org/10.1038/s41564-020-0695-z
Greenfield, S. (2000). *The private life of the brain: Emotions, consciousness and the secret of the self.* Wiley.
Haraway, D. (1984). *A cyborg manifesto*. Routledge/ Macat Library.
Haraway, D. (1991). *The reinvention of nature*. Free Association Books.
Haraway, D. (1992). The promises of monsters: A regenerative politics for inappropriate/d others'. Cultural Studies. In L. Grossberg, C. Nelson, & P. A. Treichler (Eds.), *Cultural Studies* (pp. 295–337). Routledge.

Haraway, D. (2008). *When species meet*. University of Minnesota Press.
Haraway, D. (2011). Speculative fabulations for technoculture's generations: Taking care of unexpected country. *Australian Humanities Review, 50*, 2011.
Haraway, H. (2014). Anthropocene, Capitalocene, Chthulucene: Staying with the trouble, 5/9/2014. https://vimeo.com/97663518
Haraway, D. J. (2015, May 1). Anthropocene, Capitalocene, Plantationocene, Chthulucene. *Environmental Humanities, 6*(1), 159–165. https://doi.org/10.1215/22011919-3615934
Haraway, D. J. (2016a). *Staying with the trouble: Making Kin in the Chthulucene*. Duke University Press.
Haraway, D. (2016b). *Anthropocene, Capitalocene, Chthulucene: Making string figures with biologies*. Arts, Activisms – Aarhus University YouTube https://m.youtube.com/watch?v=CHwZA9NGWg0
Haraway, D. (2018). Staying with the trouble for multispecies environmental justice. *Dialogues in Human Geography, 8*(1), 102–105.
Harris, L. D., & Wasilewski, J. (2004). Indigeneity, an alternative worldview: Four R's (relationship, responsibility, reciprocity, redistribution) vs. two P's (power and profit). Sharing the journey toward conscious evolution. *Systems Research and Behavioral Science, 21*(5), 489–503.
Higgins, P. (2012). *Ecocide, the 5th crime against peace*. https://www.youtube.com/watch?v=8EuxYzQ65H4. See the link in this lecture to Stop Ecocide, Change the Law. https://www.youtube.com/channel/UCCYoIf880oaM419JO8XUhEA
Higgins, P. (2016). *Eradicating ecocide: Laws and governance to stop the destruction of the planet*. Shepheard-Walwyn (Publishers) Ltd.
Higgins, P., Short, D., & South, N. (2013). Protecting the planet: A proposal for a law of ecocide. *Crime Law Soc Change, 59*, 251–266. https://doi.org/10.1007/s10611-013-9413-6
IPCC. (2021). Summary for Policymakers. In*: Climate Change 2021: The Physical Science Basis. Contribution of Working Group I to the Sixth Assessment Report of the Intergovernmental Panel on Climate Change* [VP Masson- Delmotte, A Zhai and SL Pirani]. https://www.wri.org/insights/ipcc-climate-report
Kaza, S. (2020). *A wild love for the world: Joanna Macy and the work of our time*. Shambhala.
Lopes, H., & McKay, V. (2020). Adult learning and education as a tool to contain pandemics: The COVID-19 experience. *International Review of Education, 66*, 575–602. https://doi.org/10.1007/s11159-020-09843-0
Macy, J., & Brown, M. (2014). *The updated guide to the work that reconnects: Coming back to life*. New Society Publishers.
Macy, J., Young-Brown, M., & Fox, M. (2014). *Coming Back to life: The updated guide to the work that reconnects*. New Society Publishers.
McIntyre-Mills, J. (2006). Consciousness, caretaking and compassion based on listening and making connections underpins systemic governance. In J. P. C. Van Gigch (Ed.), *Wisdom, knowledge and management: A critique and analysis of Churchman's systems approach* (West Churchman series) (Vol. 2, pp. xv–xxiii). Springer.
McIntyre-Mills, J. (2014). *Systemic ethics and non-anthropocentric stewardship* (pp. 270). Springer. http://www.sense.nl/gfx_content/documents/20161115_Uitgevers.xlsx. Springer, International, Cham.
McIntyre-Mills, J. (2017a). *Planetary passport*. Springer.
McIntyre-Mills, J. (2017b). Recognizing our hybridity and interconnectedness. *Current Sociology*, 1–24. https://doi.org/10.1177/0011392117715898
McIntyre-Mills, J. (2017c). Representation and accountability in Glocal governance and the 2030 development agenda: Narrowing the gap between perceived needs and outcomes. *Systemic Practice and Action Research, 30*(5), 447–469. Springer.
McIntyre-Mills, J. (2018). Policy Design for non-anthropocentric Pathways to protect biodiversity and regenerate the land. *International Education Journal in South Africa*. https://journals.co.za/content/journal/10520/EJC-117cea7615
McIntyre-Mills, J. (2019a). *Social engagement to redress the banality of evil and the frontiers of justice: Limitations of the social contract to protect habitat and why an international law to prevent the crime of ecocide matters*. Proceedings of International Systems Sciences The

63nd Annual Meeting of ISSS, Oregon State University, Corvallis, Nature's Enduring Patterns: A Path to Systems Literacy., 2nd-5th July. http://journals.isss.org/index.php/proceedings63rd/article/view/3642

McIntyre-Mills, J. (2019b). *The potential of eco-facturing to re-generate rural-urban balance through eco-villages and city hubs: Promoting social and environmental justice through vocational education and training hubs*. Proceedings of International Systems Sciences the 63nd Annual Meeting of ISSS, Oregon State University, Corvallis, Nature's Enduring Patterns: A Path to Systems Literacy, 2nd-5th July. http://journals.isss.org/index.php/proceedings63rd/article/view/3643

McIntyre-Mills, J. (2020). The Covid-19 era: No longer business as usual. *Systems Research and Behavioral Science, 37*(5), 827–838. https://doi.org/10.1002/sres.2745

McIntyre-Mills, J. (2021a). Chapter 1: Communication and culture: A multispecies Endeavour: Recognising kinship with multiple species. In J. J. McIntyre-Mills & Y. Corcoran Nantes (Eds.), *From polarisation to multispecies relationships in the age of mass extinctions* (pp. 3-26). Springer Nature. https://doi.org/10.1007/978-981-33-6884-2

McIntyre-Mills, J. (2021b). Chapter 11: McIntyre-Mills, J. Vignette: Why thinking matters: Constructivism, relationships and the performative universe. In *From polarisation to multispecies relationships*. Springer.

McIntyre-Mills, J. J., & Corcoran-Nantes, Y. (Eds.). (2021). *From polarisation to multispecies relationships in the age of mass extinctions*. Springer.

McIntyre-Mills, J., with contributions by Wirawan, R., Widiansingh, I., & Riswanda. (2021). *Social engagement to protect multispecies habitat: Implications for re-generation and food security From Polarisation to Multispecies Relationships*. Springer.

McKie, R. (2020). *Rampant destruction of forests 'will unleash more pandemics' Researchers to tell UN that loss of biodiversity enables rapid spread of new diseases from animals to humans*, Sun 30 Aug 2020 16.03 AEST. https://www.theguardian.com/environment/2020/aug/30/rampant-destruction-of-forests-will-unleash-more-pandemics

Meijer, E. (2016a, translated, 2019). *Animal languages: The secret conversations of the living world.* John Murray.

Meijer, E. (2016b). Translated, 2019. In J. Murray (Ed.), *Animal Languages: The secret conversations of the living world.* John Murray.

Mitman, G. (2019). *Reflections on the Plantationocene: A conversation with Donna Haraway and Anna Tsing June 18, 2019 updated October 12, 2019.* https://edgeeffects.net/haraway-tsing-plantationocene/

Morgan, L., Protopopova, A., Birkler, R. I. D., et al. (2020). Human–dog relationships during the COVID-19 pandemic: Booming dog adoption during social isolation. *Humanities Social Science Communications, 7*, 155. https://doi.org/10.1057/s41599-020-00649-x

Nussbaum. (2003). *Beyond the social contract: Toward global justice.* https://tannerlectures.utah.edu/_documents/a-to-z/n/nussbaum_2003.pdf

Nussbaum, M. (2006). *Frontiers of justice: Disability, nationality, species membership*. Harvard University Press.

Nussbaum, M. (2011). *Creating capabilities: The human development approach.* Cambridge Belknap Press of Harvard University Press.

Olive Veverbrandts. (2003). *Prologue: My Town in McIntyre-Mills, J. 2003. Critical systems praxis: participatory governance for social and environmental justice. The Contemporary Systems Series*. Kluwer/Springer.

Pauli, G. (2010). *The blue economy: Report to the Club of Rome*. Paradigm Publications.

Rawls, J. (1999). *The law of peoples with the idea of public reason revisited.* Harvard University Press.

Rayner, A. D. M. (2010). Inclusionality and sustainability – Attuning with the currency of natural energy flow and how this contrasts with abstract economic rationality. *Environmental Economics, 1*, 98–108.

Rayner, A. (2017). *The origin of life patterns in the natural inclusion of space in flux*. Springer.

Rittel, H., & Webber, M. (1984). *Planning problems are wicked problems, developments in design methodology*. Wiley.

Romm, N. R. A., & Lethole, P. V. (2021). Prospects for sustainable living with focus on interrelatedness, interdependence and mutuality: Some African perspectives. In *Chapter 6 in From polarisation to multispecies relationships in an era of mass extinctions*. Cham.

Roy, T. (2020). *Coronavirus restrictions see demand for pets surge as shelters issue warning to prospective owners*. https://www.abc.net.au/news/2020-04-05/demand-for-pets-surge-as-australians-stay-at-home/12118888

Selinger-Morris, S. (2020). *They've wanted a dog for years. Lockdown has finally made it happen*. https://www.smh.com.au/lifestyle/life-and-relationships/they-ve-wanted-a-dog-for-years-lockdown-has-finally-made-it-happen-20200424-p54n0n.html

Shiva, V. (2020a). *Ecofeminism and the decolonization of women, nature and the future*. https://www.youtube.com/watch?v=hVbbov9Rfjg. Mar 25, 2020 – Uploaded by San Telmo Museoa.

Shiva, V. (2020b). *Two paths to the future of food and farming – EcoFarm keynote*. https://www.youtube.com/watch?v=9vNPGwqR4TQ

Taylor, A. (2009). *Examined life: Excursions with contemporary thinkers*. New York Press.

Tsing, A. (2011). Arts of inclusion, or, how to love a mushroom (Anna Ting, for the matsutake worlds research group). *Australian Humanities Review*, 50. https://press-files.anu.edu.au/downloads/press/p111121/html/ch01.xhtml?referer=&page=2#tocanchor

Tsing, A. (2013). More-than-human sociality: A call for critical description. In K. Hastrup (Ed.), *Anthropology and nature* (pp. 27–42). Routledge.

Tsing, A. L. (2015). *The mushroom at the end of the world: On the possibility of life in capitalist ruins* (p. 352). Princeton University Press.

Tsing, A. L. (2017). *The mushroom at the end of the world*. Princeton University Press.

Ulrich, W., & Reynolds, M. (2010). Critical systems heuristics. In M. Reynolds & S. Holwell (Eds.), *Systems approaches to managing change: A practical guide* (pp. 242–292). Springer.

United Nations Declaration on the Rights of Indigenous Peoples I United Nations For Indigenous Peoples. (2007). https://humanrights.gov.au/our-work/un-declaration-rights-indigenous-peoples-1. 13th September 2007.

United Nations. (2014). *World urbanisation prospects: The 2014 revision*. https://esa.un.org/unpd/wup/publications/files/wup2014-highlights.Pdf

United Nations Office for Disaster Risk Reduction. (2015–2030). *Sendai Framework*. http://www.preventionweb.net/drr-framework/sendai-framework/

United Nations Paris Climate Change. (2015). *Conference of the Parties Twenty-first session Paris*, 30 November to 11 December 2015. https://unfccc.int/resource/docs/2015/cop21/eng/l09r01.pdf

United Nations. (2014). *World urbanisation prospects: The 2014 Revision*. https://esa.un.org/unpd/wup/publications/files/wup2014-highlights.Pdf

Urry, J. (2010). Consuming the planet to excess. *Theory, Culture & Society, 27*(2–3), 191–212. https://doi.org/10.1177/0263276409355999

Wahl, C. (2016). *Designing regenerative cultures*. Triarchy Press.

Wahl, C. (2019). Sustainability is not enough. In G. D. Dabelko & K. Conca (Eds.), *Green planet blues: Critical perspectives on global environmental politics*. Taylor & Francis Group, 2019. ProQuest Ebook Central. http://ebookcentral.proquest.com/lib/flinders/detail.action?docID=5892383

Washington University In St. Louis. (2004, April 1). Laboratory rat gene sequencing completed; humans share one-fourth of genes with rat, mouse. *Science Daily*. Retrieved January 9, 2021 from www.sciencedaily.com/releases/2004/04/040401075930.htmThe

Watts, J. (2021). *Jane Goodall: 'Change is happening. There are many ways to start moving in the right way'*. https://www.theguardian.com/environment/2021/jan/03/jane-goodall-change-is-happening-there-are-many-ways-to-start-moving-in-the-right-way?

Public Education on the Rights of Sentient Beings to a Life and Death Free of Suffering

Janet J. McIntyre-Mills

Abstract Sentience is not the preserve of human animals; it is shared by all animals. Once this is accepted, it raises questions about rights, responsibilities and where we draw the line of inclusion and exclusion when making policy decisions. It has implications for ethics, law, governance and democracy and the way in which education is framed at the primary, secondary, tertiary and vocational level. Nussbaum has stressed the need to extend the boundaries of rights and responsibilities to all sentient beings. Many indigenous first nations extend the boundary to protect totemic kin that may include organic and inorganic life, which has implications for social and environmental justice.

Keywords Suffering · Rights · Life · Education · Multispecies relationships · Respect · Capabilities

1 Introduction

The chapter makes a case that respect for multiple species should be based on our recognition of what we have in common with other beings and our interdependency. We are not exceptional insofar as we are able to communicate, make decisions, demonstrate a sense of community, show empathy or to make political calculations. What does this mean for public policy, ethics, education, governance and the way we live our lives? What sort of changes need to be made to the content, structure and process of educational systems?

J. J. McIntyre-Mills (✉)
College of Education, University of South Africa, Pretoria Central, Gauteng, South Africa

College of Professions, Adelaide University, Adelaide, Australia
e-mail: janet.mcintyre@adelaide.edu.au

J. J. McIntyre-Mills, Y. Corcoran-Nantes (eds.), *Transformative Education for Regeneration and Wellbeing*, Contemporary Systems Thinking,
https://doi.org/10.1007/978-981-19-3258-8_11

Respect for others needs to include animals, plants and the earth. It is time to rethink rights and responsibilities and to end the commodification of other species and the habitat which we all share. We can no longer pretend that sentience and the ability to suffer is the preserve of human beings. As stressed elsewhere (McIntyre-Mills, 2021a, b, c, d) communication per se does not require human voices. It can rely on text, signalling and symbols. Meijer's (2016) research on the vast array of animal communication[1] explains that sounds, movements and chemicals, for example can convey information. Animals can *communicate* and some animals have a sense of self. Meijer (2016) explains that as dogs can recognise their own urine smells and are more interested in the smells of other dogs, it demonstrates a sense of self. This sense of self was once used as a criterion to separate humans from other animals and to justify treating them as a means to support human ends.

Tsing (2013, 2015) and Haraway (1991, 2011, 2015, 2016a, b) have also stressed that many species have a form of culture or learning that they can pass on to others, furthermore we need to realise that human culture has evolved as a multispecies relationship.

Currently ethical frameworks are inadequate as they are inherently anthropocentric without giving appropriate rights to other sentient beings. A non anthropocentric approach to systemic ethics is overdue (McIntyre-Mills, 2014, 2021a, b) as stressed in chapters "Regenerative Education and Safe Habitats for Diverse Species: Caterpillar Dreaming Butterfly Being", "Nurdles and Food Security: Education for Wellbeing and Implications for Pan-African Social and Environmental Justice", "Sacred Groves of the Tolon District of the Northern Region, Ghana: Where Spirituality Meets Education for Sustainable Development", "Aa!!Venda Women and Social Enterprise: Stepwise Process to Regenerative and Sustainable Living" and "Women, Education and Socialisation: Transforming Gender and Power in the MST, São Paulo, Brasil" in this volume in which we emphasise the need for a sense of kinship with other species.

Transformative education curricula need to mainstream a respect for multiple species and it needs to highlight the disastrous consequences of our current way of life which poses an 'existential risk' (Bostrom, 2011).

The pandemic has taught us that human beings are not top predators (McIntyre-Mills, 2021a, b, c, d, e) and that business as usual (McIntyre-Mills, 2020) is no longer taken for granted. As a result of the global pandemic academic engagement research and learning are now on line activities by necessity, but it has also provided us with an opportunity to engage as zoom researchers with colleagues in many parts

[1] Meijer (2016: 74) also stresses *that* the mirror test, developed by western male academics should not be regarded as the basis for deciding on species intelligence or sense of self. She adds that Gorillas do not gaze directly at others, as it is aggressive behaviour, a further reason why this sort of test is inaccurate. Many species are able to communicate as highlighted by Meijer (2016) who stresses that bats sing, bees dance, domestic dogs, cats and horses read our behaviour and body language, parrots use language to achieve goals, primates have learned sign language and dolphins have specific sounds which equate to names.

of the world. I am struck by the theme of death and dying and the way it has been addressed across the world. The suffering and death of human beings remain centre stage in the media. The suffering and death of other sentient beings are given far less attention, but of course both are inextricably linked.

The destruction of habitat by the carbon economy, the commodification of animals as sources of food, handbags or shoes can no longer be glossed over as the pandemic provides a platform for Attenborough and Goodall to remind people that we are part of one living system. Goodall[2] (2020) accentuates that the way animals are contained, farmed and marketed contributes to the ill health of animals and human beings resulting in pandemics. De-forestation and land clearance leads to species that have never been in contact before (and thus without any resistance) being thrust together causing cross species infections.

Humanity can only be protected by defending the biodiversity in which we are embedded. The current challenges facing humanity requires a different approach to the way we live our lives, underpinned by systemic ethics (McIntyre-Mills, 2014). The rate at which the planet is urbanising will result in more habitat being at risk than ever before in human history. Industrial farming is the root cause of swine flu and bird flu – and bringing wild creatures from many different contexts together in wet markets cannot be good for animal or human health as species that have not been in contact have no immunity (see McIntyre-Mills in *From Polarisation to Multispecies Relationships*, 2021 and McIntyre-Mills, 2021e, a paper on the importance of relationality and implications for Covid-19). This is a pragmatic response to the pandemic, but a normative response is simply to recognise that human beings share sentience with other creatures, (Ackerman, 2016; De Waal, 2009; Meijer, 2016). The excuse for denying rights and responsibility for a life worth living for other animals is no longer excusable.

The pandemic has given us time to consider human mortality and what it means to suffer and to die. It also raises the question of the quality of life and quality of death for all sentient beings.

This is a topic that has been given attention by Martha Nussbaum (2006) in *Frontiers of Justice* in which she outlined the limits of the social contract and protection of citizens who have the capability and right to vote, but which denies rights to young people and those outside the mantle of citizenship, the frail and disabled and other animals. Protection of the environment is vital to protect all the inhabitants of a region, an issue raised by the *Stop Ecocide* movement[3] to implement a new Ecocide Law (Higgins et al., 2013) that builds on the work of Arthur Galston who campaigned in the wake of the Vietnam war.[4]

---

[2] https://www.cnn.com/videos/us/2020/03/19/coronavirus-jane-goodall-acfc-full-episode-vpx.cnn

[3] https://www.stopecocide.earth/dare-to-be-great The website defines the concept that is now widely cited by world leaders such as Pope Francis, Greta Thunberg and Jane Goodall.

[4] https://news.yale.edu/2008/07/18/memoriam-arthur-galston-plant-biologist-fought-use-agent-orange

The right to a life worth living requires some basic conditions outlined by Nussbaum (2011). If we consider the capabilities for human and other sentient beings they can be distilled as being based on a sense of purpose, trust and an ability to engage with others. This applies to all animals ranging from domestic, to farmed and wild animals, to use the categories outlined by Donaldson and Kymlicka (2011) in *Zoopolis*. They stress the need to give rights to animals and to ensure that these rights are upheld in law.

The next frontier of sentience and suffering is to give animals rights to 'a life worth living' (see Nussbaum, 2006 on extending frontiers of justice).

Margaret Somerville (2003, see White & Marciniak, 2019; Ryan, 2019) a bioethicist stresses that euthanasia is being proposed as bills or passed as laws in many more contexts and she has expressed reservations saying that with appropriate care, most people could die dignified deaths, Ryan cites Somerville as follows:

> The pro-euthanasia lobby aims to relieve suffering and to be compassionate and those are absolutely good motives but to do it this way is ethically unacceptable,

Ryan continues to cite her argument that:

> With our advanced research on pain and methods … we can do that without killing the person and in fact it's been shown in research; eliminating pain extends life.

White and Marciniak (2019) explain that the Voluntary Assisted Dying Act in Victoria and WA only allow euthanasia if:

> terminally ill with intolerable pain, and likely to die within six months, or 12 months for neurodegenerative diseases. It is illegal in all other states and territories.

Since writing the first draft of this chapter, the Voluntary Assisted Dying Bill for those aged above 18 years tabled in 2020 in South Australia[5] passed the lower house and will be subject to a further voting process to decide if it could become law within 2 years' time.[6]

In the *Ethical Canary* Margaret Somerville (2004: 142) recalls her personal experience with addressing the pain her father in Australia suffered. She was fortunately able to intercede on his behalf as a bioethicist, but she acknowledges that not all elderly people in unbearable pain have advocates who are able to achieve adequate pain management. She discusses the distinction between pain management to kill a patient and pain management to kill pain and the importance of achieving a delicate legal balance. She recalls how her father said she should do more to address this area of silence.

Nussbaum, however acknowledges that some forms of cancer and neurological pain are very difficult to manage and stresses that pain management if successfully managed can prolong life in some instances and in others (if necessary) it could shorten it. She also distinguishes between active and passive euthanasia but does not

---

[5] Voluntary Assisted Dying Bill 2020, Introduced by Susan Close MP.

[6] https://www.9news.com.au/national/voluntary-euthanasia-bill-south-australia-parliament-assisted-dying-closer/923c46c6-a64f-43aa-b76e-849c0e4687f7. 6:30pm May 6, 2021 accessed 5/08/2022.

engage in the issue of long, prolonged suffering that can occur for those without legal advocates. Since 1997 the Death With Dignity Act in Oregon (USA) makes euthanasia permissible *under specific circumstances*[7] and in Australia a bill is currently being proposed in Tasmania whilst Voluntary Assisted Dying (under specific circumstances) is legal in WA and Victoria[8] and in Tasmania.[9]

The documentary Laura's Choice (2021) explores the right to die. Laura was aged 90 and was afraid that if she had a stroke she would be abused in an old age home and die without dignity.[10] Her daughter and granddaughter as film makers were tasked with documenting her final journey. They had great reservations about the task set for them by Laura, but eventually decided to comply with her request. They diarised her journey and explored Laura's choice by interviewing a Christian, an Islamic cleric an Aboriginal elder and a Buddhist as well as a counsellor who worked with the dying and bereaved. Whilst all were tactful and empathetic about suffering and none passed judgement, *not one supported active (as opposed to passive)* euthanasia. They all said nature should take its course (whilst providing passive palliative care and support) was a better option. Showing elders that they *are valued* is an important aspect of care underlined by the Aboriginal elder. The Buddhist who was interviewed listened with empathy, but also made it clear that whilst it is important to die peacefully, all stages of life require being mindful. A legal expert stressed that if the family had intervened in any way (as originally requested) they could have been charged with manslaughter. Accompanying Laura to Switzerland for *a holiday* is legal which is why she needed to take a two way ticket. Suicide by rational people of sound mind is legal in Switzerland, but they have to activate the final dose themselves. At each step they need to say that they wish to proceed and the entire process is carefully governed by checks, balances and legal representatives.

The notion that suffering is acceptable for others and that the decisions are made by those who are not suffering[11] is a point made by Laura that is worth considering. But to me the issue is *who decides, why and under what circumstances* before assessing whether euthanasia is acceptable.

---

[7] https://www.oregon.gov/oha/ph/providerpartnerresources/evaluationresearch/deathwithdignity-act/pages/faqs.aspx#whatis

[8] https://end-of-life.qut.edu.au/euthanasia

[9] https://www.abc.net.au/news/2021-03-04/tasmania-house-of-assembly-passes-vad-bill/13215328

[10] Her argument is that if she had the right to medication that could end her life painlessly that she would go on living right up until the last stages knowing that she had the right and the power or control to end it. She summed it up by saying that denying elderly people the right was cruel, but she did not enter the argument of how younger people would feel if they were also bereft of capabilities and reliant on the kindness of others.

[11] The last stages of dying are not easy for those with many kinds of cancer, for example. It was harrowing to beg for more pain management for a relative. I was not the primary advocate and to be honest I think it was insufficient and so did my family members.

In the documentary Laura summed up her opinions only for the elderly leaving out the more difficult areas of prolonged suffering by others with incurable illnesses, immobility, pain and an inability to address many of the capabilities[12] deemed to make life worth living.

Pereira (2011) provides counter arguments about the slippery slope and claims that the Dutch example shows how euthanasia has been expanded to include more and more categories of suffering. What about young or younger people, those who are " just depressed" and those with debilitating conditions? Perhaps society needs to change and to realise that being inclusive and valuing people is vital.

How long should the reflection on the choice be required, to ensure that it is not just a passing shadow on a life that could be worth living?

Transformative public education needs to open up this debate to find a way forward that protects the sanctity of all life as an ideal, that recognises the need for compassion, fairness, empathy to all sentient beings. When life becomes unbearable the law should remain open to exceptions where suffering needs to be alleviated.

For human beings ending suffering entails ensuring social inclusion, giving sentience meaning in terms of the 10 capabilities:

> (1) living a life that is not cut short prematurely (2) bodily health, (3) bodily integrity, (4) sense, imagination, and thoughts, (5) emotions, (6) practical reason, (7) Affiliation, (8) links to other species, (9) Play, (10) Control over one's environment.

According to a BBC news report (14th Oct, 2020):

> The Dutch government has approved plans to allow euthanasia for terminally ill children aged between one and 12".[13] The source stresses that less than 10 children a year would be impacted by the law.

The New Zealand government has recognised animals as sentient beings since 2015[14] and recently in her second term Jacinta Ardern has stressed the need for a

---

[12] As stressed throughout this work and in several sole and joint chapters by McIntyre in 'From Polarisation to Multispecies Relationships' (McIntyre-Mills & Corcoran Nantes, 2021) and in a special essay for the ISSS bulletin (2021): "Nussbaum (2006) stresses the need for the right to a life worth living to be extended to all sentient beings. Nussbaum's (2011) ten capabilities are directly concerned with wellbeing, a life worth living and the extent to which these can be addressed. These capabilities, include: (1) living a life that is not cut short prematurely (2) bodily health, (3) bodily integrity, (4) sense, imagination, and thoughts,(5) emotions, (6) practical reason, (7) Affiliation, (8) links to other species, (9) Play, (10) Control over one's environment."

Nussbaum wrote with Amartya Sen on capabilities but in her own version she focuses on protecting "areas of freedom so central that their removal makes a life not worthy of human dignity" (p. 31)

[13] Netherlands backs euthanasia for terminally ill children under-12. https://www.bbc.com/news/world-europe-54538288, 14th October, 2020.

[14] Ivens, S. 2015 New Zealand Legally Recognises Animals as 'Sentient' Beings https://animalequality.org.uk/blog/2015/05/12/new-zealand-legally-recognises-animals-as-sentient-beings/#:~:text=New%20Zealand%20takes%20a%20monumental,%2C%20are%20%E2%80%9Csentient%E2%80%9D%20beings

Minister of Animal Welfare.[15] The reinstatement of the Minister was confirmed in November 2020.[16] But live animal export continues:[17]

> In recent weeks, the live animal export trade has again been under the global spotlight, first for abandoning animals to abject cruelty, then for a human and animal disaster almost too painful to contemplate. The world is still reeling from the heart-breaking news that an export ship carrying nearly 6,000 cattle had capsized en route from New Zealand to China, killing all animals and 41 crew members. This is the latest tragedy in a long line of devastating disasters that reveal an industry that is willing to put both people and animals at high risk in its pursuit of profit.

The capabilities for a life worth living for sentient beings (drawing on and adapting Nussbaum, 2006, 2011) include being able to express all one's animal capabilities and affiliate with one's own kind in a *species appropriate environment*. These are conditions that are denied to many animals who are treated as commodities, confined in miserable conditions and transported long distances. These are conditions that ought to be outlawed.

In terms of Nussbaum's (2006) argument the frontiers of justice need to be expanded to protect all sentient beings, namely human beings (the frail, elderly, disabled and other voiceless animals whose capabilities have long since been underestimated. Meijer (2016) has explained that animals communicate and are sentient (Proctor, 2012), whilst Gagliano et al. (2018) and Rayner (2010) demonstrate that plants are capable of making decisions and thus can have a form of consciousness, which has implications for ecological justice (Stephens et al., 2019).

This brings us to the last frontier for transformation, namely the need to recognise that the earth has rights and that all living systems of which we are a strand need to be protected (Higgins et al., 2013). This campaign continues to gain ground as we begin to face up to our interdependency which ought to be protected by systemic ethics and international laws.

James (2016) discusses the "impact of the recognition that animals are sentient in the 2013 amendment to the Animal Welfare Act 1999, examining the rationale for the change and the drivers underpinning the current legal approach to animal welfare in New Zealand." She concludes that to date the act is "largely symbolic" and that "codes of welfare need to be developed."

---

[15] https://www.worldanimalprotection.org.nz/news/why-we-need-minister-animal-welfare
Why we need a Minister for Animal Welfare23/10/2020

.. – human and animal health and survival are co-dependant – and yet we do nothing.

[16] https://www.scoop.co.nz/stories/PO2011/S00018/safe-pleased-to-see-the-role-of-minister-responsible-for-animal-welfare-returned.htm Monday, 2 November 2020, 3:19 pm Press Release: SAFE NZ

[17] https://animalsaustralia.org/features/why-we-are-determined-to-end-live-export.php

Although New Zealand banned sheep export for meat[18] it continues to transport animals at sea and they continue to suffer.[19]

Jacinda Ardern has re-instated a Minister for Animal Welfare which could provide an example of a small step in the right direction .

In South Africa two government departments, namely the Department of the Environment and the Department of Agriculture are at loggerheads about animal rights and welfare. The former has stressed that they ought not to be commodified by genetically modifying or farming them for consumption. The issue of so-called 'sustainable hunting' also needs to be addressed, according to Pinnock who cites[20] reports on animal management and stresses there is:

> … solid scientific proof that many creatures with which we share the planet are indeed sentient as individuals and can suffer. This approach is also at the heart of a raft of conservation legislation presently making its way through the British parliament and is accepted by a number of countries.

Our values and decisions shape the world in which we live. If we are to have a hope of achieving policy transformations that re-generate (Wahl, 2016, 2019) and sustain the environment, then public education needs to play a central role in protecting the future for current and future generations of life and by fostering the capabilities *we* need to enable that to occur.

References

Ackerman, J. (2016). *The genius of birds*. Scribe.

Bostrom, N. (2011). *Existential risk prevention as the most important task for humanity*. Faculty of Philosophy & Oxford Martin School University of Oxford. http://www.existential-risk.org/

De Waal, F. (2009). *The age of empathy: Nature's lessons for a kinder society*. Harmony Books.

Donaldson, S., & Kymlicka, W. (2011). *Zoopolis: A political theory of animal rights*. Oxford University Press.

Gagliano, M., Abramson, C. I., & Depczynski, M. (2018). Plants learn and remember: Let's get used to it. *Oecologia, 186*, 29–31. https://doi.org/10.1007/s00442-017-4029-7

Goodall, J. (2020). https://www.cnn.com/videos/us/2020/03/19/coronavirus-jane-goodall-acfc-full-episode-vpx.cnn

Haraway, D. (1991). *The reinvention of nature*. Free Association Books.

Haraway, D. (2011). Speculative fabulations for technoculture's generations: Taking care of unexpected country. *Australian Humanities Review, 50*, 2011.

Haraway, D. J. (2015, May 1). Anthropocene, Capitalocene, Plantationocene, Chthulucene. *Environmental Humanities, 6*(1), 159–165. https://doi.org/10.1215/22011919-3615934

[18] https://www.abc.net.au/news/rural/2018-05-08/how-new-zealand-banned-live-export-trading/9733146

[19] https://safe.org.nz/take-action/demand-ban-live-export-2/

[20] https://conservationaction.co.za/resources/reports/an-assessment-of-the-potential-risks-of-the-practice-of-intensive-and-selective-breeding-of-game-to-biodiversity-and-the-biodiversity-economy-in-south-africa/

Haraway, D. J. (2016a). *Staying with the trouble: Making Kin in the Chthulucene*. Duke University Press.
Haraway, D. (2016b). *Anthropocene, Capitalocene, Chthulucene: Making string figures with biologies*. Arts, Activisms – Aarhus University YouTube. https://m.youtube.com/watch?v=CHwZA9NGWg0
Higgins, P., Short, D., & South, N. (2013). Protecting the planet: A proposal for a law of ecocide. *Crime, Law and Social Change, 59*, 251–266. https://doi.org/10.1007/s10611-013-9413-6
James, V. (2016). *Recognising animal sentience: Including minimum standards for opportunities to display normal patterns of behaviour in codes of welfare in New Zealand*. URI: http://hdl.handle.net/10063/6335, http://researcharchive.vuw.ac.nz/xmlui/handle/10063/6335
James Tapper Sun. (2021, March 28). *16.15 AEDT*. https://www.theguardian.com/world/2021/mar/28/mps-unite-to-call-for-total-ban-on-wicked-foie-gras-in-the-uk
Laura's Choice. (2021). *ABC documentary*. https://iview.abc.net.au/show/laura-s-choice?nw=0, https://www.abc.net.au/news/2021-03-24/laura-choice-three-women-journey-with-voluntary-assisted-dying/13271708, https://iview.abc.net.au/show/laura-s-choice
McIntyre-Mills, J. (2014). *Systemic ethics and non-anthropocentric stewardship* (p. 270). Springer.
McIntyre-Mills, J. (2020). The Covid-19 era: No longer business as usual. *Systems Research and Behavioral Science, 37*(5), 827–838. https://doi.org/10.1002/sres.2745
McIntyre-Mills, J. J., & Corcoran-Nantes, Y. (Eds.). (2021). *From polarisation to multispecies relationships in the age of mass extinctions*. Springer.
McIntyre-Mills, J. J. (2021a). Communication and culture: A multispecies endeavour within a shared habitat. *Systems Research and Behavioral Science, 38*(5), 671–684 https://doi.org/10.1002/sres.2810
McIntyre-Mills, J. J. (2021b). Prologue: No longer top predator. In *From polarisation to multispecies relationships* (pp. xii–xxix). Springer.
McIntyre-Mills, J. J. (2021c). From polarisation to multispecies relationships: Re-membering narratives, chapter **10**. In J. McIntyre-Mills & Corcoran-Nantes (Eds.), *From polarisation to multispecies relationships* (pp. 173–212). Springer.
McIntyre-Mills, J. J. (2021d). Vignette: 'Knackered. We are all flesh'. In J. McIntyre-Mills & Corcoran-Nantes (Eds.), *From polarisation to multispecies relationships* (pp. 687–688). Springer.
McIntyre-Mills, J. J. (2021e). The importance of relationality: A note on co-determinism, multispecies relationships and implications for COVID-19. *Systems Research and Behavioral Science, 39*(2), 339–353. https://doi.org/10.1002/sres.2817
Meijer, E. (2016, translated, 2019). *Animal languages: The secret conversations of the living world*. John Murray.
Mitman, G. (2019) *Reflections on the Plantationocene: A conversation* with Donna Haraway and Anna Tsing June 18, 2019 updated October 12, 2019. https://edgeeffects.net/haraway-tsing-plantationocene/
Netherlands backs euthanasia for terminally ill children under-12. https://www.bbc.com/news/world-europe-54538288, 14 October 2020. https://www.researchgate.net/publication/223922921_Rethinking_Project_Management_Researching_the_Actuality_of_Projects
Nussbaum, M. (2006). *Frontiers of justice: Disability, nationality, species membership*. Belknap Press.
Nussbaum, M. (2011). *Creating capabilities: The human development approach*. The Belknap Press.
Pereira, J. (2011). Legalizing euthanasia or assisted suicide: The illusion of safeguards and controls. *Current Oncology, 18*(2), 38–45. https://doi.org/10.3747/co.v18i2.883
Pinnock, D. (2021). *South Africa at the cross roads on wildlife welfare*. https://www.dailymaverick.co.za/article/2021-06-01-south-africa-at-a-crossroads-on-wildlife-welfare-and-sentience/
Proctor, H. (2012). Animal sentience: Where are we and where are we heading? *Animals (Basel), 2*(4), 628–639. Published 14 Nov 2012. https://doi.org/10.3390/ani2040628
Rayner, A. D. M. (2010). Inclusionality and sustainability – Attuning with the currency of natural energy flow and how this contrasts with abstract economic rationality. *Environmental Economics, 1*, 98–108.

Ryan, D. 29 July 2020 (2019). https://www.catholicweekly.com.au/ethicist-margaret-somerville-compares-the-spread-of-euthanasia-legislation-to-a-slow-moving-pandemic/
Sen, A. K. (1999). *Development as freedom*. Oxford University Press.
Sen, A. (2005). Human rights and capabilities. *Journal of Human Development, 6*(2), 151–166.
Somerville, M. (2003, February 17). "Death talk": Debating euthanasia and physician-assisted suicide in Australia. *Medical Journal of Australia MJA, 178*, 171–174.
Somerville, M. A. (2004). *The ethical canary: Science, society, and the human spirit*. McGill-Queen's University Press.
Stephens, A., Taket, A., & Gagliano, M. (2019). Ecological justice for nature in critical systems thinking. *Systems Research, 36*, 3–19. https://doi.org/10.1002/sres.2532
Stokols, D. (2018). *Social ecology in the digital age: Solving complex problems in a globalized world*. Academic.
Tsing, A. (2011) Arts of inclusion, or, how to love a mushroom (Anna Ting, for the matsutake worlds research group). *Australian Humanities Review, 50*. https://press-files.anu.edu.au/downloads/press/p111121/html/ch01.xhtml?referer=&page=2#toc-anchor
Tsing, A. (2013). More-than-human sociality: A call for critical description. In K. Hastrup (Ed.), *Anthropology and nature* (pp. 27–42). Routledge.
Tsing, A. L. (2015). *The mushroom at the end of the world: On the possibility of life in capitalist ruins* (p. 352). Princeton University Press.
Wahl, C. (2016). *Designing regenerative cultures*. Triarchy Press.
Wahl, C. (2019). Sustainability is not enough. In G. D. Dabelko & K. Conca (Eds.), *Green planet blues: Critical perspectives on global environmental politics*. Taylor & Francis Group. ProQuest Ebook Central, http://ebookcentral.proquest.com/lib/flinders/detail.action?docID=5892383
White, L., & Marciniak, C. (2019). *With voluntary assisted dying not an option, Laura Henkel is heading to Switzerland to 'die with dignity'*. ABC North Coast Posted Sunday15th December 2019 at 8:41am, updated Sunday 15 December 2019 a 2:41pm.

An Assault on Situational Complexity in the Arena of Education: The Potential of Structured Dialogic Design

Alexander N. Christakis and Jeff Diedrich

Abstract The phenomenon of situational complexity emerges when groups of stakeholders are engaged in addressing wicked problems and/or designing social systems, such as education programs, health plans, transportation networks, etc. World-wide applications of the methodology of Structured Dialogic Design (SDD), with people from all walks of life, have accumulated evidence for the conceptualization of three distinct and overlapping components of situational complexity. These are: (a) objective, which is observer-independent because of its universality, (b) subjective, which is observer-dependent, and (c) intersubjective, which is observer-interdependent. We identify and investigate two critical issues inhibiting the launching of an effective assault on situational complexity: (a) how can we assault the intersubjective component by means of stakeholder inclusion, and (b) how can we adopt and implement inclusive deliberative democracy. Empirical evidence from applications of SDD, indicates that these two issues operate as a vicious cycle. The vicious cycle is responsible for the escalation of the "Global Problematique" as defined in the Club of Rome prospectus, namely of a system of strongly interacting problems, such as climate change, inequities in education, biodiversity reduction, and many more. We employ the Logosofia platform of SDD to simulate the design of a strategy for escaping "the vicious cycle Labyrinth." We argue that extensive applications of SDD in the educational arena, will enable young and old adults to become co-owners of our Problematique, and co-creators of a strategy for escaping the Labyrinth. As an exemplary application, we present a case study of an assault on the situational complexity emerging from the engagement of deaf and hard of hearing stakeholders in designing an educational program for their school in the state of Michigan, USA.

Keywords Complexity · Structured Dialogic Design · Intersubjectivity · Problematique · stakeholders · deliberative democracy · Demosofia · Logosofia · paradigm shift · School of the deaf

A. N. Christakis · J. Diedrich (✉)
Institute for 21st Century Agoras, Crete Dietrich, IL, Greece

J. J. McIntyre-Mills, Y. Corcoran-Nantes (eds.), *Transformative Education for Regeneration and Wellbeing*, Contemporary Systems Thinking,
https://doi.org/10.1007/978-981-19-3258-8_12

1 Introduction

In the early 1970s, when the American cities were burning on account of racial discrimination and injustice, a team of researchers at the Academy for Contemporary Problems, where Christakis was serving as a Fellow responsible for a program on Social Prognosis, designed an experiment on interdisciplinarity. They formed an interdisciplinary team consisting of architects, sociologists, economists, engineers, psychologists, etc., and assigned them the task of designing a hypothetical city of one million people. Christakis, together with a social psychiatrist (Jim Taylir), and a systems engineer (John Warfield), assumed the role of observing the deliberations of this interdisciplinary team to discover any barriers in their capacity to collaborate in designing the hypothetical city. After one year of observing this team, it became apparent to the observers that there existed strong barriers among the interdisciplinary team members in terms of cross-disciplinary communication and collaboration. Team members resigned because of frustration. The team leader was changed three times. The observers concluded that the interdisciplinary team was incapable of collaborating and building a consensus on account of the situational complexity embedded in the design of the hypothetical city. They discovered that the interdisciplinary team lacked the appropriate methodological support to meet the design challenge of the hypothetical city. This discovery led the Academy to launch a research program for the development of a set of consensus methods with the objective of supporting the deliberations of teams confronting complex situations emerging in designing social systems (Arnstein & Christakis, 1976; Warfield, 1976).

Prior to the interdisciplinarity experiment, Christakis was a co-founder of a revolutionary scientific initiative known as the Club of Rome (CoR). The vision of the CoR was described in the Prospectus on the Predicament of Mankind (Ozbekhan, 1970). In the prospectus, Ozbekhan conceptualized forty-nine Continuous Critical Problema (CCPs), which on account of their strong interactions create a new entity, which he named the "Global Problematique." The term Problematique was an attempt to distinguish it from ordinary problems, most of which in the past one hundred years could have been addressed independently, i.e., without taking into account their systemic connections and interactions. Due to the strong interactions among the forty-nine CCPs, the emergent entity becomes the subject of an escalated situational complexity. The connections among any set of elements, e.g., problems, actions, atoms, etc., are responsible for the creation of new entities. For example, depending on the network of connections among a set of carbon atoms, we can get coal or diamond. Similarly, the chemical reaction between oxygen and hydrogen produces water as the new entity. The CoR prospectus proposed that, in order to deal effectively with the situational complexity of the emergent Global Problematique, it is imperative to transcend interdisciplinarity and enter the domain of intersubjectivity. This domain implies the engagement of the diverse voices of stakeholders from all walks of life, as experts and producers of the CCPs, in authentic, effective, and inclusive deliberative democracy. Given that people are the producers of the Global Problematique, they ought to be included in deliberations, in order for them to internalize their role and responsibility for its resolution. It also became clear that conventional methodologies of simulation and extrapolation modelling of the data base of the current situation would perpetuate the Global Problematique instead of resolving it. In other words, the strong interactions among the CCPs required the development of new methodologies for an effective assault on its complexity. These methodologies should be capable of engaging stakeholders in consensus-building, and in co-creating visionary anticipations of alternative futures by means of successive approximations with structural ontological representations (Kakoulaki et al., 2022).

These two experiences, namely: (a) the interdisciplinary team for designing a hypothetical city ,and (b) the CoR prospectus, became the rationale and the foundation for launching a research and testing program in the 1970s, initially at the Academy for Contemporary Problems and subsequently at the Universities of Virginia and George Mason, culminating in the development of the generic/dialogic design science, and the methodology of Structured Dialogic Design (SDD) (Warfield, 1994; Christakis & Bausch, 2006; Flanagan & Christakis, 2010; Flanagan & Lindell, 2018; Jones, 2014, Diedrich & Christakis, 2021). The SDD methodology has been applied by many groups of practitioners in engaging stakeholders focusing on a variety of complex situations around the world. Examples are: (a) Tribal governance by the Americans for Indian Opportunity (www.aio.org); Broome & Christakis, 1988), (b) The Cyprus Academy for Public Administration (CAPA) in conducting Bicommunal peace-building workshops involving Turkish and Greek Cypriots for a period extending over thirty years, under the leadership of Marios Michaelides, (c) the World Health Organization (WHO) in Geneva in 1998, for the design of a plan for the elimination of the Lymphatic Filariasis epidemic in Africa by the year 2020 (Flanagan & Post, 1998), (d) drug evaluation and research by the Food and Drug Administration (FDA) (Christakis et al., 1995), (e) The School for the Blind for the state of Massachusetts in the USA (Diedrich & Christakis, 2021), (f) The issue of abandoned animals for the Demoscopio of the city of Heraklion, Greece, (Kakoulak, 2016), and many more (www.globalagoras.org). In the arena of education, there have been many applications in the USA during the decades of 2000, and 2010, for the states of Ohio (Magliocca & Christakis, 2001), and with Diedrich in Michigan beginning in 2007. Recently, Diedrich and his team of SDD practitioners, applied the methodology for the development of an educational program for the Michigan School for the Deaf, which is the case study we will discuss in this chapter in order to demonstrate empirically the potential of SDD for launching an assault on situational complexity.

2 The Demosofia Paradigm

The scientific revolution (Kuhn, 1970) for the science of dialogue started in the 1970s (Christakis, 1973). The new scientific paradigm has been called Demosofia, which in Greek means the "wisdom of the people" (Christakis, 1993). The name implies a paradigm shift from the traditional meaning of the term democracy, which in Greek meant the "power of the people", to the "wisdom of the people." The underlying premise of the Demosofia paradigm is that it is necessary (though not sufficient) to explicate the wisdom of the people in order to launch an assault on the contemporary situational complexity. Unfortunately, as we will elaborate in this chapter, wisdom is necessary but not sufficient. Moreover, because of the escalating complexity it is much more difficult today to make wise design choices collectively for a variety of social systems, as compared to those made during the golden era of democracy in the Agora of Athens, Greece (Fig. 1).

As mentioned in the Introduction, in the early 1970s a group of researchers at the Academy discovered the importance of applying science for improving the process of interdisciplinary dialogue and collaboration. After conducting several experiments on complex tasks, it became apparent to the researchers that a new scientific paradigm was needed if experts from diverse disciplines, perspectives, and experiences, were to be engaged in productive, intersubjective dialogue (Kakoulaki & Christakis, 2018; Christakis & Kakoulaki, 2021).

One of the major findings of the Academy researchers observing the deliberations of the interdisciplinary team was that their rate of progress was extremely slow. Each member would come to the interdisciplinary meetings with proposals drafted during single discipline meetings. For example, the economist would present the results of a land-use model projecting and distributing the employment of the population of the city to the year 2000. These proposals were drafted in collaboration with other economists. The single discipline meetings were productive and efficient. They were capable of integrating the "individual mental models" of every team member into "a team mental model." The dialogue among the team members of any single discipline was productive because each discipline used the scientific

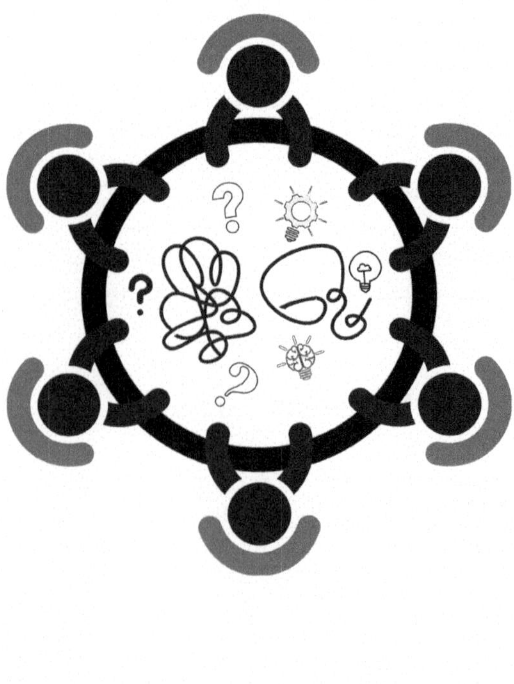

Fig. 1 The roundtable of the Demosofia paradigm for the practice of authentic and inclusive deliberative democracy in order to launch an assault on situational complexity

language of the particular discipline to communicate and build a consensus. The breakdowns in communication emerged during their efforts to engage in interdisciplinary dialogue in order to integrate the single discipline proposals into a comprehensive (systemic) design of the city. "Connecting the dots" across disciplinary boundaries turned out to be time consuming and frustrating.

The frustration of the representatives of the disciplines increased exponentially. Some members declined to participate in interdisciplinary meetings, or found excuses not to attend. The leader of the interdisciplinary team was changed three times as a result of the resignation of selected leaders. As social psychiatrist James Taylor wrote, after observing the deliberations of the team for over a year (Taylor in Arnstein & Christakis, 1976):

> *... there appears to be a pressing, well-recognized need for a kind of social intervention, the interdisciplinary team which synthesizes knowledge in order to clarify complex problems. The promise of this social invention is clear, yet in fact no workable model has emerged. The question becomes obvious: why not? What has gone wrong in existing efforts to develop 'meaningful synthesis' of 'pertinent fields of knowledge'*

Dr. Taylor, along with two researchers of the Academy, recognized that many other wicked problems – certainly the Global Problematique and problems confronting contemporary societies and organizations – could not be resolved without the integration of knowledge, expertise, and wisdom originating from diverse disciplines and perspectives. However, without empowering groups to cross the boundaries of their knowledge and experience in order to build a consensual linguistic domain, the prospects of making progress were not good. This conclusion prompted the Academy to launch a long-range research and development program focusing on the Demosofia paradigm (Christakis, 1973).

It has been suggested that a shift to the Demosofia paradigm is inevitable (Christakis, 1993), on account of the inability of organizations and the society as a whole to launch an assault on the situational complexity embedded in the Global Problematique. What is needed is a paradigm shift from the democratic ideal practiced over two thousand years ago in the Agora of the Athenian Republic to the postmodern democracy advocated by leading scholars today (McIntyre-Mills & Romm, 2018, 2021; Midgley & Rajagopalan, 2021).

3 Overview of the SDD Methodology

When solutions are not apparent,
we should turn to new processes for answers…
When consensus seems to be unattainable
Structured Dialogic Designsheds light into shades
through the spoken word
and reveals the truth of a new common language,
the shared collective knowledge
and the co-created "groundless ground" to move forward

– Maria Kakoulaki-Journalist, SDDFacilitator

In this section we present an overview of the SDD methodology by discussing: (a) the three phases of the methodology, (b) the four steps of the colaboratory of Democracy phase for generating, prioritizing, clustering, and mapping ideas, and (c) the Logosofia platform for knowledge generation and structuring (Diedrich & Christakis, 2021).

The three Phases are:

4 The Discovery Phase

During this phase, the SDD practitioners responsible for the application, engage in an inquiry which involves: (a) Conducting interviews with few selected members of the community of stakeholders in order to explore their alternative understandings and perceptions of the situation, (b) Gathering and studying relevant documentation and data (c) Selecting a Core Planning Team (CPT), which includes at least one member of the community, called the Broker, for guidance in the design of the colaboratory phase, (d) Framing one or more Triggering Questions (TQs) for focusing the deliberations of the participants during the colab, and (e) Identifying a representative group of participants to the colab, in accordance with the Law of Requisite Variety (Warfield, 1994; Christakis & Bausch, 2006), i.e., all perspectives should be present at the table.

5 The Colaboratory Phase

This phase focuses on applying the methodology of structured dialogue for engaging the group of participants in a productive, effective, authentic, and democratic deliberation. This approach to deliberative democracy emulates the Agora of the Athenian democracy in 600 BC, and is based on four distinct and overlapping steps, namely: (a) Generation of ideas in response to the Triggering Question, (b) Evaluation of the ideas proposed, by asking participants to vote on the relative saliency of the ideas at various points during the deliberation, (c) Clustering the set of ideas into affinity groups by employing the similarity relationship, and (d) Mapping a priority subset of ideas by exploring pairwise relationships of influence among them, and producing a tree-pattern displaying their influences (digraph).

6 Follow-Up Action Phase

This phase focuses on implementing the findings and recommendations of the colab. For example, if the group work during the colab has produced a consensus action plan, then specific activities are assigned to members of the group of the community, for implementation. This phase might produce the requirement for additional colabs with different stakeholders focusing on diverse, and sometimes more detailed Triggering Questions. For example, the discovery of the root cause of a complex situation, might suggest the requirement to design another colab, with another group to drill down on this root issue. This happened when working with a group of educators and parents of a school system in Flint, Michigan. The family structure emerged as the root cause for the rate of high school drop out by students. A new colab was designed focusing on discovering the reasons for the family breakdowns. Poverty emerged as the root cause of the new map, produced by the different group of stakeholders.

In many applications we have experienced a tapestry of deliberations over time, offering the community the opportunity to conceptualize and implement a continuum of iterative learnings and actions for the resolution of the complex situation (Flanagan & Christakis, 2010). An example of a continuum of applications is the case with the Center for Drug Evaluation and Research (CDER) of the Food and Drug Administration (FDA), where an SDD team worked with medical doctors and many other scientific disciplines for three years, with the objective to improve the drug evaluation and research process (Christakis, 1996).

7 The Discovery Phase

The Discovery phase involves a series of considerations and is full of nuances depending on the uniqueness of the complex situation. There is not a single formula that works for every application, but there are some general guidelines. Initially, a meeting occurs between the SDD practitioner(s) and the Broker, i.e., a person assigned to the project by the sponsor of the work to collaborate with the SDD practitioners. The sponsor is essentially the owner and the expert of the complex situation to be addressed by a representative group of stakeholders in a colab. The role of the Broker is very critical for the successful application of SDD. It is essential that the Broker has the respect and the trust of the leadership of the sponsoring organization.

During the initial meeting(s), a series of questions are addressed to assist in appreciating the complexity of the situation. Some of these questions are incorporated in the Logosofia platform because they are generic. For example, one question is to find out if other methodologies have been applied to the situation, and whether they have failed. If other efforts have not been effective, SDD should be attempted for an application, because of its track record of success. In addition, the initial discussions provide insights regarding the appropriateness of SDD for the specific situation. The reason for this is the fact that its successful application requires that the leadership of the sponsoring organization is humble, open-minded, committed, and capable of valuing the wisdom of the community (Demosofia). Participants engaging in colabs develop a strong affinity for the collective wisdom, as well as a deep investment and ownership of the definition and resolution of the situation. It follows that they would be disappointed and discouraged if their findings and recommendations were not being appreciated by the leadership (Flanagan & Christakis, 2010). It is at the Discovery phase that the SDD practitioners and the Broker assess the commitment of the sponsor to implement the results of the group work. If the sponsor ignores the recommendation of the group, engaging stakeholders will be counterproductive and detrimental to the organization to proceed with the application of SDD.

The preliminary choice of an application Archetype (Diedrich, 2010; Diedrich & Christakis, 2021), can be helpful in informing the level of effort and the depth of the Discovery phase that is necessary to ensure a successful application. In the hundreds of applications of SDD worldwide, the few failures are attributed to the lack of effective Discovery phase preparation. For example, if the sponsor is interested in engaging a group in the definition of the complex situation and not in designing a consensus action plan, then most likely a two-day colab will be sufficient. Furthermore, the choice of the Archetype can guide framing preliminary Triggering Question(s), which are essential for focusing the dialogue during the colabs. While it is helpful to identify a preliminary Archetype, it is advisable not to be constrained by the initial selection. As the Discovery phase progresses, new insights may reveal necessary adjustments, including a shift to an alternative Archetype.

Another important aspect of the Discovery phase is the formation of a Core Planning Team (CPT), consisting of five-seven members. The CPT is responsible for making critical decisions regarding: (a) Framing the TQ, (b) Finalizing the selection of the appropriate Archetype, and (c) Selecting a representative group of participants for the colab deliberations, making sure that the **Law of Requisite Variety** is not compromised (Flanagan & Christakis, 2010; Jones, 2018). The following characteristics are desirable for the members of the CPT:

- Knowledge of the situation;
- Ability to think systemically (understanding how different parts of a system can influence one another within a whole);
- Candor & Humility (willingness to share thoughts, and be open to hearing and appreciating input from others);
- Respect from colleagues.

The CPT members are thought partners, assisting in the development of the Triggering Question, identifying stakeholder perspectives and subsequently participants for the colab(s). The CPT is responsible for a thorough review of the literature. Also, interviews with selected stakeholders are conducted, culminating in a summary paper documenting the complex situation. Careful consideration to the diversity of perspectives should be given when identifying persons to interview. It is often advantageous to outsource the interviews to an individual with extensive expertise. Our experience tells us that the best qualified people for conducting the interviews are individuals trained in journalism or sociology. When conducting interviews, a general rule of thumb is, once a threshold of interviews has been reached (approximately fifteen), stop interviewing because the respondents stop generating new ideas.

Participants to the colab are identified during the Discovery phase using a rigorous process that focuses first on the diversity of perspectives, and then on selecting one or more individuals that are representatives of each perspective. For example, if a group is being selected for the design of a school district plan, the voice of the parents of the students should be represented at the colab. The success of a colab can rest on the diversity of perspectives. It is important to consider those participants who have vastly different viewpoints, namely to capture the intersubjective component of the situational complexity, for if the situation could have been resolved by like-minded individuals, it would most likely have been already resolved. A rule of thumb with regard to participants is that, while it's advantageous to have "disruptive" mindsets, those individuals should be respectful and open minded to learning from and with other participants (Christakis et al., 1999).

The Triggering Question is framed during the Discovery phase. This is perhaps the most challenging task of the colab design, as it guides the entire dialogue. A poorly crafted TQ will surely lead to a failed colab. During the framing, the initial draft should be based on input from the sponsor as well as informed by the review of literature and discussion amongst the members of the CPT. In our experience, the initial draft is never the final draft. Every word in the TQ is critical, and those nuances can make a difference between a good TQ and a bad one (Flanagan & Christakis, 2010).

Finally, logistics for the colabs are addressed during the Discovery phase. For example, the facility should be comfortable and have the capacity to accommodate a set-up as depicted in Fig. 2 (Demosofia situation room).

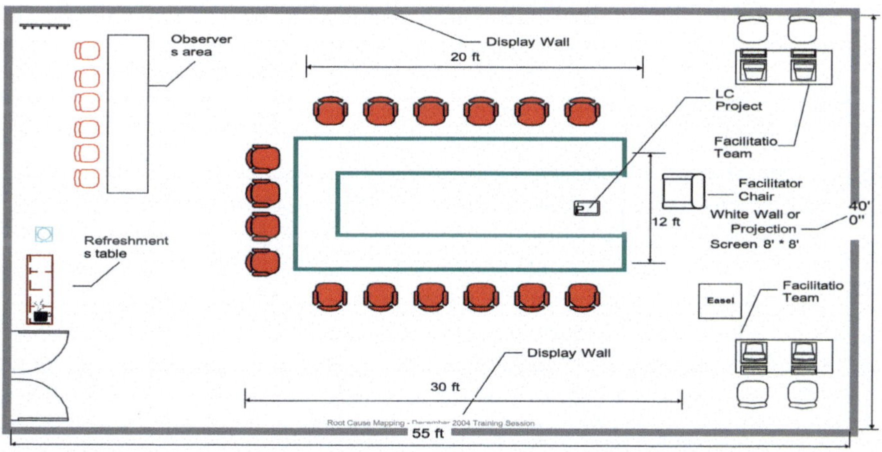

Fig. 2 The configuration of the facility (Demosofia) for the conduct of colaboratories of democracy

8 The Colaboratory of Democracy Phase

We have all experienced the benefits of dialogue when we openly and thoughtfully confront issues. We have also experienced the frustration of interminable discussion that does not lead to progress. The application of the SDD methodology, leveraging the use of the Logosofia platform in conducting colabs of Democracy, efficiently enables groups of stakeholders to dialogue and generate positive results. It accomplishes this by ensuring that the diversity, authenticity, autonomy, learning, and cognitive limitations of the participants are respected and protected during the conduct of the dialogue (Christakis & Bausch, 2006).

In the context of the Triggering Question, each colaboratory engages groups in the following steps:

(a) **Generation of Statements** – participants respond to the TQ with concise statements that capture the essence of the meaning of their ideas. The SDD Facilitator, responsible for managing the dialogue, ensures that the authenticity and autonomy of every voice is respected at the table and, through a round-robin approach, each person contributes his/her ideas. The stopping rule is that the generation of responses to the TQ stops only when participants have exhausted all contributions.

(b) **Prioritization**(initial) – Prioritization voting preferences are captured at various points throughout the colaboratory. Doing so will demonstrate: (a) the evolution of group learning, (b) the in-depth understanding of the meaning of the ideas, and (c) an appreciation for the diverse perspectives of participants regarding a complex situation. On the basis of over a thousand applications, we have discovered that the average number of ideas generated, independent of the situation and the group, is equal to sixty-four. Participants are amazed when they realize at the start of the colab the diversity of viewpoints. Reviewing all ideas generated, which are displayed for visibility on the walls of the colab facility shown in Fig. 2, participants select their top five statements on the basis of their individual judgments of saliency in the context of the TQ. Significant divergence amongst the participants is expected and observed at this stage of the process. Most participants select in their top five statements those that they have contributed.

(c) **Clarification of Statements** – the author of each statement is invited by the Facilitator to elaborate on the meaning of their concise statement to ensure all other participants fully understand the intent of the author. During the clarification of the meaning of the statements, the Facilitator protects the authenticity of every participant in terms of the intent and the language of their idea. Each author is encouraged to use their own lexicon in the clarification of their statement. Also, during the clarification of meaning step, value-judgments regarding each statement contributed by individual authors are not allowed. In other words, at this point of the process, it is irrelevant as to whether participants agree or disagree with a particular statement; only that they understand its meaning and intent counts.

(d) **Prioritization**(second) – Following the clarification of the ideas, participants are asked to select again their top five ideas. As a result of the clarification that preceded, participants begin to abandon their own ideas and to adopt statements contributed by others. It is anticipated that groups begin to converge on priorities because of understanding and learning.
(e) **Categorization** – After all ideas are clarified they are categorized inductively into affinity clusters in accordance to a relationship of significant similarity. Cluster headings are agreed upon by the group, based on the ideas within the cluster/category vs. predetermined categories, as is usually done in other processes. The categories help to reduce cognitive overload by organizing the ideas into management chunks of information. Also, they help the participants get a deeper understanding of the meaning of the ideas proposed.
(f) **Prioritization**(third) – With the aid of ideas being clustered according to similarities, participants again select their top five ideas. Their judgments of relative saliency have changed on account of: (a) their deeper understanding of meaning, and (b) their selection of their top five is supported by the organization of the ideas into affinity clusters. It is anticipated that the voting results will converge even more to a narrower subset of ideas.
(g) **Influence Mapping** – SDD utilizes the Interpretive Structural Model (ISM) consensus method (Warfield, 1976) together with the Logosofia software to yield a map that indicates the influence relationships amongst a subset of ideas included in the mapping process (as informed by the third Prioritisation). Participants engage in voting regarding the influence relationship amongst two ideas at a time, i.e. if Statement X was addressed, would that help **significantly** in addressing Statement Y? A "Yes" vote for the group requires a super-majority consensus (i.e. 75%+) amongst the group members. The participants engage in significant dialogue throughout influence mapping, with individuals sharing her/his rationale for how they voted. The new learning that occurs though this dialogue can shift opinions and votes.
(h) **Prioritization**(final) – Following the influence mapping, participants are offered a final opportunity to select their top five ideas. A tremendous amount of learning occurs during influence mapping, often resulting in new insights. Results of this round of voting may inform additional ideas that ought to be considered for adding into the influence map, in an effort to promote learning through iterations and successive approximations.

Colabs represent the correct response to the planning challenge conceptualized by Hasan Ozbekhan in his seminal paper titled "Toward a General Theory of Planning," published in 1968, in *Perspectives of Planning,* edited by a distinguished systems scientist named Erich Jantsch. Ozbekhan wrote in the chapter:

> *In my view there is no more important question in planning discourse; it is truly the heart of the matter. Let me begin by saying. Yes,* ***we can will the future, but only if change is caused to occur invaluesrather than an object's other attributes.*** *What I mean is that any change that is not a fundamental change invaluesmerely extends the present rather than creating the future. It seems to me that from this general postulate one can derive five statements which govern all planning.*

1. *Only change in the overall configuration ofvaluescan change the present situation;*
2. *Only individual will can bring about such valuechanges;*
3. *Valuechanges cannot be predicted;*
4. *Valuechanges always occur as individual ideas, or responses, or insights concerning betterment, and when they become socialized over a large part of the system we have 'progress;'*
5. *Planning is the organization of progress. Thus, the main subject of planning is the willed future.*

Colabs have demonstrated, throughout the many years of applications world-wide, the capability of satisfying all five planning requirements mentioned above.

9 Capturing New Common Knowledge with Logosofia

logosofia

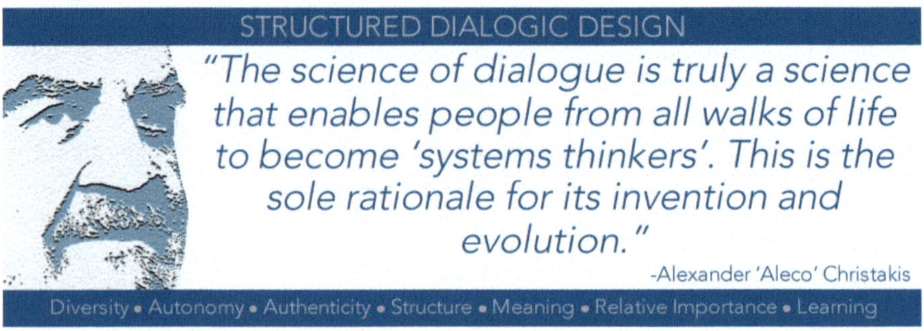

Logosofia is the software platform utilized to support the process of SDD. Logosofia tracks the votes using an Interpretive Structural Modeling (ISM) algorithm (Warfield, 1976), which is capable of significantly improving the efficiency of "connecting the dots (ideas)" by tracking connections via the ISM algorithm; However, the final product of the colab is totally dependent of the judgments of the group and not on any artificial intelligence algorithm. The reduction in the number of pairwise queries posed to the participants can reach a factor of five, which means that instead of the group having to spend a whole day to "connect the dots," it can do it in as little as two hours. The influence map often demonstrates that the top priority ideas, resulting from the voting on relative saliency, are not necessarily the most effective in terms of exerting leverage on other ideas of the subset included in the map. This phenomenon has been called the Erroneous Priority Effect, which has major implications in terms of prioritizing problems and actions (Christakis & Bausch, 2006).

The SDD process, with the help of Logosofia, leads to deep consensus, based on common knowledge of issues, and shared ownership of actions that bring desired results. Logosofia helps groups produce new knowledge and gain profound insights about their complex situation. Participants fundamentally change their opinions as they make a surprising discovery about problems and solutions: specifically, what individuals think is important when the colab starts is quite different from what the group knows is important when it is finished with the colab. The evolution in thinking, learning and knowledge is represented visually on the walls of the Demosofia facility (Fig. 2), for the group to grasp and discuss. The visuals convert stakeholder ideas into precise, easy-to-grasp categories and maps (digraphs). Participants can understand their situation from many different dimensions, and see the whole picture. Maps also help the group explain to others the rationale behind certain decisions and actions, show chains of accountability, and keep them on track with results-driven implementation (Warfield & Christakis, 1987).

9.1 Situational Complexity

Situational complexity is a phenomenon that emerges when groups of stakeholders are engaged in deliberations to define and resolve wicked problems. At a global scale, this includes climate change, the management of pandemics, the elimination of starvation, and the like. At the local scale, such complexities may include the adoption and implementation of a national health system, a master plan for a city, a regional plan for education, and others. There has been sufficient empirical evidence that the design of social systems without the inclusion of the stakeholders is both unethical and not effective (Kakoulaki & Christakis, 2018; Flanagan & Lindell, 2018; Flanagan & Christakis, 2021).

The situational complexity phenomenon is the result of the superposition of three distinct types of complexity. These are:

- **Objective complexity**, which is observer-independent, and does not take into account any action to improve the situation being addressed (e.g., astronomers observing, reporting, explaining, and measuring the motions of the stars);
- **Subjective complexity**, which is observer-dependent, and includes partially the actions to improve the situation (e.g., a doctor measuring the blood pressure of a patient, known as the white coat phenomenon; the measurement by the doctor raises the blood pressure of the patient); and
- **Intersubjective complexity**, which is observer-interdependent, and high quality observations should include fully the actions to improve the situation (e.g., the adoption and implementation of a city master plan, which most urban planners would love to have happen with minimum citizen deliberation in order to avoid dissonance).

Disciplines such as physics, chemistry, biology, and many more, have been developed by observing, studying, and developing theories and models for explaining objective complexity phenomena, such as apples falling from trees, or carbon atoms combining to form coal or diamonds. Disciplines such as psychology, medicine, anthropology, and others, have focused on studying and interpreting the interactions between observer and observant, such as the impact a psychologist exerts when she interacts with her patient. Disciplines such as urban planning, politics, journalism, social systems designing, and others, have focused on the interactions among observers. Ideally these disciplines should focus on how to improve the situation by engaging stakeholders interactively and productively in the development and implementation of policies and plans for improving the quality of their lives.

Over the past fifty years of developing and applying SDD, we have measured situational complexity by employing a metric called **Situational ComplexityIndex** (SCI) (Warfield & Cárdenas, 1994; Christakis & Bausch, 2006; Christakis & Kakoulaki, 2021). The formula for measuring the SCI is:

$$\mathbf{SCI} = (\mathbf{N} - \mathbf{7})\left[(\mathbf{V} - \mathbf{5}) / (\mathbf{V}^{\max} - \mathbf{5})\right]\left[\mathbf{K} / (\mathbf{R}(\mathbf{R} - \mathbf{1}) - \mathbf{K})\right]. \quad (1)$$

Where:

N = The number of ideas/observations proposed by the group of colaboratory participants in response to the Triggering Question (TQ);
V = The number of ideas/observations receiving one or more votes of relative saliency/priority from the group at various points during the colab;
P = The number of participants;
V^{max} = The number of participants (P) multiplied by five (5). The value of V^{max} to be entered in the formula should be smaller or equal to the value of N. If V^{max} is larger than N, then the value of N should be entered;
R = The number of ideas/observations selected for inclusion in the relational pattern displaying the network of connections among them (most often these are ideas receiving two or more votes of saliency); and
K = The number of distinct connections among the ideas/observations selected for inclusion in the relational pattern.

The formula for the SCI can be reconfigured, and simplified, by identifying three additional indices that are of particular interest to the SDD Facilitator in monitoring the progress of the group deliberations during the colab. These three indices are:

1. The Divergence (D), which is:

$$D = (V - 5) / (V^{\max} - 5) \quad (2)$$

2. The Demosensus Community Index (DCI), which is:

$$DCI = (V^{\max} - V) / (V^{\max} - 5) \quad (3)$$

Notice that when V = 5, the divergence is zero, and the Demosensus is one (1), which means that there is one hundred percent consensus among the participants in terms of relative saliency of the ideas. In this case the intersubjective component of the situational complexity has been practically eliminated since the group has converged to the five more salient ideas. On the other hand, when $V = V^{max}$, the DCI is zero, and the divergence D is maximum (100%), which means that the intersubjective component of the situational complexity is maximum.

3. The Priority Effectiveness Index (PEI), which is:

$$PEI = \mathrm{K} / \left[\mathrm{R} \left(\mathrm{R} - 1 \right) - \mathrm{K} \right] \tag{4}$$

With the use of Eqs. (2) and (4), the SCI (Eq. 1) reads:

$$\mathbf{SCI} = \left(\mathbf{N} - \mathbf{7} \right) \mathbf{D} \left(\mathbf{PEI} \right) \tag{5}$$

Studying Eq. (5), we see that the SCI is proportional to: (a) the number of ideas generated by the group (N), (b) the Divergence (D), which is a function of the number of ideas receiving one or more saliency votes (V), and the number of participants in the colab (P) which impacts the value of (V^{max}), and (c) the Priority Effectiveness Index (PEI), which is a function of K, the number of distinct connections in the map, and R, the number of ideas included in the map. It follows that the SCI is zero when: (a) the number of ideas generated N = 7, (b) when the divergence D = 0, which occurs when V = 5, and (c) when PEI = 0. which happens when K = 0, i.e., there are no connections among the ideas included in the map. Measurement for K and PEI cannot be obtained until the mapping step of the colab is implemented, which usually happens about five hours from the start of the group work. On the other hand, the measurement of D occurs from the start of the colab and continues until the end. Usually, the generation of ideas in response to the TQ is completed about one hour from the start of the colab, at which point the values of N and V^{max} become available, and hence the initial value of the divergence D can be calculated following the voting of relative saliency by the participants after the generation step.

In the three cases that the SCI is zero it happens because: (a) when N= 7, which means that the number of ideas generated by the group is very small, and there is no need for the group work to continue, (b) when V = 5, which means that the group has converged to five ideas of higher relative saliency, and hence there is consensus and no need for additional group work to decrease the intersubjective complexity, and (c) when K = 0, which means that there exist no network of connections among the subset of ideas (R) selected for inclusion in the map. In all these three cases, the situational complexity does not warrant the need to be addressed by a large team because the situation can be dealt by assigning responsibility to single individuals or small teams. On the other hand, the SCI blows up when all the ideas included in the map (R), are in a vicious cycle of mutual influence, which happens when K = R(R-1). The maximum value of SCI is obtained when K = (R-1)(R-1), which happens with one idea being outside the vicious cycle of mutual influence. We also identify the optimum SCI as the value obtained when PEI =1, which happens when K = [R(R-1)]/2. In this case, it is plausible that the pattern of influence among the subset of R ideas included in the map is linear, i.e., the influence among the ideas propagated from the bottom (root) to the branches of a tree-like pattern.

Over the fifty or so years of application of SDD in the arena (Christakis & Dye, 2008), we have observed that the average number of ideas generated (in face-to-face colabs), independent of the size of the group and the subject matter, is sixty-four (64). The average number of participants is equal to fifteen (15). The average number of ideas receiving one or more saliency votes (V), before the mapping of the subset (R), is seventeen (17). The average number of ideas included in the influence map, is fifteen (15). It would be interesting to calculate the maximum and optimum SCI for these average values of N, V^{max}, V, R, and K (for the value of V^{max} we enter the value of N, because V^{max} in this case is larger than N).

The*maximum*SCI is:

$$\mathbf{SCI} = (\mathbf{57})\left[(\mathbf{12})/(\mathbf{59})\right](\mathbf{14})\big] = \mathbf{154} \tag{6}$$

The optimum SCI is:

$$\mathbf{SCI} = (\mathbf{57})\left[(\mathbf{12}/(\mathbf{59}))\right](\mathbf{1}] = \mathbf{11} \tag{7}$$

Notice that for thevaluesof the parameters corresponding to the average case, the maximum valuefor SCI is fourteen times larger than the optimum. The first parameters that an SDD Facilitator learns at the start of a colab are the number of ideas generated (N), and the value of V^{max}. If the values of N and V^{max} are close to the values of the average case, then the Facilitator should infer that the maximum and optimal values of the SCI for the particular colab will be in the range of 154 to 11. The capacity of the Facilitator to monitor the values of V, R, K, and of SCI during the conduct of a colab is equivalent to an airplane pilot navigating the flight by means of the airplane dashboard. Specifically, if at the start of a colab with fifteen participants, the value of N is equal to sixty-four, then the Facilitator should anticipate that the values of the parameters V, and R, to be close to the average numbers mentioned above, and the SCI to be in the range between 11 and 154.

As mentioned before, there are two other indices that can be helpful to the Facilitator for monitoring progress during the colab. These are the Divergence (D), and the Demosensus Community Index (DCI). The Divergence (D) measures the degree of diversity of viewpoints among the participants, which is high at the start of the colab and diminishes as the learning and consensus among the participants evolves, and hence the level of intersubjective complexity diminishes. The structured deliberation and the evolutionary learning from the dialogue enables participants to abandon some or all of their own ideas, and adopt the ideas of others. As this happens, the DCI, which measures the degree of consensus, increases steadily. At the start of the colab, D is one hundred percent and the DCI is equal to zero. After approximately four hours of disciplined deliberation, the Facilitator can anticipate D and the DCI converge to about fifty percent, as shown graphically in Fig. 3. After about an additional four hours of group work, during which the group explores the systemic interaction among a priority subset of ideas to construct a map, the D is

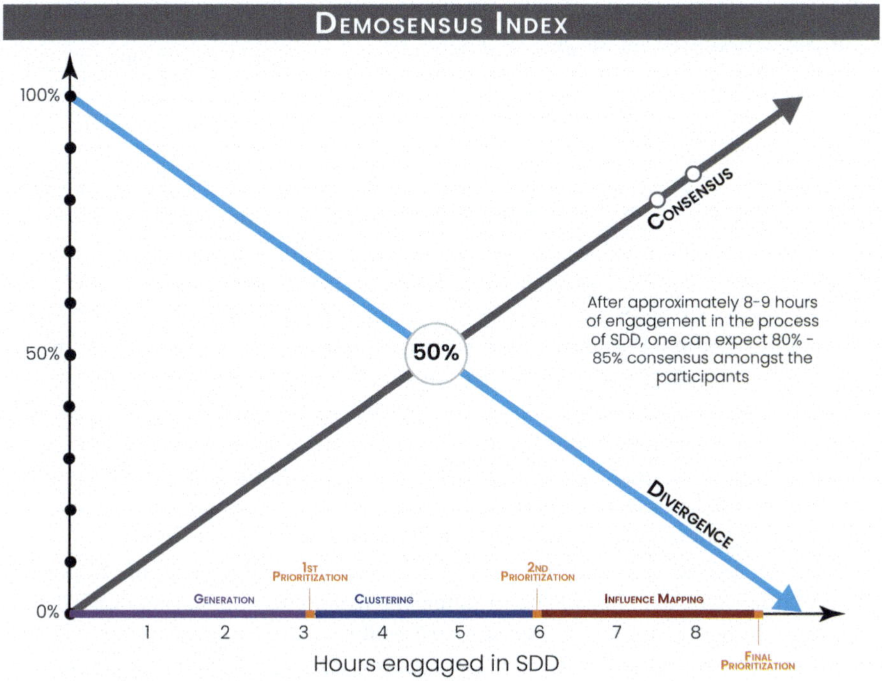

Fig. 3 Showing the evolution of the Divergence (D) and Demosensus Community Index (DCI), during the colab process as a function of time of disciplined deliberation for a group of stakeholders.

reduced to about 20%, and the DCI climbs to about 80%. As a consequence of consensus-building, the intersubjective component of the situational complexity is reduced to a minimum, as seen in Fig. 3.

It is interesting to notice in Fig. 3, that the crossing of the two lines, I.e., one for D (blue line), and the other for DCI (black line), occurs when they measure approximately 50%. This happens after about four hours of structured deliberation, as shown in Fig. 3. For the DCI to climb to 80% of consensus, it takes approximately another four hours of group work. Most governmental agencies, when they want to include the public in policy-making, allocate approximately two hours of public discussion. The majority of the two hours is spent with the managers of the agency presenting their proposals, followed by asking for comments and feedback from the public. These unstructured "public hearings," as they are called, are obviously meaningless in terms of stakeholderinclusion in making policies and becoming co-owners of the decisions. This is why we repeatedly hear stakeholders making the statement "they decide about us, without us."

Another index that may be useful to measure at the start of a colab, is the degree to which, during the Discovery phase, the Core Planning Team was successful in satisfying the Law of Requisite Variety. Variety depends on selecting a group of participants with diverse viewpoints. The proposed formula for an indicator of the Requisite Variety (RV) at the start of a colab is:

$$\mathbf{RV} = (\mathbf{N}-\mathbf{7})\left[(\mathbf{V}-\mathbf{5})/(\mathbf{V}^{\max}-\mathbf{V})\right] \tag{8}$$

Notice that RV is equal to zero, i.e., the requisite variety is not represented in the colab, when N = 7, and V =5. In both of these cases there is no need for additional group work, and the Facilitator, in consultation with the Broker, should consider terminating the coab. On the other hand, RV is at its maximum when $V = V^{max}$ (in case V^{max} is larger than N, the value of V^{max} in the formula will be replaced by N).

It should be pointed out that all the formulas presented in this section are meaningful when groups of ten or more participants are engaged in deliberations. For smaller groups, special considerations are appropriate for the measurement of SCI, because the intersubjective complexity of the situation is not significant. For example, for a single investigator applying the Logosofia platform, as we shall see in the next section of this chapter, the parameters N, V, and D, are not relevant, because the complexity is primarily subjective and not intersubjective.

Just as an airplane pilot utilizes a dashboard to navigate the plane, the Logosofia platform offers the SDD Facilitators the capability of employing a dashboard, developed recently by Diedrich & Christakis, that enables them to navigate the colab by monitoring all the indices discussed in this section. The availability of the dashboard, as shown in Fig. 4, makes the job of the Facilitator easier and more reliable, because

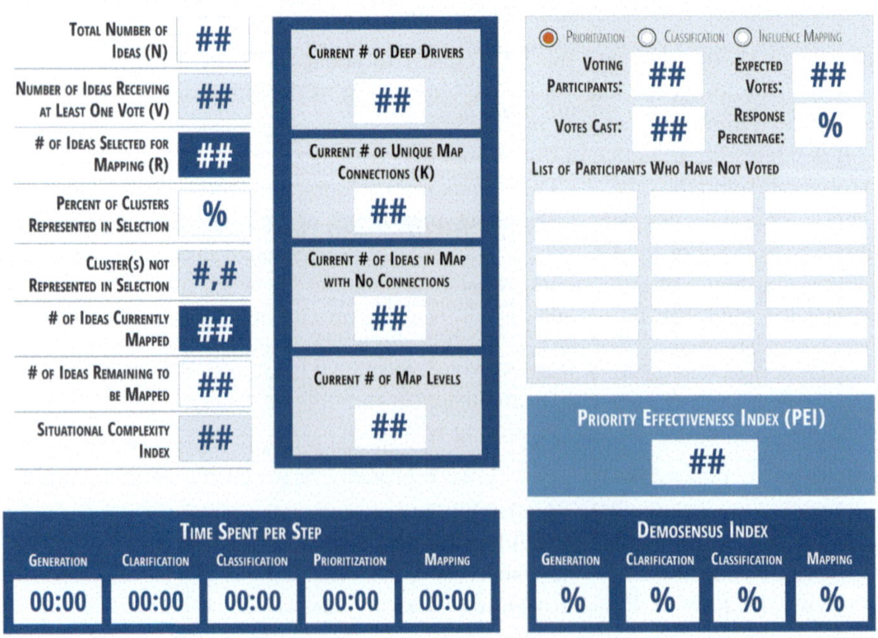

Fig. 4 Showing a picture of the Logosofia Dashboard that supports the SDD Facilitator during the conduct of a colab

it reduces the risk of bad piloting of the group work. For example, after approximately four hours of group work, the Facilitator should be able to look at the dashboard and see that the values of D and DCI are converging to around 50% for each.

10 Our Predicament of the Vicious Cycle

Unless people internalize that they
are an integral part of the GlobalProblematique,
they cannot become partners in its resolution.

– Alexander N. Christakis

In an effort to explore the relationship between situational complexity and deliberative democracy, we performed three structural inquiries by employing the Logosofia platform. The advantage of using Logosofia, even as a single inquirer and not a group, is that it enables one to efficiently and effectively explore relationships of similarity and influence among a set of ideas and to construct relational maps. In order to do this, we framed and used three distinct but interrelated Triggering Questions (TQs).

The first TQ reads:

What are requirements for an effective assault on situational complexity?

We identified ten important requirements in response to this TQ, which are shown in Table 1.

We subsequently explored the enhancement relationship among the ten requirements by answering a set of pairwise generic questions:

Suppose we were able to make progress in satisfying requirement (X), would this help us SIGNIFICANTLY in satisfying requirement (Y), in the context of the TQ?

Table 1 A set of requirements for an effective assault on situational complexity

#	Statement
1	An appropriately designed situation room (Demosofia).
2	A set of consensus methods.
3	Measurement of situational complexity.
4	A representative group of stakeholders.
5	A trained SDD Facilitator.
6	A software platform to support the process (e.g.,. Logosofia).
7	A colaboratory design team for framing the TQ.
8	Implementation of the six principles of SDD (Requisite Variety, Parsimony, Saliency, Meaning, Authenticity, Evolutionary learning).
9	A sponsor and a Broker.
10	A commitment to deliberative democracy.

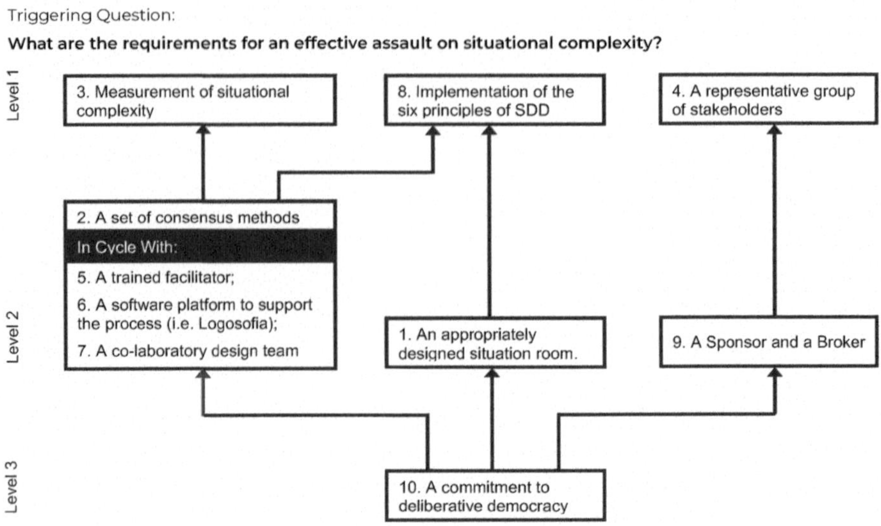

Fig. 5 A tree-pattern displaying the enhancement relationship among ten requirements for an effective assault on situational complexity

Answering this generic question produced, with the help of the Logosofia algorithm, the enhancement pattern showing in Fig. 5.

Figure 5 is a directed graph (diagraph), i.e., a network with the ideas representing the nodes and the arrows the links. The connections among the ideas represent the relationship of enhancement, which propagates upwards from Level III at the bottom, to Level I at the top. The ten requirements are positioned at three distinct levels. The one requirement (# 10 – A commitment to deliberative democracy) at the bottom (Level III), is the most influential (deep driver) because it is connected to all the others and essential to an effective assault on situational complexity. Requirement #10 enhances Level II Requirements #1, #9 and the development of consensus methods (Requirement #2), which is in a cycle of mutual influence with three other key components of the SDD methodology, (i.e., Requirements #5, #6, and #7). Three requirements are positioned at the top (Level I), and they exert no influence. The authentic inclusion of the stakeholders in democratic deliberations enhances directly the requirement of establishing a properly designed situation room (Requirement #1, i.e., the Demosofia situation room shown in Fig. 2). Ideally, the Demosofia room should be part of a Demoscopio facility, i.e., a place and a space dedicated to the application of the science of dialogic design for the practice of deliberative democracy (Kakoulaki & Christakis, 2018). Making progress in Requirement # 1, together with the cycle of four requirements at Level II, which includes such requirements as the use of consensus methods (#2), trained SDD Facilitator (#5), a software platform (#6) and a colaboratory design team (#7), lead to the proper application of the principles of authentic and productive dialogue. All these requirements contribute to a synergistic identification, measurement, and resolution of situational complexity (Requirement # 3, at Level I). Failure to appreciate the key role and contributions of the ten requirements for an assault on situational complexity, will perpetuate the existence of complex problems (e.g., the CCPs of the Club of Rome Problematique).

Our predicament is that the situational complexity represented by the Problematique cannot be addressed effectively and efficiently, unless it is owned and internalized by citizens of the planet, which cannot happen without the commitment and practice of authentic deliberative democracy. **The crucial link in Fig.5, is the one connecting the commitment to the practice ofdeliberative democracy(Requirement #10) at Level III, with the cycle of four key requirements for applying SDDat Level II (Requirements #2, #5, #6, and #7). This link implies that the application of the SDDmethodology is strongly enhanced by the commitment todeliberative democracyby a societal system.**

In light of the findings of Fig. 5, it appeared reasonable, to focus our next structural inquiry on the following question:

Why isdeliberative democracybacksliding at a global scale and, as a consequence, our capacity to deal with the situational complexityof the Glocal (Global and Local)Problematiqueis deteriorating?

In order to answer this question, especially after discovering that the deep driver of the set of requirements for an effective assault on situational complexity (See Fig. 5) is Requirement # 10), we decided to drill down focusing on the deep driver, and search for the reasons that the commitment to the practice of deliberative democracy is lacking. We focused our structural inquiry on the following TQ:

What are issues/questions that inhibit the commitment and/or practice ofdeliberative democracyat the Glocal scale?

We generated fourteen questions which act as inhibitors we need to overcome in order to reinvent deliberative democracy, and consequently launch an effective assault on situational complexity, in accordance with Fig. 5. We subsequently employed Logosofia, initially to categorize inductively (bottom-up) the set of questions into four affinity clusters. The advantage of inductive categorization is that it allows for the discovery of new categories and hence new knowledge. The four clusters are shown in Table 2.

A concise interpretation of the four clusters that emerged from the inductive clustering of the fourteen questions that act as inhibitors to the practice of deliberative democracy, based on the questions that have been assigned to each cluster, follows:

Cluster 1: PROBLEMATIQUE

The cluster labelled "Problematique" is focused on the situational complexity corresponding to the CCPs of the Club or Rome, and our incapacity to resolve them unless we commit to the practice of deliberative democracy with the authentic inclusion of the stakeholders. There exists sufficient evidence that without co-ownership of the Glocal Problematique by the stakeholders, plans for its resolution are not implementable. The MIT study, funded by the Club of Rome (CoR), title *The Limits to Growth* (Meadows et al., 1972), useful as it was in terms of predicting

Table 2 What are questions to identify Barriers to overcome in order to reinvent and practice deliberative democracy

Cluster 1:PROBLEMATIQUE	
#	**Statement**
1	Why is democracy backsliding around the world?
2	What is the contemporary Problematique with the practice of authentic deliberative democracy?
13	Is there a strategy to penetrate the democratic Problematique and approximate the ideal of an authentic deliberative democracy?
Cluster 2: ELITISM	
#	**Statement**
3	Why do politicians espouse and campaign on the practice of participative democracy and abandon it after they get elected?
7	What is the role of propaganda in misleading the choices and voting preferences of the public sector?
9	How can we diminish or eliminate the public apathy towards democracy?
11	How the dominance of the capitalist model and propaganda have compromised the voice and the preferences of ordinary people?
12	Why do elites of all kinds suppress the preferences and choices of oppressed people in conflict situations, such as Cyprus or Israel?
Cluster 3:INTERSUBJECTIVITY	
#	**Statement**
4	What is the role of the three phases of science in the practice of deliberation?
5	How can we deal effectively with the contemporary situational complexity?
6	How can we objectify intersubjectivity?
8	What is the role of the science of dialogue in rediscovering and applying authentic deliberative democracy as practiced in the Athenian Agora?
10	How can we balance observer-dependent with observer-independent data in citizen engagement? For example how can the voice of a mother of a child of special needs be heard in the context of statistical data?
Cluster 4: IDEALIZATION	
#	**Statement**
14	Why is democracy desirable in dealing with contemporary situational complexity?

the critical future, it was not capable of generating the commitment to the stakeholders for the paradigm shift. The CoR prospectus conceptualized the requirement for stakeholderinclusion as a prerequisite to the co-creation of the collective resolutions at the international, national, and regional levels of governance. A structural inquiry, conducted by Ozbkhan and Christakis in 1992, employing the set of the forty-nine CCPs of the CoR prospectus, demonstrated than no progress has been made in the resolution of the Problematique during the period of twenty years, i.e., 1970 to 1992 (Christakis, 2006). Furthermore, the lesson we should have learned from the coronavirus pandemic of the year 2020, is that national policies alone, without a global context, are not effective (Christakis, 2020). The World Health Organization (WHO), should have had the authority to act proactively and decisively for the containment of the spread of the pandemic at the global scale. If nothing else, the 2020 pandemic should have taught us a lesson regarding our vulnerability as a species.

Cluster 2: ELITISM
This cluster discusses the capacity of political, media, academic, and economic elites to control the voices, choices, and aspirations of people from all walks of life, by using propaganda and other instrumentalities of manipulation. One consequence of such manipulations is the phenomenon of apathy, especially among younger voters, regarding governance and politics. The manipulative rationality of the elites is threatening the sustainability of the planet on account of such critical issues as climate change. For example, we have experiences of the impact of these types of manipulations from application of SDD for peace-building with oppressed people in Cyprus and Palestine (Diedrich, 2010). These oppressed groups were capable of conceptualizing symbiotic existence on their lands, and proposed action plans for implementation; however, their recommendations were not adopted by the political elites, because they were not to their benefit. The political leaders talk the talk of peace, but do not walk the walk. The documentary titled "Dialogue beyond borders" (Kakoulaki, 2016), offers evidence of the hypocrisy of politicians, at local, national, and international levels who campaign on platforms espousing democratic participation and abandon their promises after being elected. It appears that the elites take advantage of the dominance of the paradigm of conventional First Phase science, together with the difficulty in objectifying the intersubjectivity component of situational complexity, to promote polarization by embracing the myth of objectivity, so that they can perpetuate oppression, conflict, and dissonance, and hence their continuity of control and power (McIntyre-Mills & Corcoran-Nantes, 2021).

Cluster 3: INTERSUBJECTIVITY
The Intersubjectivity cluster focuses on the intersubjective component of situational complexity. The role of, and complementarities among the three phases of science, i,e., objective (First Phase), subjective (Second Phase), and intersubjective (Third Phase), are clarified in terms of enabling groups of stakeholders to objectify intersubjectivity (Christakis & Kakoulaki, 2021). The distinction among the three phases of science is discussed rigorously in (de Zeeuw, 1996; Christakis & Bausch, 2006). Briefly speaking, in the First Phase of science the observations are designed to be objective, i.e., the observations are independent of the observers (objective complexity), **and no actions are included in improving the situation.** In the Second Phase, the observations are dependent on the observer (subjective complexity), **and actions to improve the situation are partially included.** In the Third Phase, the observations of the observers are interdependent (intersubjective complexity), **and the actions to improve the situation are fully included.** The intersubjective component of situational complexity is clearly connected to the Third Phase of science, especially in terms of the actions to be taken to improve the situation. In the section discussing the concept of situational complexity, we have presented the formula for measuring SCI (Eq. 1). As mentioned before, we have accumulated evidence that it takes a group approximately eight hours of structured deliberations for reducing

significantly the intersubjective component of situational complexity, and as consequence, enable the group to converge to a consensual design and/or action plan (See Fig. 3).

Cluster 4: IDEALIZATION

West Churchman, the great systems thinker of the 20th Century, said that "idealism is the best form of pragmatism." This statement is absolutely true in the case of the democratic ideal. As humankind is evolving toward the critical role of knowledge and wisdom for its sustainability, the role of deliberative democracy is becoming increasingly critical. The observed backsliding of democracy represents an existential threat for humankind on account of the absence of accountability and co-ownership of the Problematique. The role of education in terms of helping young and old adults to internalize their role and responsibility in collectively addressing situational complexity cannot be underestimated. Co-ownership of our predicament will only happen if young and old adults learn together through authentic, deliberative democracy. We must all learn together how to balance individualism and collectivism for social and environmental justice (McIntyre-Mills & Romm, 2018; McIntyre-Mills & Christakis, 2021).

After internalizing the meaning of the four clusters, as presented above, we employed the Logosofia platform and a generic question to construct an enhancement pattern among the fourteen questions/barriers to the commitment to practice authentic and inclusive deliberative democracy. The generic question we used reads:

Suppose we were able to make progress in overcoming Barrier (X), would this help us SIGNIFICANTLY in overcoming Barrier (Y), so that we can commit to the practice ofdeliberative democracy?

By responding to the generic question in a pairwise mode, we constructed a map showing the pattern of enhancement among the fourteen barriers, as shown in Fig. 6.

It is interesting to note that the deep driver in Fig. 6, is Barrier # 6, focusing on "How to objectify intersubjectivity." In the section discussing situational complexity (See Fig. 3), we have demonstrated the gradual diminishing of the intersubjectivity component of situational complexity during a colab by employing the SDD methodology. It follows that applications of the SDD are fundamentally the means for dealing with all the other thirteen questions/barriers displayed in Fig. 6. On the other hand, progress in resolving the questions located at Level I, i.e., the backsliding of democracy (# 1), and the lack of a strategy for penetrating the Problematique ($\neq$ 13), depends on making progress on resolving all the other twelve questions at the lower levels of Fig. 6. It follows that, the capability of employing SDD, and hence address effectively the issue of objectifying intersubjectivity, will enhance our capacity for overcoming practically all of the other twelve issues identified in the map, with the exception of # 14, at Level IV, which relates to the idealization of democracy as the means to deal effectively with situational complexity.

Combining the interpretations of Figs. 5 and 6, a narrative of our predicament of a vicious cycle, in terms of our capacity to address the situational complexity, emerges as follows:

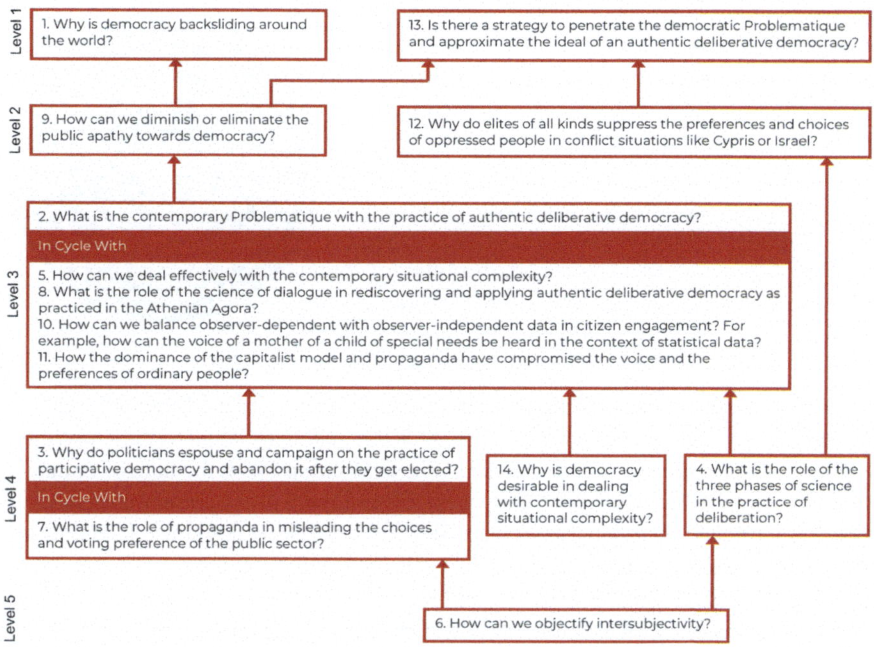

Fig. 6 A tree pattern displaying the enhancement relationship among the fourteen questions/barriers to the commitment and/or practice of deliberative democracy

> *In order to launch an effective assault on situational complexitywe need the application of SDDwith theinclusionofstakeholders, in a properly designed Demoscopio capabilitywith aDemosofiafacility (as shown in Fig.2). However, as shown in the map of Fig.5, displaying the requirements for an assault on situational complexity, such a capabilitywill become plausible only when elites of all kinds make a serious commitment todeliberative democracy. But as we see in the map of Fig.6,deliberative democracyis plausible only through the application of SDD, which is essential in order to diminish the intersubjective component by means of conducting Co-laboratories of Democracy. In other words,deliberative democracydepends on applying SDDwith theinclusionofstakeholders, and SDDdepends on the commitment to authenticdeliberative democracy. It follows that without breaking this vicious cycle, humankind will not be able to address effectively the situational complexityof such global issues as climate change, accelerated reduction of biodiversity, and the rest of the CCPs of the GlobalProblematique. This narrative captures the essence of the vicious cycle of our predicament at this juncture of the evolutionary process of humankind.*

11 Escaping the Labyrinth of Our Predicament

The third structural inquiry we chose to conduct focused on the challenge of designing a strategy for breaking away from the predicament of the vicious cycle, by becoming partners in the resolution of the Glocal Problematique. To address this challenge, we will imagine ourselves as being prisoners that are trying to escape

Fig. 7 A drawing of the Minoan Labyrinth maze (Drawn by the Cretan artist Dimitri Bounakis in 2021)

from the "Labyrinth of the Minoan mythology." According to the Minoan myth, the Athenian hero Thyseus, at the request of the King of Athens, entered the Labyrinth, killed the Minotaur, and was able to find an exit to escape with the help of Ariadne, the daughter of King Minos of Crete, who fell in love with Thyseus. Ariadne gave Thyseus a thread when he entered the Labyrinth, so that he could use it to find his way out. An artistic rendering of the Minoan Labyrinth, created by a Cretan artist and poet, Dimitri Bounakis, is shown in Fig. 7.

We will imagine that humankind is trying to escape from the Labyrinth of our vicious cycle predicament, and will employ the Logosofia platform to generate ideas in response to the following TQ:

"What areguiding principlesfor overcoming the barriers encountered in escaping from the Labyrinth of our predicament, and as a consequence securing the sustainabilityof our planet?"

The list of guiding principles we generated for escaping the Labyrinth, with clarification of meaning, follow in Table 3.

We constructed an enhancement pattern among the twelve guiding principles in order to discover those with the maximum leverage in developing a strategy for escaping our predicament. The generic question we explored in pairwise queries is:

"Suppose we implement guiding principle (X), would this help us SIGNIFICANTLY in implementing principle (Y), in our effort to escape the Labyrinth of our predicament?"

This pattern is shown in Fig. 8.

The first thing to notice in Fig. 8, is that the driver (the root) is the principle of **Love**, which enhances all the other principles in the map. On the other hand, at Level I, we notice the principle of **Purpose**, which does not enhance any other principle, but is supported by all of the others. This makes sense by the explanation we gave to this principle. i.e., if we manage to escape the Labyrinth, we should be able to attain the purpose of life, which is human development and happiness. It is also interesting to notice that at Level II, we have the principles of **Peace** and **Action**, both of which are enhancing the principle of Purpose at Level I, and are enhanced by the principle of **Justice**, which is located at Level III. The principle of Justice should be

Table 3 Guiding principles for escaping the Labyrinth of our predicament with clarification of meaning

#	Statement
1	**Humannature** The dominance of the concept of the "rational man," advanced by the economic rationality of being, is questionable and counter-productive in terms of explaining human behaviour and action. Human beings are a complicated mix of thoughts and feelings that are complementary and not mutually exclusive. Furthermore, there is ample evidence that values such as empathy, collaboration, altruism, love, have played a major role in the context of the evolutionary origins for a good society (Christakis, 2019).
2	**Planning:** The theory of planning advocated by Hasan Ozbekhan in his seminal chapter titled "Towards a General Theory of Planning," (Ozbekhan, 1968), is the appropriate clarification of the meaning of this principle. As mentioned earlier in the section discussing the Colaboratory Phase of SDD, the colab approach is capable of satisfying all the necessary and sufficient conditions of Ozbekhan's theory of planning. For example, the futures-creative Archetype, which will be discussed in the case study application of SDD at the last section of the chapter, focuses on value-base changes and the will to co-create desirable futures. Furthermore, our experience with the practice of SDD for social system design and planning, has demonstrated beyond doubt that in confronting complex situations, the plans for action are not implementable effectively and ethically, without the authentic commitment of the community of stakeholders (Metcalf, 2015; Banathy, 1996). For example, plans for education and health systems, in countries such as Greece, Cyprus, and the USA, designed by politicians without listening to the voices of the people, are doomed to failure in the long run. Even plans for dealing with the crisis of the pandemic of 2020, are being challenged because they did not seek the consensus of those being affected by the crisis.
3	**Evolution:** In order to escape from our Labyrinth, humanity must transition from "evolutionary consciousness" to "conscious evolution" (Banathy, 2000). This transition can occur only by inclusive, authentic, deliberative democracy and "dialogue beyond borders," as presented in a documentary produced and directed by Maria Kakoulaki (2016). The documentary is based on interviews with Palestinians, Israelis, and Turkish and Greek Cypriots. The documentary is based on the application of SDD, for designing a vision and an action plan for symbiotic existence by these oppressed people on their lands.
4	**Peace:** The attainment of peace requires a dynamic balance between individual wellness (tranquility) and social and environmental justice. Peace without justice is neither desirable, nor attainable and/or sustainable. We have designed and conducted many bicommunal peace-building colabs between Turkish and Greek Cypriots for a period extending over twenty years (Laouris & Michaelides, 2018; Romm, 2010). These colabs are referred to in the conflict resolution literature as "Track II diplomacy," because they are focused on engaging stakeholders for saliency, priority, and trust. However, the consensual results and recommendation of such colabs, in the majority of cases, are not adopted by elites, as mentioned earlier, for a variety of reasons, primarily for maintaining control and power.

(continued)

Table 3 (continued)

#	Statement
5	**Truth:** In complex societal situations, truth in social system design and planning, emerges as a consensual linguistic domain from deliberative dialogue among the members of a community of stakeholders. This truth is subject to evolutionary learning through successive approximations and iterations. During colabs, we promote flexibility of thought and structure. It is preferable as the deliberation evolves to embrace the consensual truth and "be right" so to speak, even if this means that one is inconsistent with oneself, than to "be consistent and wrong." Evolutionary learning through iterations is a fundamental principle of SDD. In addition, doing the right thing, and doing it right, is by far superior to doing the wrong thing, and doing it right. Doing the right thing is a measure of effectiveness, doing it right is a measure of efficiency.
6	**Justice:** Social justice is attained when human beings, individually and collectively, are enabled to fulfill their destiny, including their aspirations for wellness, emancipation, and self-actualisation. Balancing individualism and collectivism for social and environmental justice (McIntyre-Mills & Romm, 2018) is attainable provided we combine reason with vision (Bausch & Christakis, 2015).
7	**What counts:** "Not everything that can be counted, counts, and not everything that counts, can be counted." -Albert Einstein. The dominance of quantitative measures at the expense of qualitative aspirations and emotions, is counter-productive in designing social systems (Flanagan & Lindell, 2018). For example, not including non-monetary costs in performing Cost/Benefit analyses focusing on economic development, has been disastrous in terms of environmental protection and energy preservation. The mechanistic, social engineering approach, based exclusively on observer- independent data and statistics relevant to the First Phase of science, will not capture the subjective and intersubjective components of situational complexity. For example, in a colab focusing on designing an action plan for helping deaf students, which is the case study we will discuss later, the experiences and feelings of the deaf stakeholders cannot be represented solely by statistical data, but require the intersubjective data base captured from the voices of the students, parents, and teachers.
8	**Meaning:** In accordance with the principle of Requisite Meaning of SDD, deep understanding of ideas occurs when stakeholders engage in exploring relationships of similarity, saliency, priority, influence, etc., among diverse observations and perspectives. The Principle of Requisite Meaning, attributed to the American philosopher Charles Pierce (Pierce, 2012), states that the depth of understanding is enhanced significantly by the exploration of relationships. In SDD colabs, we have repeatedly observed that, after the generation and clarification of ideas steps, participants believe that they understand all the ideas generated; only to realize their confusion or misunderstanding when they are asked to make pairwise judgments of similarity or influence relationships during the next steps of the colab process.

9	**Languaging:** In complex situations, stakeholders cannot "walk the talk, unless they talk the walk." The talk must be authenticity, inclusive, coherent, and collective (Flanagan & Christakis, 2010; Flanagan & Lindell, 2018). The notion that talking, as opposed to taking action, is a waste of time is counterproductive, especially if the talk is productive. authentic, and effective. We have learned from applications of SDD that people are actually desperate to be heard, so they are elated when the opportunity to interact and articulate their views is offered to them (See participant reflection after the application of SDD with the school of the deaf of the state of Michigan). It should be appreciated that the intangible outcome of a colab is the discovery of a consensual linguistic domain among the participants, which becomes the "groundless ground" for follow-up action. Furthermore, the statement by Descartes "I think, therefore I am," should be replaced by "I interact, therefore I am," even if the interaction is virtual as we are forced by the pandemic.
10	**Action:** As mentioned earlier, in the complex situations of our contemporary societies, doing nothing is better than doing the wrong thing, especially if one does the wrong thing right. Doing the right thing implies effectiveness in approximating the objectives, and doing it right implies efficiency. Instead of approaching the exit from the Labyrinth we can get deeper in trouble by choosing the wrong pathway. The Minoan goddess Ariadne, because of her love for Theseus, was capable of anticipating the danger of her loved one not being able to escape from the Labyrinth, and gave him the thread to ensure his exit strategy. In essence, the exit strategy was a combination of anticipation based on love. Adoration of the Minoan goddesses of the planet, the mothers, daughters, and sisters, who have been historically abused, might give us a hint on how to escape our Labyrinth.
11	**Love:** Love is the essence of our existence. A life that did not experience love, is a life wasted. The meaning of love in this guiding principle is compassion, empathy, and altruism toward our fellow human beings. The American Indians, a community of people who have adopted and are implementing SDD for tribal governance, have a relevant statement, namely "Walk a mile in somebody else's shoes." Street wisdom advises us "to learn to take other people from where they are at, and not from where we want them to be."
12	**Purpose:** The purpose of life is happiness and human development. Human settlements should be designed to enhance this purpose (Doxiadis, 1968). Unfortunately, because of the lack of planning and anticipation this is not the case with modern cities. The rationale and reason for escaping the Glocal Labyrinth is to attain human development in an evolutionary context. We should escape from being prisoners of the vicious cycle, individually and collectively, in order to fulfill the purpose of life.

Fig. 8 Displaying the enhancement pattern among the twelve guiding principles to be adopted and implemented in order to overcome the barriers of escaping the Labyrinth of our vicious cycle predicament

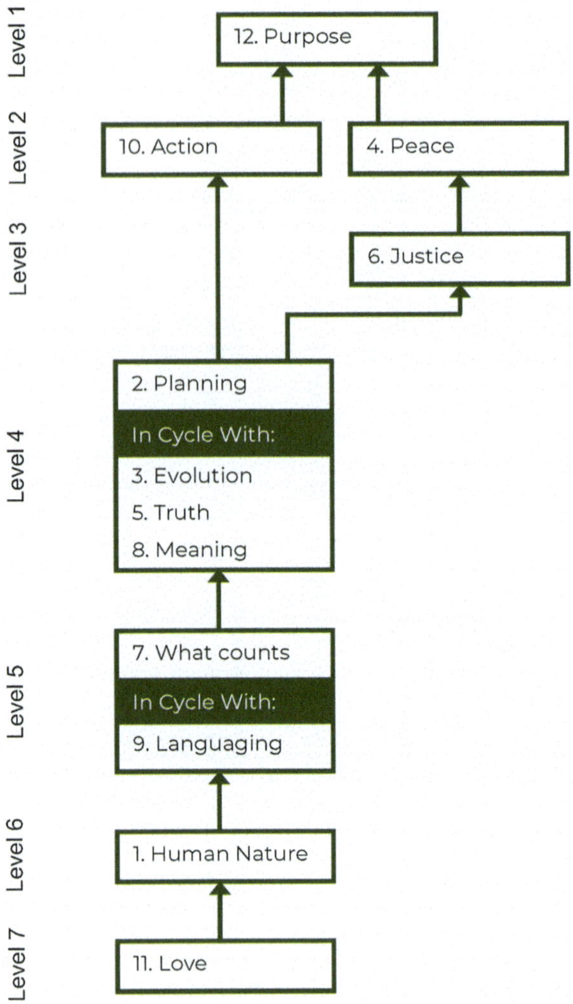

of higher priority and of stronger leverage than the principle of **Peace**, because **Peace** without **Justice** is not sustainable in the long run. Finally, we notice in Fig. 8, a cycle which contains the principles of **Planning**, **Evolution**, **Truth**, and **Meaning**, all of which are in a cycle of mutual enhancement. These four principles are mutually enhancing because they are all implementable concurrently in accordance with the General Theory of Planning of Ozbekh. It is hard to imagine humankind escaping the Labyrinth without the capacity to implement these four principles. In fact, these four principles are at the core of the development of the science of generic/dialogic design (Christakis et al, 1987; Warfield, 1994).

The main lesson from this structural inquiry with the metaphor of the Labyrinth is that escaping the predicament of our vicious cycle will require the appreciation of all twelve principles, and an understanding of the role and leverage each principle exerts in a successful escape strategy. It is almost certain that without love and compassion for all human beings and for all other living species, good planning, and justice for all, our survival and sustainability at the local and global scales is threatened (McIntyre-Mills, 2014).

The advantage of discovering this narrative from a structural modeling inquiry is that it lends more rationality and credibility compared to a narrative based strictly on prose modeling. Imagine young and old adults engaged in a learning experience by responding to the TQ of the Labyrinth metaphor inquiry, and structuring their ideas by means of the open and focused dialogue of the Logosofia platform. It would be highly desirable if something equivalent or similar to the Labyrinth inquiry presented here was being applied in the educational arena, so that young people would be sensitized early on about the vulnerability of our sustainability at large and small scales. In the next section, we will enter the educational arena and discuss the application of SDD with a community of people attending the Michigan School for the Deaf in the USA.

12 School for the Deaf Application

In 2015, the Michigan Department of Education (MDE), Office of Special Education, embarked on an effort to reimagine the Michigan School for the Deaf (MSD). The stated goal was to improve support to the MSD, and increase their capacity to help districts across the state to better serve students who are Deaf and hard of hearing, but are not attending MSD itself. This particular series of three colabs was very challenging to design and conduct by the SDD team of practitioners of the MDE; however, it affirmed the power of SDD, to address the intersubjectivity component of situational complexity due to the divergent viewpoints, to build a consensus, yielding a commitment by a community of stakeholders ready to take on the challenge of redesigning the MSD.

13 Discovery Phase

An important part of the Discovery phase is to ensure that the sponsor is well aware of the benefit of the SDD process, but also the responsibilities and vulnerabilities it encompasses. People in leadership positions, particularly in traditional, hierarchical organizations, are often expected to know the answers without input from stakeholders. Seeking input from those with less authority may actually be frowned upon. However, an effective leader understands intersubjective complexity and the need for shared ownership in order to make a plan implementable and sustainable (Magliocca & Christakis, 2001). A good SDD practitioner can sense the challenge of balancing the voices of the stakeholders with the authority of the leader of an organization, and will never attempt to oversell the SDD process – doing so will inevitably lead to failure. The sponsor of this application of SDD, Teri Rink, Director of the MDE, Office of Special Education, recognized the intersubjective complexity of the situation and understood the need to engage diverse stakeholders to co-construct a collective pathway for navigating the escape from the Labyrinth of the MSD. Ms. Rink was familiar with the SDD process, as she had sponsored previous colabs for different issues in the educational arena. Because of her previous experience, the role of the Broker (Jeff Diedrich) was minimal for this application. The sponsor chose to be a participant in the colab deliberations, recognizing she wielded no more power or influence in the colabs than any other participant.

A Core Planning Team (CPT) was established during the Discovery phase aligned with the characteristics and capacity described in the "Overview of the SDD Methodology" section of this chapter. The CPT included individuals who were Deaf/hard-of-hearing, a critical consideration that lent credibility to the effort. This also helped ensure appropriate accommodations were in place to enable every individual to fully participate in the colabs.

The tentative plan was to engage participants from across the state in the "Futures-Creative" archetype, consisting of three face-to-face colabs (Diedrich & Christakis, 2021). The first colab was focused on creating a *Vision*, the second on *Barriers/Challenges* for achieving the vision, and the third on *Actions* to overcome the identified barriers/challenges and approximate the vision. Several iterations of the triggering questions were drafted during the Discovery phase, fine-tuning it in meetings of the CPT. After the CPT was satisfied with the triggering questions, it was tested externally with seven people. The results of the tests validated the triggering questions.

Careful consideration was given to the stakeholder perspectives. Beyond students, staff, and family perspectives, the CPT also sought to ensure participants represented different and often competing philosophies regarding the education of Deaf/hard of hearing students. These competing philosophies contributed to the intersubjective complexity; however, without the requisite variety of perspectives, the effort would ultimately fail. In many processes, requisite variety is compromised in favor of convening like-minded individuals. Also included were individuals representing external agencies (i.e., Office of Civil Rights), business, etc. While easier to facilitate a group of like-minded individuals, (lacking requisite variety), there is substantial evidence that the design and overall solution will be underconceptualized. The structure of the SDD methodology makes engaging and managing people with differing ideas and philosophies not only possible, but advantageous.

Interviews were conducted during the Discovery Phase to gain additional insights and inform the approach. Results of the interviews revealed historical mistrust toward the MDE. This mistrust was, in part, due to decisions regarding the design of MSD facilities without input from stakeholders. Another contributing factor was a long-standing rumour of MDE closing the MSD. In addition, there was dissonance within the Deaf and hard-of-hearing community regarding interventions and philosophical approaches to meeting the needs of students who are Deaf/hard-of-hearing.

As with many oppressed or marginalized groups, there are many assumptions and generalizations that perpetuate the inability to escape from the predicament of the Labyrinth. Disregarding the voices of the stakeholders does not address the inter-subjective complexity. Interviewees, while skeptical of the motives and questioning the sincerity of the intent of the MDE to authentically involve stakeholders, agreed with the need to engage stakeholders. Ultimately, the interviews substantiated the need for an approach that fostered shared understanding and ownership of the situational complexity.

Logistics required additional considerations for this colab, namely accommodations to ensure all participants could fully engage and contribute during the colabs. Four certified sign language interpreters were utilized throughout each colab, as was remote "Communication Access Realtime Translation" (CART) services – which provided instant translation of the spoken word into English text using a stenotype machine. In addition to the room layout of the Demosofia space (see Fig. 2), lighting should have been a critical consideration, but was initially overlooked.

14 Colaboratory of Democracy Phase

Most organizations base their vision on present trends, utilizing current data and models to derive a projection of an extrapolated future – only slightly different from the current situation. The Future's-Creative archetype of SDD encourages participants to dream, free of all constraints, of an idealized, visionary future, that is vastly different from the current situation and its extrapolation. Liberating oneself from constraints is challenging – though people complain about constraints, there appears to be a certain level of comfort in having constraints. Freeing oneself of constraints requires an atmosphere that encourages some degree of risk-taking, which implies that participants make themselves vulnerable. SDD creates a safe space to render oneself vulnerable and imagine a new data-base; however, it requires a degree of trust.

The dialogue was launched in an atmosphere of heavy skepticism and mistrust, directed primarily towards the MDE and, to a lesser extent, each other. This led to questioning the legitimacy of the SDD methodology and the role of the facilitation team, making it vitally important that the process be followed with great integrity, e.g., ensuring that the facilitator remained content neutral and process focused, as well as protecting the authenticity of every voice at the table.

Compounding the challenge of trust, participants shared concerns over the room lighting, indicating it was difficult to see the interpreters. Lighting is often a factor in the Demosofia facility, but the facilitation team failed to recognize the critical importance of lighting relative to participants observing interpreters. The facilitation team purchased LED lights to replace the fluorescent tubes in the ceiling, as well as supplemental LED lighting during the first extended break of the day. Participants were grateful for the effort and noted a slight improvement, though they continued to indicate lighting was a problem. It is interesting to notice how the comfort of the facility utilized for conducting structured dialogue is critical for high quality observations (see Fig. 2), something which is ordinarily disregarded.

Day one ended with slightly less mistrust and the Facilitator (Diedrich) was optimistic that day two of the first colab would begin with a lower degree of skepticism regarding the process and motives of the sponsor and the facilitation team. Unfortunately, day two quickly revealed the optimism was unfounded, but the facilitation team moved forward, trusting the SDD process.

In the following section we will present the summary of the results obtained from each one of the three colabs, with primary emphasis on the metrics corresponding to the measurement of situational complexity. Because the process steps are similar for each colab with the exception of the Triggering Question, we will elaborate only on the vision colab and only present the TQs and the maps produced for the barriers and the action colabs. For more details the interested reader should visit the link:

https://mdemsd.wordpress.com/

15 Summary of Results from Each Colaboratory

15.1 Co-laboratory 1: Vision

The launching of the first colab included an overview of why reimagining the MSD was an important issue, a review of the timeline, a brief overview of the SDD process, participant expectations, and introduction of the Triggering Question. The Triggering Question for the first co-lab was:

What do you visualize an ideal model of what the Michigan Department ofEducation's School for the Deaf 'ought to be' to deliver an extraordinary educational experience that leads to successful career and college-ready student outcomes?

16 Statement Generation and Clarification

Participants generated seventy-four statements in response to the Triggering Question. The clarification step, during which the authors of the ideas explained their meaning for the understanding, and only for understanding, by the other members of the group took approximately four hours.

16.1 Categorization

Following the clarification of meaning, statements were classified into thirteen affinity clusters based on significant similarity. The clusters of similar ideas were discovered inductively, i.e., by looking at two ideas at a time, and voting on their affinity in the context of the TQ. The thirteen clusters that were discovered and labelled by the group were:

Cluster 1: Talent Recruitment Selection (five ideas)
Cluster 2: Language Competency (five ideas)
Cluster 3: Early Language Acquisition (four ideas)
Cluster 4: Student Experience (ten ideas)
Cluster 5: Leverage Expertise (seven ideas)
Cluster 6: Collaboration (nine ideas)
Cluster 7: Evaluate Impact (two ideas)
Cluster 8: Options (two ideas)
Cluster 9: Access to Curriculum (eight ideas)
Cluster 10: Awareness/Visibility (five ideas)
Cluster 11: Leadership (four ideas)
Cluster 12: Family Engagement (three ideas)
Cluster 13: Infrastructure (ten ideas)

16.2 Prioritization

At the time of this co-laboratory, (i.e., in the year 2015, the capability for Logosofia to record multiple prioritizations was not available), participants engaged in only a single round of prioritization. This occurred after classification of the vision statement and at the end of day one. Of the seventy-four ideas, thirty received at least one vote. Given the skepticism of participants, the voting or relative saliency result signified a growing appreciation for the ideas of other participants, thereby reducing the intersubjective complexity significantly, because the divergence Index D is equal to 0.36 (See Table 4). The top priority statements were:

1. Vision #27: Birth to 3 opportunities for families on MSD campus (8 Votes)
2. Vision #36: MSD will be the center of expertise on deaf education in Michigan (8 Votes)
3. Vision #18: Broaden the capacity of MSD to support the educational needs of all students who are deaf and hard-of-hearing statewide (7 Votes)
4. Vision #39: MSD shall see Deaf leaders in the administrative level (7 Votes)
5. Vision #53: Conduct the hiring processes of teachers of the deaf with a better collaboration with the MSD with the MDE to be a part of the review and hiring process (7 Votes)
6. Vision #58: To set up a new campus for MSD (7 Votes)

Table 4 Colab series metrics

colab	(N) # of Ideas	(V) # of Ideas that received a vote	(R) # of Ideas in a Relational Pattern	(K) Relationship Cardinality – # of distinct inter-connections	(D) Divergence in terms of relative importance voting	(PEI) Priority Effectiveness Index	(SCI) Situational Complexity Index
VISION	74	30	16	117	0.36	0.95	23.09
BARRIERS	88	32	15	73	0.33	0.53	14.04
ACTIONS	77	35	16	67	0.42	0.39	11.30

17 Influence Mapping

Based on the prioritization, a subset of sixteen ideas receiving two or more votes was selected for influence mapping. The mapping process invited participants to engage in pairwise decision-making regarding significant influence of idea X over idea Y. Participants were invited to share their rationale for their vote regarding the influence relationship of a pair of ideas. The dialogue during influence mapping is critically important to the learning that occurs during a colab. Participants deepen their understanding and appreciation for alternative perspectives while also deepening their own understanding by explaining their rationale.

The process of SDD is, while manageable, cognitively demanding. The burden for those participants who were Deaf/hard-of-hearing increased exponentially. Those individuals had to pay attention to the Facilitator, the interpreters, CART, and to each other, all with less-than-ideal lighting. The Facilitator had to be cognizant of and empathetic to the increased burden on the group, and was perhaps a bit too slow in recognizing this need. That being said, participants fully engaged, sharing their thoughts with and appreciating one another, changing their minds, and learning together.

The resulting influence map yielded many cycles of mutual influence among the subset of sixteen ideas. These cycles contribute to the overall complexity of the situation and can be seen in Fig. 9.

This map is reasonably linear with four distinct levels and three cycles containing twelve of the sixteen ideas selected for inclusion in the influence map. Statement #53, focusing on the requirement of hiring teachers for the school of the deaf with a dominant role by the MSD with the consultation of the MDE. The root statement influenced fourteen of the total of sixteen ideas included in the map, with the exception of statement #12 at Level III. On the other hand, at Level I we have visionary statements, such as building a new campus for the MSD (#58), and the use of best learning practices for the MSD students. The cycle at Level III, which contains six statements of mutual influence, is primarily focused on the attributes of the teaching staff, which is clearly instrumental in attaining the visionary statement located at Level I. All the visionary statements are meant to inspire the group of participants to think about those actions which, if adopted and implemented, will approximate the attainment of the visionary future for the MSD. The results for the metrics corresponding to this colab, including the SCI, are shown in Table 4.

The top five priority effectiveness statements are:

1. Vision #53 – Conduct the hiring processes of teachers of the Deaf with a better collaboration with the MSD with the MDE to be a part of the review and hiring process (7 votes)
2. Vision #3 – I want a state-of-the-art ASL/English bilingual early childhood program for Deaf children for the age of 0–3 encompassing the entire state of Michigan (4 votes)
3. Vision #18 – Broaden the capacity of MSD to support the educational needs of all students who are Deaf and hard-of-hearing statewide. (7 votes)

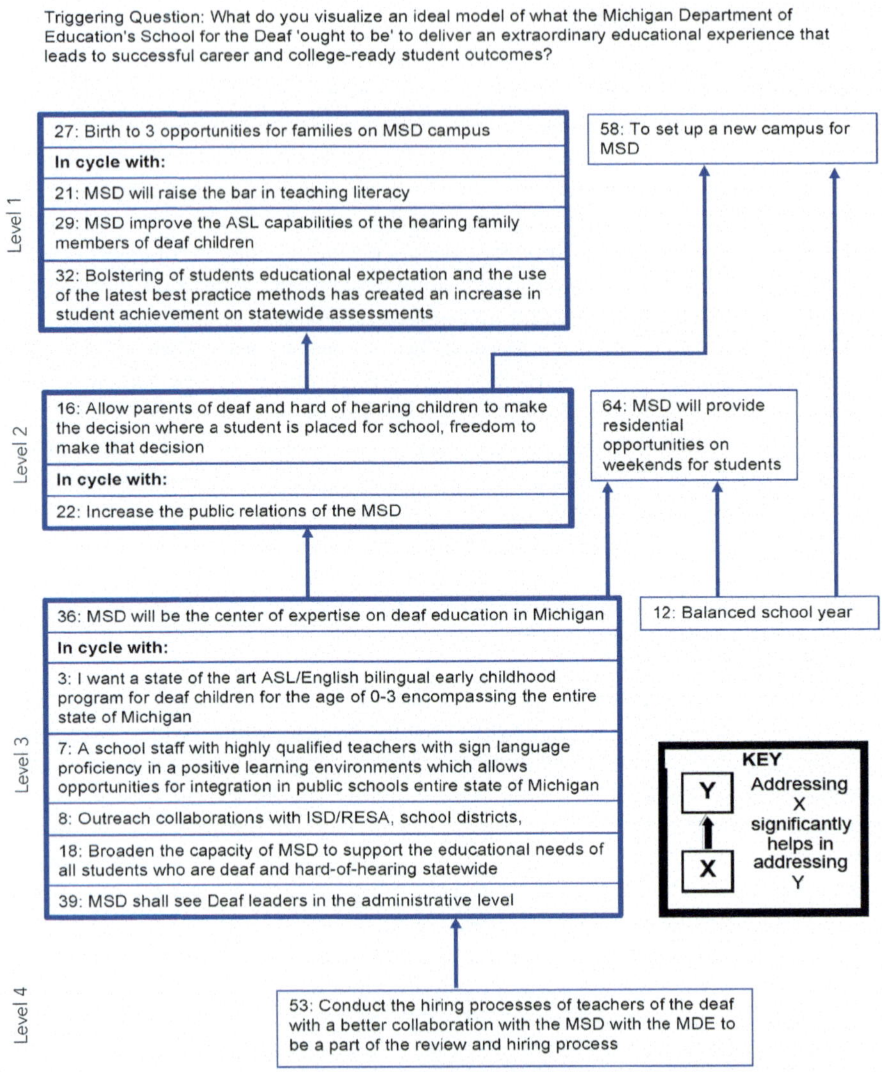

Fig. 9 Influence map showing the enhancement relationship among sixteen vision statements

4. Vision #39 – MSD shall see Deaf leaders in the administrative level (7 votes)
5. Vision #12 – Balanced School Year (5 votes)

If we study the top five saliency priorities identified earlier, and compare them with the top five effectiveness priorities determined from Fig. 9, we will see that Visions #53, #18, and #39 of the latter five are included in the subset of the former five. In the SDD methodology (Christakis & Bausch, 2006), a lack of significant correlation between these two subsets, i.e., the saliency and the effectiveness results) has been repeatedly observed and has been identified as the Erroneous Priorities Effect. Organizations that implement strictly the priorities determined on the basis of relative saliency run the risk of allocating resources to the least effective priorities. In this case, the large cycle at level III makes it challenging to draw this conclusion though it should be noted that Vision statements #3 and #12 were not represented in the top 5 votes on relative saliency, and should be considered in the design of an action plan because of their priority effectiveness.

18 Colaboratory 2: Barriers/Challenges

The Triggering Question for the second colaboratory was:

What are barriers to realizing the Michigan Department ofEducation's School for the Deaf ideal model, as defined in colab I, to deliver an extraordinary educational experience that leads to successful career and college-ready student outcomes?

18.1 Statement Generation, Clarification and Prioritization

Participants generated eighty-eight statements in response to the Triggering Question. The clarification step took approximately four hours again. The number of statements that received one or more votes was equal to thirty-two, resulting in a divergence measure D is equal to 0.33 (See Table 4). The number of statements that were selected for inclusion in the influence map was fifteen.

The top priorities in terms of relative saliency are (statements that received six or more votes):

1. Barrier #27: The funding protocol (how we fund) the MSD (9 Votes)
2. Barrier #33: Underutilization of resources, extend collaborations with all agencies or entities to expand possibilities for students (7 Votes)
3. Barrier #36: Conflict between philosophies/ methodologies in Deaf education (7 Votes)
4. Barrier #1: Lack of support for implementation of 0–3 ASL- English bilingual programs (6 Votes)
5. Barrier #25: Current facility location, size and design are all limiting factors (6 Votes)
6. Barrier #61: A new type of leadership structure on the campus. (6 Votes)
7. Barrier #68: Inability of parents of Deaf and hard-of-hearing children making the decision to attend MSD (6 Votes)
8. Barrier #77: Lack of sensitivity and understanding of Deaf needs (6 Votes)

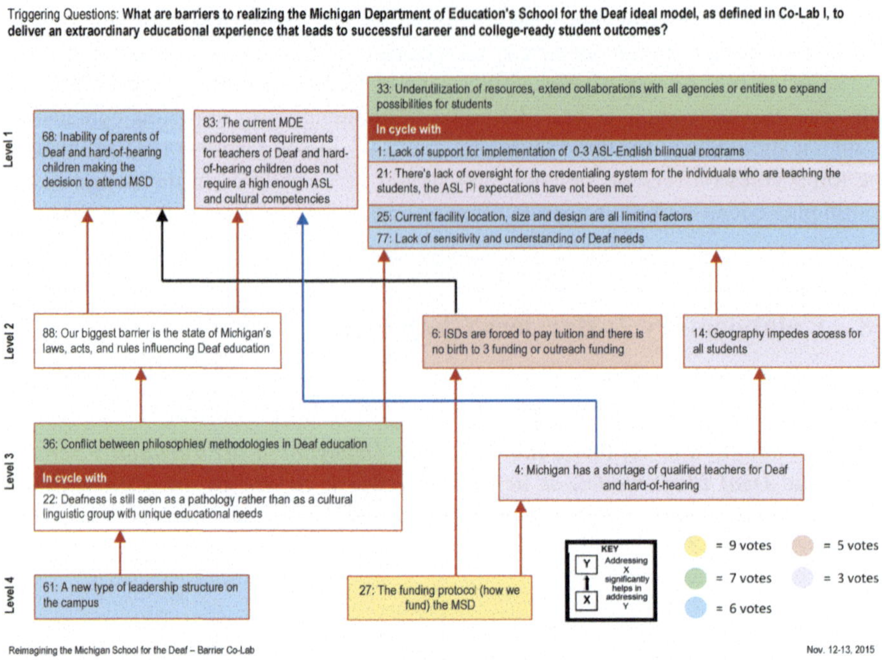

Fig. 10 Influence map showing the enhancement relationship among fifteen barrier statements

18.2 Influence Mapping

Employing again the stakeholders in a deliberation that focuses on a generic question, the subset of fifteen statements was mapped in accordance with the voting judgments of the group. The resulting influence map can be seen in Fig. 10.

The metrics resulting from this colab appear in Table 4. We notice that this map has four levels with two barriers at the deepest level. There exists one cycle with five barriers at Level I, and another cycle with two barriers at Level III. Usually, the emergence of cycles in a map represents a new entity, because of the couplings and strong interactions among the barriers included in the cycle, that require special consideration in terms of the capacity of dealing with them effectively.

The top five barriers in terms of priority effectiveness are:

1. Barrier #61 – A new type of leadership structure on the campus (6 Votes).
2. Barrier #27 – The funding protocol (how we fund) the MSD (9 votes)
3. Barrier #4 – Our biggest barrier is the state of Michigan's constitution (4 votes)
4. Barrier #22 – Deafness is still seen as a pathology rather than as a cultural linguistic group with unique educational needs (3 votes)
5. Barrier #36 – Conflict between philosophies/ methodologies in Deaf education (7 votes)

Comparing the top priorities in terms of relative saliency, with the top five in terms of priority effectiveness, we see that only Barriers #27 and #61, of the latter subset are included in the former, indicating that the Erroneous Priority Effect (EPE) applies for the Barriers/Challenges colab. An action plan designed without careful consideration to the top five effectiveness barriers identified from the map of Fig. 10, will probably fail to approximate the vision map (Fig. 9), by overcoming the barriers map. The essence of the Futures-creative Archetype of SDD is to enable a group to co-create a vision (Fig. 9), followed by the identification of barriers/challenges that are inhibitors to the attainment of the vision (Fig. 10), and finally in the third colab designing an action plan that creates the plausibility of penetrating the wall of barriers in order to approximate the vision (see short video by Jeff Diedrich created for a colab with a group of Israelis and Palestinians).

19 Colaboratory 3: Actions

Again, using the same process, the Triggering Question for colaboratory 3 was:

What specific ACTIONS, if adopted and implemented, can overcome the barriers to approximate the ideal vision leading to MDE's School for the Deaf ability to deliver an extraordinary educational experience that leads to successful career & college-ready student outcomes?

19.1 Statement Generation, Clarification and Prioritization

Participants generated seventy-five statements in response to the Triggering Question. The clarification step took approximately four hours. The number of statements that received one or more votes was equal to thirty-five. The number of statements that were selected for inclusion in the influence map was sixteen. The divergence measured after the clustering was equal to 0.42 (See Table 4).

The top priorities in terms of relative saliency are:

1. Action #4 – Engage MDE leadership and state budget office to work with the legislature to change MSDs funding so local districts are not charged. (9 votes)
2. Action #22 – Move to a new state owned facility with sufficient capacity and proper design in a desirable, optimal location (6 votes)
3. Action #41 – Promote and adopt a balanced calendar (6 votes)
4. Action #43 – Enable parents to enroll their children who are Deaf or hard-of-hearing in the MSD (6 votes)
5. Action #60 – Require culturally Deaf leaders in administrative levels. (6 votes)

The resulting influence map can be seen in Fig. 11.

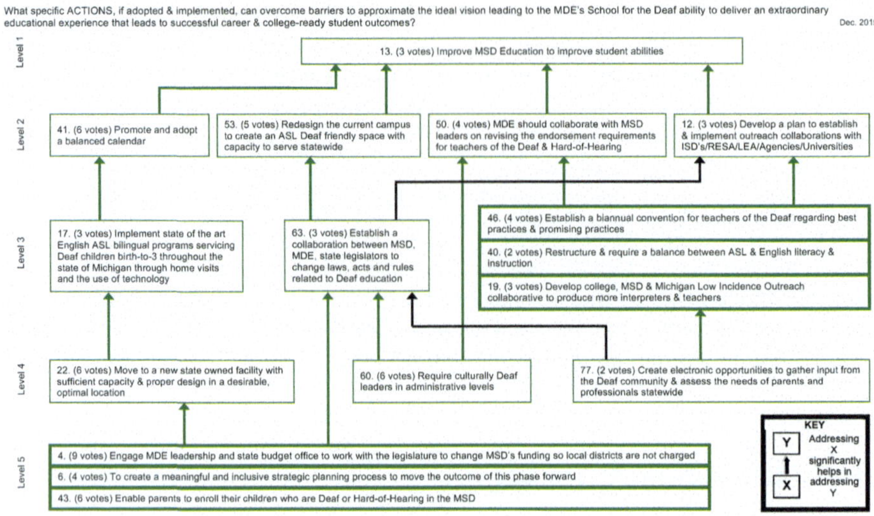

Fig. 11 Influence map showing the enhancement relationship among sixteen action statements

The Action map has five levels with a cycle of actions (#4, #6, #43) at the deepest level. There exists a second cycle with five barriers at Level III. As noted, the strong interactions often represent a new entity that requires special consideration to implement them effectively.

The top five actions in terms of priority effectiveness are:

1. Action #4 – Engage MDE leadership and state budget office to work with the legislature to change MSDs funding so local districts are not charged. (9 votes)
2. Action #6 – To create a meaningful and inclusive strategic planning process to move the outcomes of this phase forward. (4 votes)
3. Action #43 – Enable parents to enroll their children who are Deaf or Hard-of-Hearing in the MSD (6 votes)
4. Action #77 – Create electronic opportunities to gather input from the Deaf Community and assess the needs of parents and professionals statewide. (2 votes)
5. Action #60 – Require culturally Deaf leaders in administrative levels. (6 votes)

Comparing the top five priorities in terms of relative saliency, with the top five in terms of priority effectiveness, we see that Actions #4, #43, and #60 of the latter subset are included in the former. However, Action #6 received four votes regarding relative saliency, and Action #77 received only two. Yet, as we see in Fig. 11, both exert substantial leverage and effectiveness. Traditional methods would have likely overlooked these actions in favor of more popular actions, leading to the erroneous priorities effect. A strategic action plan designed without careful consideration of the top five effectiveness actions identified from the map of Fig. 11, will probably fail to approximate the vision map (Fig. 9).

20 Reflections

The sponsor (Teri Rink) of the MSD application, recognized from the outset the value of engaging the community of stakeholders; however, neither she nor the Facilitator (Jeff Diedrich) anticipated the challenge of launching an assault on this particular situational complexity. Nevertheless, it turned out that, in spite of the high level of the MSD complexity (as we shall see from Table 4), the SDD process proved capable of enabling the community to reducing the SCI of the MSD to only 7% of the maximum for the vision colab, and to 2% for the action colab, by means of: (a) enforcing the content neutrality of the Facilitator, (b) ensuring every voice is heard and is equal at the table, and (c) focusing on protecting the authenticity of every participant.

With every hour of this series of colabs, incremental increases in trust was observed. Participants learned, perhaps for the first time, that the authenticity of their experiences and of their sign language will be honored. They began to trust each other, they began to trust the process, they began to trust the facilitation team, and they began to trust the Michigan Department of Education (on the issue of trust see Christakis & Kakoulaki, 2021). By the conclusion of the third colab, participants who represented a requisite variety of philosophies, experiences, and perspectives, became a community that collectively owned the responsibility to redesign the Michigan School for the Deaf. While it was one of the most challenging series of colabs for Jeff Diedrich to facilitate, it was also one of the most rewarding. The methodology of SDD encouraged: (a) deep self-examination of one's own thinking, (b) a willingness to abandon some/all of their own ideas in favor of ideas produced by others, (c) consensus on how to improve the MSD, and (d) a shared ownership and commitment in doing the right thing (Vision map), and doing it right (Action map). Best of all, it created a community – with new-found friendships and respect as a result of the process of structured deliberation.

20.1 Situational Complexity Index (SCI)

We will now offer some reflections on the values of SCI obtained as a result of conducting the three MSD colabs in the context of the futures-creative Archetype of SDD. The values of all the relevant indices (metric), as displayed in the screen of the Logosofia Dashboard for each of the colabs are shown in Table 4.

For the first colab, focusing on envisioning the ideal MSD future as represented by the map of Fig. 9, the actual SCI is equal to 23. From the vision map, together with the values for the parameters N, D, and R, shown in Table 4 (the values for D and R correspond to the start of the mapping step, see Fig. 2), the maximum value of SCI is equal to 362, which would be obtained if K was equal to 225. The optimum is equal to 24, obtained when K is equal to 120. That means that the actual SCI measured after the completion of the mapping step, corresponds to 7% of the maximum, and 100% of the optimum. Essentially, the SDD process has enabled the group to reduce the situational complexity of the MSD vision from its maximum value of 362 to the optimum value of 24, a non-trivial accomplishment. Even though twelve ideas from a total of sixteen selected for mapping are included in three distinct cycles in Fig. 9, the pattern of the ideas offers the group the capacity to identify the most effective priorities, and to draw distinctions between statements with strong influence at Levels III and IV, and weak at Levels II and I. Furthermore, the identification of effective priorities enables the community to avoid implementing erroneous priorities. Also, the map of Fig. 9, appears to resemble a linear tree-pattern from bottom up, because the number of distinct connections K = 117, which is practically equal to half of the maximum number of connections (K) for an R = 16, (i.e., all sixteen ideas are in a cycle of mutual influence, and the value of K is equal to 240). This is the reason that the SCI in this case is equal to 100% of the optimum value of SCI. The group work during the vision colabs has managed to help the stakeholders identify effective priorities and draw distinctions among the original set of ideas (i.e., N = 74), generated in response to the Triggering Question. In essence, the map of Fig. 9, becomes the foundation and the magnet for developing a strategy and an action plan for overcoming the barriers identified in the second colab and approximate the ideal MSD future.

For the colab focusing on the barriers, and the map of Fig. 10, the actual SCI is equal to 14, while the maximum SCI is equal to 374, and the optimum is equal to 27. Hence, for this colab the actual SCI corresponds to 4% of the maximum, and 52% of the optimum. Six ideas, from a total of fifteen, are in cycles in Fig. 10. For this case the actual K = 73, and the maximum corresponding to R =15 is equal to 210, so the value of K is 35% of the maximum. Again, for this colab the group work has managed to reduce the SCI by 96%. The maximum value of SCI is obtainable when all the ideas in the map are connected to each other, something that is conceivable unless a group of stakeholders discovers the actual connections by means of mapping, and as a consequence reduce the SCI from its maximum conceivable value, as happened in this case.

For the third colab producing the action map (Fig. 11), the actual SCI is equal to 11, the maximum equal to 441, and the optimum equal to 29. The actual SCI is 2% of the maximum and 38% of the optimum. This map also has six ideas in cycles from a total of sixteen. For this map the actual K = 67, and the maximum corresponding to R =16, is equal to 240, so for this case, K is 28% of the maximum. The reduction of SCI from its maximum value corresponding to the actions colab is 98%. The significant reduction of SCI for this case is attributed to the smaller values of both the K and the PEI, as compared to the values for the other two colabs. Also, by looking at the pattern of Fig. 11, one can see that it is much simpler in terms of the number of links and crossings among the sixteen action ideas, as compared to Figs. 9, and 10.

The Logosofia Dashboard was not available at the time of the MSD application. However, from the data reported in Table 4, we can make some inferences regarding the utility of the Dashboard for navigating future colabs. The calculations performed so far, offer some insights of the type of indicators to be displayed on the Dashboard, so that it will be useful to the Facilitator. Specifically, once the values of N, D, and

R are determined, which happens after approximately four hours of group work, the Dashboard should display the maximum and optimum value of SCI for the Facilitator to observe. During the mapping process, the Dashboard should display the actual SCI measures as the percentages of the maximum and optimum SCI, in order to help the Facilitator pilot the colab. For example, if the percentage of the actual SCI approaches the optimum value, the Facilitator should slow down the speed of the mapping process. The reason for this is to avoid the plausibility of the group not drawing distinction between strong and weak interactions during the pairwise voting queries. If this happens, the group constructs spurious cycles in the mapping process, which is the equivalent to bad weather for an airplane pilot.

We intend to utilize the Dashboard in the conduct of colabs in the future in order to gather data and be able to gradually improve the quality of SDD facilitation. As an example, in the case of the vision colab for the MSD application, it is conceivable that, if the Dashboard was available to Jeff Diedrich, he would have slowed down the pace of the mapping process to avoid getting twelve of the sixteen ideas in three cycles in the map of Fig. 9. As we can see from Table 4, the value of K being equal to 117, is almost fifty percent of the maximum value of K, which is obtained when all sixteen ideas are in a cycle of mutual influence. **To summarize, the experienced Facilitators should be watching the valueof the divergence D during the generation, prioritization, and clustering steps of the process (See Fig.2), and seeing if the valueof D is converging to a valueof 50% or less after clustering. During the mapping step, the Facilitator should be watching the valueof PEI as it unfolds, and seeing if it approaches the valueof 100% of its optimum value, (i.e. PEI = 1). In case the PEI is close or larger than one, as it did for the vision map of the MSD application, and if also the number of levels in the map is small (see Fig.9), the Facilitator should slow down the subsequent voting on the influence between pairs of ideas.**

21 Participant Reflections

It was my first time meeting Teri. I did wonder whether she or Jeff had hidden agendas for the workshops but I kept my mind open. I felt a lot better that meetings were proctored or facilitated in a certain way and now we as a team are saying in our 6th meeting day that our meetings went well.

I believe we should continue to support each other as a team and I cannot wait

–Gary B.

Director, NTID Center on Access Technology

At the beginning of this meeting I thought oh, this is not going to work. Because of past experiences. And I thought MSD is going to close because nobody is going to want it to do well! Or be successful. But now, after 6 days, I'm thinking wow, this is hard and it's my school and this is so interesting how this is becoming a successful process. And I think it will be, we can achieve what we want to do. And I appreciate all of you coming for this whole 6 days and taking our school seriously. I want to thank all of you for everything – your philosophies and opinions and, just everything.

–J. C.

Student

One thing that was interesting, part of the dialogue that we had, that one person's statement could influence a person's vote. I appreciated this dialogue procedure. It's not easy to go through. It's thought provoking and, a lot of self-analysis, introspective thinking has to happen for this to be a successful process. We come to the table with diverse philosophies and thought processes in general and opinions and this dialogue was a successful process. And if you look at the 6 days and the way that the work maps out, we are talking funding and philosophies and student outcomes as a priority. The days here together – we know that we want the school to succeed and that's a common goal and thread that we all have here.

–***Annie U.***

Division On Deaf, DeafBlind, and Hard-of-Hearing (DODDBHH) Director, Michigan Department of Civil Rights

I feel a need to reiterate the experiences felt by Stevie Naeyaert that we, as Deafstakeholdersof MSD, have always faced the issue of being marginalized with the planning and decision making of MSD over the years. Through this process, the collaboration that have been experienced with the different lens of Deaf and hearing participants from the north of the Upper Peninsula to south of Detroit, have shown that each and every one of you have brought forth respectand consideration of one another's viewpoints. Through this, the lid has been lifted off this marginalized issue that we, the Deafstakeholdershave experienced, and provided hope for a positive collaboration of both communities for the sake of MSD.

– ***Camille J-L.***

Teacher

Being part of these structured dialogues have brought forth renewed optimism for the future of Michigan School for the Deaf (MSD) through the alliance that has been evident with insights from both deaf and hearing participants. Even more so, I was pleased to see our present MSD students become part of the team and gain experiences that will carry with them as they pave the way of becoming positive, contributing deaf leaders.However, it is incumbent upon us, as deaf and hearing participants, to maintain having autonomous and authentic dialogues after this opportunity to conquer any skepticism that positive change at MSD is imminent and will become a reality. There is hope and we need to continue to be visible. For allowing me to be part of this vital process, I thank you.

–***Natalie G.***

Teacher

22 Conclusions

In this chapter, we have conceptualized situational complexity as a phenomenon that emerges when groups of stakeholders attempt to collaborate for the definition and resolution of wicked problems. The strong interactions among a set of problems creates a new entity, which was identified in the Club or Rome prospectus as the Global Problematique. Launching an effective assault on the situational complexity corresponding to the Glocal (i.e., Global and Local) Problematique, and/or designing complex social systems, such as the case of the Michigan School for the Deaf (MSD), makes it an imperative for leaders and followers to adopt and implement genuine deliberative democracy. However, we have accumulated empirical evidence, since the interdisciplinarity experiment of the 1970s, that deliberative democracy is not implementable effectively and authentically, even with the best intentions, unless it is supported by the science of dialogue and the application of the methodology of SDD. On the other hand, applications of SDD with the inclusion of stakeholders, depend critically on the commitment to deliberative democracy by leaders, such as Teri Rink in the case of MSD. This phenomenon of the cyclical interdependence between assaulting situational complexity and practicing genuine deliberative democracy, produces a vicious cycle, which we have called metaphorically the Labyrinth of our predicament. Designing a strategy for escaping the Labyrinth requires the adoption and implementation of such guiding principles as Love, Planning, Conscious Evolution, Justice, Peace, etc., in order to fulfill the purpose of life for all human beings, which is happiness and human development.

The application of the SDD methodology for the school of the deaf provided tangible evidence of the potential of SDD to build trust and consensus for a challenging community, such as the MSD, by engaging the stakeholders in colabs so that they significantly reduce the measure of the SCI. Furthermore, it accomplished this goal for a community of stakeholders for which there has been historically major skepticism and mistrust. This particular application of SDD also provided evidence and insights on how inappropriate and unethical it is to design action plans for the future of MSD without listening to the voices of the Deaf/hard-of-hearing people. As Native American have been trying to teach the "White Man" for generations, it is about time to learn "to walk a mile in somebody else's shoes." At the conclusion of the colabs, a small and diverse team was compiled to draft a strategic plan to set into motion the outcomes of the colabs. The results of the colabs continue to inform and inspire the MSD community. For example, stakeholder input was sought when determining the desirable characteristics of a new principal for the MSD when the position became vacant. The Deaf community was involved in the interview process and provided feedback regarding the candidates. The position was ultimately filled by an individual who is culturally Deaf. In addition, the MDE has engaged legislators to examine the funding structure of MSD and conversations are ongoing.

We anticipate the emergence of major challenges regarding public policy issues in the scientific domain unless we pay attention to the phenomenon of situational complexity. For example, recently the Greek parliament passed legislation with the intent to protect the medical doctors (epidemiologists) who, as members of a task force, were responsible for making public policy for preventing the spread of the pandemic. The legislation protects the epidemiologists from any legal actions against them by Greek citizens. The concern is that citizens, who might disagree with these policy measures adopted and implemented by the Greek Government, would choose to sue the doctors. We label this legislative action "The Galileo Effect," because it reminds us of what happened to Galileo, four hundred years ago, after he advanced the heliocentric explanation of planetary motion. We recognize, of course, that the pandemic and the planetary motion are drastically different occurrences, in that the situational complexity explained by Galileo belongs to the domain of First Phase science objectivity. And yet, the dominant dogma at the time of Galileo, represented primarily by the Catholic church, took three hundred years to accept Galileo's observations and theory. Our hope is that the situational complexity phenomenon confronting the MSD, and many other communities, will be considered sooner rather than later, by means of the authenticinclusion of stakeholders in SDD applications for designing their futures.

References

Arnstein, S., & Christakis, A. N. (1976). *Perspectives on technology assessment*. Israel Press.

Banathy, B. H. (1996). *Designing social systems in a changing world*. Plenum.

Banathy, B. H. (2000). *Guided societal evolution: A systems view*. Kluwer Academic/Plenum Publishers.

Bausch, K., & Christakis, A. N. (2015). *With reason and vision: Structured dialogic design*. Emergence Publications.

Broome, B. J., & Christakis, A. N. (1988). A culturally sensitive approach to tribal governance issue management. *International Journal of Intercultural Relations, 12*(2), 107–123.

Christakis, A. N. (1973). A new policy science paradigm. *Futures, 5*(6), 543–558.

Christakis, A. N. (1993). The inevitability of demosophia. In I. Tsivacou (Ed.), *A challenge for systems thinking: The Aegean seminar* (pp. 187–197). University of the Aegean Press.

Christakis, A. N. (1996). A people science: The CogniScope system approach, systems. *Journal of Transdisciplinary Systems Sciences, 1*(1), 1–9.

Christakis, A. N. (2006). A retrospective structural inquiry of the predicament of humankind. In J. P. van Gigch & J. McIntyre-Mills (Eds.), *Volume 1: Rescuing the enlightenment from itself*. Springer.

Christakis, N. A. (2019). *Bluprint: The evolutionary origins of a good society*. Little, Brown.

Christakis, N. A. (2020). *Apollos' arrow, the profound and enduring impact of coronavirus on the way we live*. Little, Brown.

Christakis, A. N., & Bausch, K. C. (2006). *Co-laboratories of democracy: How people harness their collective wisdom to create the future*. Information Age Publishing.

Christakis, A. N., Dye, K. M. C., & Shearer, W. L. (1999). Collaboration through communicative action: Resolving the systems dilemma through the cogniscope. *Systems: journal of transdisciplinary systems science, 4*(1–2), 9–32.

Christakis, A. N., & Dye, K. M. (2008). In P. Jenlink & B. H. Banathy (Eds.), *CogniScope: Lessons learned in the Arena, chapter: Dialogue, dialogue as a collective means of design conversation* (pp. 187–203). Springer.

Christakis, A.N. & Kakoulaki, M. (2021–forthcoming). Objectifying intersubjectivity through inclusion for a scientific [r]evolution (Chapter 35). In J. McIntyre & N. Romm (Eds.), *From polarisation to MultiSpecies relationships: Re-generation of the commons in the era of mass extinctions*. Springer.

Christakis, A. N., Whitehouse, R., & Conoway, D. (1995). Redesigning the good review practices of the food and drug administration, CWA Ltd Report.

de Zeeuw, G. (1996). *Second order organizational research, working papers in systems and information sciences*. University of Humberside.

Diedrich, J. (2010). *Act beyond borders video*. https://www.youtube.com/watch?v=E7NSjhZno80

Diedrich, J., & Christakis, A. N. (2021). *The Co-Laboratory of Democracy Archetypes: Engaging Stakeholders in Deliberative Democracy to Respond Proactively to Diversity. In From Polarisation to Multispecies Relationships* (pp. 451–470). Springer.

Doxiadis, C. A. (1968). *Ekistics: An introduction to the science of human settlements* (pp. 27–31). Oxford University Press.

Flanagan, T., & Christakis, A. N. (2010). *The talking point: Creating an environment for exploring complex meaning*. Information Age Publishing.

Flanagan, T. R., & Christakis, A. N. (2021). *The Talking Point: Creating an Environment for Exploring Complex Meaning* (2nd ed.). IAP.

Flanagan, T., & Lindell, C. (2018). *The Coherence factor: Linking emotion and cognition when individuals think in groups*. Information Age Publishing.

Flanagan, T., & Post, D., (1998). Eliminating Lymphatic Philariasis by the year 2020, CWA eport for the World Health Organization.

Jones, P. (2014). Systemic design principles for complex social systems. In G. Metcalf (Ed.), *Social systems and design* (Volume 1 of the Translational systems science series) (pp. 91–128). Springer.

Jones, P. H. (2018). Contexts of co-creation: Designing with system stakeholders. In P. Jones & K. Kijima (Eds.), *Systemic design: Theory, methods and practice* (Volume 8 in Translational systems sciences series) (pp. 3–52). Springer.

Kakoulaki, M. (2016). *Dialogue beyond borders documentary based on interviews Israel, Palestine, Cyprus*. https://www.youtube.com/watch?v=SKwXw6hFpAA

Kakoulaki, M., & Christakis, A. N. (2018). Demoscopio: A demosensual [r]evolutionary Eutopia. In J. McIntyre & N. Romm (Eds.), *Balancing individualism and collectivism for social and environmental justice*. Springer.

Kakoulaki, M., Fkanagan, T. R., & Christaksi A. N. (2022–forthcoming). An approach for co-creating visionary anticipations. In G. Midgley & D. Cabrera (Eds.), *Routledge volume on systems thinking*.

Kuhn, T. (1970). *The structure of scientific revolutions*. University of Chicago Press.

Laouris, Y., & Michaelides, M. (2018). Structured democratic dialogue: An application of a mathematical problem structuring method to facilitate reforms with local authorities in Cyprus. *European Journal of Operational Research, 268*(3), 918–931.

Magliocca, L. A., & Christakis, A. N. (2001). Creating transforming leadership for organizational change: the CogniScope System approach. *Systems Research and Behavioral Science: The Official Journal of the International Federation for Systems Research, 18*(3), 259–277.

McIntyre-Mills, J. (2014). *Systemic ethics and non-anthropocentric stewardship*. Springer.

McIntyre-Mills, J. J., & Christakis, A. N. (2021). *Social and environmental justice. In From Polarisation to Multispecies Relationships* (pp. 283–307). Springer.

McIntyre-Mills, J., & Corcoran-Nantes, Y. (2021). *From Polarisation to Multispecies Relationships*. Springer.

McIntyre-Mills, J., & Romm, N. (Eds.). (2018). *Balancing individualism and collectivism for social and environmental justice* (Springer Planetary passport). Springer.

McIntyre-Mills, J., & Romm, N. (Eds.). (2021). *From polarisation to MultiSpecies relationships: Re-generation of the commons in the era of mass extinctions*. Springer.

Meadows, D. H., Meadows, D. H., Randers, J., & Behrens, W. W., III. (1972). The limits to growth: a report to the club of Rome (1972). *Google Scholar, 91*.

Metcalf, G. S. (2015). A Constructivist Perspective on Banathy's Conversation Methodology. *Constructivist Foundations, 11*(1).

Midgley, G., & Rajagopalan, R. (2021). Critical systems thinking. *Systemic Intervention, and Beyond.*https://doi.org/10.1007/978-981-13-0370-8_7-1

Ozbekhan, H. (1968). The Triumpf of Technology:> Can< Implies> Ought<, Massachusetts Inst. of Technology (Colonial Press, Inc.) 1968.

Ozbekhan, H. (1970). *The predicament of mankind. quest for structured responses to growing world-wide complexities and uncertainties*. http://quergeist.net/Christakis/predicament.pdf

Pierce, C. S. (Author) & Buchler, J. (Ed.) (2012). *Philosophical writings of peirce*. Dover.

Romm, N. (2010). *New racism: Revisiting researcher accountabilities*. https://doi.org/10.1007/978-90-481-8728-7

Warfield, J. N. (1976). *Societal systems: Planning, policy, and complexity*. Wiley.

Warfield, J. N. (1994). *A science of generic design* (2nd ed.). The Iowa State University Press.

Warfield, J. N., & Cárdenas, A. R. (1994). *A handbook of interactive management*. The Iowa State University Press.

Warfield, J. N., & Christakis, A. N. (1987). Dimensionality. *Systems Research, 4*, 127–137.

Rethinking Teacher Education Practicum for Transformative and Regenerative University – Community Partnerships

Pinkie L. Mabunda

Abstract One of the critical features of Teacher Education programmes across the globe is the incorporation of the practicum component to ensure that students obtain adequate classroom experience that provides opportunities for merging theory and practice. This chapter provides various critical perspectives from a range of countries based on the transformative and regenerative discourses on teacher professional standards in practice. I offer some selected cases through which I show the important role of acknowledging the context or situatedness of learning. The salient lessons emerging from the studies conducted are that the practicum can transform and regenerate the individual and collective change agencies and attitudes and practices of student teachers, mentor teachers and other school role players, whilst the actual implementation of the practicum strengthens university–school and community partnerships. While the chapter shows the importance of teacher professional standards as being pivotal to promoting quality we argue that without the focus on context they cannot on their own promote transformation and regenerative practices. In such cases, these standards may serve as constraining if they are not sufficiently broad in their conceptualisation.

Keywords Teacher education · Teaching practicum · Teaching standards · Transformative education · Regenerative education · Community partnerships

1 Introduction

One way in which governments and regulatory bodies aim to ensure consistent preparation, readiness and 'quality' of teachers is through the implementation of professional standards. However, a problem with this approach is that quality is an

P. L. Mabunda (✉)
University of South Africa, Pretoria, South Africa
e-mail: mabunpl@unisa.ac.za

J. J. McIntyre-Mills, Y. Corcoran-Nantes (eds.), *Transformative Education for Regeneration and Wellbeing*, Contemporary Systems Thinking,
https://doi.org/10.1007/978-981-19-3258-8_13

ambiguous and contested term, which can be reductionist and superficial when measured against the achievement of predetermined outcomes. Moreover, quality is also a highly contextualised concept (Ryan & Bourke, 2018 in Call, 2018) and its role in context can often be lost unless the standards specifically acknowledge teaching in context. While it must be acknowledged that professional standards vary in content, form and purpose, they generally outline key knowledge and skills that teachers are required to acquire. However, caution must be exercised over the usefulness of standards that encompass 'everything' teachers need in their toolbox (Santoro & Kennedy, 2016). The contradiction that standards are open to interpretation yet also narrowly defined has characterised contemporary discussions on professionalism as entailing democratic professionalism which upholds the values of equity and social justice (Douglas, 2017).

Historically, professional standards have been promulgated as a fundamental means of enhancing the quality of teaching.[1] While the use of such standards has become normalised in education, they are however constantly questioned on their necessity, format, content and application (Call, 2018).

Moreover, there are widespread calls for their reforms in initial teacher education programmes, which are normally based on two notions, namely: that successful school education increases student achievement due to being based on the quality of teachers and the instruction[2] as opposed to the amount of school funding; and that teachers are recognised as the single most important factor when it comes to improving students' performance within schools.

A common criticism is that professional standards put forward a reductive model where teaching is seen as an instrumental process that reduces professional practice to a narrow set of behaviours or, as Sachs (2003) cautions, their use as a government-imposed 'regulatory framework' which promotes a singular view of teaching (Torrance & Forde, 2017). Such a framework "advocate standardisation, conformity, control, certainty, predictable outcomes and appear to offer little space or hope for creativity" (Turvey & Lloyd, 2016: 246)

While standards do not always have the status of defining teachers' practice and achievements, they invariably have a 'backwash' effect in defining teacher development programmes and thereby defining the attitudes and behaviours of teachers. They also assume a developmental function as a tool for teachers' own self-evaluation and are often used as part of a system-wide strategy for professional learning (Kleinhenz & Ingvarson, 2007).

According to Glasswell and Ryan (2017), 'the pervasiveness of mandatory reflective practice in teacher standards has been regarded as a means of elevating teachers' work to match that of other professions'. Nevertheless, they argue that standards are frequently regarded as too narrowly conceived and context-free to

---

[1] Standards can also have a developmental function as a tool for teachers' own self-evaluation or be used as part of a system-wide strategy for professional learning (Kleinhenz & Ingvarson, 2007).

[2] Generally, however, instead of showing an appreciation of teachers and providing resources for teachers' professional development, teachers are blamed for poor quality and performance (Treagust et al., 2015).

serve a role in terms of improving practice. Under circumstances such as this, they argue that it is not surprising that the impact of such standards on schools and teachers' work remains restricted, with little evidence to indicate that the standards are capable of producing a transformative effect. Such arguments have intensified calls to broaden standards in order to encapsulate notions of context (Cummings & Dyson, 2007: 18). Parr et al. (2015: 145) argue that decontextualised teacher standards limit the boundaries of thinking and knowing, prescribe who the teachers ought to be and provide limited spaces for interconnectedness possibilities.

A further criticism of the lack of contextual conceptions in teacher standards stems from the view that education systems themselves feed into unsustainable societies. Teachers need practices that will allow them to survive not only in these environments, but also in environments that will permit them to succeed in re-enlivening the landscapes they exist within (Smith et al., 2020). Typically, teacher standards have not included standards that reflect the contextual systemic and ecological competences.

This chapter thus critiques professional teaching standards and offers perspectives and guidelines from specific cases on how transformative and regenerative discourses could be infused in professional teaching standards for meaningful teaching practicum experiences of student teachers.

2 Global Perspectives on Professional Teaching Standards

Schools have been subject to a persistent policy focus on their performance in keeping with 'modernisation' standards. This focus on standards revolves around an expectation that schools will improve their students' achievements by teaching better, through developing more effective leadership and by following the recommended practices as per national strategies. However, it has become increasingly recognised that something more than a single, narrow focus on standards is necessary (Cummings & Dyson, 2007: 3). Among other things, professional teaching standards define high-quality or accomplished teaching and are among the main tools whereby policy makers and education authorities in numerous nations desire to make teaching practice 'less variable, more reliable and increasingly effective' (Mulcahy, 2011). Certain supporters of professional standards for teacher highlight the capability of standards to provide 'a basis for deliberation and reflection' within learning environments and a framework for career progression of teachers. Others state that standards serve the purpose of upholding public trust in teaching and teachers, which is a necessary requirement for any profession (Santoro & Kennedy, 2016).

As already indicated above, standards serve a variety of functions including accountability measures, which contribute to the regulation of the profession. In addition, the authors are explicit in terms of the knowledge and skills required by teachers and furnish a framework for the curricular of initial teacher education along with graduate teacher professional development. Notably, they serve to communicate publicly the crux of teaching within a particular national context. However, what they fail to communicate is how the standards are enacted in practice, thereby revealing a common policy-tension (Santoro & Kennedy, 2016; Altaf & Saeed, 2019). Accordingly, the standards-based approach has continued to be adopted, which has led to teacher education reform proposals which emphasise the training of teachers to follow the standards (Liu & Ball, 2019).

With the above in mind, this section reviews certain perspectives on professional teaching or teacher standards based on the typical initial teacher education and specific teaching practicum literature within the contexts of specific countries. The aim of this chapter is not to evaluate the various policy backgrounds and documents and does not present a detailed comparison of the standards. Instead, this section presents an overview of the primary themes in the debates on standards and features some of the contrasts, challenges and issues that role players are confronted by in implementing the standards, along with the possibilities they present to teacher education and the teaching practicum. Some general observations are presented at the end of the section.

Commencing with the Scottish standards, these provide a career-long development framework in contrast to some of the more concise standards, for example the English Teachers' Standards and New Zealand's Practising Teacher Criteria. These latter standards convey a basic level of competence required to practise; however, they do not include an extended or broader role as teachers progress in their careers. The concise design of these standards inhibits an extended definition and exemplification of revised constructions of professional practice. These standards are therefore not designed to communicate the complexities of practice realities (Forde et al., 2016), and instead present key concepts that need to be understood and enacted in practice (Torrance & Forde, 2017).

The Australian National Professional Standards for Teachers includes the key components of quality teaching' and refer to what teachers should know and be able to do at the various stages of their careers, which are described as 'Graduate, Proficient, Highly Accomplished and Lead teachers'. The primary components of the standards are as follows: professional knowledge (content and pedagogical knowledge), professional practice (within schools) and professional engagement (within professional and school communities). Since newly graduated teachers are expected to meet the professional standards of 'Graduates', each initial teacher education programme provider is required to demonstrate how its initial teacher education programme supports the education students in meeting these professional standards upon graduation (AITSL, 2017). The Accreditation of Initial Teacher Education Programmes in Australia have clearly identified national teacher education programme standards and accreditation procedures in order to 'improve teacher quality through continuous improvement of initial teacher education' and 'accountability of providers for their delivery of quality teacher education programmes' (Treagust et al., 2015).

The English 'Teachers' Standards' document presents official guidance published by the Secretary of State for Education and is owned explicitly by the government. In analysing the recently revised national teachers' standards in the United

Kingdom, it is clear that they fail to provide sufficient attention to the broader aspects of learning to teach, such as acquiring an understanding of educational theory and the ability to problematise educational values.

There is also concern that certain standards 'do not require students to be critically engaged in their practice or to consider disciplinary-based knowledge', which suggests that to meet the teacher standards and learn the techniques necessary for performance are all that is necessary to realise adequate teacher development. Questions have therefore arisen about where and how new teachers are able to build their understanding of educational values, and whether these are now playing a diminishing role in teacher formation (Douglas, 2017).

With regard to South African teachers, who come from diverse backgrounds, experiences and qualifications, they frequently find themselves working in schools without access to a common language of practice or a shared understanding of professional teaching practices. In order to address these challenges, the South African Council of Educators (SACE) recently published (in 2020) professional teaching standards, in which the failures and successes of past policies to support teacher professionalism are considered. While significant, it is important to note here is that the development of professional standards for teachers should not be considered as a 'magic bullet' that is capable of solving the complex problems within the education system (Kimathi & Rusznyak, 2018). The South African education system is still hampered by inequality, a high learner drop- out rate, inconsistent teaching quality and poor learner performance. The 10 professional teaching standards published by SACE delineate what is expected of teachers in an endeavour to develop a common understanding of professional and ethical teaching practices. They also describe in broad terms what an educator must know and be able to do in order to provide quality teaching and learning opportunities for all learners in the diverse context of South African schools. These standards apply to pre-service and in-service teachers in all the school phases, subject specialisations, job descriptions and institutions; however, 'they are to be interpreted in the context, role and career stage of each individual educator' (SACE, 2020).

Similarly, in New Zealand, graduates of all accredited programmes are mandated to meet the Aotearoa New Zealand Graduating Teacher Standards (GTS), which exemplify what a pre-service teacher should know and be able to do at the culmination of their tertiary education. The GTS provides a framework for building 'a shared understanding of what it is an accomplished teacher knows and does ... representing the opening of professional practice to debate, discussion and improvement' (Tait-McCutcheon & Knewstubb, 2018).

Generally, schools serving disadvantaged areas have been subject to a relentless policy focus on performance and standards within the context of modernisation and marketisation. This 'standards agenda' revolves around an expectation that schools will improve their students' attainments by teaching better, developing more effective leadership and by following the practices recommended by national strategies for the various phases of schooling (Cummings & Dyson, 2007). The common conception of standards prevalent in education research and education policy, a conception that may serve to conceal the circumstances (social, material, discursive and political) which give rise to them, as well as erode the 'invisible work' that teachers and learners do in order to sustain them (Mulcahy, 2011: 95). Instead, standards should be used as a means of discussing the developmental and situational aspects of the teacher's role, as opposed to focusing predominantly on only meeting required teachers' standards – a mode which does not inspire student-teachers to develop a broader vision or consider the context of their work (Douglas, 2017).

In the study by Glasswell and Ryan (2017), in only two of six standards documents reviewed (Scotland and New Zealand) were specific examples of reflective practice identified, including reference to reflection as underpinning the transformative functions of education, along with involving the critical consideration of teachers' values, or the values as embodied by the education systems. In a study by Smith et al. on the social and political landscape of reform in initial teacher education in the United Kingdom, they argue that 'drawing on the wisdom of local indigenous practices alongside innovative approaches to learning, authentic regenerative practices can be brought to life. Revitalisation of teacher professional standards is key to this endeavour' (2020: 92, 95). Although the 'standards agenda' remains more or less in place, schools are now beginning to explore a broader set of activities and responsibilities. Whether or not these may contribute to the raising of attainment in the long term is yet to be seen; however, they cannot be reduced to this in the short term. Under these conditions, schools must therefore attempt to balance what appear to be competing demands. For instance, if the focus is largely on community-oriented work, will attainments decline? Conversely, if schools fail to address their community issues, do they have any chance of raising attainment? (Cummings & Dyson, 2007: 3).

While there are numerous competing theories on how to produce teachers who are equipped for the classroom, best practices suggest that a more agile and reflexive approach would be better than prescriptive approaches. Critical reflection on quality teaching stems from a strong base of appropriate knowledge and skills, but these do not define what quality teaching is comprised of. It is through informed deliberation and reflexive agility in practice that quality is far more likely to be achieved, which has no definitive 'standard' or blueprint (Ryan & Bourke, 2018: 182). In a study of the Pakistani education system, teachers expressed disappointment that the Standards were not sufficiently contextualised for their education system, they were not adequately aligned with the needs of their classroom and confirmed that the standards require amendments (Shaukat & Chowdhury, 2020). Universities, schools and communities have a responsibility to empower student teachers to be reflexive not only about theory and teaching practicum programmes but to encourage and offer them opportunities to critically question socio-political assumptions, practices and set standards in order to promote equality (Smith 2013).

According to Liu and Ball (2019), critical reflection for transformative teaching and learning in the classrooms of individual student teachers carries the potential to transform practice in the broader teacher education community for both social and educational equity.

3 Transformative and Regenerative Education and Development

In this first section, I provide a rationale for propagating transformative and regenerative approaches, rather than for describing, measuring and evaluating the quality of the teaching practicum process from the perspectives on professional teaching standards. The second section presents specific case studies that exemplify how transformative and regenerative development discourses and approaches can facilitate and enhance teaching practicum in the context of the 'real world'.

While the teaching practicum is intended to enhance 'real-world' training for pre-service teachers, increased awareness exists that student teachers may learn negative behaviours and attitudes from schools and mentors, and that the current teaching practicum systems might possibly reinforce bad practices or inequity. It is thus necessary that teacher development programmes produce teachers with the requisite skills, dispositions and knowledge required to teach a diverse array of learners in diverse contexts including race, ethnicity, gender, sexual orientation, socioeconomic status, exceptionalities, language, religion and geographical area which accommodate the differing abilities, experiences, talents and prior learning of the learners, in addition to the values (language, culture, family and community) they bring to the school contexts (Liu & Ball, 2019).

However, whether this knowledge and understanding required to educate teachers in the 'social and political contexts of the curriculum, of pedagogy, of educational purposes, of the structures of schooling and education, and the effects these have on reproducing and widening racialised, gendered, social class-based inequalities' (Hill, 2007: 214) is possible within the standards agenda is extremely doubtful. Previous studies have critiqued the 'technicised' and 'reductive nature' of the standards (for example: Beck, 2009; Furlong, 2005). It is thus essential that student teachers are critically reflexive within their broader social and eco contexts, which will in turn involve a historical critique of the cultural, political, social and economic landscape in order that students 'can resist being unknowingly inculcated into politically motivated equality practices' (Smith et al., 2020). In practice, it would require teachers to understand the way in which the socio-economic inequalities play themselves out in the context of schooling with children facing poverty, hunger and inadequate health services (Mabunda & McKay, 2020).

The philosophical arguments underpinning a transformative pedagogy can be seen in the works of numerous renowned authors such as Freire and Giroux. As an approach, transformative pedagogy aims to capacitate student teachers to examine which educational, moral and political commitments assist to guide their progress towards becoming professional teachers. Moreover, the pedagogy encourages reflective thinking, critical citizenship, social consciousness and a disposition for social justice. The possibility of transformative practice appears dependent on an array of social, political and material factors which constitute the teacher education landscape. The configuration of these factors is what enables or disables any transformation from taking place. On the whole, every effort to enact a transformative pedagogy as a form of professional learning stems from within a particular cultural and political arena, where various forms of student experience and subjectivity are produced and mediated in a manner that enables students 'to engage in critical activities that may challenge the status quo, reconstruct social-political-historical knowledge, question dominant ideologies and make public the histories of those marginalized, disenfranchised and/or disaffected' (Ovens, 2017).

According to Farren (2019: 41), teacher education through the socio-cultural context of the practicum, course content and transformative pedagogy 'should aim to support student teachers in developing a more encompassing professional identity as practitioner-researchers and leaders with moral consciousness'.

A radical egalitarian initial teacher education curriculum capacitates students to look inwards at their own racial positioning to understand how their 'prior life experiences, beliefs and assumptions, ... act as powerful filters through which they interpret teaching, students and communities'. However, without such a curriculum, the status quo will thus be reproduced, the result being that educational equity will remain little more than a dream of some, whilst most student teachers and teacher educators will remain content and secure in the knowledge that they are doing their best for every pupil within a system they believe is designed to promote equality. Taking into account the global and national events in 2020, a question that arises is this: What changes are in need of being made to teacher practices in order to further these agendas? (Smith et al., 2020). The notion of regenerative practice addresses the need for renewal, which connects professional learning with lifelong learning as well as supporting teacher development as a process of flourishing.

The concept of regeneration encompasses an approach that is both rooted in the past whilst orientated towards the future. At both an individual and societal level, regeneration is possible 'through going deeper, in our relationships with ourselves, with nature and through interpersonal and inter-relational experiences that become the embodiment of transformational learning in teacher development' (Smith et al., 2020). It is necessary to work together with numerous stakeholders who are able to contribute diverse means of knowing, including non-anthropocentric approaches informed by an understanding of nature and other life forms. With regard to personal praxis, this necessitates learning the ability to be inclusive through enabling other people to feel that they are heard and respected by giving them time and developing the ability to communicate respectfully across diverse participants spanning diverse cultures, experiences, interests and discipline-based knowledge (McIntyre, 2018).

According to Muller (2018: 104), regenerative development starts with profoundly reforming our way of doing things by holistically integrating six fundamental pillars, which include:

> social strengthening by community organization, empowerment and development to allow local communities to recognize and reverse their negative behavioural patterns and enhance the positive ones;
>
> necessary bond for community life, where local knowledge, values and traditions are shared within family, friends and the community as a whole, giving meaning to these terms, where self-esteem is recovered and enhanced, allowing community proudness to facilitate collective action and the true achievement of a harmonious relation with all; and
>
> rethinking and redesigning current political structures so they reflect true participatory democracy and inclusiveness, without the influence of money and power.

In order to regenerate teachers' identities as agents of change situated within an interconnected ecology, transformational practice is necessary (Smith et al., 2020). Müller (2018: 140) argues that:

> graduates will have a higher ability to go far beyond other professionals, resulting in effective positive change within their organizations, by showing strong differentiation with positive impacts in the well-being of society and the environment, while increasing the economic benefits…. A good starting place is to un-learn many of the things that were taught to us. We need to develop a much stronger critical analysis of what is considered good for development. We also need to re-evaluate how we do science and manage knowledge….

Transformational or profound change requires deeper social connectivity and deeper understanding which develop through practice. Practitioners are regarded as actors who are able to shape the construction of practice through who they are, how they act and the resources that they draw upon. In this way, knowledge is thus 'embedded in practice and embodied in people, is socially constructed and culturally embedded' (Jakubik, 2018, p. 206, 389 in Glasswell & Ryan, 2017). In the study by Arko-Achemfuor et al. (2019: 8) of academic practitioners collaborating with communities towards social and ecological transformation, the authors consider knowing to be linked to action and argue that this is in keeping with the transformative and indigenous paradigmatic views, which tie 'knowing specifically to an ethical action, with those involved in the inquiry process becoming agents of justice-oriented interventions'. In the same way, wisdom is built through wisdom practices, and teachers, 'as agents within an ecology, hold the key to wisdom through varied reflective practices on their experiences and relationships and the cultivation of environments, such as learning communities, that support this growth'. Marshall and Thorburn (2014) contend that the "perceptual" or "situational insight" can empower the "moral agent" to perceive and interpret the features of practice through a developed sense of discernment guided by practical wisdom.

Collaboration with others for reflective purposes was evident in every country's standards, and while some made more of this than others, it is evident that the prevailing view is that teacher learning can and does occur as teachers work together in communities of practice. As Warwick notes (Glasswell & Ryan, 2017: 20), there is a powerful case for considering reflection as a socially-mediated process in which interactions with others create opportunities for open discussion and drive forward the development of shared understandings and increased awareness.

The instantiation of the significance of supportive colleagues in guiding teacher development in the standards documents appears to be a valuable counter to the view that teachers work or indeed reflect autonomously (Glasswell & Ryan, 2017). Romm (2018) makes a case for the consideration of communities of research (COR) as a practice, where the prime concern to COR professionals is that 'justice-infused interventions become part of the purpose of organising/facilitating inquiries'. Inquiry practices along these lines are oriented to identify and address (as opposed to running the risk of ignoring) historically marginalised people as part of a process of developing what Midgley et al. (2018: 771) refer to as 'meaningful engagement of communities'.

4 Selected Exemplar Cases

In this section, I refer to three cases of teaching practices that aptly demonstrate that the development of quality in teacher education is a much broader notion than any standards can prescribe. Ryan and Bourke's (2018: 182) argument is critical in this regard. They point out that a solid skills base in pedagogical content knowledge is necessary but insufficient to ensure quality practice. Rather, they argue that: "*quality is far more likely to be achieved through informed deliberation and reflexive agility in practice, which has no definitive blueprint or 'Standard'* (Boxes 1, 2 and 3).

Box 1: The South African Rural Teacher Education Project (Islam, 2012)

My first example draws on the South African Rural Teacher Education Project as described by Islam (2012), who explains how a school – university partnership for teacher preparation involving two rural schools in the Vulindlela area, a deep rural district in the province of KwaZulu-Natal, and the University of KwaZulu-Natal. The teacher education project was implemented in the context of post-apartheid South Africa and aimed, *inter alia*, to show how experiential learning was significant for professional development and to debunk racial reifications of the students through their exposure to schools in varied contexts. This exposure also deepened their theoretical understandings of race and class as well as of poor rural communities, but possibly more importantly, it also gave students a scaffold to challenge their own conscious and unconscious biases. During the practicum, students were required to reflect on their daily experiences and to record these in their reflective journals and discuss their experiences in their focus groups.

Islam (2012) points out that students reflected on their own intercultural experiences arising from living in the rural area and co-teaching at the same schools. His case study demonstrates how the reifications of race and ethnicity need to be unravelled in teacher education programmes. The location of students for a period of time in a rural project showed how these ideologies played out in practice and in ways that extend beyond what is stipulated in professional teacher standards.

The case study also shows the importance of communities of practice and shows the way in which inter identity exposure in the context of broad and varied situations can enable student teachers to transcend the narrow standards that are often imposed. This form of practicum plays a critical role in the development of tacit knowledge that is so critical for the adaptability and agility required for teaching.

Box 2: The North East England Case (Cummings & Dyson, 2007)
This case study also showed the need for a broader conceptualisation of the roles of teachers in 'relation to the neighbourhoods they serve' in order for schools to play a role in community regeneration.

The study sought to investigate how the needs of these areas for regeneration were understood by different stakeholder groups, how leaders within schools and external stakeholders understood the contributions that schools might make to regeneration, and what the constraints upon and opportunities available to these schools were. The primary schools in the study had developed close working relationships with each other as a 'family' of schools in which parents, families and wider communities worked together.

The study identified three models that were useful in analysing the role of schools in relation to area regeneration, namely: 'community resourcing', 'individual transformation' and 'contextual transformation' models.

Evidence of impact suggested that each of these models had important small-scale and short-term impacts, but that none of them offered a means of addressing the underlying problems of disadvantaged areas and indeed those of disadvantaged learners.

As with the previous case study (Islam, 2012), the evidence suggests firstly that individuals have greater need for developing the contextual or situational understandings of teachers as well as the need for teachers' 'individual transformation'. Islam (2012) makes a case for the inclusion of situational understandings in teacher development. This form of 'knowledge' is proposed as offering an essential backdrop for teaching in South Africa by the Department of Higher Education and Training (2015: 11), which describes situational knowledge as the:

> knowledge of the varied learning situations, contexts and environments of education (classrooms, schools, communities, districts, regions, countries and globally) ... and includes an understanding of the complex and differentiated nature of the South African society... in confronting the diverse challenges faced by children in schools and the communities they serve, for example HIV and AIDS, poverty and the lingering effects of apartheid, dealing with diversity, promoting inclusivity and environmental sustainability.

Box 3: The Australian Case Study (Smith et al., 2020)
An implicit aim of this study was to connect the *local* with the *global* through identifying practices and sensibilities which could guide a regenerative approach that expands and enriches reflective practice. Teachers are hereby framed as agents of change in educational evolution, located within the

(continued)

Box 3 (continued)

ecologies of schools, communities and the natural world. This reconceptualisation of education systems as ecologies recognises that teachers practice not merely within physical environments, but also within social and mental environments. By the deployment of ecological principles, symbiotic, interconnected and relational landscapes are fostered which nourish and support teacher practice, as opposed to power-based frameworks, thereby allowing teachers to become more mature, authentic and wise practitioners. The authors present three layers of regenerative practice, namely 'being with ourselves', 'being with nature' and 'being with each other' as essential to a holistic approach which can be aligned with and supported by teacher professional standards. The authors also considered how 'local wisdom', specifically of indigenous Australians, could be used authentically in the regeneration of shared landscapes through recognising the importance of 'place'. Ecology models of education were therefore explored to ascertain the nature of regenerative practices and how these may be integrated into teachers' professional development. Attention was then placed on teaching practice and the theories and policies that undergird it, highlighting 'teacher agency and a nuanced discussion surrounding reflective practices'. The construct of 'regenerative practice' emerged conceptually as an effective way to respond to the challenges brought about by COVID-19.

Cummings and Dyson (2007) reflect on the broad-based community engagement as requiring a broad research base and argue that:

> research has a part to play both in untangling some of the conceptual muddles of professional standards and in searing out empirical evidence for policy directions. Inevitably, this means that education researchers will need to begin collaborating with a range of other researchers, not just from the education but other disciplines such as sociology and economics. (2007: 19)

5 Implications and Conclusion

What lessons can be learned from the three cases cited above? All three cases confirm the need for reconceptualisation of teacher professional standards to ensure that they are less constraining and include a broader contextual understanding, which relies on communities of practice and reflection to bring about social transformative pedagogy. They also call attention to the contextual backdrop of the school and the wider educational system.

While it is imperative to align the standards to national education policies, the cases indicate the need for a combination of 'transformative and regenerative principles, approaches and concepts to guide implementation strategies' in order to prevent (Liu & Ball, 2019) teaching practices and even entire teacher programmes from reinforcing negative aspects of school and society. This will be necessary for shifting teaching education from individual attitudinal changes to actual changes in teachers' practices.

Ryan and Bourke (2018: 181) conclude that "*For quality teaching to take place, this cannot be delimited to a list of standards*."

Instead, regenerative practice entails 'expansive learning' which encourages epistemic change at the deepest level, where 'the learner questions the validity of tasks and problems' and 'makes an effort to change the context that posed the problem'. Deliberating on the outcomes of practice is what opens regenerative potential (Smith et al., 2020).

Clearly, universities and educators and other community agencies have an important role to play in regeneration. The prime challenge for those designing and leading teacher education programmes is to ensure leadership and practitioner enquiry are not reduced to sets of standards. Instead, practices based on influence and agency on the part of the universities and teachers should be central to genuine re-professionalisation of teaching (Torrance & Forde, 2017: 123; Christie & Kirkwood, 2006). As argued by Mabunda and McKay (2021), transdisciplinarity and a widened ontological lens offers possibilities for improved and agile reflective and transformative practices.

To conclude I cite McIntyre Mills (2020: 834), who sums up the importance of transformative teacher practices. She argues for:

> …the opportunity to apply participatory action learning and research as a component of each of the core subjects could create the context for a new approach to regenerative education. The thematic patterns across many ways of knowing could be a key learning outcome in capstone action research projects that would strive to achieve systemic integration. To sum up, primary, secondary and tertiary education nodes and networks need to be linked, spanning face-to-face engagement and supported by capacity building through online training. The mission ought to enable the development of regenerative social, economic and environmental systems. The current education system churns out certificates and degrees for jobs within the neoliberal economy. But, supporting "business as usual" will not address social and environmental injustice.

References

AITSL (Australian Institute for Teaching and School Leadership). (2017). Australian Professional Standards for Teacher. https://www.aitsl.edu.au/docs/default-source/national-policy-framework/australian-professional-standards-for-teachers.pdf

Altaf, F., & Saeed, M. (2019). Teachers' and education managers' perceptions and practices regarding national professional standards for teachers: A multiple case study. *Global Regional Review, 4*(1), 326–334.

Arko-Achemfuor, A., Romm, N. R., & Serolong, L. (2019). Academic-practitioner collaboration with communities towards social and ecological transformation. *International Journal of Transformative Research, 6*(1), 1–9.

Beck, J. (2009). Appropriating professionalism: Restructuring the official knowledge base of England's 'modernised' teaching profession. *British Journal of Sociology of Education, 30*(1), 3–14.

Call, K. (2018). Professional teaching standards: A comparative analysis of their history, implementation and efficacy. *Australian Journal of Teacher Education, 43*(3), 93–108.

Christie, D., & Kirkwood, M. (2006). The new standards framework for Scottish teachers: Facilitating or constraining reflective practice? *Reflective Practice, 7*(2), 265–276.

Cummings, C., & Dyson, A. (2007). The role of schools in area regeneration. *Research Papers in Education, 22*(1), 1–22.

Department of Higher Education and Training. (2015). *Policy on the minimum requirements for teacher education qualifications* (2nd ed.). Government Gazette.

Douglas, A. S. (2017). Raising the standard: Contradictions in the theory of student-teacher learning. *European Journal of Teacher Education, 40*(2), 157–170.

Farren, P. (2019). 'Transformative pedagogy' in language teacher education. In *Second language acquisition – Pedagogies, practices and perspectives*. IntechOpen.

Forde, C., McMahon, M., Hamilton, G., & Murray, R. (2016). Professional standards for teaching in Scotland: Sustaining professional learning. *Professional Development in Education, 42*(1),19–35. https://doi.org/10.1080/19415257.2014.999288

Furlong, J. (2005). New labour and teacher education: The end of an era. *Oxford Review of Education, 31*(1), 119–134.

Glasswell, K., & Ryan, J. (2017). Reflective practice in teacher professional standards: Reflection as mandatory practice. In *Reflective theory and practice in teacher education* (pp. 3–26). Springer.

Hill, D. (2007). Critical teacher education, new labour, and the global project of neoliberal capital. *Policy Futures in Education, 5*(2), 204–225.

Islam, F. (2012). Understanding pre-service teacher education discourses in communities of practice: A reflection from an intervention in rural South Africa. *Perspectives in Education, 30*(1), 19–29.

Jakubik, M. (2018). Practice ecosystem of knowledge co-creation. *International Journal of Management, Knowledge and Learning*, (2), 199–216.

Kimathi, F., & Rusznyak, L. (2018). Advancing professional teaching in South Africa: Lessons learnt from policy frameworks that have regulated teachers' work. *Education as Change, 22*(3), 1–25.

Kleinhenz, E., & Ingvarson, L. (2007). Standards for teaching: Theoretical underpinnings and applications. *Teaching Standards and Teacher Evaluation, 1*, 46–71.

Liu, K., & Ball, A. F. (2019). Critical reflection and generativity: Toward a framework of transformative teacher education for diverse learners. *Review of Research in Education, 43*(1), 68–105.

Mabunda, P., & McKay, V. (2020). An analysis of newspaper depictions of schooling: A participatory study for the development of a teacher education curriculum. *Systemic Practice and Action Research*, 1–20.

Marshall, A., & Thorburn, M. (2014). Cultivating practical wisdom as education. *Educational Philosophy and Theory, 46*(14), 1541–1553.

McIntyre-Mills, J. (2018). Recognizing our hybridity and interconnectedness: Implications for social and environmental justice. *Current Sociology, 66*(6), 886–910.

Mabunda, P.L., & McKay, V. (2021). Educational Curriculum and Multispecies Relations. In: McIntyre-Mills, J.J., Corcoran-Nantes, Y. (eds) *From Polarisation to Multispecies Relationships*. Contemporary Systems Thinking. Springer, Singapore. https://doi.org/10.1007/978-981-33-6884-2_16

McIntyre-Mills, J. (2020). The COVID-19 era: No longer business as usual. *Systems Research and Behavioral Science, 37*(5), 827–838.

Midgley, G., Johnson, M. P., & Chichirau, G. (2018). What is community operational research? *European Journal of Operational Research, 268*(3), 771–783.

Mulcahy, D. (2011). Assembling the 'accomplished' teacher: The performativity and politics of professional teaching standards. *Educational Philosophy and Theory, 43*, 94–113.

Müller, E. (2018). Regenerative development in higher education: Costa Rica's perspective. In *Higher education in the era of the fourth industrial revolution* (pp. 121–144). Palgrave Macmillan.

Ovens, A. (2017). Putting complexity to work to think differently about transformative pedagogies in teacher education. *Issues in Teacher Education, 26*(3), 38–51.

Parr, G., Bulfin, S., Castaldi, R., Griffiths, E., & Manuel, C. (2015). On not becoming 'a mere empirical existence': Exploring 'who' and 'what' narratives in pre-service English teachers' writing. *Cambridge Journal of Education, 45*(2), 133–148.

Romm, N. R. (2018). Reflections on a multi-layered intervention in the South African public education system: Some ethical implications for Community Operational Research. *European Journal of Operational Research, 268*(3), 971–983.

Ryan, M., & Bourke, T. (2018). Spatialised metaphors of practice: How teacher educators engage with professional standards for teachers. *Critical Studies in Education, 59*(2), 167–186.

SACE (South African Council for Educators). (2020). *Professional teaching standards*. SACE.

Sachs, J. (2003). Teacher professional standards: Controlling or developing teaching? *Teachers and Teaching, 9*(2), 175–186.

Santoro, N., & Kennedy, A. (2016). How is cultural diversity positioned in teacher professional standards? An international analysis. *Asia-Pacific Journal of Teacher Education, 44*(3), 208–223.

Shaukat, S., & Chowdhury, R. (2020). Teacher educators' perceptions of professional standards: Implementation challenges in Pakistan. *Issues in Educational Research, 30*(3), 1084–1104.

Smith, H. J. (2013). A critique of the teaching standards in England (1984–2012): Discourses of equality and maintaining the status quo. *Journal of Education Policy, 28*(4), 427–448.

Smith, S. E., Mason, J. C., & Bowden, M. (2020). Local wisdom in regenerative teacher practices. *Australian Journal of Teacher Education (Online), 45*(9), 92–107.

Tait-McCutcheon, S., & Knewstubb, B. (2018). Evaluating the alignment of self, peer and lecture assessment in an Aotearoa New Zealand pre-service teacher education course. *Assessment & Evaluation in Higher Education, 43*(5), 772–785.

Torrance, D., & Forde, C. (2017). Redefining what it means to be a teacher through professional standards: Implications for continuing teacher education. *European Journal of Teacher Education, 40*(1), 110–126.

Treagust, D. F., Won, M., Petersen, J., & Wynne, G. (2015). Science teacher education in Australia: Initiatives and challenges to improve the quality of teaching. *Journal of Science Teacher Education, 26*(1), 81–98.

Turvey, A. M., & Lloyd, J. (2016). From preservice to early-career English teacher in the UK: negotiating powerful myths. *English Teaching: Practice & Critique, 15*(2), 236–250. https://doi.org/10.1108/ETPC-02-2016-0036

Online Learning and the Pedagogy of Resilience, Agency and Protest: Lessons from the COVID-19 Experience

Veronica McKay and Pinkie Louisa Mabunda

Abstract The global Covid-19 pandemic continues unabated with educational systems operating in crisis mode in an attempt to keep learning and teaching going through the various levels of lockdowns and the shutdowns of educational institutions. The ramifications of the pandemic cut across all sectors of society thwarting endeavours to achieve the Sustainable Development Goals. This chapter presents a critical overview of some of the challenges that impacted on the education sector. Notwithstanding these challenges, the chapter tells a positive story of students forming new digital communities that provide ways of overcoming the lack of a social presence and who show resilience in the face of hardship. Amidst the near calamity is a positive story of goodwill, ubuntu and resilience. The chapter reflects on the development of collective agency and resilience engendered in the digital space, enabling students to navigate the crises and supporting the momentum of learning. It focuses on the way in which the monitoring of the social media enables institutional responsiveness in catering for more than 400,000 university students and concludes with some reflections on education beyond the pandemic.

Keywords Digital divide · Online learning · Resilience · Protest · Covid-19 · Ubuntu

1 Introduction

The global Covid-19 pandemic continues unabated with educational systems and most education institutions operating in crisis mode in an attempt to keep learning and teaching going through the various levels of lockdowns and the shutdowns of educational institutions across 190 countries and affecting 1.6 billion school learners and more than 300 million higher education students (UNESCO, 2021).

V. McKay (✉) · P. L. Mabunda
University of South Africa, Pretoria, South Africa

J. J. McIntyre-Mills, Y. Corcoran-Nantes (eds.), *Transformative Education for Regeneration and Wellbeing*, Contemporary Systems Thinking,
https://doi.org/10.1007/978-981-19-3258-8_14

Showing no sign of it abating (apart from fluctuations in the numbers of affected people), the pandemic wreaks havoc on all aspects of our lives, regardless of where on we are on the planet, and irrespective of who we are. The ramifications of the pandemic cut across all sectors of society, thwarting endeavours to achieve the Sustainable Development Goals (SDGs) to which all nations committed as part of the global agreement concluded in 2015 by the United Nations (2015a). The year 2015 also marked the conclusion of the Sendai[1] Framework for Disaster Risk Reduction Framework (United Nations, 2015b) that aims at improving preparedness for disaster responses and promoting a "Build Back Better" post-disaster recovery and reconstruction approach. Although this framework was timely in its formulation, the world was clearly unprepared for the Covid-19 pandemic. This chapter focuses on the interface between the global pandemic and the SDGs and considers of some endeavours to build back better in the education sector. It also shows the intersections of the pandemic with poverty and on learning losses.

Most notably, the Covid-19 pandemic had a severe impact on SDG 4 of the United Nations 2030 Agenda (United Nations, 2015a) disrupting endeavours to "ensure inclusive and equitable quality education and promote lifelong learning opportunities for all" and disrupting efforts for inclusivity in offering universal primary and secondary education, early childhood development and universal pre-primary education, as well as efforts to expand technical, vocational and higher education. These goals are regarded as indispensable not only for the achievement of SDG 4 but also for the other 16 SDGs, specifically health and well-being (SDG 3) and gender equality (SDG 5) (English & McKay, 2022).

2 A Disruption of Global Proportions

UNESCO (2021) refers to the pandemic as the most severe "global education disruption in history" as teaching and learning ground to a near halt with the closures of educational institutions, impacting on more than 190 countries, 1.6 billion learners, 300 million higher education students and 100 million teachers and school personnel (UNESCO, 2021).

It is contended that the pandemic has caused much more than just a year of disruption in the education sector. Rather, its impact will have long-term and lasting effects on the most vulnerable and marginalised—impacting on the physical and mental health of students, teachers and school personnel. UNESCO (2021) estimates that 24 million children and youths are at risk of dropping out and as a result of learning loss a further 100 million will fall below the minimum proficiency level in reading which is a foundational skill for all learning. In the higher education

[1] "The Sendai Framework for Disaster Risk Reduction 2015–2030" was adopted at the Third UN World Conference in Sendai, Japan, on 18 March 2015.

space, UNESCO (2021) shows that all countries needed to rely on digital means but there was a clear socio-economic disparity with all low-income and lower-middle-income member states reporting the need for improvement in digital infrastructure to support online teaching and learning. Clearly the impacts of the pandemic were more acute when intersecting with poverty (McKay, 2022).

To corroborate this view of the interface of poverty and Covid-19, a recent South African study shows that approximately 750,000 children had not returned to school in 2021 when the Covid-19 lockdown levels were lowered. School dropout was cited as a problem that was more prevalent among lower socio-economic households, and was found to be more acute in rural areas among households which had run out of money for food as a result of unemployment and a lack of salaried income (Shepherd & Mohohlwane, 2021). These researchers (2021: 12) state that "the effect of the COVID-19 pandemic, therefore, may have taken engagement in [South African] schooling back two decades" to what it was at the end of apartheid. Hunger and not having enough money for food, going hungry or relying on food donations was reported as a critical challenge by more than 40% of higher education students (DHET, 2021).

The global economic downturn resulted in public education budgetary cuts at a time when increased funding is needed across the educational sector (UNESCO, 2021). With virtually no preparation, teachers were required to teach virtually, adjusting curricula and assessment methods in an endeavour to mitigate learning losses and to prevent learner attrition as a result of the institutional lockdowns. In many poor schools the time lost by rotational timetables that were put in place to reduce the number of learners per classroom where rotation as a means of "social distancing" was necessary because of the high learner-to-teacher ratios in poor schools that are generally overcrowded (Shepherd & Mohohlwane, 2021).

The pandemic also had an impact on food security which was exacerbated by the global closure of educational institutions. While the school closures were implemented as a way "to save lives" this conversely gave rise to wide-spread food insecurity. The question of "how to provide meals to children eligible for free or reduced lunch at school" among low-income school learners and university students who were residential, was resounded across countries (Morgan, 2020) with the lack of institutional feeding giving rise to increased hunger and high levels of parental stress in low-income households.

According to the United Nations World Food Programme, 154 million children globally missed meals as a result of Covid-19-related school closures.

Learner hunger was clearly visible in South Africa where the schooling system normally provides meals for 10 million children per day as part of the large-scale National School Nutrition Programme (NSNP). Household food insecurity increased with the closure and partial closure of schools, a problem exacerbated as breadwinners lost their employment and their children were unable to access or only had erratic access (owing to the rotational timetable) to meals provided by the Department of Basic Education's (DBE) nutrition programme.[2]

The impacts of the resultant food insecurity were far-reaching. Shepherd and Mohohlwane (2021) contend that just over three million adults reported that children their households had experienced hunger suggesting high rates of adult and child food insecurity. This, they identified as significant sources of adults' worry and depression more so because the economic hardships such as have occurred since the start of the Covid-19 pandemic have brought to bear pressures on families including on their ability to fulfil the basic needs of food and shelter.

UNESCO (2021) reports that the closures of education institutions reversed the progress made over the years towards gender equality in education. In addition to affecting learning, countries reported increased teenage pregnancy indicating that the shutdowns had exposed girls and women to increased gender-based violence both inside and outside of their homes. The focus on education for girls and women has historically been a priority for UNESCO and there has been a constant call for nation states to emphasise women's education, to eliminate the gender gap in the access to education, and to promote the continuation and progression of girls in education. The lockdown had a severe impact on education but also cut access to vital services for protection, nutrition, health and well-being thereby impacting on SDGs 3, 4 and 5 (English & McKay, 2022).

3 The Digital "Haves and Have-Nots"

Digital learning emerged as the default mode of mitigating the impacts of school and university closures. Despite the acceleration of digitisation across the world, the pandemic starkly revealed digital inequalities—showing that not all young people show characteristics of the Tic-Toc generation who are well connected, digital natives. In fact, onset of the Covid-19 pandemic and the need to shift to remote learning showed that high numbers of youths are not digitally savvy, that both the youths and indeed their teachers are digitally marginalised. Zimmerman (2020) argues that this marginalisation is not a dichotomy between the digital haves and the have-nots. Rather he points out that access is multifaceted and across a spectrum that includes students having access to a shared a device, or having only a mobile device, or having a device with no or low internet connection. He further refers to the gaps in digital competencies pointing out that young people who do not have digital access at home (a socio-economic indicator) were likely to have poorer digital skills than their peers.

Transitioning from contact face-to-face offline learning to online learning as the dominant mode during the global pandemic revealed many new and complex

[2]The DBE was subsequently mandated by a High Court ruling on 17 July 2020 to reinstate the NSNP even while the schools were closed.

challenges. Almost overnight teachers were expected to be proficient in teaching online while learners needed matching proficiencies as well as digital access. In practice, teachers were not able to use the same material as they did in face-to-face teaching[3] and were faced with challenges arising from the rapid introduction of learning management platforms, changing curricula and changing modes assessments, changing ways of mediating learning and changing ways of invigilating assessments which often necessitated the use of proctoring. It required educators, irrespective of their training, to enter the "brave new world" of online teaching in spite of the many challenges of digital inequality. In many cases it meant that the digital have-not learners were left behind. Moreover, schools in lower socio-economic areas usually have fewer resources than their more endowed counterparts, thus precluding a considerable number of children from participating in online learning. With parental abilities also factored into the challenge, the low adult literacy rates in developing countries reduced parental abilities to "home teach" learners, making online learning less effective for poorer learners assuming that they had access to technology.

Zimmerman (2020) points out that the rapid shift to online learning brought about by the pandemic has catalysed the world's biggest educational experiment in history with nearly 2 billion learners locked out of their learning institutions and attempting to navigate online learning. The 'corona experiment' will significantly influence what education and learning will look like in the twenty-first century.

Jarke and Breiter (2019: 1) refer to the "increasing datafication" through the pandemic stating that it brought about new means to "measure, capture and describe the social reality of the education sector which is hardest hit by datafication". They opine that the collection of data at all levels of educational systems (individual, classroom, school, district region, country) has brought about associated fears with respect to privacy being compromised by the resultant surveillance, control, power relations and digital inequalities.

Efforts to provide data for students at the University of South Africa (Unisa) also gave rise to fears of datafication and surveillance. As learners increased their presence on the university's learning platforms their concerns over privacy increased. They were suspicious that their presence was used for student profiling (beyond learning analytics). This led to a legal probe in South Africa into whether the university, after providing students with 30 GB of data, was monitoring student devices without their knowledge. McKay dismissed these allegations in an interview with Mungadze (2020):

> 'We have no reason to do that; we only offered them data so that they can write their exams. What they do with data is not for us to monitor,' says McKay. McKay dismissed allegations that UNISA and mobile network operators were snooping and actively monitoring the online activities of students who were allocated free data under COVID-19 relief efforts.

---

[3] McKay (2019a) refers to the way in which the South African Department of Education's school workbooks were designed using a distance education mode. These workbooks proved to be of value during lockdown for home teaching and later for home learning during rotational school attendance.

Clearly, the politics, pedagogies and practices emerging from the great Covid-19 experiment calls for a critical and theoretically informed analysis of what is emerging as twenty-first century education. It has also demonstrated that while digital connectivity is important, it cannot fix social inequality.

4 The Political Economy of Pandemic Pedagogy

Pandemic pedagogies have become the focus of the education technology (ed-tech) industry and technology businesses have rapidly increased their marketing of products to support online learning. Even "free open-source platforms" find their way into institutions coming with huge price tags which are eclipsed by the institution needing to buy all the plug-ins and add-ons to make the system work. They further come with 'lock-in' contracts for 'after-care' provided at exorbitant costs and provided by their own employees in ways that will inevitably dislodge the existing capacity within institutions.

The crisis has provided the educational-tech industry with one of the biggest business opportunities to vigorously market their products to education institutions, and to grow its market with learning institutions scrambling to procure platforms, digital invigilation tools and artificial intelligence — presenting the industry with increased opportunities for increasing profit-making, and also for increasing its influence over the practices of education.

Hillman et al. (2019) refer to this as the "platformisation" of education which leads to a dependency situation where the dominant technical platforms mediate and maintain learning. Their political economy analysis of educational platformisation suggests the need for serious caution with educational-tech businesses installing more than platforms. These platforms act as intermediaries between educational institutions and their students with the platformisation displacing academics while giving business the opportunity to not only sell its wares but inevitably impacting on pedagogies and content, recentralising decentralised systems through platforms all in the name of mitigating the disruption caused by Covid-19.

One cannot ignore the geopolitical dimensions with technology industry from the US and China veering for commercial advantage through penetrating the educational domain, as tech-platforms are being empowered alongside national and international policy (Knox et al., 2020).

Moreover, with the Covid-19 pandemic comes the hypothesis of "broken education" which offers an opportunity to ed-tech a massive market to fix by selling "untested solutions which sometimes have little to do with proper teaching and learning philosophies", and which, as they become rooted in institutions, develop a dependency relationship where institutions henceforth need to pay for maintenance and for additional add-ons and increased 'work-arounds' to the ostensibly open source and free services provided as education institutions become more dependent (Teräs et al., 2020).

We, as academic managers at Unisa, write this chapter immersed in our own struggles endeavouring to ensure the continuation of learning for 400,000 students during the pandemic. We now turn to our case study or reflect-piece in which we present from an action-oriented point of view.

5 The University of South Africa

Unisa is the largest comprehensive open distance learning institution in Africa, and with approximately 400,000 students it is one of the world's 30 mega-institutions. The University enrols over one-third of all of South Africa's higher education students. It is statutorily a distance education institution with 31 campuses across South Africa which serve largely as administrative and student support hubs and which provide library facilities and computer laboratories specifically for students who lack computer and/or internet facilities at their homes.

For the past 10 years, the institution has attempted to move from the more paper-based and post-office-dependent teaching towards a greater commitment to online teaching and learning, assessment and student support. However, the transition was marked by fits and starts as the burden of transition was not well received by the student body who come from backgrounds of poverty and who largely emanate from poor communities and fee-free schools (that provide school feeding as already referred to) and households' dependent on social (welfare) grants. The University has a strong commitment to social redress and after the demise of apartheid, the university is home to a large majority of African students who dwell in rural areas,[4] urban townships[5] and informal settlements[6]— environments with very high poverty rates. The majority of these students require financial assistance to study and approximately half of the Unisa students are supported by the Government's National Student Financial Assistance Scheme.

6 Remote Learning Solutions the Unisa Experience

The pandemic compelled makeshift adaptations using remote methods to rescue teaching and learning across the education sector often with limited resources, rapidly needing to grow cloud capacity and include the need to provide free mobile data for students. This case study is intended to show the way in which the University needed to rapidly shift from a blended mode (Mabunda, 2010) of learning (with

[4] Areas usually without piped water or sanitation, electricity and tarred road infrastructure.

[5] Separate townships were established for Africans, Indians and people of mixed race. While the urban townships are underdeveloped when compared with urban suburbs where white people lived, they nevertheless offer more services than the informal settlements, and would be more organised than informal settlements.

[6] Informal settlements are unplanned and unauthorised residential areas. The informal dwellings are usually shacks made from corrugated iron sheets. These communities have poor quality housing, inadequate access to safe water and sanitation, and insecure residential status.

some campus-based contact tuition) to becoming fully a fully online learning institution as a way to save the academic project. We reflect on the way in which the socio-economic divide had an impact on learners' lack of access to connectivity, devices and adequate learning spaces and the way in which this was compounded by the lack of e-readiness of learners and teachers. We also show the way in which the shift to online modalities required the teaching and learning portfolio to innovate ways of teaching and assessing using high-, low- and no-tech solutions.

Although socio-economic inequalities were always apparent, the pandemic exposed the fault lines, with South Africa's Gini coefficient among the countries with the highest levels of inequality in the world. Inequality manifests itself through a skewed income distribution, unequal access to opportunities, and regional disparities. Low growth and rising unemployment have contributed to the persistence of inequality. South Africa is known as one of the most unequal countries in the world, reporting a per-capita expenditure Gini coefficient of 0.63.

Over the past two years, the pandemic has required a number of changes to be made in line with the health protocols of the various levels of lockdown and campus closures which kept staff and students isolated at home. The greater majority of Unisa students are digitally disadvantaged without laptops/tablets and connectivity and were severely impacted on by the national "need to save lives and save the academic year". Numerous protests and student meetings were held to demand that the university and government provided learning devices for disadvantaged students. Student leaders made a case for the provision of mobile data to at least enable students to use their mobile phones and tools for learning in the absences of broadband and laptops. Students with disabilities were among the most affected needing adaptive devices and supportive teaching tools that were not readily available thereby threatening inclusivity.

The inequalities in South Africa have been the centre of various student struggles for many years with the #FeesMustFall movement rejecting the neoliberal education system and forging a new collective identity using an intersectional lens to advocate for free and decolonised education (Chinguno et al., 2017: 16).

7 Digital Inequality and #FeesMustFall

The Covid-19 pandemic exposed the inequalities in higher education and the digital divide in the country and showed that the inequalities were far more extensive than the #FeesMustFall campaign initially expressed. In response, the #FeesMustFall vision was expanded to include the focus on the digital divide with the need for technological resources in this period being akin to a form of cultural capital, a means for survival and an essential tool for learning to continue.

Although Unisa provided students with 30 GB of free data, they had to devise ways of working using their mobile phones as the device of last resort. This meant that students could handwrite their examinations and assignments, scan and PDF them using a scanner app on their mobile phones, and thereafter upload their PDFs onto the university platform for assessment.

Unisa as a distance education provider has been grappling to shift from a blended mode of teaching to online provision for many years. Blended in this sense refers to the use of both physical and virtual resources—using a range of technologies in combination with face-to-face interaction. Educators had to modify curricula and to adapt lesson plans to carry on with instruction using high-, low- and no-tech solutions regardless of their preparedness for the pandemic-induced changes to their educational practices. With this came the need for continued training on remote teaching and for using available technologies and alternative pedagogies for online, blended and offline learning during the closures and beyond.

8 An Analysis of the Media as an Action-Oriented Methodology

We have elsewhere elaborated on the analysis of media analyses as an emerging methodology in educational research showing that such analyses as the potential for immediacy and for action-oriented impact (Mabunda & McKay, 2020). We show (in Mabunda & McKay, 2020) that the analyses of media is a growing field of education research that draws on the public discourse about education issues and which uses the the public media to influence, shape and inform education management or policy formulation. We argue that the public media offers a way for immediacy in enabling a better understanding of educational matters and for enhancing the engagement of educationstakeholders. Baroutsis (2019: 6–7) refers to the growing body of education research that actively explores media representations of different aspects of education.

In similar vein, Coe and Kuttner (2018: 1) point out that the media plays a significant role in the education policy, influencing the way in which issues are prioritised and understood and they argue that researchers need to consider "how much attention news media give to education, which topics are covered, and how topics are addressed and what elements might guide action oriented responses of management".

In the case in point, in this chapter, we relied on the social media for informing our own actions as education managers. Following Kemmis (2009: 468), we use our analyses of the media for action-oriented purposes and using our analysis of the media as the practice of action-oriented research that "happens in sayings, doings and relatings".

As we argue in Mabunda and McKay (2020), the media represents a voice of civil society particularly with regard to social issues in education, providing a backdrop for conceptualising problems, planning and policy. Although we do not underestimate the role of the reader as an active participant in engaging with the interpretation of media messages, we recognise that the media is a vehicle for articulating the voices of civil society and we are cognisant of the role that these voices might play in guiding management decisions, defining agendas, informing policy development and highlighting a number of critical contextual and situational aspects. Rheingold (2008: 101) in fact refers to the media in the public sphere as a realm of social life in which public opinion can both be formed and be used to mediate political expressions. He draws Habermas's notions of the public sphere as a democratic space open to citizens wishing to disseminate information with society through channels of communications that are available, and he argues that media democracy entails a "more equal distribution of economic, social, cultural, and information capital, which would lead to a more informed citizenry, as well as a more enlightened, representative political discourse". Habermas (1974) defined the public sphere as a conduit for political expression in which the intimate connection between a web of free, informal personal communications reside as the foundations of a democratic society (Mabunda & McKay, 2020).

In this sense, the Habermasian emancipatory interest with knowledge being constituted as and guided by reflection is action oriented (McKay, 2018). As McKay and Romm (2008: 153) argue, "active action" research enables decision makers to become more attuned to public considerations to provide opportunities for interventions and changes. Following this argument, this chapter shows how reflection on the media gives rise to praxis.

How is this relevant for the situation of the Covid-19 pandemic? The separation between students and their lecturers and the challenges in remote working made it necessary to multiply sources of feedback about the students' challenges. With 430,000 students it was necessary for us as university management to keep our finger on the pulse—and the social media provided the main source for monitoring the mood of the student body.

9 Reading the Mood on the Social Media

With the lockdown levels and the rapidity at which the academic calendar was being adjusted, the social media enabled the University to tune into the way students received the changes and the challenges and indeed what their desires and needs were. The social media was therefore mined each day to identify any challenges that students faced and matters that needed further exploration. The charts in Fig. 1 indicate the moods and emotions of the students.

Amidst the fear and uncertainty, the social media represented the students' areas of foci and showed their emotions. The quantitative findings depicted areas to hone and probe or to triangulate by interviews with the Student Representative Council. Further data sources were used to interpret the findings and to guide us as university managers when making crucial decisions on the way in which to manoeuvre the pandemic.

One of the areas of insight we gained from our analyses of social media was the way in which the students supported one another by providing self-help

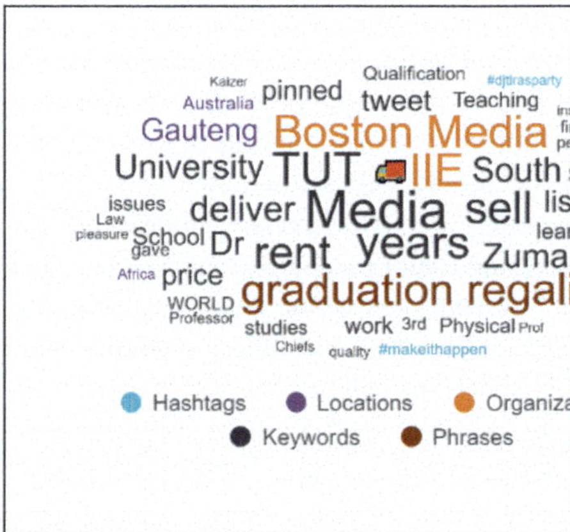

Chart 1: Social media word cloud for the week identified trends from the web.

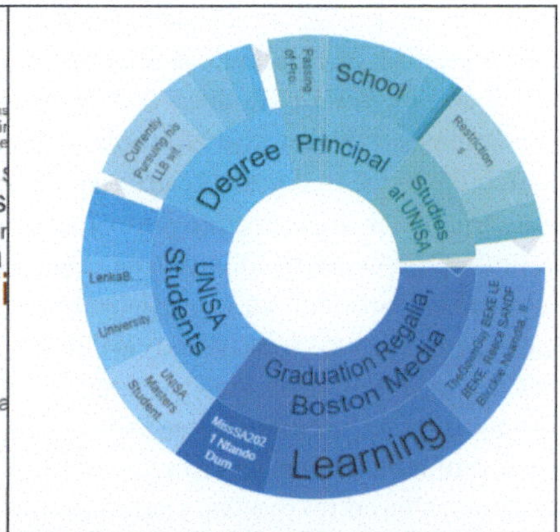

Chart 2: Topic Wheel
The topic wheel, and the workcloud highlight trending topics through the week.

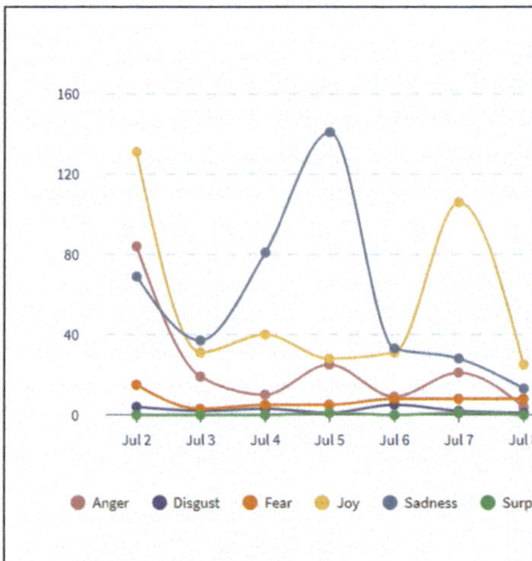

Chart 3: Emotions volume

The emotions chart showed daily emotions expressed by the students. As indicated, the anger shown on 2 July meant that immediate probing and immediate action were necessary. On probing, students required more time to complete their assignments. An extension later in the day turned the emotions. The sadness depicted on the 5 July was highest after releasing information on the death of the Dean of Law.

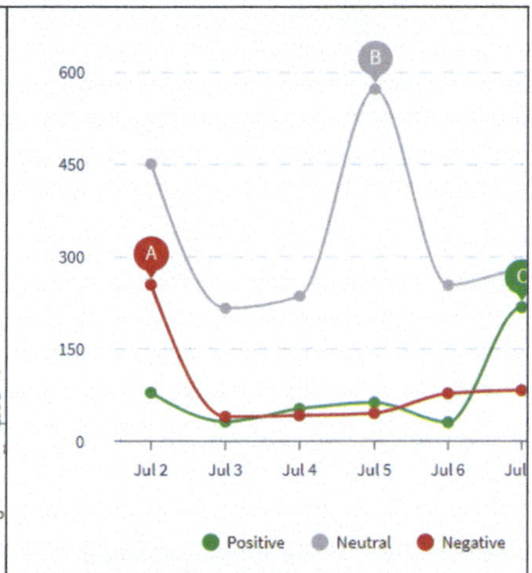

Chart 4: Sentiment volume

The sentiment chart taken from Facebook, Twitter and LinkedIn suggested a neutral mood for most of the week, apart from the negativity expressed on 2 July.

Fig. 1 Moods and emotions of students expressed on social media

information, providing guidance through the new online eco-system, answering administrative or academic questions and providing emotional support. Our newly developed Power BI Business Intelligence was further used to provide intelligence on student behaviour and students' needs. For example, we found that during the course of 2020, under the restrictive Covid-19 Level 1 lockdown, nearly 9 million assignments were uploaded onto the university platform and a further 3 million examinations were written online. This was done under extreme circumstances and depicted a resilience and grit showing that students rose to the challenge of online learning in the face of adversity. With the digital divide, it meant that students often needed to travel from rural villages to distant towns where they could receive a mobile internet signal so that they could upload their work onto the relevant platform. Students living in villages and townships "crowd funded" to pay for shared modems, shared textbooks, shared laptops and shared mobile phones.

10 Data and Devices as an Offshoot of #FeesMustFall

As the tweets in Fig. 2 show, data and devices featured very regularly in the social media. This we show subsequently as being an extension of the #FeesMustFall which was now broadened to include internet data and devices as a form of capital. As indicated in Fig. 2, tweets 1–8 and 12 highlight the urgency of addressing digital divide pointing the the lack of data and devices with both being portrayed as essential for learning under lockdown circumstances.

What was prominent and possibly unprecedented was the number of students offering to help those less fortunate —offering to donate data, textbooks or laptops to students who did not have the means.

As one of the students in tweet 9 says, "this is a family." The engagement of students shows the responsiveness of students to each other's plight in the spirit of ubuntu. A new virtual engagement (now crossing race, class and gender) with students helping one another and appealing to the university management for moral understanding of the many challenges students experienced. The tweets show that students were able to invoke the spirit of ubuntu, collectivity and cooperation through their engagements in a virtual world; it may be argued that under the most difficult circumstances, this was possibly the only reason why in the midst of the pandemic, the academic year was saved.

The solidarity on the social media and the personal testimonies of sharing achievements and through their virtual (at home) graduations are shown in tweets 10 and 11 also serving as peer and ingroup motivation.

The media and student reports highlight that community resilience is not an individual matter. In the education context of the pandemic, it relied on our communal strength (the students, the lecturers and the university management) where resilience was now a communal asset generated through the responses on the online public forums. It required cooperation and responsiveness from university managers, teachers and students, in leading and organising students and academics. It

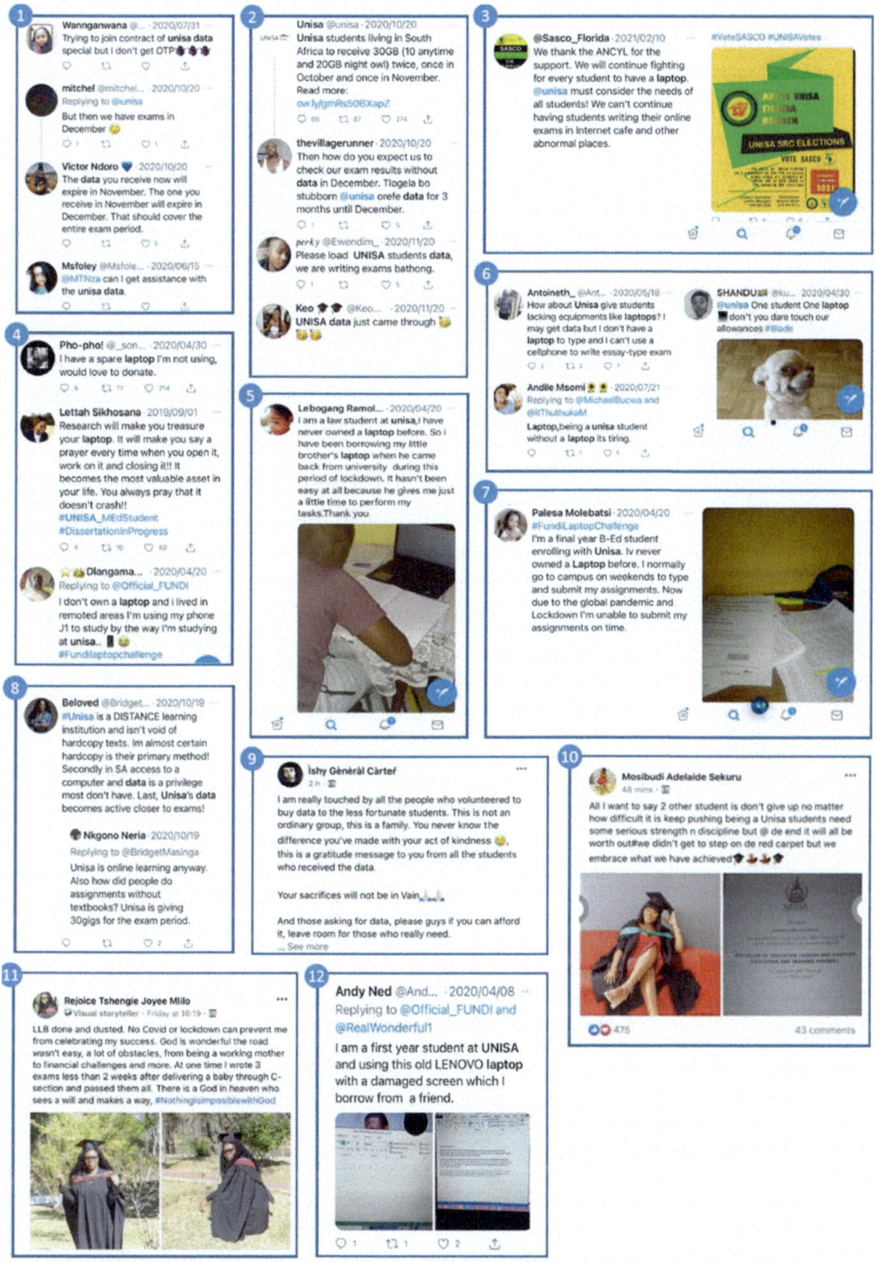

Fig. 2 Student tweets

required university management to keep a finger on the pulse of the organisation—with the social media providing a critical beat.

The assistance and cooperation and the expression of ubuntu and compassion shown, contributed to the sense of community where in the void of separation, it enabled students and academics to feel a sense of belonging to a community. The internet and the online space was not just about personal learning. It was now a place where people could meet, get support and it opened ways individual and collective agency could generate a new form of social capital that supported the momentum for learning while recognising that the digital divide was a force to be reckoned with. Mabunda and McKay (2020) also show the ways in which the online space was used for protest and resistance during the lockdowns and the ways in which female learners have challenged patriarchy and racism. (It was also a space for funerals, weddings and the collective Jerusalema dance prodiced by musicians Kgaogelo Moagi and Nomcebo Zikode as a source of motivation during the hard lockdown.)

As part of the twenty-first century skills that have emerged as critical, learners have needed to acquire skills which include critical thinking, creativity, collaboration, and communication along with the digital skills to use and navigate online learning resources. Girardi and Scheg (2021) outline strategies for student engagement in the online environment. The most significant of these is to create an online community that connects learners and enables them to overcome isolation in the absence of social presence and see themselves as part of a larger community, to connect and collaborate with others and to use each other as a sound board for problem-solving.

Much of what is shown in Fig. 2 reflects students' using the digital space to air their concerns with their challenges to actually access to the digital space. (These were selected tweets for this purpose.) Their views on data were taken seriously in guiding the university management to procure 30 GB of data for 430,000 students for 3 months. With close on 40 million gigabytes procured for one segment of the academic year, students were able to use their mobile phones as hotspots (or by using the scanning app) to upload handwritten documents onto the platform. Mobile learning was the lowest common denominator during the pandemic when students lacked other means of connecting.

11 Pedagogies Beyond the Pandemic

In his book, *The cruel pedagogy of the virus*, Boaventura de Sousa Santos (2020) reflects on the impacts on the world plunged into crisis by the virus with the pandemic, which he argues, revealed the deep social, economic, political, cultural, environmental, and educational inequalities (Lopes & McKay, 2020; Mabunda & McKay, 2021).

Henry Giroux (2021) expands on this in his book, *Race, politics, and pandemic pedagogy:Education in a time of crises*, and argues that in the midst of the storm, the pandemic has acted as a catalyst for reflecting on what kind of society the world wants in future and on what role that education might play in engendering a new society. He offers a theoretical and pedagogical lens for understanding education as critical to the post-pandemic renewal and to "building back better" by arguing that education can lead to a more compassionate, just and equitable society. He argues that critical pedagogy is central to producing citizens who are informed, politically aware, and willing to produce a culture that keeps democracy alive, since democracy is only as strong as the people within it. In this regard, he considers the role of education as a political tool for human agency that creates the possibilities for radical change. Giroux thus argues that in a post-Covid-19 pedagogy, a critical and reflective understanding underscores empowerment and provides the knowledge and agential potential, and he extends the parameters of agency which guide the individual to embrace a social and collective agency.[7]

In the South African context, the focus of education as a tool for democracy lies in tracing poverty and inequality to their roots. The renewal of education focuses strongly on decolonial struggles and with the resurgent decolonial struggles of the twenty-first century as reintroduced with the #FeesMustFall and which was at the centre of the #RhodesMustFall movement that swept South African universities. Ndlovu-Gatsheni (2018: 223) explains this as being an "ideological amalgamation of radical black feminism, black consciousness, Fanonism and pan-Africanism with demands that can be categorised into free, quality, decolonised education; an end to sexism, patriarchy and racism". In pedagogical terms, this implies the decommissioning of all offensive colonial or apartheid iconographies (Ndlovu-Gatsheni, 2018), the importance of using indigenous African languages in teaching (Hanemann & McKay, 2019) and indigenising learning, research and research ethics (Chilisa, 2009). Ndlovu-Gatsheni (2018: 238) further contends that the university of the future will "banish epistemicides, linguicides, culturicides, racism, sexism, patriarchy, tribalism, xenophobia and classism to become a home of everyone".

In similar vein, Le Grange (2020) refers to "the epistemic violence inflicted on black students and staff by a higher education system . . . based on Western models" and sees the pandemic as timely for dealing with the cognitive colonisation and the appropriation of the curriculum and opening for redressing the inequalities laid bare by the pandemic, no least, the digital divide where he argues, the "platform pedagogy" morphs into "platform capitalism". He points to the foregoing challenges that are interwoven in complex ways in the way in which the Covid-19 pandemic played out in emergency education across the globe with inequalities (digital divide) and the economy (platform-capitalism).

Emanating from the pedagogy of the pandemic, Le Grange (2020) eloquently proposes a resolution of ubuntu currere as a concept that might open possibilities for imagining the different curriculum that enables (post)human and decolonial perspectives to meet each other halfway arguing that ubuntu currere captures at least

[7] McKay (2019b) explores the collective agency as opposed to individual agency and shows its role in effecting collective resilience and bringing about communal change.

some dimensions of both (post)human and decolonial perspectives He proposes that it offers an amalgam of the African notion of ubuntu and a conception of curriculum drawing on the Latin meaning of curriculum "to run" and ubuntu or "botho" an African concept embodying the way in which we ought to relate to each other, to become fully human through being in community with others.

The interconnection embraced by ubuntu takes expression in generosity, caring, a morally compassionate and practically enabling state (Assié-Lumumba, 2017: 14), and it is this which Ramose (2015) presents as departing from Western individualism and any "-ism" by embracinges a "wholeness" —a relatedness and an interconnectedness that cannot be conceptualised by Western thinking. Msila (2013: 495) further elaborates on what ubuntu means and points out that at the centre of ubuntu is oneness of community, communication with each other.

Hlatshwayo et al. (2020: 4) also contend that the notion of of ubuntu currrere offers an emancipatory alternative transforming the education curriculum (McIntyre-Mills et al., 2014). They argue for an ubuntu currere conceptual tool that is dialectical, inclusive, and multivocal and which will avert "epistemic closure". We agree with these authors that in the South African context, contemporary curriculum practices cannot be divorced from the broader calls emanating from the 2015 to 2016 students' movement for decoloniality and transformation in the academy without challenging the contested history and institutional differentiation that characterises the sector. Ubuntu currere thus provides a framework for centring African thought, for dismantling cognitive and epistemic violence and for dislodging the grand Western narratives that peripheralised African thought. Herein lies our views of pedagogy beyond the pandemic.

References

Assié-Lumumba, N. D. T. (2017). The ubuntu paradigm and comparative and international education: Epistemological challenges and opportunities in our field. *Comparative Education Review, 61*(1), 1–21. https://doi.org/10.1086/689922

Baroutsis, A. (2019). Mapping the field of education research and media. In A. Baroutsis, S. Riddle, & P. Thomson (Eds.), *Education research and the media: Challenges and possibilities* (pp. 1–24). Routledge. https://doi.org/10.4324/9781351129114-1

Chilisa, B. (2009). Indigenous African-centered ethics: Contesting and complementing dominant models. In D. M. Mertens & P. E. Ginsberg (Eds.), *The handbook of social research ethics* (pp. 407–425). Sage. https://doi.org/10.4135/9781483348971.n26

Chinguno, C., Kgoroba, M., Mashibini, S., Masilela, B. N., Maubane, B., Moyo, N., Mthombeni, A., & Ndlovu, H. (2017). *Rioting and writing: Diaries of Wits fallists*. Society, Work and Development Institute.

Coe, K., & Kuttner, P. (2018). Education coverage in television news: A typology and analysis of 35 years of topics. *Aera Open, 4*, 1–13. https://doi.org/10.1177/2332858417751694

Department of Higher Education (DHET). (2021). *Social impact of the COVID-19 pandemic on youth in the Post School Education and Training Sector in South Africa*. DHET.

English, L. & McKay, V. (2022). SDGs 3, 4, 5: Educating for health, literacy and gender. *Covergence, 43*(1), 31–40. ISSN 2791-1675

Girardi, T., & Scheg, A. G. (Eds.). (2021). *Theories and strategies for teaching creative writing online*. Routledge. https://doi.org/10.4324/9781003019848

Giroux, H. A. (2021). *Race, politics, and pandemic pedagogy: Education in a time of crisis*. Bloomsbury Academic. https://doi.org/10.5040/9781350184466

Habermas, J. (1974). The public sphere: An encyclopaedia article (1964). *New German Critique, 3*, 49–55. https://doi.org/10.2307/487737

Hanemann, U., & McKay, V. (2019). Learning in the mother tongue: Examining the learning outcomes of the South African Kha Ri Gude Literacy Campaign. *International Review of Eudctaion, 65*(3), 351–387. ISSN0020-8566. https://doi.org/10.1007/s11159-019-09782-5

Hillman, T., Bergviken Rensfeldt, A., & Ivarsson, J. (2019). Brave new platforms: A possible platform future for highly decentralised schooling. *Learning, Media and Technology, 45*(1), 7–16. https://doi.org/10.1080/17439884.2020.1683748

Hlatshwayo, M. N., Shawa, L. B., & Nxumalo, S. A. (2020). Ubuntu currere in the academy: A case study from the South African experience. *Third World Thematics: A TWQ Journal, 5*(1–2), 1–17. https://doi.org/10.1080/23802014.2020.1762509

Jarke, J., & Breiter, A. (2019). Editorial: The datafication of education. *Learning, Media and Technology, 44*(1), 1–6. https://doi.org/10.1080/17439884.2019.1573833

Kemmis, S. (2009). Action research as a practice-based practice. *Educational Action Research, 17*(3), 463–474. https://doi.org/10.1080/09650790903093284

Knox, J., Williamson, B., & Bayne, S. (2020). Machine behaviourism: Future visions of "learnification" and "datafication" across humans and digital technologies. *Learning, Media and Technology, 45*(1), 31–45. https://doi.org/10.1080/17439884.2019.1623251

Le Grange, L. (2020). The (post)human condition and decoloniality: Rethinking and doing curriculum. *Alternation Special Edition, 31*(2020), 119–142. https://doi.org/10.29086/2519-5476/2020/sp31a7

Lopes, H., & McKay, V. (2020). Adult learning and education as a tool to contain pandemics: The COVID-19 experience. *International Review of Education*. https://doi.org/10.1007/s11159-020-09843-0

Mabunda, P. L. (2010). Information and communication technologies for teaching and learning: Challenges and implications for ODL universities. *Progressio, 32*(2), 222–244.

Mabunda, P. L., & McKay, V. (2020). An analysis of newspaper depictions of schooling: A participatory study for the development of a teacher education curriculum. *Systemic Practice and Action Research*. https://doi.org/10.1007/s11213-020-09550-9

Mabunda, P. L., & McKay, V. (2021). Educational curriculum and multispecies relations. In J. J. McIntyre-Mills & Y. Corcoran-Nantes (Eds.), *From polarisation to multispecies relationships* (pp. 369–389). Springer. https://doi.org/10.1007/978-981-33-6884-2_16

McIntyre-Mills, J., Kedibone, G. M., Arko-Achemfuor, A., Mabunda, P., & Njiro, E. (2014). Participatory approach to education: An action learning approach at the University of South Africa. *Participatory Educational Research, 1*(2), 106–132. https://doi.org/10.17275/per.14.14.1.2

McKay, V. (2018). Through the eye of a fly: Action research as a support for the South African national literacy campaign. *Systemic Practice and Action Research, 31*(4), 375–393. https://doi.org/10.1007/s11213-017-9431-x

McKay, V. (2019a). Fit for purpose: Using a distance education approach to support underperforming schools in South Africa. *Progressio, 41*(1), 1–26. https://doi.org/10.25159/0256-8853/4597

McKay, V. (2019b). Communities in action: The participation of communities in two South African literacy campaigns. In I. Eloff (Ed.), *Handbook of quality of life in African societies, international handbooks of quality-of-life* (pp. 22–43). https://doi.org/10.1007/978-3-030-15367-0_2

McKay, V. (2022). The impact of the Covid-19 pandemic on education: Intersections with poverty. Covid perspectives across Africa. In: Fymat, A., Room, N and Kapalanga, J. (Eds), *Society for the advancement of science in Africa* (pp. 244–264). Tellwell Publishers. 978-0-2288-7440-9.

McKay, V. I., & Romm, N. R. A. (2008). Active research towards the addressal of HIV/AIDS in the informal economy in Zambia: Involvements in social development. *Action Research, 6*(2), 149–170. https://doi.org/10.1177/1476750307087050A

Morgan, H. (2020). Best practices for implementing remote learning during a pandemic. *Clearing House, 93*(3), 135–141. https://doi.org/10.1080/00098655.2020.1751480

Msila, V. (2013). Stephen Biko's philosophy and its pedagogical implications in South Africa. *Creative Education, 4*(8), 492–496. https://doi.org/10.4236/ce.2013.48071

Mungadze, S. (2020, 30 July). Academic urges student discounts from telcos, device-makers. *IT Web*. Retrieved from https://www.itweb.co.za/content/nWJad7begNdvbjO1

Ndlovu-Gatsheni, S. (2018). Rhodes must fall. In S. Ndlovu-Gatsheni (Ed.), *Epistemic freedom in Africa* (pp. 221–242). Routledge. https://doi.org/10.4324/9780429492204-9

Nxumalo, F. (2020). Situating indigenous and black childhoods in the Anthropocene. In A. Cutter-Mackenzie-Knowles, K. Malone, & E. Barratt Hacking (Eds.), *Research handbook on childhood-nature: Assemblages of childhood and nature research* (pp. 535–556). Springer. https://doi.org/10.1007/978-3-319-67286-1_37

Ramose, M. (2015). Ecology through ubuntu. In R. Meinhold (Ed.), *Emerging from cultures and religions of the ASEAN Region* (pp. 69–76). Konrad-Adenauer-Stiftung.

Rheingold, H. (2008). Using participatory media and public voice to encourage civic engagement. In L. Bennett (Ed.), *Civic life online: Learning how digital media can engage youth* (pp. 97–118). MIT. https://doi.org/10.1162/dmal.9780262524827.097

Santos, B. de S. (2020). *The cruel pedagogy of the virus*. Almedina.

Shepherd, D., & Mohohlwane, N. (2021). *The impact of COVID-19 in education: More than a year of disruption*. Coronavirus Rapid Mobile Survey (CRAM) WAVE 5. National Income Dynamic Study.

Teräs, M., Suoranta, J., Teräs, H., & Curcher, M. (2020). Post-Covid-19 education and education technology 'solutionism': A seller's market. *Postdigital Science and Education, 2*, 863–878. https://doi.org/10.1007/s42438-020-00164-x

UNESCO. (2021, 19 March). *One year into COVID-19 education disruption: Where do we stand?* Retrieved from https://en.unesco.org/news/one-year-covid-19-education-disruption-where-do-we-stand

United Nations. (2015a, September). *2030 Agenda for sustainable development: Seventeen development goals*. United Nations Summit: New York.

United Nations. (2015b). *The Sendai framework for disaster risk reduction 2015–2030*. United Nations.

Zimmerman, J. (2020, 10 March). Coronavirus and the great online-learning experiment: Let's determine what our students actually learn online. *Chronicle of Higher Education*. Retrieved from https://www.chronicle.com/article/coronavirus-and-the-great-online-learning-experiment/

Transformative Exhibitions and Artistic Practices: Vision and Culture in the Educational Environment of the Museum

Rania Jaber and Yvonne Corcoran-Nantes

Abstract Museums have slowly transformed from colonial and national repositories to gateways that offer "global" platforms which aim to invite interaction between people from different ethnic backgrounds and socio-cultural contexts. Here we consider case studies from different settings including: an exhibition that took place at the National Museum in Lebanon in 2018 entitled "Chou Hayda?" which interfaces a pluralistic and collective approach to heritage and the socio-political past in Lebanon, along with the permanent show "Identity: Yours, Mine, Ours" at the Immigration Museum in Melbourne that looks at the question of cultural heritage and multiculturalism. Both exhibitions question ideas of citizenship and belonging as well as thinking about heritage and multiculturalism from different geopolitical angles and demonstrates how museums can play a central role for disseminating knowledge and challenging notions of ourselves and others.

Keywords Museum exhibitions · Heritage and culture · Transformative education · Artist · Artworks · Exhibitions

A first encounter with museums for most people is often connected to the perfunctory educational school trip. Whereby, the young and impressionable is introduced to physical representations of art and history sourced locally and globally to offer visual insight and a tacit appreciation of archaeology, art and sculpture as the physical retainers of the past. Silent witnesses of tragedy and joy, of injustice and triumph and insight into culture and racial difference. The curator is the keeper of

---

The fieldwork research which appears in this chapter was undertaken by Dr Rania Jaber.

R. Jaber
Center for Arab and Middle Eastern Studies, American University of Beirut, Beirut, Lebanon

Y. Corcoran-Nantes (✉)
College of Humanities, Arts and Social Sciences, Flinders University, Adelaide, Australia
e-mail: yvonne.corcoran-nantes@flinders.edu.au

J. J. McIntyre-Mills, Y. Corcoran-Nantes (eds.), *Transformative Education for Regeneration and Wellbeing*, Contemporary Systems Thinking,
https://doi.org/10.1007/978-981-19-3258-8_15

knowledge, the oracle, and the child or young adult is expected to draw on the received knowledge. As Peter Sellars, in an address to museum educators in 2009, observed:

> Unfortunately, even many of the funders of education programs want to see that kids are put through predetermined hoops that will demonstrate that they have learned something. They want quantifiable results from a museum visit when in fact, the most beautiful thing about a museum experience is that there is the space to actually discover rather than be told. And those are the times in your life that you remember. Not when someone told you, but when you discovered. And nothing on earth is like the look on a child's face when they discover. That moment of discovery is a moment of empowerment, a moment of vision, a moment of beauty (2010, 104).

The idea of interactivity or interpretation is a situation beyond the habitual question and answer session. The rarefication of the museum as the font of knowledge and an esoteric experience has remained across generations and geographies. Alternative interpretations remained unspoken, the interrogation of difference and its subjective meanings were occulted and unrequited in the museum setting

Thankfully, museums have slowly transformed from colonial and national repositories to gateways that offer "global" platforms which aim to invite interactions between people from different ethnic backgrounds and socio-cultural contexts. We can think of museums as universal yet contextual and intersectional by asking the following questions: what is the mission for museums and how is an engagement with art a transformative educational experience for audiences? What are the museum's motives for presenting certain artists from specific cultures and why? How do exhibitions and art works contribute to education and inclusion within and across different cultural communities? Which artists and artworks are included to convey certain ideas, histories, and perspectives? These ideas will be explored by looking at specific exhibitions in various settings. The case studies from different settings include: an exhibition that took place at the National Museum in Lebanon in 2018 entitled "Chou Hayda?" which interfaces a pluralistic and collective approach to heritage and the socio-political past in Lebanon, along with the permanent show "Identity: Yours, Mine, Ours" at the Immigration Museum in Melbourne[1] that considers the question of cultural heritage and multiculturalism. Both exhibitions question ideas of citizenship and belonging as well as reflecting on heritage and multiculturalism from different geopolitical angles which reveals how museums can play a central role in disseminating knowledge and challenging notions of ourselves and others.

In 2018, on an early summer afternoon in June, I headed to the National Museum of Beirut to attend the opening of the exhibition *Chou Hayda? (2018)* by Annabel Daou. I had never been to the National Museum prior to the commencement of Daou's project which began a year earlier. In November 2017, the artist had sent an invitation, through social media, emails, word of mouth, asking for participants to

[1] Sadly, at the time of writing, due to COVID-19 the exhibition is temporarily closed but an interactive website has been constructed conveying the themes and stories which form part of the exhibition https://museumsvictoria.com.au/immigrationmuseum/resources/identity/

join her in creating an audio guide for the National Museum of Beirut. The artist had invited local individuals and groups living in Beirut, from university students, artists, activists to shop owners and those who ran the local bakery making the classic *man'oushe*[2] – to enter the museum space and collectively participate in creating an audio guide for the National Museum of Beirut. Most of those who attended had never been inside the museum before.

Visitors were required to sign up for an allocated arrival time and to participate in the creation of the audio guide. Upon arrival, visitors would be met outside of the museum's main entrance which consists of a grand staircase and large temple-like columns. During the session that I attended on a Saturday afternoon in 2017, we were received by the artist in front of the museum. After she introduced herself and explained the project, Daou guided us to an abandoned sarcophagus in the small garden outside of the museum. The unkempt and abandoned structure was clearly not cared for and it was overgrown with shrubs and greenery.

Daou directed everyone's attention to the structure and asked them what they saw before she began to elicit responses as to why the statue had been placed outside of the museum. Thus, alluding to the delineation between objects which generally exist outside museums and those objects that are deemed 'worthy' of being housed within the museum. This encouraged participants to begin thinking about the objects by considering how each piece was created, by whom, and the potential trajectory of the object before arriving at the museum. Other questions to consider include: why the object was forbidden from entering the museum's front doors? Which objects have been neglected? And which ones were prized and valued? Beginning with an object's narrative allowed visitors to focus their attention and thoughts on the material composition and significance of each exhibit that one encounters within the museum space and those outside of the museum.

The exploration began as the group was directed from one sculpture to the next. As we huddled around Daou, she tendered a question with regard to each of the objects in question, the participants responded and this process was recorded, with the use of a microphone, by one of Daou's collaborators from the Temporary Art Platform (TAP). In fact, the title of the exhibition 'chou hayda?' translates, in English, to 'what is this?' Thus, participants were encouraged to observe, analyse, and reflect on what they were seeing. Daou asked questions as she pointed to various objects encouraging everyone to think *with* and *through* each one. Each question and reflective answer were recorded. Exhilaration grew as the participants realised that their voices would be included in the final product. In fact, every participant contributed their voice, a narrative, and a new perspective to addressing

[2] It is made from dough that is rolled out to create a flat round surface and is considered staple food found in most local bakeries in Lebanon. The traditional *man'oushe* is either *zaa'tar* or *kechek*. *Zaa'tar* is made from dried thyme, roasted sesame seeds, and sumac, which are mixed with olive oil while *kechek* is made from burghul and strained yoghurt, or *labne*, and mixed with olive oil. Each mixture is placed on top of the kneaded dough before it is put into the traditional *furn* or oven. Often, they are baked in batches and sold for a reasonably cheap price, making the *man'oushe* a fast and tasty street food.

artifacts and statues that Daou had selected. They were participants in a creative interactive process offering new reflections and interpretations of silent, historical objects.

Prior to initiating the development of the audio guide, the artist had spent considerable time at the museum. She was commissioned by TAP to create an intervention in the National Museum of Beirut. Daou had previously collaborated with TAP in 2016, on an artistic project entitled *Works on Paper (2016)* which comprised of a series of interventions in four Lebanese papers (Assafir, Al Akhbar, The Daily Star, and L'Orient Le Jour).[3] Artists were commissioned to reflect on the circulation of news and the materiality of the newspaper. In her work, Daou had invited members of the public (an ad had been included in the newspaper) to send a photograph of the palms of their hands.[4] This was a continuation of her fortune telling series whereby she offers silent readings for participants in exchange for a fee.[5]

However, in 2017, the collaboration was more encompassing and specific to one location in Beirut. During that time, Daou immersed herself ethnographically in the museum space by getting to know the objects and closely scrutinising the artifacts, murals, and various material included in the museum. She silently observed the objects and the spaces they occupied inside the museum and began to craft a series of questions with which to engage participants and bring the works to life. Without researching or reading about the history of each object, Daou initiated a new and different perspective by evoking various characteristics such as noticeable marks, fragments, missing pieces or absent parts of statues, or by questioning what could be added to the sculpture or statue under scrutiny. What stories might each object reveal? Based on the questions created by the artist, it is evident that objects were imagined as historical witnesses.

This exhibition relied on participants. It could not have been produced using a singular voice or authoritative narrative as is often experienced when one enters a museum. It also depended on the possibility that objects could contain transformative stories. What if an object had several stories? Several lives? Or what if they had more than one storyteller and more than one perspective? When visitors enter museums, they are often told what to see, believe, look at, what to value – what to observe along with the story and history behind each exhibit. The experiences portrayed and conveyed in this chapter are interactive and expansive placing outcomes and knowledges in collective hands wherein no one is the oracle and where intersectional voices and knowledges coalesce and reside.

---

[3] The selected newspapers are published in different languages (Arabic, English, and French). For instance, Assafir and Al Akhbar are published in Arabic, The Daily Star in English, and L'Orient Le Jour in French.

[4] The advertisement in the Daily Star reads as follows: In this ongoing project, artist Annabel Daou invites members of the public to partake in an intimate exchange in which they present their palms and receive their fortunes transcribed onto paper. For the current iteration of this project, 12 participants will have their fortunes published anonymously in this paper on May 28th.

[5] For more information about *Fortune,* please see Daou's website: https://annabeldaou.com/FORTUNE-1

1 Methodology

Research for this chapter was conducted in part through the ethnographic practice of being a participant observer. Rania participated in the creation of the audio guide and attended the opening of the exhibition on June 18, 2018. She also relied on secondary resources such as newspaper accounts about the National Museum of Beirut, scholarly publications, and reviews. As previously mentioned, Rania was present at the opening of the exhibition *Chou Hayda (2018)* and took part in the recording of the audio guide by participating in the event. More specifically, she also contributed by answering some of the questions asked by Daou when prompted – answers which were recorded and edited and subsequently turned into the audio guide found on the website for the project *Chou Hayda*.[6] Consequently, Rania's voice and some of the answers she provided are included in, and are part of the resultant audio piece. Since then, Rania has visited the exhibition several times.

During the opening event, visitors were encouraged to download the audio guide on their phones and follow the map to locate each object. The occasion was well attended as visitors wandered about in the spacious museum setting. A talk by the organiser and head of TAP, Amanda Abi Khalil introduced the project and offered an account of the exhibition along with a conversation between Abi Khalil and Daou followed by a trailer the organisers had created in order to explain the process and aims of the work.[7] The video played in a small theatre close to the main entrance of the museum and was left to play on repeat for future visitors.

Concerning the exhibition *Identity: Yours, Mine, Ours (2011)*, although Rania did not experience the show first-hand, she conducted research using various scholarly papers, reflections, and documents. Such scholarly attention makes the show stand out as an event that questions various collective and individual identities. Furthermore, Rania's experience of this exhibition has been mediated through online resources which include the website for the museum[8] along with several videos and artistic material which have been accessed online. What is striking about this exhibition is the number of scholarly responses to the show – some refer to the exhibition as a game-changer, as 'unapologetic' (McKernan, 2020) as instilling significant changes through one's understanding and knowledge of 'others,' while others claim that the show offers a return to discussing a critical and widespread concept of 'racism' when tackling cultural diversity. For Rania, as someone who grew up in Adelaide, the topic of migration and constructed identities as well as stereotypical roles in multicultural settings is integral to her own migratory experience.

Although *Chou Hayda (2018),* is more recent and little has been written about the show thus far, the exhibition lends itself to re-thinking and re-imagining heritage

[6] The website for *Chou Hayda* can be accessed on: https://www.chouhayda.com/. The audio guide can be either downloaded through this link or accessed at the museum.

[7] The trailer can be accessed on this link: https://www.chouhayda.com/media

[8] https://museumsvictoria.com.au/immigrationmuseum/whats-on/identity-yours-mine-ours/

practices and past histories through offering a similar connection between 'self' and 'others'. The opening question to *Identities (2011)* is "What does it mean to belong and not belong in Australia?" Likewise, one of the questions asked in *Chou Hayda (2018)* with reference to one of the sculptural objects of a figure with arms across the chest and head looking up echoes similar notions of national belonging. The question asked is "Are you Lebanese?" According to the description of this statue:

> Ever since this figure appeared in the world, somewhere within the borders of what is now this country, the question of nationality has hovered around him. While it is hard to grasp exactly what nationality means to him, there is something in the way he clutches what may be a bird in his hand that gives one the sense that he is holding onto something that could fly away at any moment and that he cannot let go. Perhaps he is afraid waiting for something before he can relax, but then again there may be a completely different explanation that is far less anxiety-inducing.

The description evokes the nation's historical past and the complex question (Are you Lebanese?) regarding cultural identities and pasts. The responses are often ample, diverse, and divergent. The following section will provide a brief history of the museum along with the colonial history of the republic of Greater Lebanon which was established in 1920.

2 History and Significance of a National Museum

The inauguration of museums focusing on contemporary and modern art in Lebanon has proliferated over the last few years. For instance, Sursock Museum, a museum of contemporary and modern art, re-opened in 2015 after an eight-year closure. Another case in point is Beit Beirut (or the House of Beirut in English), a museum dedicated to the memory of the civil war which has also been reconstructed. The building once known as the Barakat building, named after the family who owned it, used to be an apartment building built in the 1920s and is one of the few buildings that survived the civil war.

The building suffered severe damage due to its location on the green line dividing Beirut into two zones: east and west. The apartments served as the principal hideout for snipers because it offered a clear vision of anyone crossing the divide. Plans to demolish the edifice were hampered by a campaign to preserve the building as a surviving monument of the war. Hence, the structure was refurbished in a way to maintain the traces of destruction and the ruins replete with bullet holes along with the surviving shoot-out holes used by snipers. The museum is meant to serve as a memorial to the war, however, there are clear disagreements concerning what the building should be used for and what objects should be stored there (The Economist, 2017). The National Museum of Beirut is located in the same area as Beit Beirut, on Rue Damas which cuts across the dividing line between east and west Beirut that existed during the civil war from 1975 until 1990. The National Museum closed down during the war years and reopened again in 1999 after undergoing renovation and restoration of damaged and lost artifacts. According to Joseph

Pharès (2003) the actual idea of a national museum was planted in 1919 when Raymond Weill, a French officer, had purportedly received antiquities 'discovered' on Lebanese soil. Although it is unclear who had found the antiquities and why they were delivered to Weil in his Beirut apartment, what is clear is that the area which was to become Lebanon contained a plethora of archaeological findings that were amassed and eventually housed in a museum. Construction of the building began in 1930 and took seven years to complete (1930–1937), followed by five years of assembling and instituting a collection to exhibit. The official inauguration by the president of the Lebanese republic, at that time Alfred Naccache, took place on May 27, 1942.

The building was constructed in an 'Egyptian Revival' style (Pasquini, 2019) which resembles a temple. During the civil war years, most of the items were placed in crates and stowed away in both the museum's basement and vaults in the Central Bank in Beirut (Young, 1995). However, items such as statues that were too big to move were placed in concrete caskets (Young, 1995). According to Elaine Pasquini (2019), after the museum reopened, it harboured a collection that went from the Bronze age up until the Ottoman era. For experts, the collection signals an important reflection on past civilizations such as Phoenician heritage along with the Roman and Persian Empires (Pasquini, 2019). Of course, Lebanon's past is a delicate issue since there are various narratives and ideologies that are dissonant with one another.

In fact, Lebanon has existed as a country with internationally recognized borders since 1920 when it was declared the State of Greater Lebanon under the French mandate. The State of Greater Lebanon was transformed into the Lebanese Republic in 1926 (Salibi, 1988, 27). Since the inception of the republic, Lebanon has been described as an 'artificial creation of foreign imperialism' (Salibi, 1988, 31). Long before Lebanon was established as a country by the French mandate, numerous empires (from Roman and Byzantine to Ottoman and French) had contributed to the cosmopolitan status of the country. There are significant residual traces that remain to this day, from the Phoenician ruins to Roman and Mamluk past influences, which have been preserved at various sites around Lebanon such as Batroun, Baalbeck, and Anjar. The country is currently home to three languages: English, Arabic, and French, as well as colonial architectural styles, some of which are still evident.

During the French mandate from 1920 to 1943, Beirut was constructed to resemble a French colonial city. Consequently, during post-independence, from the French mandate in 1943, and prior to the civil war that erupted in 1975, Lebanon was frequently mythologised as the 'Paris of the Middle East' and the 'Switzerland of the East'. Such connotations, however, were short-lived and failed to provide an antidote to the country's turbulent past and politically unstable future. At the turn of the twentieth century, prior to becoming a French mandate, Lebanon was still under

Ottoman rule. However, there was growing dissatisfaction and rebellion against the regime, and a rise in nationalist thinking to create a sovereign nation.[9]

Accordingly, the fissures dividing different group ideologies led to a civil war in 1958 and again in 1975. In a short-lived civil war in 1958, politicians were divided between those who favoured patriotic leanings towards Lebanon against the Arab nationalists. According to historian Kamal Salibi (1988), these divides have produced the greatest tensions amongst the population in the country between two different loyalties of Arab nationalism (or pan-Arabism) and Lebanism (advocating a particular Phoenician and Mediterranean history). Fast forward to the end of the civil war in 1990, the official slogan used to announce a shift from a state of conflict to a state of peace was "*la ghalib la maghlūb*" which translates to 'no victor, no vanquished' (Haugbolle, 2005, 191). Therefore, the state of forgetting became the resolution to ensuring peace.[10] The civil war of 1975 erupted as social and political divisions produced what has been described as a religious and sectarian war dividing the city into two sides: east and west (Salibi, 1988). Clearly, there were various political parties and groups involved. Ultimately, the divide spread to towns and villages in the mountains, which led to significant displacement of the population. The general amnesty granted to all participants was an invitation to forget the war. Therefore, amnesia became a useful political method to evade pinpointing the blame on the individuals who were still in power after the war ended.

No coherent narrative emerged of the 'events' as the civil war was often referred to during that period, and the generic term was accepted which obliquely downplays the reality of a fully-fledged and morbid war. Indeed, many artists and writers have argued that the war never ended, and that Lebanese society is enduring a 'protracted war' that has never been resolved (Sadek, 2016). The willingness to forget war time atrocities has denied Lebanese people closure, the right to publicly grieve, mourn, and accept what happened. Sune Haugbolle (2005, 192) argues that the 'unfinished nature of the end of civil war' has had a negative impact on the population's ability to deal with the memory of the violent conflict. However, it can also be argued that people preferred to forget the war, even though it formally never ended. Recent work has addressed the role of memorials. Larkin and Parry-Davies (2019) in their work on memorials and memory-making in Lebanon, focus on museums dedicated to commemorating events connected to the civil war in Lebanon. According to Larkin and Parry-Davies (2019) their case study of Lebanon's memory museums sits within a new genre of museology that is exemplified by what the authors term as 'dark tourism' which emphasises 'the decentering of knowledge from rigid pedagogical narratives in favour of the participatory and affective experience of

[9]The public hanging in 1914 in the heart of Beirut of several individuals who publicly opposed the Ottoman Empire is a crucial event in Lebanese history. It created a contested identity for nationalists that was immortalised in a monument depicting the lives of individuals who believed in a sovereign nation.

[10]The slogan was also used in another, although brief, civil war in 1958. The war in 1958 had been the result of conflicting ideologies and long-standing clashes based on sectarian animosities between different sects.

viewers.' In truth, Beit Beirut has no clear structure or program. The space has no clear guidance and lacks administration (Larkin & Parry-Davies, 2019). Although the building has housed a series of interventions, according to Larkin and Parry-Davies (2019) the building remains empty. They claim that the museum is undergoing an 'impasse' which is indicative of contemporary politics of war memory in Lebanon – since political parties and civil society have no neutral agenda (Larkin & Parry-Davies, 2019, 92). Their work focuses on three museums including Beit Beirut which is in the same area as the National Museum of Beirut. In fact, both constructions are located on the demarcation line – the designated dividing line during the civil war.[11] This emphasises the dominant role of memory-making in this area, with its dividing history, but also demonstrates the role of transformative education, ideas, exhibitions as crucial tools to discuss events from the past and the present.

3 Changing Institutions

Museums are critical public institutions that are predicated on social and political hierarchies as well as exclusion practices depending on the narrative (Levitt, 2015). James Clifford (1997, 192) describes museums as "contact zones" or as spaces of 'an ongoing, historical, political and moral *relationship*—a power-charged set of exchanges, of push and pull.' As stated by Eilean Hooper-Greenhill (2000), there has been a shift from modernist museums as locations of authority to what she terms the 'post-museum' as a place of mutuality (xi). In fact, the change in thinking about museums through this lens began in the 1990s. The move towards a decentred approach to knowledge production involves the use of oral histories, testimonies and collective participation (Larkin & Parry-Davies, 2019). This is evident in both exhibitions addressed in this chapter. *Chou Hayda (2018)* challenges the pedagogical roles of displays that are assembled in museums by creating contemporary and persuasive conversations through objects that have been relegated in status to artifacts and archaeological discoveries. As attested by Hooper-Greenhill (2000) the pedagogical role of displays is not often researched nor is it properly acknowledged. Furthermore Hooper-Greenhill (2000), claims that museum pedagogy is structured through the narratives assembled by museum displays as well as the methods used to communicate such narratives. *Chou Hayda (2018)* alters the narratives through an audio intervention. The fact that the work is conducted in Arabic also challenges global art world domination and highlights a decolonial method for creating artistic works that envision a different future for the archaeological objects on display.

In her book on global museums, Peggy Levitt (2015) explores nationalism through looking at cultural institutions. She conceives of the concepts of

[11] The reason why the segregation between east and west Beirut is referred to as the green line, is because the area was occupied by snipers therefore few people crossed this zone during the war years, which resulted in overgrown green shrubs and foliage.

'nationalism' and 'cosmopolitanism' as a continuum. For example, some museums see the world through a local lens, while others bring a more cosmopolitan approach to looking at local nationalism. On the other hand, Ien Ang (2016) argues that the relationship between the two should be seen as one of 'enduring friction, creating a major challenge for museums around the world to tell stories suited for our mobile, global times.' According to Ang (2016), there is a tendency to segregate 'the story of immigration from the main national story – and to marginalise it – which is a common representational strategy in the museum landscapes of many countries, including avowedly immigrant nations such as Canada and Australia' The exhibition *Identities (2011)* presents critical tools for thinking about how museums can incorporate questions regarding what it means to be a citizen and who belongs in a nation.

Now in the case of Lebanon's institutional frameworks and structures, when the civil war ended, the contemporary art scene in the 1990s and early 2000s was booming with artists returning to Beirut who began experimenting with new media. These artists formed communities and created artworks that addressed the period and memory of war, which was still very fresh in people's minds. The first public engagement took place in 1995 in Sanayeh Gardens, a garden space in the Hamra area of Beirut. Academics and critics such as Wilson-Goldie (2007), Rogers (2007) and Sadek (2016) have written quite extensively about this event, which generated an organic significance in the creation of a public space and interest in the arts. This period coincided with a growing interest in global art (Rogers, 2007).

The first contribution by artists after the war ended was not only to criticize local politics and the public space, but to focus on Western representations of Lebanon and to reject coherent narratives of the civil war itself (Abdelhady, 2011). Historically speaking, constructions of national identity have been based on political narratives such as a commitment to a notion of Arab nationalism or a specific Lebanese national identity. According to Hanan Toukan (2010, 7) international support for local initiatives began in the late 1990s and early 2000s, to work with home-based initiatives by Beirut-based artists and cultural organisations that had been active since the early 1990s. Similarly, Jessica Winegar (2008) and Maymanah Farhat (2009) documented a growing interest in artworks from the Middle East region after September 11, 2001. In 2002, the Canadian art magazine *Parachute* devoted an issue to the art scene in Lebanon. In the same year, the well-established art space Ashkal Alwan, led by Christine Tohme, made its debut art intervention in the streets of Hamra, engaging the public with works by Lebanese artists. It was an exciting time for both artists and audiences to engage with artworks in public spaces. However, prior to the year 2000, there were very few international curators who were interested in looking at what artists were creating in Lebanon (See Rogers (2007), Muller (2013), Sadek (2016)).

Therefore, since the late 1990s, there has been an increase in curatorial interventions in Beirut regarding the global art market. In their introduction to the topic of collecting practices in the Arab world, the authors Sonja Mejcher-Atassi and John Pedro Schwartz (2012) highlight the necessary questions one must ask when considering the activity of collections surrounding museum cultures. The main challenge delineated by both authors is how to think:

> ….critically about the transmission of Western assumptions, methods and goals while adapting the insights of a tradition most commonly associated with the West, and to combat the Orientalist tendency to essentialize the region's culture and society as timeless and identical (Mejcher-Atassi & Schwartz, 2012, 1).

Similarly, Lina Tahan (2010) states that the National Museum of Beirut is a colonial institution. According to Tahan (2010), the focus of the archaeological collection at the National Museum of Beirut is dedicated to a Phoenician heritage and artifacts. She argues that this focal point offers a skewed, colonial interpretation of a vast array of historical and political identity. In fact, the debate over Lebanon's national identity is also evident in collections at the Archaeological Museum at the American University of Beirut (which was established in 1868), as well as the National Museum of Beirut. These debates also translate to the world of contemporary and modern art museums as well as art scenes in Beirut. Hanan Toukan (2010, 10), claims that identity debates regarding cultural practice reflect Lebanon's relentless contention with multiple identities and various narratives, which demonstrate that cultural production and representation 'whether for a local or global audience, inevitably become domains of contestation'. Furthermore, this is not merely internal. Toukan (2010, 10) argues that outside players make their presence felt through funding, visions and discourses, which includes a geopolitical confluence of political identity.

In a more recent interview, Christine Tohme (2015, 321) claims to be following the footsteps of other 'non-traditional institutions' from Lebanon's past history. She asks:

> Are we really interested in spaces that do not reflect anything around us-that do not resonate with the flux of a political and social situation that is changing constantly? How do you respond to this constant change in an institution that only presents art as things on the wall, that envisions institutional space as a zone of static, manicured perfection? (Tohme, 2015, 322).

An exhibition like *Chou Hayda (2018)* addresses these concerns by directly involving participants in various encounters and scenarios.

Emerging in the context of a significant growth in art institutions, galleries, and museums, the Temporary Art Platform is a program that aims to work with public spaces and current issues. Temporary Art Platform (TAP) is described as:

> ….a context-responsive curatorial platform founded on para-institutional modes of operation responding critically within shifting and dynamic situations. We are concerned with the development of social practice in Lebanon, the region, and the Global South through modes of participation, multi-disciplinary formats and decentralized locations.[12]

Based on the projects created so far, TAP appears to invite debate and dialogue surrounding critical subjects such as the lack of public space in Beirut. TAP was

[12] Information about TAP can be found on the Temporary Art Platform website: https://temporaryartplatform.com/what-we-do/

founded in 2014 and creates partnerships with public and private institutions aiming to foster transformations and public interventions. They work beyond the art world to advocate for environmental as well as policy changes. They identify crucial issues to discuss through the following way:

> We identify a topic, a need, a site, a community, a theme, a question that we want to put at the core of our investigation, and then we define the format of the project and its methodology.[13]

They aim to produce knowledge about social practice, public art, as well as administrative and legal frameworks. Past projects include: the Dalieh project whereby TAP invited artists to intervene in the site by creating artworks about the Dalieh project. This project specifically caters to the privatisation of the coastal land in Lebanon. The Dalieh area is one of the few public spaces remaining in Beirut and one of the only entry points to the sea. The coast is largely overrun by private resorts, often in violation of zoning requirements. In 2017, TAP worked with The Civil Campaign to Protect the Dalieh of Raouche, to highlight the encroaching theft of an 'open-access shared space' for city dwellers.[14]

3.1 Chou Hayda

The project *Chou Hayda (2018)* was sponsored by both TAP and *Mathaf Mathaf. Mathaf Mathaf* is a collaborative commissioning project between TAP and the Association for the Promotion and Exhibition of the Arts in Lebanon (APEAL) which is responsible for establishing the upcoming Beirut Museum of Art.[15] The word *mathaf* translates to museum in English. However, in Lebanese vernacular, *mathaf* also refers to the area where the National Museum of Beirut is located. In fact, the geographical location of the National Museum of Beirut will be in close proximity to the up-and-coming Beirut Museum of Art (BeMA). The two museums will face each other, offering a contextual as well as geographical dialogue between a historical archaeological museum and a museum of modern and contemporary works. Therefore *mathaf mathaf* also refers to the potential connections between two distinct museums in the same area. It will be interesting to see how the past and future of the city will be told through the proximity of these two museums along with Beit Beirut.

When visitors enter the National Museum of Beirut, they are confronted with sculptures, artifacts, statues and the silences of past histories and wars. They are met with the knowledge that the museum has endured violence, closures, disruptions,

[13] This is taken from the TAP website: https://temporaryartplatform.com/what-we-do/

[14] For more information about this campaign, please visit the organisation's website: https://dalieh.org/

[15] Mathaf Mathaf" accessed at: http://www.bema.museum/bema/main-menu/outreach/upcoming-programs/mathaf-mathaf

and had once been a hide-out for snipers. The statement for the exhibition *Chou Hayda (2018)* reads as follows:

> In November of 2017, people from across Beirut came to the National Museum and gave their voices to a number of objects in the collection. The people spoke to and for and about these objects from the past, and, in doing so, they revealed fragments of the present. They did not attempt to disclose a particular historical narrative. Nor did they attempt to create a fiction. They did not lie and they did not try to tell the truth.

Rather, each participant told their own stories and when future visitors listen to the audio guide, they will encounter a plethora of responses and narratives embedded in each exhibit included. At times, answers coalesced regarding certain objects. For instance, one of the first objects that spectators were asked to respond to includes two bulls joined in the middle with each side facing away from the other (resembling a Janus-faced expression). As stated on the *Chou Hayda* website by Daou, these bulls have been in opposition for 2500 years 'They tower above their viewers, daring those below to climb up and mount them. Everyone has a different idea about the direction in which they will sit once they reach the top.' Daou asked participants: "which side are you on?" A range of responses were offered and each one was recorded – some examples include on the left, in the middle, on the right, both sides, etc. The question posed "which side are you on?" demands viewers to reflect on their role and involvement in interpreting objects that are often distant and disconnected from one's own subjective experiences. For those listening to the audio guide, they are also encouraged to reflect on their answers and to potentially transform their experiences of different objects.

Thinking about the movement of sculptural pieces and how meanings change, Daou has invited participants to interact with the statue. The social life of objects has been thoroughly studied by anthropologists, interested in how objects become imbued with certain meanings and values. Building upon the 'social life of things' by Arjun Appadurai (1986), more recent studies have analysed the various intersections enacted as objects circulated beyond object biographies which often appear to be too linear. In fact, Alexander A. Bauer (2019) focuses on the 'object's itinerary' by addressing the object's materiality, agency, networks and circulation as well as power structures. According to Bauer (2019, 366),

> ....thinking about objects as having itineraries rather than biographies collapses the temporal distinctions between past and present (and future) and helps us to fully consider objects' present entanglements as central to their story.

This is useful when thinking about the various narratives that each object offers viewers in the present, despite the object's itinerary before arriving at the National Museum in Beirut. In addition to considering objects as itinerant, we can also consider objects as witnesses to past wars and conflicts through the evident residual marks that remain.

Furthermore, thinking of objects as witnesses has artistic and scholarly resonance. For instance, an artwork by Mona Hatoum immediately comes to mind, the work titled *Witness (2009)* is a miniature rendition of the statue of the martyrs located in Martyrs Square in downtown Beirut (in the central area of the city which often attracts large crowds of protests). Hatoum's figure alludes to the visible scars that caused damage to the actual statues, a reminder of brutal fighting and the history of the war. The miniature replica reproduces what statues, like the buildings that had been hit by bullets and shells, must have silently witnessed.[16]

Thinking of objects as itinerary witnesses adds further layers to the encounters between viewers and objects. According to Pasquini (2019), the sculpture of the two bulls is a white marble fragment of a capital in the shape of a bull's head which dates back to the fourth century BCE. 'The sculpture was excavated from the Temple of Eshmun in Sidon in 1967, while the life-size bull's head was stolen from a warehouse in Byblos in 1981 during Lebanon's civil war and ultimately emerged in the private collection of Colorado residents Lynda and William Beierwaltes.' In 2016, while the sculpture was on loan to the Metropolitan Museum of Art, a curator confirmed that the sculpture was one of several artifacts that were stolen from Lebanon during the civil war. After informing the Ministry of Culture in Lebanon and following a lawsuit in the U.S., the sculpture was repatriated to Lebanon in 2017. The statue came back just in time to be included in the exhibition, adding further narratives to the object's journey across borders.

In another section of the museum, there is a lone statue. The statue is a woman whose face has been eroded or destroyed. Daou points at the sculpture and asks participants: what have you seen? Therefore, participants pretended to be in her shoes and act as witnesses, bringing a stone statue's stories to life. What story would she tell? Another statue that appears to be a head – a jar, with visible cracks covering its fragile face also provokes similar questions. Daou asks visitors: Can you keep a secret?

The questions asked make visitors reflect and use their voices to speak from a personal conviction therefore shifting the power from a voice of authority to individual and local voices of citizens who are often deprived of power. In fact, viewers respond to the objects in their current state. For example, at one-point Daou focuses on three keys housed in a glass cabinet. She asks participants: What Door Can You Open? The responses to this question were almost unanimous as views expressed similar accounts regarding political hierarchies and corruption in Lebanon. The question seemed to resonate with many participants who responded by stating they would open the doors of the house of parliament. The potential to open closed doors to access restricted and heavily guarded places that consistently keep people at bay became an eery premonition of the October 17th uprisings in 2019, that sparked massive nation-wide protests in Lebanon. Whereby as the protests grew, offices of parliament in downtown Beirut were safe-guarded and secured from protestors.

In their reflection on *Chou Hayda (2018)*, Daou, Nadim Mishlawi and Amanda Abi Khalil focus on the ways in which the multiple voices recorded play an integral role in 'knowledge production within museum/institutional contexts and its power to shift received notions of authority and access within and outside the institution'

[16] Mona Hatoum: Witness (2010) accessed at: https://universes.art/en/nafas/articles/2010/tips/mona-hatoum-beirut

(2019, 367). Furthermore, asking questions in Arabic is a profound commitment to inclusionary practices in the art world in Lebanon since most shows are intended for multilingual audiences. In fact, relying on knowledge that is produced and created by locals in Arabic vernacular challenges knowledge production in a global art world which is often dominated by other languages.

The fact that the questions and responses are all in Arabic demonstrates a quest to redirect the growing art spaces in Beirut to include locals. In her study of the art world in Cairo, Jessica Winegar (2006) reflects on how local artists often felt excluded from events and art talks held in Cairo because most of those events, including art talks, were presented in English. In this instance, the audio guide is in Arabic, but it is still an immersive experience for visitors who do not understand Arabic – it allows for an experience that has the potential to be translated from local languages and cultural ideas rather than the other way round – from global to local movements.

4 From National Museum to Immigration Museum

The permanent exhibition *Identities: Yours, Mine, Ours (2011)*, opened on May 9, 2011 at the Immigration Museum in Melbourne. Upon entry into the exhibition space, visitors are confronted with an immersive video installation playing on a screen at the end of the hallway – as viewers walk up to the exhibition rooms. The video entitled *Welcome (2011)* by Lynette Wallworth portrays a human scaled video comprised of a series of people standing at the entrance of the museum looking into the camera – or what appears to be looking directly at an audience. These strangers seem to be looking at visitors in an either welcoming manner based on their gaze and body language or standing with arms crossed with a disdainful look on their faces. As the visitor turns the corner to enter the exhibition, one comes face to face with a ruptured mirror, reflecting the visitor's own face. The show is divided into three sections: First Impressions, People Like Us, and People Like Them. Based on a review by Eureka Henrich (2011), the main theme of belonging is emphasized through the interpretation of the show as well as its advertising campaign. Henrich (2011) states that when the exhibition opened in 2011, large banners with black and white photo portraits broadcasting the question "I belong, do you?" were draped outside the museum and appeared on bill posters that were located on walls in various places around Melbourne. She also claims that unlike other museums which tend to focus on 'national' celebration the exhibition under question, focuses on 'how we constitute ourselves.'

The description of the exhibition on the Immigration Museum website states that the show focuses on:

> How our cultural heritage, languages, beliefs, and family connections influence our self-perceptions and our perceptions of other people – perceptions that can lead to discovery, confusion, prejudice and understanding.[17]

Yet who does the term 'our' refer to? Is it certain cultural groups or all audiences? As announced on the website, the exhibition aims to communicate these questions regarding who belongs in Australia and how one's cultural background influences perceptions about others by using personal stories, objects, images, and interactive multimedia experiences which encourages viewers to reflect and engage with the questions posed. In turn, visitors are encouraged to share their own stories and reflect on what 'diversity' means to them.

According to the Exhibition Evaluation Report (Walton et al., 2016) the target audiences consisted of secondary school students addressing curriculum units tackling identity, belonging, and ethnicity.[18] Henrich (2011) also claims that the target audience was principally the youth. The evaluation aimed to understand the public role of museums in 'countering racism and promoting positive attitudes and acceptance toward people from diverse racial, ethnic and cultural groups' (Walton et al., 2016, 7). One significant factor demonstrated by the evaluation is that although immersive, interactive and affective engagements allowed for knowledge acquisition of experiences of racism, cultural identities and ideas of belonging, the learning process was insufficient. Rather, what enabled a more reflexive and empathetic understanding of racism and diversity that ventured beyond reading about contemporary issues and experiences, was the introduction of other mediums of communication and engagement such as audio-visual and interactive approaches (recorded as multi-modal experiences) in the museum exhibition.

For the curator, Moya McFadzean (2019) the museum must:

> ….be constantly engaged in the 'dialogic encounter of narratives' to avoid the binary assumption of a relationship that positions the Museum as the producer and its public audience as the consumer, with all the implications of power and single authoritative voice that this implies. Multiple voices no longer just focus on conventional narratives of a linear journey of migration and settlement but reflect on larger questions of personal identity formation, individual and collective belonging and prejudice, transient experiences and mobilities.

She writes about the pivotal role of memory for sharing past histories (through oral histories) and cultural heritage. However, the role of museums is to also re-evaluate 'who is doing the remembering and how that memory is presented' (2019, 266). This is insightful and compelling when considering migrant memory. Australia is considered to be one of the receptive western nations to immigration (Markus, 2014) and has a long history of migrants who have maintained transnational

---

[17] The description is taken from the exhibition's website which can be accessed on the following link: https://museumsvictoria.com.au/immigrationmuseum/whats-on/identity-yours-mine-ours/

[18] The Evaluation Report is a result of research conducted through Deakin University within the Alfred Deakin Institute for Citizenship and Globalisation, researchers from the University of Melbourne and partner organisations including Museum Victoria (Immigration Museum and Melbourne Museum) and the Victoria Health Promotion Foundation.

connections with their respective homelands but have also forged new identities that are constantly changing. Despite being a multicultural society, Ghassan Hage (1998, 17) argues that national identity is based on what he terms 'internal orientalism.' In fact, the growing visibility of migrant and ethnic minorities in the public sphere continues to 'engender indifference, hostility, prejudice, uncomfortable feelings of anxiety and fear in everyday spaces' (Mansouri & Lobo, 2016, 6).

According to Kylie Message (2014), the exhibition engages audiences in the changing debates around cultural diversity – from the macro/structural level to everyday experiences that are often heterogenous. Andrea Witcomb (2007) suggests thinking of museums as 'communities' which is helpful in developing policy frameworks and curatorial projects. All in all, it is fair to say that museums play a crucial role in the formation, discussion and implementation of important social issues that engender racist attitudes and feelings of discomfort. Ang (2005) states that cultural diversity is a predicament for museums in integrating diverse communities while maintaining a divide between cultural objects and "art" since museums could be considered as occupying an alienating space by ethnic minorities. Furthermore, in another study, Noble and Ang (2018) notice that the impact of cultural diversity on cultural consumption by different ethnic groups is quite complex. Firstly, ethnic groups can never be homogenous, instead groups are already significantly diverse. Secondly, the authors identify a need for further research by posing more questions and conducting further investigation regarding ethnicity and cultural consumption.

The exhibition offers a plethora of experiences to start conversations and create encounters. In fact, part of the aim of the Immigration Museum is to create knowledge about diverse cultures and individuals. According to Mckerman (2020, 7), a central component of the museum is to recognise the 'humanity' of migrants and refugees who are often dehumanised by the media. Elias et al. (2020) consider the framing of multiculturalism and interculturalism to be a bifurcated approach for successful diversity governance. In fact, intercultural frameworks encourage bi-directional and multidirectional engagements cross- culturally (Elias et al., 2020). With the rise in nationalism and antagonism against diversity, exhibitions such as *Identity* (2011) highlight these antagonisms and evoke a necessary dialogue regarding shared histories, plurality, and diversity. While *Identity* aims to challenge racist attitudes towards non-White Australians, *Chou Hayda (2018)* invites locals to create their own narratives about museum objects.

Both exhibitions are focused on inviting interactions and participation for local audiences to change how one views objects from the past (in Chou Hayda) or how one views diverse cultural groups (Identity). *Chou Hayda (2018)* is constructed as a participatory event as well as an exhibition inside a museum. Those who listen to the audio guide are hearing multiple voices and perspectives, they are also encouraged to reflect on their own perceptions and answers. Similarly, Mckernan (2020, 10) emphasizes: 'The challenge of Identity is as much about an encounter with ourselves as it is about an encounter with the Other.' These encounters take place through oral histories and interactive scenarios. Witcomb (2019) considers three kinds of pedagogy in oral history exhibitions in museums, which are: pedagogy of looking, pedagogy of feeling, and pedagogy of listening. According to the author, oral histories offer encounters between differences or what she terms 'us' and 'them'. She claims that oral histories are not only about listening and learning from those telling their stories, but also involves implicating oneself in someone else's experiences or what she terms a 'pedagogy of feeling'. Witcomb (2019, 213) states:

> Oral histories need to be presented in ways that stage an encounter with both the voice and face of the person giving the oral history, rather than through a written quote. This is because of the importance of mimetic communication which establishes infectious connections between people, and which demands attention.

Thus, engaging with experiences and lived accounts prompts transformational viewing and educational possibilities.

5 Conclusion

This chapter offers new insights into the significant role of exhibitions and artistic interventions in the construction of new conversations about citizenship, location, and globalization. Although museums have modified their artistic, curational and 'educational' strategy, certain limitations remain. In Lebanon we have witnessed a persistent and non-traditional lack of government support, which allows projects and platforms to create a new strategy for producing knowledge that caters to local audiences. The exhibitions discussed here also ask pertinent questions such as: who are locals? Who are citizens? What languages are integral to one's identity and the ability to communicate their perspectives and heritage? In fact, one of the most urgent questions that museums should be asking is: what constitutes a nation? *Identity* (2011) attempts to question who is included in that view of nationhood. This is especially pertinent since most cultural identities are constantly changing and are a result of past, present and future encounters. Cultural identities are also dependent upon migrants, circulation of objects, ideas, food, and global economic powers and politics. As Sellars states:

> This is why the arts exist. Art is about going into those difficult spaces and trying to describe those things that we don't know how to describe. And part of the reciprocal gaze is the challenge. Having the permission to say, well, what isn't there? Why? What are we also looking for? What are the questions this work is not able or willing to answer? We can demonstrate that a visitor does not only have to receive a masterpiece but is in a position to challenge it. Through reciprocity we demonstrate equality (2010, 109).

It is the knowing and unknowing that takes us on the journey to knowledge/s. Here, several questions were explored regarding how museums contribute to education and inclusionary practices within and across different cultural communities and knowledges. *Chou Hayda* (2018) offered an entry point through intervening in the linear history of viewing objects, by inviting participants to be an 'authority' and offer a resounding voice for selected objects. Daou invited local communities within the vicinity of the museum who may have never set foot inside the space to the show. The production of an audio guide provided the entry point for locals, offering a sense of accessibility and engagement. In *Identity* (2011), the target audience was secondary school students who were invited to engage with racist scenarios and expand their understanding of cultural heritage and comprehend differences. The various objects and videos used provided a platform for viewers to engage, reflect, and document their own experiences. The show offered space to create a transformative experience encouraging self-reflexive responses by the attendant viewers. Thus, the show contributes to educational and inclusionary practice within and across different cultural communities.

As demonstrated here, both exhibitions have grown out of local ambitions. In *Chou Hayda (2018),* the TAP, an active platform creates a point of mediation between locals, civil society, and artistic interventions. The organisers are keen to invite locals to engage with public spaces. Daou's intervention in the museum's collections allows for an evocative, new way of engaging with archaeological artifacts. In fact, in both exhibitions dialogue was instigated regarding cultural heritage and past histories. Even more so, self-reflection was highly encouraged and expected. If we think of museums as contextual and intersectional, the requirement is to create a rendezvous with art that inspires a transformative experience. For instance, student responses for the *Identity* (2011) exhibition demonstrated a capacity for self-reflection (Schorch, 2015). Both exhibitions provide effective ways of exploring deeply enriching questions in addressing who we are and what kind of narrative each of us carries. Similarly, objects hold a significant narrative value and meaning. Asking the question "Chou Hayda" which translates to "what is this?" in English, is a way of inviting viewers to share their own perspectives and narratives through historical objects that have a distinct past. *Chou Hayda (2018)* offers audiences multiple perspectives in a way that encourages encounters, narratives and new interpretations. The answers and responses to the questions posed by Daou, are all recounted in Arabic. The intervention subverts the dominance of English in the global art world and presents a decolonial strategy to include local voices, stories, and visions. The strategies used by both exhibitions offer a resounding impact for local and cross-cultural communities in both local and global settings. A transformative educational experience where the knowing becomes the knowledge and the knowledge precipitates knowledges.

References

Abdelhady, D. (2011). *The Lebanese Diaspora*. New York University Press.

Abi Khalil, A., Daou, A., & Mishlawi, N. (2019). 'Sounding the Museum: A Shared Reflection on the Chou Hayda (What is this?) Intervention at the National Museum of Beirut', *Curator: The Museum Journal., 63*(3), 367–376.

Ang, I. (2005). The predicament of diversity: Multiculturalism in practice at the art museum. *Ethnicities, 3*(5), 305–320.

Ang, I. (2016). What are Museums for? The enduring friction between nationalism and cosmopolitanism. *Identities, 24*(1), 1–5.

Appadurai, A. (1986). *The social life of things: Commodities in cultural perspective*. Cambridge University Press.
Bauer, A. A. (2019). Itinerant objects. *Annual Review of Anthropology, 48*, 335–352.
Clifford, J. (Ed.). (1997). *Routes: Travel and translation in the late twentieth century*. Harvard University Press.
Elias, A., Mansouri, F., & Sweid, R. (2020). Public attitudes towards multiculturalism and interculturalism in Australia. *Journal of International Migration and Integration* [Open Access]. https://doi.org/10.1007/s12134-020-00784-z
Farhat, M. (2009). Imagining the Arab world: The fashioning of the 'War on Terror' through art. *Callahoo, 32*(4), 1223–1231.
Hage, G. (1998). *White nation: Fantasies of white supremacy in a multicultural society*. Routledge.
Haugbolle, S. (2005). Public and private memory of the Lebanese Civil War. *Comparative Studies of South Asia, Africa and the Middle East, 25*(1), 191–203.
Henrich, E. (2011). Identity: Yours, mine, ours review. *reCollections, 6*(2). https://recollections.nma.gov.au/issues/vol_6_no_2/exhibition_reviews/identity
Hooper-Greenhill, E. (2000). *Museums and the interpretation of visual culture*. Routledge.
Larkin, C., & Parry-Davies, E. (2019). War Museums in postwar Lebanon: Memory, violence, and performance. *Nationalism and Ethnic Politics, 25*(1), 78–96.
Levitt, P. (2015). *Artifacts and allegiance: How Museums put the nation and the world on display*. University of California Press.
Mansouri, F., & Lobo, M. (2016). Introduction social inclusion: Exploring the concept. In M. Lobo (Ed.), *Migration, citizenship and intercultural relations: Looking through the lens of social inclusion* (pp. 1–10). Routledge.
Markus, A. (2014). Attitudes to immigration and cultural diversity in Australia. *Journal of Sociology, 50*(1), 10–22.
McFadzean, M. (2019). Purposeful memory-making: Personal narratives of migration at Melbourne's immigration Museum. In K. Darian-Smith & P. Hamilton (Eds.), *Remembering migration: Oral histories and heritage in Australia* (pp. 255–270). Palgrave Macmillan Memory Studies.
McKernan, A. (2020). 'A little more courageous': Cultural recognition and histories of migration and diversity at the immigration Museum, Melbourne. *Museum Management and Curatorship, 35*(1), 2–14.
Mejcher-Atassi, S., & Schwartz, J. P. (2012). *Archives, Museums and collecting practices in the Modern Arab World*. Routledge.
Message, K. (2014). Returning to racism: New challenges for Museums and citizenship. In L. Gourievidis (Ed.), *Museums and migration: History, memory and politics* (pp. 44–66). Routledge.
Muller, N. (2013, June 27). The many Metamorphoses of Mounira Al Solh: Mounira Al Solh in conversation with Nat Muller. *Ibraaz*. http://www.ibraaz.org/interviews/82
Noble, G., & Ang, I. (2018). Ethnicity and cultural consumption in Australia. *Continuum, 32*(3), 296–307.
Pasquini, E. (2019). Beirut Museum reclaims Lebanon's ancient history. *The Washington Report on Middle East Affairs, 38*(6), 63–63.
Pharès, J. (2003). The National Museum of Lebanon in Beirut. *Museum International, 55*(3-4), 38–43.
Rogers, S. (2007). Out of history: Postwar art in Beirut. *Art Journal, 66*(2), 8–20.
Sadek, W. (2016). *The ruin to come: Essays from a protracted war*. Motto Berlin.
Salibi, K. (1988). *A house of many mansions: The history of Lebanon reconsidered*. I.B. Tauris.
Schorch, P. (2015). Experiencing differences and negotiating prejudices at the immigration museum Melbourne. *International Journal of Heritage Studies., 21*(1), 46–64.
Sellars, P. (2010). Museums, educators and social action. *The Journal of Museum Education, 35*(1), 103–110.

Tahan, L. G. (2010). Challenging colonialism and nationalism in Lebanese Archaeological Museums. *Near Eastern Archaeology, 73*(2/3), 195–197.
The Economist. (2017, August 10). *A Museum of memory in Beirut gets off to a Troubled Start.* https://www.economist.com/news/books-and-arts/21726053-no-one-can-agree-what-beit-beirut-or-what-objects-should-be-display-there
Tohme, C. (2015). Nouveaux Niches. *Artforum International, 53*(9), 320–323.
Toukan, H. (2010). On being the other in post-civil war Lebanon: Aid and the politics of art in processes of contemporary cultural production. *Arab Studies Journal, 18*(1), 118–161.
Walton, J., Paradies, Y., McFadzean, M., Molloy, J., Sproul, L., Priest, N., Kowal, E., Mansouri, F., Kelaher, M., & Meehan, C. (2016). *Identity: Yours, mine, ours exhibition research project evaluation report.* Museum Victoria.
Wilson-Goldie, K. (2007). The war works: Videos under Siege, online and in the Aftermath, again. *Art Journal, 66*(2), 68–82.
Winegar, J. (2006). Cultural sovereignty in a global art economy: Egyptian cultural policy and the New Western interest in art from the Middle East. *Cultural Anthropology, 21*(2), 173–204.
Winegar, J. (2008). The humanity game: Art, Islam, and the war on terror. *Anthropological Quarterly, 81*(3), 651–681.
Witcomb, A. (2007). 'A place for all of us?' Museums and communities. In S. Watson (Ed.), *Museums and their communities* (pp. 133–156). Routledge.
Witcomb, A. (2019). Oral history and first-person narratives in migration exhibitions: Tracking relations between "Us" and "Them". In K. Darian-Smith & P. Hamilton (Eds.), *Remembering migration: Oral histories and heritage in Australia* (pp. 203–217). Palgrave Macmillan Memory Studies.
Young, P. (1995). Beirut's past comes to life. *History Today, 45*(10), 4–6.

Part III
Learning in 360 Degrees

Sacred Groves of the Tolon District of the Northern Region, Ghana: Where Spirituality Meets Education for Sustainable Development

D. Addae and Janet J. McIntyre-Mills

Abstract The paper discusses the potential of indigenous customs in the Tolon district of the Northern of Ghana to support government legislation to promote the country's sustainable development agenda. It is estimated that throughout the district sacred groves are protected by the belief in the power of Jaagbo. The case study of the Tolon District is discussed with reference to Douglas' notion of sacred and profane and the implications for fostering Bateson's ecological mindset through public education that builds on indigenous values and customs and mainstreams the sense of awe and respect for the systemic role that forests play in protecting the wellbeing of living systems.

Keywords Sacred · Profane · Wellbeing · Systemic

1 Introduction and Statement of the Problem

Sustainability for the purposes of this paper is defined in terms of living within the limits of a sustainable ecosystem that is able to re-generate resources for all the inhabitants of a region. It is defined in non-species centric terms and based on a sense that these limits are sacred.

D. Addae
Department of Adult Education and Human Resource Studies, University of Ghana, Legon, Accra, Ghana

Department of Adult, Community and Continuing Education, University of South Africa, Tshwane, South Africa
e-mail: daddae@ug.edu.gh

J. J. McIntyre-Mills (✉)
College of Education, University of South Africa, Pretoria Central, Gauteng, South Africa

College of Professions, Adelaide University, Adelaide, Australia
e-mail: janet.mcintyre@adelaide.edu.au

J. J. McIntyre-Mills, Y. Corcoran-Nantes (eds.), *Transformative Education for Regeneration and Wellbeing*, Contemporary Systems Thinking,
https://doi.org/10.1007/978-981-19-3258-8_16

The paper explores the public education opportunity to build on the traditional customs. Without forests for photosynthesis, carbon capture, maintaining water table, climate change will continue to accelerate. Drawing on Arko-Achemfuor (2021 forthcoming), the paper provides an outline of Ghana's environmental conservation regulations and highlight the significant complementary role of indigenous values and customs in conserving forests in the country. We then examine how Jaagbo – a twin deity and a well demarcated sanctuary for these gods – more than any other factor, has helped to safeguard the forest from human destruction. The degradation of other sacred groves in surrounding areas as a result of human activities (Dorm-Adzobu et al., 1991) has accentuated the spiritual significance of the Jaagbo shrine and forest. Based on this, we examine specific taboos, rituals, customs and practices which have helped to sustain the conservation of the 'Jaagbo forest'. Subsequently, we discuss, how through an indigenous research methodology which borrows greatly from Bateson's (1972) idea of metalogue – a conversation between the authors and also by the first author with the participants about their understanding of their relationship with nature and to weave an understanding of their interactions to achieve a better multispecies balance.

With the help of the various narratives from a key informant from *Tolon* – the study site – we discuss how their traditional beliefs serve as a buffer for preserving the Jaagbo forests. Finally, we discuss the potential for mainstreaming indigenous values in a range of contexts from public education, primary, secondary, tertiary and vocational levels.

The first author reflects:

> In 2020, I received an invitation from Janet to contribute a chapter to a book she was developing titled, 'Transformative Education for Regenerative Development: Pathways and Perspectives.' This call couldn't have come at a more difficult period for me. The rise in COVID-19 cases had me pondering over several questions:
>
> How can life be sustained? How can the multiple species survive a massive global tsunami resulting from climate change? How did the world reach this current state of ecological degradation? Given my interest in indigenous knowledge, what indigenous practices, beliefs and customs are helping to protect what is left of the environment? How can transformative regenerative education help to heal or regenerate the world? We can no longer deny the fact that the current human demands for environmental services have serious environmental consequences, while we often forget that 'we are part of a social system, and our experiences shape our values which in turn shape our economies and the environment for better or worse' (McIntyre-Mills & Romm, 2019, p. 5). Luiz Marquez (2020) in his book Capitalism and the Environmental Collapse exposes how man's search for immediate economic satisfaction has led to unimaginable assault on the 'ecosystems of Earth, which in fact support humanity and all of life, and without which we would perish' (Scott, 2020, p. v). This, Marquez argues, is contributing to an environmental decline unfavourable to the multispecies of the Earth. Lamarck (1744–1829) – one of the foremost naturalists – made this observation of and prediction about the causal link between industrial civilization and environmental collapse:
>
> By his egoism too short-sighted for his own good, by his tendency to revel in all that is at his disposal, in short, by his lack of concern for the future and for his fellow man, man seems to work for the annihilation of his means of conservation and for the destruction of his own species. In destroying everywhere the large plants that protect the soil in order to secure things to satisfy his greediness of the moment, man rapidly brings about the sterility

> of the ground on which he leaves, dries up the springs, and chases away the animals that once found their subsistence there. He causes large parts of the globe that were once very fertile and well-populated in all respects to become dead, sterile, uninhabitable, and deserted. (…) One could say that he is destined to exterminate himself [and other species], after having rendered the globe uninhabitable. (Marquez, 2020, p. 5)

This classical statement and similar comments have been made recently nauseam, but it is an inescapable reality facing humanity.

We need to ensure that our environment does not only become sustainable – satisfying the needs of mankind today without necessarily compromising the quantity and quality of natural resources available to future generations – but is able to regenerate – feeding forward support for future generations through restoring the social and environmental fabric … it refers to nurturing young people, their community and the environment on which they depend (McIntyre-Mills, 2019, p. 140) – and thrive. We conceive both sustainability and regeneration as responsibilities bestowed upon human beings by virtue of our existence – we have to ensure that this earth will continue to support multi-species existence. The choice of the title for this chapter, therefore, is informed partly by the above questions and more importantly by an encounter I had with a colleague regarding a particular forest reserve in Ghana and how the beliefs, customs and practices of the indigenous people have helped to preserve the patches of forests.

This indigenous system has survived centuries of unadulterated influence by other religious systems notably Islam and Christianity – which are widely practiced in the country. There is no doubt that 'when two different religious systems exist in one community, the tendencies of conflicts of socioreligious, ideological, physical, and even political characters will arise between them'. And indeed in many communities in Ghana, this conflict is often subtle and covert; the various religious systems have co-existed on parallel lines based on mutual respect and non-congruent relationship. Traditional religion in Ghana shares commonalities all the 'features of world traditional religion, including its beliefs, sacred myths, oral qualities, strong appeal to the hearts of adherents, high degree of ritualization, and possession of numerous participatory personages such as officiating elders, kings, priests, and diviners' (Okeke et al., 2017, p. 2).

2 Forest Preservation in Ghana: A Confluence of Legal and Traditional Regulations

Ghana is recognized as one of the most advanced tropical African countries in terms of established forest policies, legislations, forest inventory, and management planning systems (Boadi et al., 2017). Dating back to 1909, the country's long and cherished history of formal forest management practices is manifested by the existence of 283 forest reserves under the Ministry of Lands and Natural Resources (Boadi et al., 2017). Furthermore, the forest reserve system which was established in the 1920s and 1930s is one of the most extensive in sub-Saharan Africa, yet much of the protected forest, especially in the reserve has been degraded (Dorm-Adzobu et al., 1991).

Notable legislations such as the Environmental Protection Act, 1994; Forest and Wildlife Policy, 2011; and Forest Protection Act, 2002 are all part of the government's grand Sustainability Agenda. In addition to these notable state efforts at preserving the environment in general and forests in particular, local conservation has also played a role as it has been an important part of the cultural ethos in many remote rural and indigenous communities (Singh et al., 2017). Local conservation relies on indigenous norms reinforced by taboos which prohibit certain types of interactions with nature. Compliance with these taboos rests in the belief in the supernatural as couched in the local worldviews or knowledge systems which forms the basis for safeguarding the way traditional institutions function (Aniah & Yelfaanibe, 2018). Dorm-Adzobu et al., (1991) argues that these traditional regulations (e.g. taboos) established centuries ago for using and protecting the forests are well observed by local people, preventing bushfires, farming, grazing and most hunting and gathering afforest resources from degrading the integrity of the forest ecosystem. Popularly known as sacred groves, these forests are based on the cultural ethos which

> ascribes nature a sacred place and contains a system of taboos and adherences that preserves portions of the biosphere. ... Through these sacred groves there is potential to recognize the role of culture in biodiversity conservation. (Barre et al., 2009, p. 27)

Within the African Traditional Religion, certain forests – known for their biodiversity – are believed to be the residing places for several supernatural beings. These places are revered for their sanctity and some are even institutionalised through the development of shrines as representation of the gods, and Priest(s) to serve as mouthpiece of their mouthpiece. All things regarded as sacred are deemed clean and prohibits certain perverse acts. This underscores the reverence the local people associate with those places. Local people's interactions with their environment are guided by norms, taboos and values. While these measures unconsciously help to ensure preservation of forests by providing the values, worldviews, or environmental ethics that shape the way in which different societies interact with biological diversity and nature in general, Berkes (2001) argues that religious traditions have little to say specifically about biodiversity (Negi, 2010).

But in fact, religious beliefs, practices and customs have played a major role in promoting conservation of various patches of forests in Ghana. Furthermore in Jens Koehrsen's (2015) paper titled, Does religion promote environmental sustainability? – Exploring the role of religion in local energy transitions, the interconnection between religion and environmental sustainability was explored. Prominently discussed in the paper was the issue of value dissemination – even though the degree of influence could not be established – as a cardinal function of religion in sustainability efforts – what was referred to as subliminal influences (Koehrsen's (2015).

3 Sacred Groves

Sacred groves are found in many places across Ghana. Gordon (1992 estimated there are 2000–3200 sacred groves in Ghana, approximately 80 percent of which are located in the southern half of the country (Nganso et al., 2012, p. 38). Sacred groves as noted by Nganso et al. (2012) serve vital ecological and sociocultural functions by preserving virgin forests that provide refuge for rare and useful local biodiversity, and sources of herbs for medicinal, social and religious purposes (p. 1). These are 'sacred places where trees and plants were allowed to grow undisturbed' (Nayar, 1987, p. 4) and are an embodiment of the spiritual makeup of the indigenous peoples. They are believed to the house communities' god(s) and ancestral spirits hence their social reverence. Hagan (1998) and Grim (2000) maintain that sacred groves derives from the desire of local people to live in harmony with numerous spirits that are believed to be associated with the natural environment, especially forests (Sarfo-Mensah et al., 2010, p. 2). While the significance of many sacred groves in the country is tied to the belief in a supernatural being, others are largely associated with certain 'holy' animals.

> For instance, 'the Buabeng Fiema sacred grove (or monkey sanctuary) has a strong association with 'sacred' monkeys ... and the fear associated with retribution from the god believed to own the primates, have partly served to protect the monkeys and preserve their habitat' (Sarfo-Mensah et al., 2010, p. 18).

In the Northern part of Ghana (made up of Savannah, Upper West, Upper East, North East, and Northern Regions), many sacred groves are deeply rooted in the belief in certain deities who are believed to protect the people from all adversities, and so their resting place (the forests) should not be desecrated with human activities. In these areas, Aniah et al. (2014) observes that traditional institutions comprises chiefs, elders, Tindaana (earth priests), soothsayers, diviners and sorcerers, rainmakers, traditional healers (including herbalists) as well as clan/family heads (Aniah & Yelfaanibe, 2018). These people are accorded immense respect due to their unique traditional positions in the community and serve as gatekeepers in the community's quest to respect the sanctity of the homes of their gods. Through sets of beliefs, practices, customs, values and traditions, these people help foster culturally-inspired conservation of forests due to the belief of their spiritual significance in the lives of the people. As Campbell (2005) notes, a powerful reason for the perceived value of sacred groves is their preservation of forests that may predate those of the surrounding, usually more modified landscape.

Through sets of beliefs, practices, customs, values and traditions, the indigenous people unconsciously foster culturally-inspired conservation of forests due to the belief of their spiritual significance in the lives of the people. As Campbell (2005) notes, a powerful reason for the perceived value of sacred groves is their preservation of forests that may predate those of the surrounding, usually more modified landscape.

4 Methodology

"Zoom anthropology" is an ironical term coined by the second author when reflecting with Corcoran Nantes, that in some ways it enabled her to bring together people from many parts of the world to discuss shared concerns and to connect where possible if data associated costs and access to the internet made it possible.

The first and second author engaged in a series of *emailed* conversations whilst the first author made attempts to recruit a key informant for the study. The first author combined safe engagement with the selected key informants about how the belief in a spiritual being perpetuates forest conservation behaviours among indigenes of a community in the Northern part of Ghana and how it weaves into the larger Sustainable Development Goals (SGDs). The discussions give only a brief insight as the distance between Accra and Tolon (almost a 12 h drive) made it difficult to connect during the pandemic. A further consideration is the distrust that the information (if not suitably general) could be misused by developers and the paper draws attention to the need for knowledge management across cultures and disciplines (Gibbons et al., 2010; Hoppers, 2013).

5 Case Study of Tolon and Jaagbo Sacred Grove

The Tolon District of the Northern Region of Ghana is home to several clusters of indigenous vegetation revered for their spiritual significance. The main vegetation in the district is grassland, interspersed with guinea savannah woodland, characterised by drought-resistant trees such as acacia, (Acacialongifolia), mango (Mangifera), baobab (Adansoniadigitata Linn), shea nut (Vitellariaparadoxa), dawadawa, and neem (Azadirachtaindica) (Government of Ghana, 2014, p. 19).

In the various communities in the district, culture and nature are inseparable in the worldview of local indigenous people. To ensure regenerative development in communities in the area, certain patches of forests are protected by government and/or traditional regulation. It is worth noting that sacred groves all over the country host biodiversity of various categories (birds, insects, mammals, plants, etc.) at higher levels than neighbouring unprotected land, and many sacred groves are home to ecosystems native to Ghana that no longer exist outside the groves (Nganso et al., 2012, p. 4). However, recent infrastructural developments in the district have not only resulted in the wanton destruction of the environment but also threaten the very survival of sacred groves in the area.

This case study draws on a series of conversations with a key informant in the Tolon about the significance of how the belief in *Jaagbo* – a twin deity and the name of the shrine located in the patches of forests in the community – has helped preserve the Jaagbo sacred grove. The name of the grove is derived from the name of the twin gods believed to be residing in the grove and the set of beliefs helps to protect the grove and the animals within it from human destruction.

Tolon is one of towns constituting the Tolon District of the Northern Region of Ghana and also serves as its administrative capital. The district shares borders with North Gonja District to the West, Kumbungu District to the North, Central Gonja District to the south and with Tamale Metropolitan to the East. The district is home to several clusters of indigenous vegetation revered for their spiritual significance. In the various communities in the district, culture and nature are inseparable in the worldview of the indigenous people. To ensure regenerative development in communities in the area, certain patches of forests are protected by government and/or traditional regulation. It is worth noting that sacred groves all over the country host biodiversity of various categories (birds, insects, mammals, plants, etc.) at higher levels than neighboring unprotected land, and many sacred groves are home to ecosystems native to Ghana that no longer exist outside the groves (Nganso et al., 2012, p. 4). However, recent infrastructural developments in the district have not only resulted in the wanton destruction of the environment but also threaten the very survival of sacred groves in the area.

Tolon is however one community where human activities – including agriculture which is the predominant occupation of people – have not affected patches of forests considered sacred to the people. While the Islamic faith is practiced among many people in the village, it has not affected their indigenous belief systems which have guided their personal and communal lives for many centuries, nevertheless one informant stressed that both Christianity and Islam could be used to justify abandoning the belief in the indigenous customs that protect the forest.

Located in the Tolon district is a popular ecotourism site called the Jaagbo sacred grove. For centuries this grove has had cultural significance in the lives of many people living in the Tolon area. The Grove is approximately 11 kilometres square and is predominantly savannah reverie forest or woodland with a stream running East West direction serving as source of water for some communities fringing it (Tolon District Assembly, 2014, p. 40). It is a land area of untouched forests which shelters varied medicinal, endangered and mysterious plants. The site is rich in diverse flora and fauna notably serving as a home for crocodiles (Jaagbo crocodile pond). As early as the 1990s, over 220 plant species were discovered in the Jaagbo sacred grove (Schaaf, 2003). It is a taboo to cut down trees in the forest for logging, or hunt for animals in the forest. However, people could harvest parts of the trees and other specific plants for medicinal purposes. To do this one must seek permission through the Tindana (land owner or an intermediary between man and the gods) from the deity through the priest. The Tindana cannot act unilaterally in this instance; instead he seeks permission on behalf of the person in need by providing an offering to appease Jaagbo. Wahabu (2013, p. 6) states that "according to oral tradition, before the settlement of the Mole-Dagomba and the establishment of the institution of chieftaincy in this part of the country, the people lived an acephalous or head-less life with the Tindana playing the role of symbolic head who seeks answers to the challenges of his people through the gods". The Tindana's role in the cultural structure of the Mole-Dagomba cannot be understated; in fact they are seen as the custodians of the land – land has both social and religious significance in the life of the community. The traditional beliefs of the people of Tolon recognise a Supreme Being known as Naawuni (God) who made tingbani nama t1ri (land and all that exists in it). Naawuni then made daadama (human beings) and made them stewards over all His creation. Tigbani (land) has spiritual connection with man because it is the source and 'sustainer' of life. Below the Supreme Being are Nature gods who inhabit several objects such as rocks, trees, rivers, etc. Man relies on natural resources for sustenance, without which man cannot survive.

The belief system stresses the spiritual significance of Jaagbo who is the god of Tolon. According to oral tradition, somewhere in the 15th century, communities in the Tolon area experienced two epidemics. On consulting the soothsayers in the area, it was revealed that the people had neglected Jaagbo. Subsequently, when the necessary rituals and sacrifices were made by the people to the gods, the diseases disappeared from the communities. The Jaagbo Sacred Grove and Shrine, apart from serving the spiritual needs of the local people, is also an ecotourism site generating resources for the community and it has become a place of study for tours by the schools in the area because of its diverse plants and animals, especially the bird sanctuary (Tolon/Kumbungu District Assembly, 2006 in Wahabu, 2013, p. 5).

5.1 Customs, Rituals and Practices

Traditional African cultural practices are generally built into ways of conserving and protecting natural resources against overexploitation through the use of taboos and totemic affiliation with localities and wild flora and fauna species (Kideghesho, 2009, p. 87). Taboos and totems are indications of 'don'ts' instituted to guide the actions of local people in their interactions with the environment. Taboos and totem enjoins local people from harvesting certain plants and killing and consuming identified animals. Their reverence derives from the culture of the people. Bulow (1991) asserts that 'in some ways almost all of the taboos are religious in nature since they are part of a right way of living' (p. 191). They serve as warning signs for community members to desist from certain acts which are deemed by the community as abhorrent. As has been observed by Boafo et al. (2016) community taboos are considered necessary management strategies for reducing over-harvesting of critical provisioning services as well as restoring endangered animal and plant species.

Customs and rituals are specific social behaviours, practices, and ceremonies performed on a regular basis by individuals or specialized people within the study communities (Boafo et al., 2016). They are meant to connect the individual/group to the spirits. Gupta and Agrawal (2017) contend that spirituality helps in seeing the individual importance and inevitability of each segment of economy, environment and companionship.

The pervading belief is that the Jaagbo is a god of protection for all people in the Tolon area. In times of need, the god manifests himself to people in the form of other creatures (e.g. as human beings, animals, etc.). Typically, this traditional knowledge has been transmitted orally from generation to generation, often in the form of stories, songs, folklore, and proverbs, as well as direct training of youth by elders (Parrotta et al., 2016, p. 2). Traditional knowledge transmission reflect the various taboos (chiha), customs, rituals and practices that define the very essence of people's place in the spiritually significant land they occupy. From birth, the child is taught to appreciate the belief system which reinforces an awareness of and compliance with certain social rules that guide behaviour which enculturates younger generations into the worldview of members of the community.

6 Metalogue with the Key Informant

There was an attempt to engage the Tindana and other community leaders in the area on the topic of interest. However, the first author was informed by the Tindana that he and other community leaders in the area had been directed by the Chief of Tolon not to discuss anything pertaining to Jaagbo with outsiders because they feel they have been exploited for information in the past. However, the transcript below offers a brief, grounded understanding of the importance of Jaagbo to the local people of Tolon:

David: How significant is Jaagbo to the people of Tolon?

Respondent: Jaagbo is very important not only for the people of Tolon but Dagbon as a whole. In times past, Jaagbo formed an integral part of the system of sacred grooves that were seen as "protecting" the people of the local areas as well as Dagbon in terms of protection and general wellbeing in health, fortune, procreation, rainfall, etc. Procreation is an interesting one. It is believed that couple who struggle to have children can go and make special request or appeal, through the right custodian, to Jaagbo for children. When this happens the first child that the couple have is named Jaagbo, and it is believed that this child has strong connection to the spirit of Jaagbo throughout their lives. I have a cousin sister name after Jaagbo!!! Just as any sacred groove in Dagbon, Jaagbo also serves an important function of dispute resolution, crime control as well as any social vice. This function is activated by simply asking litigants to swear by Jaagbo if they are telling the truth in whatever it is that is being litigated upon.... [The] general custom regarding groves in Dagbon is reverence towards the grove. Such that nothing is done to disturb the peace and sanctity of the spirits or ancestors.

David: In what ways does the belief in Jaagbo help to preserve the 'Jaagbo forests'?

Respondent: Like I said … the cardinal act in grove management in Dagbon is that no one should do anything to disturb the peace of the spirit and ancestors. Therefore, given that the forests in the grove provides home.. and ecosystem for the spirit, destroying the forest harms the spirits and ancestors. So, it is prohibited for people to cut down or destroy the vegetation in the groves. Therefore, the belief is a simple symbiotic relationship whereby the people preserve the forest, which in turn "nourishes" the spirits and ancestors and the latter ultimate provide the various benefits to the people as I outline in my response to the first question.

Janet: This response is very similar to the custodians of the sacred Tathe Forest in Venda who believe that they are related to the forest through their totems and that this makes the forest and the creatures within it, part of their family. eating a totem is more than a taboo, it is a violation of trust and family relationship with other sentient beings. In fact it goes further than just organic life, it also includes the inorganic life as the two are part of one cycle.

This appears to be confirmed because when David asked about specific taboos prohibiting the harvesting of forest resources in the Jaagbo forests the respondent explained that there is no one taboo, it is a whole way of life, not just "specific taboos that prohibit harvesting of forest resources within Jaagbo".

David: In what ways has contact with other cultures affected belief in Jaagbo?

Respondent: Other cultures? I will say foreign religions: Islam and Christianity. I will say that these two religions are the biggest threat to the belief system surrounding sacred groves in Dagbon. Those Dagbamba who strongly profess Islam or Christianity would say they do not belief in Jaagbo or sacred grooves generally. It means that these groups of people will most likely not know or will violate any taboos meant to help manage the grove.

Janet: Interestingly some branches of Christianity and Islam focus on the sanctity of all life. Those who follow St Francis of Assisi or the mystics in Islamic faith as well as some current leaders who have declared a Fatwa (ban) on the cutting of the forest in East Kalimantan. I heard this when I attended the Annual Council of Islamic Studies Conference (Balikpapan, East Kalimantan 2014) stressed the need to protect the habitat of indigenous animals and to preserve agricultural land, because urbanisation poses a threat to food security.

The theme of sustainability was raised by David who asked how the Jaagbo forest could be sustained for the next generation. The respondent explained that:

> The vegetation in and around Jaagbo is tied to the belief in Jaagbo. Therefore, the sustenance of the vegetation depends on the sustenance of belief in Jaagbo. I must say that besides Islam and Christianity, urbanisation and associated land markets also pose a threat [to]the existence or sustenance of Jaagbo. Experience, with "Kpalvogu" (another important groove very close to Tamale), has shown that urbanisation pressures can easily destroy Jaagbo. Therefore, a practical way to preserve Jaagbo for posterity for the chiefs who allocate customary land in Dagbon to not be seen to attempt allocating land near Jaagbo for *any development*. Doing so, like it happened with Kpalvogu, will open the floodgates for increased human activity that ultimately will deplete Jaagbo of its vegetation.

7 Discussion

As detailed in chapter "Nurdles and Food Security: Education for Wellbeing and Implications for Pan-African Social and Environmental Justice" of this volume Ostrom's (1992) local design principles[1] can be undermined by big corporations and therefore protecting the ecological fabric of life needs to be buttressed by international law (see Higgins et al., 2013 and chapters "Fluid, Organic Thinking and Relationality: Implications for Education and International Relations" and "Systemic Praxis and Education to Protect the Commons" of this volume).

A further protection for forests is added by a sense of a priori norms that are rooted in Indigenous culture and religion. In Venda (see Romm & Lethole, 2021) and Lethole (2021, personal communication) referred me to the Gaia Foundation[2] and the award they presented to indigenous leader Mphatheleni Makaulule who works in VhaVenda with *Makhadzi* (senior women custodians) to restore indigenous values and culture. The Gaia Foundation explains that These women are:

> custodians of *sacred* natural sites such as indigenous forests, rivers, springs and wetlands. (Author's emphasis)

They are also

> custodians of traditional knowledge about seeds and soils – including planting, harvesting, seed selection, storage and saving. Together they are now recuperating local seed varieties....[They] .... stand firm against some of the threats to their land, culture and livelihoods, such as mining projects in the region.

The Gaia Foundation explains:

> Mupo means bowl of life all of creation, including the cosmos. It is a name, a principle and a practice.

---

[1] Forsyth and Johnson (2014) explain that:

> Ostrom's main point was that individuals will be more likely to create and conserve the commons when they have credible and reliable information about the costs and benefits of resource decisions and (crucially) when they have an opportunity to decide the rules of the game.

They explain that she drew on her research to identify design principles:

> ...that would improve the effectiveness and sustainability of common property regimes: clear resource boundaries (i.e. knowing physical and ecological properties of the resource); clear rules of membership (knowing who is entitled to use the resource); congruence between rules of provision/appropriation and local conditions; arenas for 'collective choice'; mutual monitoring; 'graduated' sanctions; mechanisms for conflict resolution (i.e. ensuring that resource users are able to monitor and sanction other resource users); and finally, a state that is willing to recognize (or at least not challenge) local rights of organization (ibid.: 90).

[2] https://www.gaiafoundation.org/mphatheleni-makaulule-awarded-global-leadership-award-in-new-york/

The notion of *sacred* and *profane* are concepts popularised by Mary Douglas in "Purity and Danger" (1978) in which she elaborates the importance of perceptions of the sacred and the beliefs in taboos which if ignored or disobeyed would result in danger for the individual and the group.

Purity refers to acceptable thinking and behaviour within the boundaries of local customs. Dirt or pollution refers to the unacceptable and dangerous. Belief systems can serve to protect people and their environment. For example in East Kalimantan the second author was told in 2014 when she gave a paper on critical thinking and multicultural education at the Conference of Islamic Scholars how the forest and creatures were protected in terms of Islamic Law. She was asked what I thought of the fatwa that had been declared against those who destroyed the forest and how important it is to protect orangutans. I agreed and explained how this was very similar to approaches taken in the societies studied by Mary Douglas. Sanctions guide perceptions, decisions and behaviour.

It is worthwhile reflecting on what modern, Western societies perceive to constitute risk. We face an "existential risk" (Bostrom, 2011) as a result of climate change and as we lose habitat, increased displacement and loss will impact food security.

The need to protect the global commons from so-called 'free riders' will become all the more important as we face the next decade. According to Jane Goodall the 2020 pandemic underlines the need to protect habitat as destruction of habitat leads to species that have never previously been in contact being exposed to pathogens for which they have no immunity. Many of the pandemics that human beings have faced in recent decades have resulted from the containment and abuse of farmed animals such as pigs (Swine flu), chickens (bird flu) and cattle (so-called Mad Cows disease caused by feeding cattle offal), whereas Aids has been linked with hunting bush meat and Covid 19 has been linked with trafficking animals and keeping them in confined close contact with species that have never previously been in contact, such as bats and pangolins (Goodall, 2020).

Why is any of this relevant to a discussion of beliefs that protect sacred groves?

The answer is because natural habitat is an essential part of the global commons. According to Nobel Peace Prize winner Elinor Ostrom (2009):

> It was long unanimously held among economists that natural resources that were collectively used by their users would be over-exploited and destroyed in the long-term. Elinor Ostrom disproved this idea by conducting field studies on how people in small, local communities manage shared natural resources, such as pastures, fishing waters and forests. She showed that when natural resources are jointly used by their users, in time, rules are established for how these are to be cared for and used in a way that is both economically and ecologically sustainable [3]

"Governing the commons" (Ostrom, 1990) is one of the greatest challenges, but it involves more than contractual arrangements amongst those who make a shared living from fishing or managing a forest resource, it needs to protect vast tracts of

[3] Elinor Ostrom – Facts. NobelPrize.org. Nobel Media AB 2021. Tue. 9 Mar 2021. https://www.nobelprize.org/prizes/economic-sciences/2009/ostrom/facts/nable.[27]

commons that are used not for human direct economic benefit, but which are indirectly essential for human survival.

8 Beliefs and Perceptions for Transcendence and Transformation

According to Kenneth Boulding (1956) human evolution will require transcendence. In his so-called skeleton of science he mapped complexity from the inorganic to the organic and the increasingly complex designs that human beings can create. But he also alluded to the layer of complexity added by values and perceptions. If we are to evolve we will need to understand that out transcendence will require a transformation in values

Rituals and taboos that remind people of the sacred can play a vital role in creating a sense of respect for other species. It reminds human beings that they are a strand in the web of life and thus they learn humility.

Ostrom's focus is on 8 design principles based on the idea that the common pool would be delimited to those with rights to use it. Unfortunately in the real world no matter what local agreements are reached outside powers, such as large multinationals can make governance deals with the state and the market that override the local.

Thus local written and unwritten sanctions need to be supported at a post national and regional level if the commons are to be protected.

A priori norms and values can be supported by A posteriori requirements that govern actions and are sanctioned in laws that protect the commons. One such law is the Ecocide Law (Higgins et al., 2013) that has been proposed and accepted by a few nations.

The perception that destroying the commons will cause Existential Risk needs to be explained so that there is increasing acceptance of the need to accept Ecocide Laws. Even if the law is accepted nationally by more nations this would be a first step

Distrust of nations for one another has led to stalling on laws to halt climate change. Nations were of course founded as governance structures for war and trade. Post national structures will need to be created to protect human security.

Water, food and energy security are vital determinants for life across the evolutionary scale. They cannot be protected simply within the boundaries of one nation state. The perception of shared risk need to be developed locally, nationally and post nationally.

9 Potential for Mainstreaming Indigenous Values Through Education and Law

Nussbaum's (2011) ten capabilities are directly concerned with wellbeing, a life worth living and the extent to which these can be addressed. These capabilities, include: (1) living a life that is not cut short prematurely (2) bodily health, (3) bodily integrity, (4) sense, imagination, and thoughts,(5) emotions, (6) practical reason, (7) Affiliation, (8) links to other species, (9) Play, (10) Control over one's environment. In Nussbaum's version, the Capabilities Approach focuses on "the protection of areas of freedom so central that their removal makes a life not worthy of human dignity" (p31).

Public Education to protect the forest is supported by the traditional belief in customary obligation and stewardship based on strong emotional connections with the forest and all the living systems it supports. It would be valuable to link Indigenous Knowledge Systems(IKS) explicitly with the Sustainable Development Goals.

9.1 Ways to Translate Educational Capabilities as A Priori Norms *and* A Posteriori Measures *of Assessment*

IKS support a values shift in understanding that the commons are not merely a resource that needs to be managed and that it also provides the fabric of life on which all species depend. Just as human beings need shelter and an ecological niche, so do other species.

The value of studying, documenting and recognising the wisdom and value of the notion of sacred and profane is in helping human beings to know their place within a living system.

The wisdom of traditions that inspire respect and shape perceptions of awe have a place in protecting the fragile balance of individual and collective needs.

Multiple species require safe habitats within ecological niches. A web of living systems is needed to support human life and human beings need to recognise their interdependency and their stewardship role.

Domestic and farm animals require safe habitat and decent living conditions that they share living alongside human beings.

The liminal creatures who share our suburban gardens or cities require protection and

The wild creatures require corridors and spaces that remain wild habitat as Donaldson and Kymlicka (2011) stress in their treatise on rights. Without these diverse niches for other species, human beings are at risk as biodiversity and resilience go hand in hand (Scott, 1998). Monocultures in conceptual and geographical terms (Shiva, 2012a, b, 2020) have been discussed by Vandana Shiva in a range of publications on the need to protect seed diversity and the environments in which they thrive.

10 Conclusion

The paper explores the public education opportunity to build on the traditional customs that promote environmental conservation. Historically, human beings have been faced with the question of how best to manage the environment to ensure its sustainability. As has been stated by Pope John Paul II:

> The earth will not continue to offer its harvest, except with the faithful stewardship. We cannot say that we love the land and then take steps to destroy it for use by future generations.

As faithful stewards, we need to recognise that without forests for photosynthesis, carbon capture, maintaining water table, climate change will continue to accelerate. As a result, we need to ensure that our actions, even though globally uncoordinated, ultimately lead to long-term protection of the ecosystem. Evidently, the exigencies of our time requires that approaches for addressing global environmental challenges such as climate change, food and water security, land erosion, and biodiversity loss need to acknowledge that natural resources are part of complex and highly dynamic social-ecological systems (Berkes & Folkeet et al., 1998 in Delgado-Serrano et al., 2017). However, current economic and infrastructural development models have failed to incorporate environmental externalities which are exerting increasing pressures on natural resources at local and global scales (Delgado-Serrano et al., 2017). In Africa, forests are being 'harvested' on a massive scale to satisfy complex human demands. Blessed with many natural resources – oil, gold, diamonds, bauxite, manganese, iron, cobalt, uranium, etc. – many African countries see these resources as the panacea to their development woes. Therefore, necessary conditions have been created for these resources to be exploited by both local and international companies. This has led to unimaginable levels of environmental destruction with adverse effects on human lives. However, there is a realisation in recent times of the importance of forests for the socioeconomic development of societies; ameliorating global warming, providing income, recycling water and oxygen, and serving as homes to most of the world's plants and animals (Asante et al., 2017). Resultantly, several sweeping measures have been instituted in many African countries to conserve the environment. Largely, 'forests in most countries are managed through formal scientific methods, which include the use of silvicultural principles enshrined in policies, regulations, and international conventions, while relegating the indigenous system that entailed cultural beliefs and practices to the background' (Asante et al., 2017, p. 2). Effective forest management in Africa cannot solely be attributable to a single approach (Western-inspired), instead other less respected (indigenous) approaches have helped protect patches of forests in many communities. Kendie and Ghartey (2000) argue that the wider mistrust of traditional values in supporting natural resource protection has negatively affected humanity's pursuit of environmental sustainability (Asante et al., 2017). This is because the environment is shaped by the culture and vice versa and that 'the extent to which people participate in natural resources management depends on the approach used by conservation institutions' (Mariki, 2013, p. 3).

In the past three decades however, there has been a profound realisation among academics and policymakers of the value of indigenous beliefs and practices in preserving indigenous forests. Ruiz-Mallén et al. (2014) and Wily and Dewees (2001) note that this these village-scale reserves and other locally managed protected areas as potentially more effective environmental conservation tools than large, state-driven parks: less prone to apathy or resistance (Kelly, 2020, p. 203). This is even more profound in Ghana, where several groves (estimated to be around 2000–3200) nationwide have been acknowledged as 'sacred' to their indigenous people. Amoako-Atta (1998) and Gordon (1992) report that nearly all the forest reserves in Ghana have close links with sacred groves and/or sociocultural ties with the local communities' zone (Boadi et al., 2017).

One of the most defining features of all African societies is that of cultural diversity through which the common concern for protecting the environment needs to be fostered by exploring *common concerns* for current and future generations of life and an understanding of interconnectedness and co-dependency on nature. By seeing nature as part of one's family, indigenous knowledge systems have an important message for the next generation.

Bibliography

Acheampong, E. O., Macgregor, C. J., Sloan, S., & Sayer, J. (2019). Deforestation is driven by agricultural expansion in Ghana's forest reserves. *Scientific African, 5*. https://doi.org/10.1016/j.sciaf.2019.e00146

Algraini, S., & McIntyre-Mills, J. (2019). Education policy and governance of secondary schools in Saudi Arabia: A critical review informed by Nussbaum's capabilities approach. In *Democracy and governance for resourcing the commons: Theory and practice on rural-urban balance* (pp. 295–369). Springer.

Amoako-Atta, B. (1998). Preservation of sacred Groves in Ghana: Esukawkaw Forest Reserve and its Anweam Sacred Grove. *Paris UNESCO Division of Ecological Sciences South-South Co-operation Programme, 40*, 3.

Aniah, P., & Yelfaanibe, A. (2018). Environment, development and sustainability of local practices in the sacred groves and shrines in Bongo District: A bio-cultural study for environmental management in Ghana. *Environment, Development and Sustainability, 20*, 2487–2499.

Aniah, P., Yelfaanibe, A., & Abindaw, B. A. (2014). Impact of climate variability on smallholder households and indigenous coping strategies in Bongo District. *International Journal of Development Research, 4*(3), 693–699.

Arko-Achemfuor, A. (2021 forthcoming). Balancing the interests of wildlife and humans resulting in sustainable ecotourism: The case of Boabeng-Fiema monkeys' sanctuary, Ghana in McIntyre-Mills, J.J and Corcoran-Nantes. From polarisation to multispecies relationships. Springer.

Asante, E. A., Ababio, S., & Boadu, K. B. (2017). The use of indigenous cultural practices by the Ashantis for the conservation of forests in Ghana. *SAGE Open*. https://doi.org/10.1177/2158244016687611

Barre, Y. R., Grant, M., & Draper, D. (2009). The role of taboos in conservation of sacred groves in Ghana's Tallensi-Nabdam district. *Social & Cultural Geography, 10*(1), 25–39. https://doi.org/10.1080/14649360802553194

Bateson, G. (1972). *Steps to an ecology of mind*. Ballantine.

Berkes, F. (2001). Religious traditions and biodiversity. In S. A. Levin (Ed.), *Encyclopedia of biodiversity* (pp. 109–120). Elsevier. https://doi.org/10.1016/B0-12-226865-2/00231-5

Berkes, F., & Folke, C. (Eds). (1998). *Linking social and ecological systems: Management practices and social mechanisms for building resilience*. Cambridge University Press.

Boadi, S., Nsor, C. A., Yakubu, D. H., Acquah, E., & Antobre, O. O. (2017). Conventional and indigenous biodiversity conservation approach: A comparative study of Jachie sacred grove and Nkrabea forest reserve. *International Journal of Forestry Research, 2017*. https://doi.org/10.1155/2017/1721024

Boafo, Y. A., Saito, O., Kato, S., Kamiyama, C., Takeuchi, K., & Nakahara, M. (2016). The role of traditional ecological knowledge in ecosystem services management: the case of four rural communities in Northern Ghana. *International Journal of Biodiversity Science, Ecosystem Services & Management, 12*(1–2), 24–38. https://doi.org/10.1080/21513732.2015.1124454

Bostrom, N. (2011). *Existential risk prevention as the most important task for humanity.* Faculty of Philosophy & Oxford Martin School University of Oxford. http://www.existential-risk.org/

Boulding, K. (1956). General systems theory – The Skeleton of science. *Management Science, 2*(3), 197–208. Retrieved March 9, 2021, from http://www.jstor.org/stable/2627132

Bulow, E. (1991). *Navajo taboos*. Buffalo Medicine Books.

Campbell, O. (2005). Sacred groves for forest conservation in Ghana's coastal savannas: Assessing ecological and social dimensions. *Singapore Journal of Tropical Geography, 26*(2), 151–169.

Delgado-Serrano, M., Mistry, J., Matzdorf, B., & Leclerc, G. (2017). Community-based management of environmental challenges in Latin America and the Caribbean. *Ecology and Society, 22*(1), 4. https://doi.org/10.5751/ES-08924-220104

Donaldson, S., & Kymlicka, W. (2011). *Zoopolis: A political theory of animal rights.* Oxford University Press.

Dorm, A. C., Ampadu-Agyei, O., & Veit, P. G. (1991). *Religious beliefs and environmental protection: The Malshegu sacred grove in Northern Ghana*. WRI Washington, DC, USA and Acts Press, Africa Centre for Technology Studies, Kenya.

Douglas, M. (1978). *Purity and danger: An analysis of the concepts of pollution and Taboo*. Routledge and Kegan Paul.

Forsyth, T., & Johnson, C. (2014). Elinor Ostrom's legacy: Governing the commons and the rational choice controversy. *Development and Change, 45*, 1093–1110. https://doi.org/10.1111/dech.12110

Gaia Foundation. (n.d.). https://www.gaiafoundation.org/mphatheleni-makaulule-awarded-global-leadership-award-in-new-york/. Accessed 23 Mar 2021.

Gibbons, M., Limoges, C., Nowotny, H., Schwartzman, S., Scott, P., & Trow, M. (2010). *The new production of knowledge: The dynamics of science and research in contemporary societies*. SAGE Publications Ltd. https://doi.org/10.4135/9781446221853

Goodall, J. (2020). https://www.cnn.com/videos/us/2020/03/19/coronavirusjane-goodall-acfc-full-episode-vpx.cnn

Gordon, C. (1992). Sacred groves and conservation in Ghana. *Newsletter of the IUCN SSC African Reptile & Amphibian Specialist Group, 1*, 3–4.

Government of Ghana. (2014). *Ghana statistical service 2010 population and housing census*. District Analytical Report. Tolon District. Available from: www.statsghana.gov.gh/docfiles/2010_District_Report/Northern/TOLON.pdf

Grim, J. A. (2000). *Indigenous traditions and ecology. 2000 forum on religion and ecology*. Harvard Environmental Information Centre.

Gupta, K., & Agrawal, R. (2017). Sustainable development and spirituality: A critical analysis of GNH index. *International Journal of Social Economics, 44*(12), 1919–1939. https://doi.org/10.1108/IJSE-10-2015-0283

Hagan, J. E. (1998). *The Kogyae strict nature reserve*. The World Bank/WBI's CBNRM Initiative.

Higgins, P., Short, D., & South, N. (2013). Protecting the planet: A proposal for a law of ecocide. *Crime Law Soc Change, 59*, 251–266. https://doi.org/10.1007/s10611-013-9413-6

Kendie, S. D., & Ghartey, Y. B. (2000). *Culture and development occasional*. University of Cape Coast.

Kideghesho, J. R. (2009). The potentials of traditional African cultural practices in mitigating overexploitation of wildlife species and habitat loss: Experience of Tanzania. *International Journal of Biodiversity Science & Management, 5*(2), 83–94. https://doi.org/10.1080/17451590903065579

Koehrsen, J. (2015). Does religion promote environmental sustainability? – Exploring the role of religion in local energy transitions. *Social Compass, 62*(3), 296–310. https://doi.org/10.1177/0037768615587808

Mariki, S. B. (2013). Conservation with a human face? Comparing local participation and benefit sharing from a National Park and a State Forest Plantation in Tanzania. *SAGE Open*. https://doi.org/10.1177/2158244013512665

Marquez, L. (2020). *Capitalism and environmental collapse*. Springer. https://doi.org/10.1007/978-3-030-47527-7

McIntyre-Mills, J. (2014). *Systemic ethics and non-anthropocentric stewardship* (p. 270). Springer. http://www.sense.nl/gfx_content/documents/20161115_Uitgevers.xlsx.

McIntyre-Mills, J. (2015). Multicultural education: Challenges and opportunities for participatory education research: From clash of civilisations to co-creation and co- determination. *Participatory Educational Research (PER), 2015*(I), 101–131.

McIntyre-Mills, J. (2019). Dynamic weaving together strands of experience: Multiple mixed methods approaches to resilience and re-generation based on intra-, inter- and cross-disciplinary approaches. In J. J. McIntyre-Mills & N. R. A. Romm (Eds.), *Mixed methods and cross disciplinary research: Towards cultivating eco-systemic living* (pp. 15–58). Springer. https://doi.org/10.1007/978-3-030-04993-5_2

McIntyre-Mills, J. J., & Romm, N. R. A. (Eds.). (2019). *Mixed methods and cross-disciplinary research: Towards cultivating ecosystemic living. Springer*. https://doi.org/10.1007/978-3-319-58014-2

Miller, G., Jr. (2004). *Sustaining the earth: An integrated approach*. Brooks/Cole.

Nayar, P. K. B. (1987). *Religion, mythology and the ecosystem*. Government of India.

Negi, C. S. (2010). The institution of taboo and the local resource management and conservation surrounding sacred natural sites in Uttarakhand, Central Himalaya. *International Journal of Biodiversity and Conservation, 2*(8), 186–195.

Nganso, B. T., Kyerematen, R., & Obeng-Ofori, D. (2012). Diversity and abundance of butterfly species in the Abiriw and Odumante sacred groves in the Eastern Region of Ghana. *Research in Zoology, 2*(5), 38–46.

Nussbaum, M. (2011). *Creating capabilities: The human development approach*. The Belknap Press.

Odora Hoppers, C. (2013). Community engagement, globalisation, and restorative action: Approaching systems and research in the universities. *Journal of Adult and Continuing Education, 19*(2), 94–102. https://doi.org/10.7227/JACE.19.2.7

Okeke, C. O., Ibenwa, C. N., & Okeke, G. T. (2017). Conflicts between African traditional religion and Christianity in eastern Nigeria: The Igbo example. *SAGE Open*. https://doi.org/10.1177/2158244017709322

Ostrom, E. (1990). *Governing the commons: The evolution of institutions for collective action*. Cambridge University Press.

Parrotta, J., Yeo-Chang, Y., & Camacho, L. D. (2016). Traditional knowledge for sustainable forest management and provision of ecosystem services. *International Journal of Biodiversity Science, Ecosystem Services & Management, 12*(1-2), 1–4. https://doi.org/10.1080/21513732.2016.1169580

Romm, N. R. A., & Lethole, P. V. (2021). *Prospects for sustainable living with focus on interrelatedness, interdependence and mutuality: Some African perspectives. Chapter 6 in From polarisation to multispecies relationships in an era of mass extinctions*. Springer.

Ruiz-Mallén, I., Newing, H., Porter-Bolland, L., Pritchard, D., García-Frapolli, E., Méndez-López, E., De la Peña, A., Sánchez-González, M. C., & Reyes-García, V. (2014). Cognisance, participation, and protected areas in the Yucatan Peninsula. *Environmental Conservation, 41*(3), 265–275.

Sarfo-Mensah, P., Oduro, W., Antoh-Fredua, E. & Amisah, S. (2010). *Traditional representations of the natural environment and biodiversity conservation: Sacred groves in Ghana.* FEEM Working Paper No. 87. https://doi.org/10.2139/ssrn.1665065

Schaaf, T. (2003). UNESCO's experience with the protection of sacred natural nites for biodiversity conservation. The importance of sacred natural sites for biological conservation-Paris (UNESCO)-proceedings of the international workshop, Kunming and Xishuangbanna (pp. 13–20).

Scott, J. (1998). *Seeing Like a State.* Yale University Press.

Shiva, V. (2012a). *Monocultures of the mind.* Third World Network.

Shiva, V. (2012b). *Making peace with the earth.* Fernwood Publishing.

Shiva, V. (2020). Ecofeminism and the decolonization of women, nature and the future. Mar 25, 2020 – Uploaded by San Telmo Museoa.

Singh, S., Youssouf, M., Malik, Z., & Bussmann, R. (2017). Sacred groves: Myths, beliefs, and biodiversity conservation—A case study from Western Himalaya, India. *International Journal of Ecology, 2017.* https://doi.org/10.1155/2017/3828609

Stewart, J. E. (2020). Towards a general theory of the major cooperative evolutionary transitions, *Biosystems, 198,* 1–12. ISSN 0303-2647. https://doi.org/10.1016/j.biosystems.2020.104237. (https://www.sciencedirect.com/science/article/pii/S030326472030126X)

Tolon District Assembly. (2014). *Draft district medium term development plan (2014 – 2017).* District Planning Coordinating Unit.

Tolon/Kumbungu District Assembly. (2006). *Tolon/Kumbungu district assembly profile.* http://tolonkumbungu.ghanadistricts.gov.gh/

Wily, L. A. & Dewees, P. (2001). *From users to custodians: Changing relations between people and the state in forest management in Tanzania.* The World Bank, Policy Research Working Paper Series.

Wahabu, S. (2013). Exploring local belief systems for the benefit of land restoration: A case study of the Jaagbo Shrine restoration in the Tolon/Kumbungu District of Northern Region, Ghana. United Nations University Land Restoration Training Programme [final project]. http://www.unulrt.is/static/fellows/document/wahabu2013.pdf.

AA!!Venda Women and Social Enterprise: Stepwise Process to Regenerative and Sustainable Living

Patricia Lethole, Janet J. McIntyre-Mills, Rudolf Wirawan, and Mphatheleni Makaulule

Abstract The sacred lake and forest in Venda are threatened by the development of a coal mine. The paper makes the case why building a mine would be counterproductive for water and food security and that alternative forms of energy are feasible in South Africa. Furthermore, a green cyclical economy could provide a source of income for women through setting up an eco-village co-operative growing organic produce.

Keywords Public education · Sacred sites · Cyclical economy · Environmental protection · Village co-operative

---

Mphatheleni Makaulule: Indigenous Knowledge Project coordinator of environmental heritage with communities around Vhembe District/ Venda, The water podcast series | JoJo
The defender of sacred sites and water Faces of Africa – The Defender of the Sacred Sites – YouTube Zwa Maramani: Fundudzi | 11 March 2020 – YouTube.

P. Lethole
University of South Africa, Pretoria, South Africa

J. J. McIntyre-Mills (✉)
College of Education, University of South Africa, Pretoria Central, Gauteng, South Africa

College of Professions, Adelaide University, Adelaide, Australia
e-mail: janet.mcintyre@adelaide.edu.au

R. Wirawan
University of Adelaide, Adelaide, Australia

M. Makaulule
Indigenous Knowledge Project Coordinator of Environmental Heritage with Communities Around Vhembe District/Venda, Thohoyandou, Venda, South Africa

J. J. McIntyre-Mills, Y. Corcoran-Nantes (eds.), *Transformative Education for Regeneration and Wellbeing*, Contemporary Systems Thinking,
https://doi.org/10.1007/978-981-19-3258-8_17

1 Introduction, Statement of the Problem and Rationale

This small pilot was conducted at the height of the Covid Pandemic by setting up a face to face and virtual community of practice (Wenger, 1999; Wenger et al., 2009) spanning the University of South Africa and University of Adelaide in line with the UNISA ethics protocols (Reference 2021/06/09/90001745/22AM)[1] and supported in part by UNISA community engagement team including Professor NRA Romm.

The group focused on shared concerns, namely poverty, climate change, destruction of biodiversity and habitat. We used a combination of WhatsApp, email zoom meetings and face to face fieldwork conducted by Pat Lethole and Mphatheleni(Mphathe)[2] to foster a shared vision and a commitment to help one another and the community, defined as "all the inhabitants within a region" (Higgins et al., 2013) in order to protect people, plants, animals and their shared habitat. A recognition of our hybridity and interconnectedness provided the starting point for the collaboration (McIntyre-Mills, 2017; Romm & Lethole, 2021).

Rudolf and Janet explained that In Indonesia in West Java a civet cat who lives in the forest eats the coffee beans and their faeces fertilise the forest. All the species work together when we work with nature. Through confining and caging the civet cat for Luwak coffee – a cruel practice – we work against nature, because the forest is not fertilised, so eventually the forest becomes less resilient as it is not nourished. This leads to loss of habitat for many species, erosion, loss of ground cover and the eventual lowering of the water table. Crushing nature, or mastering her does not lead to success, it leads to the destruction of the web of life. This is what has been forgotten when we focus only on quick profits at the expense of people, plants and animals.

1.1 The Makhadzi – Senior Women Custodians

Pat explained to us when we met that "Aa!!" is a Venda phrase for greeting that is used by women. Thus, some greetings are gender sensitive. For men is ndaa! greeting other human beings. "Zwavhudi "means it is good or magnificent' in Venda, but it can also be used to say "good bye".

[1] All team members are named and a member of the ethics committee linked with a registered Tshidzivhe Bamboo Primary co-operative Limited 2021/600766/24.

[2] See ethics form approval number Ref: 2021/06/09/9000174522AM.

The women who form the leaders of the co-operative are Makhadzi or custodians of the local sacred forest in Venda who wish to protect the forest, the creatures and their habitat in line with their sacred Indigenous Knowledge.

The only male member in the co-operative is the traditional chief of the village who is also in charge of the management and protection of the indigenous functioning of the sacred forest within the region. The other members(each of them, being founding members of the co-operative and the local community) have specific totems with which they identify and which form their identity. The totems link them with the natural world they inhabit and links people with specific totems to protect wild animals, the sacred forest, trees and lakes. For instance, Mphatheleni's totem is *nguluvhe ya* Luonde (warthog of the Sacred mountain of Luonde) which might be interpreted as wild pig of Luonde (place/ sacred place of Luonde mountain) and Pat Lethole is (Ndou ya Vhutanda) an elephant of Vhutanda (see Romm & Lethole, 2021). Pat is also a custodian of the Vhutanda sacred forest within the Vhutanda territory.

Mphatheleni draws on (Ross, 2017: 111) and uses her own experience to explain that as a member of this community the totem "holds the spiritual connection and health of a person from birth, when growing up until death."

She explained that:

> "*Mutupo* means a totem of a *Vhongwaniwapo* or *Vhangona* (indigenous) clan amongst the Vhavenda, and those clans are identified by Zwifho (Sacred Natural Sites). According to totemism, the owners of the totem need to protect either a forest or animal to which they are identified. A totem is usually an animal or reptile and their wellbeing, and the welfare of their environment are linked through their responsibility to both the forest and their specific totem.
>
> Thus, it is not permissible for Mphathe to eat pork or I (Lethole) to eat elephant meat and it is also my responsibility to protect the Vhutanda sacred forest."

Mphatheleni continued by explaining that the meaning of *Mutupo* "refers not merely to the totem animal, it also refers to the 'the well-being of the whole of life in the *Zwifho* (Sacred Natural Sites), which in turn relates to the *tshikhodo* (praise) of the Clan'".

Mphatheleni draws on Ross (2017, 12, 27, 91, 115) in order to explain further that the "well-being of the of the clan members is linked (through their totem) to their *responsibility* to protect sacred sites (Zwifho)."

Mphatheleni describes her role as a traditional custodian who needs to:

> "show vhuthu towards others – this is caring for others based on gratitude towards trees, plants, nature and all others, people, a dog a cat, to my cow, I have to show feeling, warmth to others so they have a sense of belonging … …what you feel you would like for yourself, you must do for others…. This is the basis for the stewardship enterprise for the Thathe village."[3]

---

[3] https://drive.google.com/file/d/1Tmm2bBNG_ykrHLYVa_T8b8je0Pd8EjNN/view?usp=sharing

Pat shared in an email (after listening to the recording of our conversation):

> The concept [of ubuntu] is reciprocal because the role players work together for the benefit of all of them including trees, animals and all non- living things. Thus, there is an element of sharing ideas, assisting one another as a team

Pat groups organic and inorganic life into two categories but shows how they are mutually dependent. Living systems go through a cycle of being organic when alive and then providing inorganic matter to support life.

2 Unemployment and Business as Usual Versus Employment Through Re-generative Work

In South Africa unemployment remains one of the greatest challenges alongside climate change, the pandemic and growing dissatisfaction. According to a Reuters article (2021): [4]

"South Africa's unemployment rate jumped to a record high in the fourth quarter of last year, the statistics agency said on Tuesday, as the economy was further battered by the COVID-19 pandemic.

Statistics South Africa said the unemployment rate stood at 32.5% in the October-December quarter, meaning 7.2 million people were unemployed, up from 30.8% in the previous three months.…The figure was the highest since the survey began in 2008…."

The same report continues that if those who are "discouraged from finding work are included then the figure would be around 42.6% who are unemployed." It continues by stressing that employment even back in 2018 the International Labour Organisation had estimated that "60% of the global population" relied on the informal sector for survival.[5] This would be much higher in the wake of the unfolding pandemic.

The potential to work with nature through ecofacturing (Pauli, 2010) and to develop the potential of new digital technologies is important for empowering those who are currently marginalised.

[4] https://www.reuters.com/article/us-safrica-economy-unemployment-idUSKBN2AN0S1 February 23, 2021.

[5] https://www.ilo.org/global/about-the-ilo/newsroom/news/WCMS_627189/lang%2D%2Den/index.htm

2.1 Regeneration with Makhadzi Custodians

The development of mines, such as coal mines in Venda, *business as usual* and industrialised style farming methods are re-considered with the suggestion of alternatives that could protect the environment and the heritage of Venda by working with the Makhadzi, or traditional women leaders.

As a result of collaborative research with colleagues at UNISA Janet met Pat Lethole who was at the time meeting with Norma Romm. Together they shared ideas around promoting opportunities for women in Venda along the lines of building up a community of practice to support a circular green economy that could help to re-generate opportunities for those who are currently in need of opportunities in line with the United Nations Sustainable Development Goals.

Janet suggested that the greater the level of participation in social, economic and environmental decision making – linked with green social enterprise and supported by a community of practice – the better the outcomes in line with UN Sustainable Development Goals, such as Goal 1 to address poverty, Goal 3 (gender empowerment), goal 10 (structural inequalities) and Goal 17 to build partnerships.

2.2 The Risk of Coal Mining

In Venda coal mining is a risk to the water table and communities have been saying 'no' to coal in Venda since 2012. The Gaia Foundation has the potential to support women to advocate against coal and Mphathe drawing on a report explains that :

> "Environmentalists have already questioned the water and carbon footprints of South Africa's coal power stations.... [and]. The coal mining process is ... water intense and ... pollutes soil and water supplies as it spreads underground and flows into rivers and streams."

Mphathe was part of an action group supported by the Gaia Foundation that opposes coal mining. The arguments against coal mining are the need to protect Lake Fundudzi and Thathe Sacred forest in Venda. The need for women to be able to earn their own living from sustainable enterprises could help in preventing the mine.

3 Design and Methods

The Multiple Mixed Methods research aims to set up a community of practice with colleagues in South Africa to address poverty (in line with the UN Sustainable Goal 1) and to enhance the agency of women (SDG 3) by addressing inequalities (SDG10) by working with young people through developing partnerships (SDG 17).

The hypothesis is that groups can manage shared resources in such a way that the commons (habitat) for multiple diverse species can be maintained whilst simultaneously setting up a viable circular economy that enables those –who are currently marginalised and at risk of displacement – to establish their own income streams within the co-operatively run social enterprises that are supported as learning organisations/communities.

4 Mupo and the Local Context of Regenerative Leadership by Women

In terms of Venda custom, the father's sister was regarded as an elder with authority. This has been sidelined in recent years, according to Matshidze (2013: 14)[6] in her thesis on women and leadership in Venda. She highlights that in terms of the South African constitution that traditional leadership is important and that other legislation on "Traditional Leadership Government Framework Act and the Traditional Court Bill" also provides opportunities for empowering women further at the local, provincial and national level. In other parts of Africa such as Ethiopia, for example women held senior positions in the past (Woldegies, 2016) but empowerment needs to be based on restoring a more balanced society in which patriarchy does not silence women or limit their opportunities.

Lethole (2021, personal communication) referred the team to the Gaia Foundation and the award they presented to indigenous leader Mphatheleni Makaulule who works in Venda with Makhadzi (senior women custodians) to restore indigenous values and culture. The Gaia Foundation explains that these women are:

> custodians of sacred natural sites such as indigenous forests, rivers, springs and wetlands. (author's emphasis)

[6] "Chapter 12 of the South African Constitution Act 108 (1996) thus specifically acknowledges the institution of traditional leadership and its place and role in the system of democratic governance. It provides for the continued authority and functioning of such leaders in accordance with traditional law, within the broader legal framework, and for traditional leaders to participate at local government level. Further the Constitution also established a Council of Traditional Leaders. Other pieces of legislative enactments that reflect commitment on the part of government towards the strengthening of traditional leadership system include the Traditional Leadership Government Framework Act and the Traditional Court Bill. The primary objective of the 2003 Framework Act was to establish and recognise traditional councils at a local level, in order to align traditional rule and traditional councils more strongly with the principles of democracy and hopefully to remove some of the negative perceptions associated with traditional rule and traditional councils. Under this law, the premier of each province has power to influence the establishment of a traditional council in an area which he/she has recognised as a traditional community. Furthermore, the Act acknowledges a position for traditional leadership, not only within local government, but at the provincial and national levels as well."

They are also

> custodians of traditional knowledge about seeds and soils – including planting, harvesting, seed selection, storage and saving. Together they are now recuperating local seed varieties….[They] …. stand firm against some of the threats to their land, culture and livelihoods, such as mining projects in the region.

The Gaia Foundation explains:

> "Mupo means bowl of life all of creation, including the cosmos. It is a name, a principle and a practice. "Women have played and continue to play an important role in Venda. Mphathe was nominated for a leadership award."

The approach to ubuntu leadership is summed up in the phrase " I am because we are" (Ngunjiri, 2016). The relationship of ubuntu is extended to all living systems which requires moving away from business as usual by taking small steps towards re-generative living which requires more collective responsibility – not only for one another but for multiple species (see Macy et al., 2014; Kaza, 2020; McIntyre-Mills and Corcoran Nantes, 2021; Wadsworth, 2010; Wahl, 2019). Regenerative living is about moving away from individualism towards a more balanced way of life supported by managing resources from below and above to ensure that international goals are supported locally by engaged communities. Pat stressed that (Photos 1 and 2):

> "The preliminary meeting is very critical because it will assist us to pave the way forward. I also like you to know that we (Pat and Mphatheleni Makaulule) have already started to engage some chiefs, and it is bearing fruit. ….. Please be informed that the bamboo project is still conceived in my mind. There is this place in Venda where bamboo just grows like wild fire. The community just cuts it and throws it away":

Photos 1 and 2 Types of local bamboo. (Photo supplied by Pat Lethole)

The following is an extract from Letter of Support sent to the South African National Research Foundation by Pat Lethole to support the project (Box 1):

Box 1: Background on Indigenous Leaders and the Project

Mphatheleni "Mphathe" Makaulule is the winner of the 2013 Global Leadership Award, founded Mupo Foundation in 2007, established registered Dzomo la Mupo (Voice of the Earth)[7] as a community Based Organisation in 2015. For almost 30 years, Mphatheleni has been recording the indigenous knowledge of local elders. In 1998, Mphatheleni finished her teaching Degree from University of Venda, then went straight back to her communities to learn from elders about the culture: forests, rivers, rituals and in that learning she worked with school children and teachers to transfer the indigenous knowledge. In 2000, she established Luvhola Cultural village, and in 2007 she founded Mupo Foundation and co-founded Dzomo la Mupo ("Voice of Mupo"), a community-based organisation, based on the work she was doing with communities - including working with the VhaVenda – to revive their confidence and stand firm against some of the threats to their land, culture and livelihoods, such as destruction of sacred sites, threats of mining projects in the region. Dzomo la Mupo is the voice of rivers, mountains, insects, wildlife. They are the voice of clouds, stars and cosmos. They are the voice of Mother Earth and her communities. In 2006, Mphathe met with the African Biodiversity Network and The Gaia Foundation who have been working together to seek African solutions to the ecological and socio-economic challenges that face the continent. Mphathe worked with Gaia Foundation from August 2007 till 2014, at which point over-representations from Gaia Foundation did not work well with local communities. Presently Mphathe is volunteering with Dzomo la Mupo to carry on this same vision and activities. The protection of sacred natural sites is at the core of the work and Mupo are part of the growing global network of communities and their allies saying Zwifho are "No Go Zone Areas" and instead "Yes to Life, No to Mining! Mphathe is also pursuing her Master's degree at University of Venda on African Studies focussing on indigenous knowledge systems….

Pat Lethole has met Janet McIntyre in 2019 when she visited the University of South Africa to engage in community research with Norma Romm and Akwasi Arko-Achemfuor. They discussed the potential for the development of social enterprises based on bamboo cultivation and offered to introduce the participants who engage in sustainable social enterprises in West Java.

Janet helped to support community research in Manyeledi and Vryberg whilst scoping the potential for this project. Pat Lethole and Norma Romm contributed a chapter to a volume published by Springer called: "From

(continued)

[7] https://www.thedzomolamupo.org/home/

Box 1 (continued)

Polarisation to Multispecies Relationships" (2021). I am aware of the track record of the team members at University of South Africa in the field of social enterprise, Indigenous knowledge, community development and gender mainstreaming.

Due to the partnership across University of South Africa, the University of Adelaide and Universitas Padjadjaran, the team will be able to collect data that is directly relevant to the project enabling the creation of practical training on sustainable social enterprises and policy guidelines that not only can reach the participating social enterprises but can continue to be employed after the completion of the project.

The ability of the training materials to be provided using online and print based media will enhance access. By including women and young people's voices the research will address ways to empower and reveal their issues and frustrations in setting up a viable social enterprise that can serve as an inspiration create a learning community.

This project will signal that South Africa is serious in addressing ways to increase employment and gender mainstreaming and equality to ensure better societal outcomes. I believe that the objectives of the proposal are well thought out and argued. The strategies in achieving the stated objectives are robust. I look forward to the findings and believe that the research will lead to further work and contributions.

I highly recommend this proposal as I strongly believe it will enhance employment opportunities in Venda.

Regarding Pat and Mphathe's meeting, it is deeply rooted in the friendship of their fathers (Mr B.R Manyatshe (Pat's father) and Mr. R. A Makaulule (Mphathe's father) Pat summed up as follows: "We met on face book after losing my son after a gross negligence from the hospital. I felt that I had a void which needed to be filled by the Makaulules. I wrote on the messenger on the 11 July 2020 requesting for her contact details. From that day we started to share the information on organic farming and the potential to use bamboo (not wood) to protect the forest. We worked with Mphathe and the ommunity to establish Community Engagement. Though, the conception of the Community Engagement was a little bit skewed, Janet helped to enable it to be more cooperative. It is important to note that we met face to face on the 16 April 2021 for the first time, preparing for field research in Sundani village and we ended up also going to Tshidzivhe village as mentioned above. From the day we started to communicate until now, our conversation is on how best we can improve the lives of communities by facilitating their financial independence and protecting their natural environment."

5 A Project Rooted in Friendship, Gratitude and Healing

In the work that connects website[8] and her work book "Coming alive again" Joanna Macy (2014) stresses that the spiral for creating pathways towards healing begins with a sense of gratitude for our connections with others which helps to override grief and loss. The creation of a sense of community is vital for finding the courage and energy to move towards the regeneration of people and places.[9] This resonates with the community custodianship shown by Mphatheleni, a Makhadzi who protect their spiritual Mupo heritage (Box 2):

Box 2: Mphatheleni Reflects

Mphatheleni reflects:

"That was on 11th February 2021, late afternoon.... The voice said, 'Do you know who is speaking with you? '....

The respondent said "ni khou amba na nwana wa Vho Manyatshe" meaning you are speaking with Vho Manyatshe's daughter."

"I screamed, shouted with happiness because I know, I once received a Facebook messages about my father and Vho-Manyatshe. ..... She explained about her son and I quickly remembered that I listened to the radio news[10] and how that news affected many people around. I told Pat that the day when she spoke on the radio (about her son's passing), I felt so much affected, however I never knew that, that was about my relative....That conversation connected me and Pat, and I introduced Pat to my siblings.

The following day I went to my mother and told her. My mother when responding to the phone to Pat, she said that Vho Manyatshe was my teacher at school and he was a friend to her father. Pat's father was also a friend to my maternal grandfather. As a result of this information and meeting with Pat, we revived family friendship and renewed our relationship.

(continued)

[8] Macy, J. The work that connects website https://workthatreconnects.org/network/get-involved/

[9] An International Systems Sciences women's zoom circle set up a workshop drawing on Joanna Macy's groupwork on the cycle of regeneration. As a result of signing up to the zoom meeting (which I missed as it was midnight) I signed up to her website "The work that connects to learn more about her engagement and praxis. I discovered Joanna Macy's *Work of our Time* through reading Kaza's (2020) edited collection and was struck by synergies and parallels in our thinking about the importance of narrative for healing and how reflection can help people to imagine alternative scenarios. Her analogy of turning resonates with the Mahayana notion of making a difference through our relationships with others and to turn away from 'business as usual' by taking daily steps towards living in ways that promote wellbeing. Macy interprets Mahayana Buddhism's exhortation to respect Mother sentient beings as respecting mother earth. Her groupwork encourages a spiral approach from personal gratitude for what we have received, to reflection on how greed and anger result in most of the problems facing the world.

[10] The South African news highlighted the neglect of a young man (Pat's son) in a hospital during the pandemic.

Box 2 (continued)

One other exceptional role which Pat played in, my life, is that gratitude process has been happening since our interaction. It is so deep and profound that words cannot explain it. My meeting with Pat, is part of the healing process through our father's friendship. I was also in the same time going through some personal problems in the family. Pat became like a psychological counsellor, and I remember she would tell me words such as 'identity crisis, gratitude, connectivity'. We talked about the way Vhavenda culture is rooted in identity in the clan, gratitude to one another and connection with nature.

From February Pat and Myself call each other frequently with the aim of prompting sisterhood (mukomana na murathu-big sister and younger sister).

The Bamboo project took us even to revive the connectivity to my grandmother's clan, The Sundani village. If we refer to the notes on the Sundani visit, there is explanations on how do I relate to Sundani village and how Sundani elders acknowledge with pride Pat's father, as Vho-Manyatshe, who was well known.

Again, when we go to Tshidzivhe, a community with which I worked deeply with the members there, Pat's visit was embraced with happiness, but the meeting was also surrounded with the pain [caused] by the passing of Pat's son. Tears fell, and some feeling of healing was sensed. I hope I would not be wrong if healing can be acknowledged to accompany the Bamboo project as another way to pay tribute to Pat's son ..."

6 First Steps: Key Points from Zoom Meeting 28th March, 2021[11]

The meeting was attended by Mphatheleni (Mphathe),[12] Mulalo Tshikalange,[13]Pat Lethole,Norma Romm and Rudolf Wirawan.

> Mphathe explained that: "Bamboo grows well in Venda. This is interesting and motivational, because even though I do not know if there is special species of bamboo, but around here we have thin bamboo and other types (as detailed below)... we have many spots of bamboo which people do not use, people do not know how bamboo can be used.... For women who are just sitting and not knowing what they can do.... Many [of the] projects government is imposing [are not sustainable]...for example injecting chickens with chemicals so that they can grow in one month.....People cut forest ... [to plant] for chemical[ly grown] vegetables .. but bamboo is friendly for [the]environment ...."

[11] Meeting recording: https://drive.google.com/file/d/1zu2u-JBNeP06tBQ_Y2BIIOA7Idx4OpAd/view

[12] https://www.gaiafoundation.org/mphatheleni-makaulule-awarded-global-leadership-award-in-new-york/

[13] **Beekeeper & Pollinator Activist**: "NYOC Organics is an apiculture enterprise focusing on bee colony management, apiculture training & development, production of local raw honey and beeswax-derived organic products. We are dedicated to the conservation of bees and ethical production of organic bee products harvested from our own hives. Services Offered: Raw Honey. Beeswax and Propolis Balms. Bee Removals. Beekeeper Training"

Rudolf explained that the right bamboo needs to be planted in the right location so as not to do more harm than good, but it can be helpful as a source of fuel and building materials for housing.

Mphatheleni (Mphathe): " I like bamboo for electricity for constructing houses, because rivers are getting dry, because people are taking stones away and sand to build the houses…"

Rudolf: "Bamboo or mangoes…just let nature do her work… this is working with nature … the earth, rain and sun continues to work and they do not come to say we do not pay the bill…We just need to nurture nature – it gives life . …."

Janet: "Perhaps we could do brainstorming to ask women what they want to do – plant vegetables ? We can work virtually with women to find out what they want to do and set up a community of practice which means working together. Maybe bamboo ? Maybe planting vegetables and using bamboo pipes to water them? Maybe the bee hotels with the bee farmers? Perhaps the women could team up with people with building skills to make bamboo houses or a shop or cafe ? It seems Venda has the right conditions for eco-tourism ? Perhaps shops or a café selling products? The bamboo cafes could sell arts and crafts made by the women. It could be an excellent way to make a circular economy to make money for the women living in the community so that they can help their families. This could be an example so other people are inspired to find ways to work with the co-operative on behalf of the whole community….The discussion and presentation by Rudolf has given plenty of food for thought to work in a circular economy making money with the community co-operative."

"We welcome the opportunity to work with you – even a small project to inspire a handful of women….. "

The participants agreed that: "Bamboo can grow …[it could] improve holding ground water and replace firewood …"

It was also agreed that by helping the community it could prevent some of the mining companies targeting the community for development. The point was made that a coal source had been found in the area and instead of establishing a coal mine an alternative source of sustainable green enterpris would be better …

Finally it was agreed:

> "Yes, let us start small and 5 years from now starting to reap …..Yes let us start small but also raise awareness so that they do not have to mine."

Rudolf then explained that the sawdust from bamboo can be used like a candle as a light and that bamboo biochar forms a type of coal that can be used for an energy supply. It can also be used to filter water… (Box 3).

> Janet: "Small success stories around food, energy and water could be summarized to lobby and to work with NGOs and Government. We hope to do work with other communities – some want to work with vegetables, goat's cheese, bees and wild life. One community in Vryberg is working with the dis-Abled to support growing and selling organic vegetables. It is good to give women an opportunity but we also need to foster ways to support mainstreaming gender equality and opportunities for all who are marginalised….When the ethics process is finalized we can invite the women to talk about how they can work on developing a co-operative. We can invite women to Zoom with us and to take care because of Covid."

Box 3: The Visit to Sundani Village

The field research 17 April 2021

was facilitated and documented[14] by Mphatheleni Makaulule together with Pat.

Community visit at Ha- Sundani village.

"The purpose was to hold a conversation with the community about capacity building in social enterprises. Before this meeting Mphatheleni had already held a discussion with the acting chief and made the necessary arrangements for setting up meetings with the community as explained in her previous report. On the 17 April 2021, it was easy to meet the community because they were waiting for the bamboo project. The Focus Group conversion was conducted as mentioned below:

The twenty six participants comprised twenty females and six males. Their age ranges from 20 to 46 years old. One male elder who participated is 76 years old.

During the conversation, the males were participating fully whilst the women were just looking without contribution. I tried to probe the women to speak but it was in vain. I asked Mphathe why the women were so quiet. She mentioned that it was her first experience to be involved with the Sundani village. Her experience is with other communities and they [also]started this way. She explained that according to their culture they are not supposed to raise their voice in the presence of male counterparts because it can be interpreted as "u thanyela dubo"(being clever about unnecessary things). They also believe that they need to be submissive as mentioned in the bible.

Mphathe mentioned that she had to send the community members to the wilderness for training purpose. I (Pat) suggested that the Sundani community might be [ready] for training in the wilderness and also encouraged to engage with progressive Tshidzivhe community.

The bamboo is available, but I (Pat) suggest that we [also] grow a bamboo nursery in this community. The acting chief promised that will avail space for growing bamboo."

Focus Group Conversation with Acting Chief

"The acting chief raised awareness of unemployment in the village and poverty to be dealt with by ourselves (the team). She also mentioned that every household needs to be prepared to accommodate the opportunity that arises. However this business proposal will be discussed with the whole community on 02/05/2021. Mphathe will be given the feedback."

(continued)

[14]They translated the English questions to Venda Research questions, wrote the report and made the video.

Box 3 (continued)

Question 1. What do you think of the results of cutting trees, removing stones and sand from the rivers?

"They mentioned that destruction of nature by cutting trees, caused winds that can destroy houses and removing of stones and sand compromise the life of fish. Thus, life as a whole will be disturbed."

What can be done by community to prevent this?

"They mentioned that it is the responsibility of the community leader and its people to enforce rules and regulations which can safeguard the natural resources like Marula trees. As a results they intend to introduce strict rules to protect that."

Question 2. If I lived here, how many types of bamboo will I see?

"It was mentioned that they are three types of bamboo (lutanga, the thin one. Musengele the thick one and thanga the very thin one). The problem is that some are dying and some got burned during the wild fire."

Question 3. What do you think of making business from bamboo, clay,traditional bangles?

"They indicated they are prepared to plant bamboo on the sides of Letaba river in which they can plant a fowl run for their free range chicken. They have a wood crafter who is capable of making wooden spoons and different types of sculptures."

Question 4. What do you think what are the importance of Bamboo

"They said that the bamboo is very useful because it captures mist from the space which creates wetlands that increase water table. In the olden days it was used to build houses and cover them with cow dung."

Question 5. How can chicken farming be combined with bamboo farming

"It was indicated that there is a lot of free range chicken in their community. Almost all households do have these chicken but they never thought of making business with those chickens."

5.1. What do you think are the needs of chickens and their interests?

"They mentioned that chickens likes to run around, to lay eggs and also that the cock likes to sing in the morning standing on the poles. They mentioned that free range chickens needs a bigger space for moving around, for climbinga ladders, and nests for laying for eggs. So it means bamboo can also be used to build nests for eggs and stands. Pat explained that the group could consider building an eco- village which could comprise houses(owned by different households) built from bamboo. Bamboos can also generate electricity and charcoal."

(continued)

Box 3 (continued)

Closing remarks as a response from the Acting chief.

"She was thankful for the visit … She was also thankful for the possible employment which could be brought by working with universities …. She was adamant that cutting of trees like Marula and others will be stopped immediately... She promised that there would be better management of sand collections and that bamboo could be planted where the damage had already been done. She emphasized that we should consider that the project has already been welcomed in Sundani village."

Tshidzivhe Village

Date: 18 April 2021
Venue: Chief kraal

"They were 12 members of community which comprised three males(aged 50-60) and nine females (aged 26-60). The community is very organised. Mphatheleni trained them to work as a team which comprises team leaders of different categories. For example one group is dealing with indigenous trees and herbal plants. Another group deals with the protection of sacred sites.

Knowledge is also invested in youth, because intergenerational learning about indigenous knowledge (involves youth learning from elders about biodiversity and food security)."

Informal conversation

"We discussed the natural resources around the Tshidzivhe community. They mentioned that they have bees. L. shared with them the benefits of bee farming. So the knowledge about bees was discussed and the community were fully participating. The participants are well trained in how to work in groups. There is also limited patches of bamboo plants in the area. It was suggested that they establish nurseries for bamboo in appropriate places. The land to plant bamboo and to build with bamboo will be negotiated with the chief who will be involved in the project."

7 Making Progress: Zoom Meeting 7th June 2020[15]

The meeting was attended by Mphatheleni, Pat, Rudolf, Janet and Dr. Enid Pitsoane joined and was introduced to the others. She explained: "My preferred nickname is 'Bunkie' everyone calls me this and stressed that she was happy to help the project and had been invited by Pat to join as a member of the team. Dr. Pitsoane is a counsellor and has skills in conflict resolution that could be useful in the context of the

[15] https://drive.google.com/file/d/13QUeWtc3llgniSqpDSLZaGovvTVeHITZ/view?usp=sharing

research we agreed we could all learn from one another. Pat explained that there was conflict in one of the villages that had expressed interested in being involved and although this conflict had nothing to do with the proposed project it made her realise that it would be a good idea to ask someone with counselling skills to join the project. We agreed this was a good idea and welcomed Bunkie to the project."

Mphatheleni explained that they had successfully registered The Tshidzivhe Bamboo Co-operative Ltd..[16] She continued by sharing that:

> "of the five members only four are residents of Tshidzivhe village, three of the women members are dedicated to protecting the sacred Thathe forest. The chief and the other three women have relationships to the clan of Netshidzivhe and participate in protocols for the protection of sacred forests, however the three women come from other sacred forests and are not custodians of the Thathe sacred forest. Mphathe, as a member of the co-ooprative belongs to other sacred sites, however she is intrumental in the protection of many sacred sites through Dzomo la Mupo, a community based organisation that she founded. One of the custodians is also the wife of the local chief."

Maphateleni then explained drawing on her reading of Ross (2017: 12, 27, 91, 115) that:

> "In the Luvenda language sacred forests are called Zwifho and that she had held conversations with elders who are the custodians of the sacred indigenous forests and other natural sites including mountains, lakes ponds, waterfalls and caves."

Maphateleni continued by explaining that:

> "The *sacred sites* were created by *Ri sa mu di* or *Nwali* (a Creator or God) since the beginning of the creation of the universe. Ross (2017) explains that *Zwifho* are not human made and cannot be defined as *Zwiendeulu* (royal graveyards) or game parks. *Zwifho* thus means quite literally "to give and be given." Accordingly *Zwifho* can be said to be the root of our life and well-being. There is no written record which can say when those places become *Zwifho*, and who chose the people or the Clan to be the Custodians of the *Zwifho*. This has been respected and passed on by generations of our ancestors through our oral tradition, which is older than the written tradition, and must be respected as such,[17] because sacred *Zwifho* is a place where *Vhadzimu* (ancestors' spirits) are found. It is where *thevhula* and *u phasa* (ancestral rituals) are performed by a Clan chosen by *Ri sa mu di* or *Nwali* (Creator or God). When we do *thevhula* and *u phasa* (ancestral rituals), we ask for *shothodzo u ri shango li lale* (harmony and well-being, peace and health for all).....We the Custodian Clans uphold our responsibility to follow the order which originated with *Tsiko* (Creation of the Universe) that our ancestors respected and showed us the way to continue performing *thevhula* and *u phasa* (ancestral rituals) and respecting the ecosystem in the *Zwifho*. We the *Vhongwaniwapo* or *Vhangona* (indigenous) of Venda cannot live another life than to follow our *thevhula* or *u phasa* (ancestral rituals). …A *Zwifho* is not a place to be interfered with by any other activities or for entertainment. Since *Tsiko* (Creation of the Universe) the *Vhavenda* (Venda tribe of South Africa) respected these sites for many generations. The *Zwifho* are *No Go Areas*, even harvesting is not allowed. Now, despite a democratic government, the *Zwifho* are being disturbed and threatened, for example by tourism, development, plantations and mining. *Zwifho* is about sacredness not commercialisation….The

[16] Name of Co-operative is Tshidzivhe Bamboo Primary Co-operative Limited Reg no: 2021/600766/ 24.

[17] https://www.thedzomolamupo.org/home/

Commission for the Promotion and Protection of Rights of Cultural, Religious and Linguistic Communities (CRL)[18] and IUCN Sacred Natural Sites Guidelines for the Protected Area Managers 2008 clarify that there are different forms of Sacred Sites which should be recognised. The *Zwifho* or Sacred Natural Sites in Venda are not human made and cannot be replaced. The *Zwifho* carry the life and well-being of all communities of *Mupo* (all of Creation which is not human made)."

7.1 *Organising the Focus Group Discussions on Bamboo and Social Enterprise*

We agreed that when the focus group discussions are held, that each of the members of the focus group will register with their totem names, not their actual names. This will ensure confidentiality. Pat shared that her totem is Elephant and if more people shared her totem they would be elephant 2 or 3. Other local totems are kudu and wild pigs.

Janet explained that she was indeed familiar with the concepts of totems and that this Indigenous Knowledge System (Odessa Hopper, 2013) is vital for protecting wild life and their habitat. Totems guide people's thinking and practice in relation to one another and the environment in many parts of the world. In Australia, for example totems define kinship, affiliation to the land and the creatures who live within a region with the protection of their custodians. Holding totems sacred involves protecting one's namesake and not using it as a source of food. This helps to prevent over- harvesting or hunting specific flora or fauna and thus it helps to protect biodiversity.

We also agreed that the planned webinars would be a team effort and interpreted by Pat and Mphateleni who would focus on how bamboo could help to re-generate the local social, economic and environmental fabric of Tshidzivhe Village. We agreed that we could focus on a circular green economy and link all the messages by connecting their role as custodians with climate change, habitat protection, the United Nations Sustainable Development Goals, Gender Mainstreaming and Indigenous Knowledge Systems.

Rudolf stressed that the co-operative needs to serve the community and it needs to ensure that the *right* local bamboo is grown in the right location.

Clumping bamboo could be planted around the perimeter of running bamboo. Running bamboo should be restricted to areas that would not threaten the natural environment. Running bamboo must be used very carefully as a source for energy and regularly harvested for making biochar, a form of energy supply which could

[18] Commission for the Promotion and Protection of Rights of Cultural, Religious and Linguistic Communities, 'Sacred Sites Protection and Promotion of the Rights of Ownership, Access and Use', Vol 1 (2008), pg. 11.

provide the community with fuel. By harvesting it regularly the running bamboo could be controlled effectively and the sacred forest could be protected, because an alternative source of fuel could be provided for everyone. The clumping bamboo could be harvested for making houses or furniture as can the running bamboo. The first Webinar could make this point and we could discuss the implications with the women and those in the community who choose to work actively as part of the new social enterprise.

The notion of social enterprise will be discussed in the first webinar stressing that the aim is ***not*** to enrich a few people at the expense of others and the environment. The aim is to enable the five women custodians to empower and enable the whole community (comprising the people, plants and animals) to live harmoniously and well. In this way the community can develop wellbeing stocks (Stiglitz et al., 2010) for this generation and the next and play an active role in supporting the prevention of environmental destruction. The social enterprise (adapting Higgins et al., 2013) definition will define community as protecting "all the inhabitants" served by the co-operative to ensure that "peaceful enjoyment by the inhabitants" is promoted and not diminished. Thus social enterprise will be purposefully focused on promoting the wellbeing of people, plants, animals and their habitat whilst working with nature, rather than against it (Pauli, 2010).

The wider context of the bill to protect biodiversity is further explained by Greene (2019) who cites Higgins's[19] definition of ecocide as follows (Box 4):

> "1. Acts or omissions committed in times of peace or conflict by any senior person within the course of State, corporate or any other entity's activity which cause, contribute to, or may be expected to cause or contribute to serious ecological, climate or cultural loss or damage to or destruction of ecosystem(s) of a given territory(ies), such that peaceful enjoyment by the inhabitants has been or will be severely diminished. 2. To establish seriousness, impact(s) must be widespread, long-term or severe."

8 Conclusion

According to the fieldwork facilitators, the Sundani community needs some empowerment as the women still lack the confidence that is more evident in the Tshidzivhe. Potentially the two groups could assist one another and the Sundani community will be invited to pair with the other community.

It was suggested that the community facilitators develop a business plan with each of the two communities and that this be discussed with a local South African bamboo grower to discuss potential co-operation. A win-win solution is being

[19] *Ecocide Law*, MISSION LIFEFORCE, https://www.missionlifeforce.org/ecocide-law [https://perma.cc/L326-S4KA].

Box 4: The Website The Many Uses of Bamboo Have Been Summarised on Behalf of the Co-operative in a Website Created as Follows:
https://www.wirasoftfoundation.org/en_GB/web/ecovillage/venda

Rudolf explained that the website could be a resource for training and resourcing the community. He stressed that he would add resources and adapt it for them.

The website provides details on bamboo and links to some of the inspirational speakers on bamboo and its many uses. It also links with the work of Vandana Shiva on the importance of protecting local biodiversity and seeds.

1. Bamboo for social enterprise, social and environmental justice. Bamboo is good for energy so the forest is protected.
   Different types of bamboo
   Working with the environment
   Indigenous Knowledge Systems
2. Empowering the community and running the co-op
   Ubuntu, gender mainstreaming
   The circular economy
3. The eco village and the many linked enterprises

explored with advice from an organic farming support network PGS SA https://www.pgssa.org.za/ and enabling access to land so that the women grow their skills to manage a successful organic co-operative.

References

Flood, R. L., & Romm, N. R. A. (2018). A systemic approach to processes of power in learning organisations: Part 1 – Literature. *Theory, and Methodology of Triple Loop Learning the Learning Organization, 25*(4), 260–272.

Greene, A. (2019). The campaign to make ecocide an international crime: Quixotic quest or moral imperative? *Fordham Environmental Law Review, 30*(3), See https://ir.lawnet.fordham.edu/cgi/viewcontent.cgi?article=1814&context=elr

Higgins, P., Short, D., & South, N. (2013). Protecting the planet: A proposal for a law of ecocide. *Crime, Law and Social Change, 59*, 251–266. https://doi.org/10.1007/s10611-013-9413-6

Kaza, S. (2020). *A wild love for the world: Joanna Macy and the work of our time*. Shambhala.

Odora Hoppers, C. (2013). Community engagement, globalisation, and restorative action: Approaching systems and research in the universities. *Journal of Adult and Continuing Education, 19*(2 Autumn), 94–102. https://doi.org/10.7227/JACE.19.2.7

Macy, J., Young-Brown, M., & Fox, M. (2014). *Coming Back to life: The updated guide to the work that reconnects*. New Society Publishers.

Matshidze, P. E. (2013). *The role of Makhadzi in traditional leadership among the Venda*. Submitted in fulfilment of the requirements for the degree of doctor of philosophy in the faculty of anthropology and development studies at the University of Zululand.

McIntyre-Mills, J. (2017). *Recognising our hybridity and interconnectedness: Implications for social and environmental justice*. Current Sociology. https://uk.sagepub.com/en-gb/journals-permissions
McIntyre-Mills, J. J. & Corcoran Nantes, Y. (2021). *From Polarisation to Multispecies Relationships in the Age of Mass Extinctions*. Springer.
Ngunjiri, F. W. (2016). "I am because we are": Exploring Women's leadership under Ubuntu worldview. *Advances in Developing Human Resources, 18*(2), 223–242. https://doi.org/10.1177/1523422316641416
Pauli, G. (2010). *The blue economy: Report to the Club of Rome*. Paradigm Publications.
Romm, N. R. A., & Lethole, P. V. (2021). Chapter 6: Prospects for sustainable living with focus on interrelatedness, interdependence and mutuality: Some African perspectives. In *From polarisation to multispecies relationships in an era of mass extinctions*. Springer.
Ross. K. B. (2017). *Traditional terrains: Land, Gender and cultural biodiversity preservation in Venda, South Africa*. PhD thesis, Department of African American and African Studies, Michigan State University.
Stiglitz, J. Sen, A. & Fitoussi, J. P. (2010). *Mis-measuring our lives: why the GDP doesn't add up*. The New Press.
United Nations. (2007). *Declaration of the rights of Indigenous peoples*. https://www.humanrights.gov.au/publications/un-declaration-rights-indigenous-peoples-1
United Nations. (2003). Human Development Index. In *A compact among nations to end poverty*. Oxford University Press.
United Nations. (2014). *World urbanisation prospects: The 2014 revision*. https://esa.un.org/unpd/wup/publications/files/wup2014-highlights.PdfUnited
Nations Paris Climate Change. (2015). *Conference of the Parties Twenty-first session Paris*, 30 November to 11 December 2015. https://unfccc.int/resource/docs/2015/cop21/eng/l09r01.pdf
United Nations. (2017). *Sustainable development goals*. https://www.un.org/development/desa/publications/sdg-report-2017.html
United Nations Office for Disaster Risk Reduction. (2017). http://www.preventionweb.net/files/55465_globalplatform2017_edings.pdf
Wadsworth, Y. (2010). *Building in research and evaluation. Human inquiry for living systems*. Allen and Unwin.
Wahl, C. (2019). Sustainability is not enough. In G. D. Dabelko & K. Conca (Eds.), *Green planet blues: Critical perspectives on global environmental politics*. Taylor & Francis Group, ProQuest Ebook Central. http://ebookcentral.proquest.com/lib/flinders/detail.action?docID=5892383
Woldegies, B. (2016). Research highlighting options for gender empowerment in Amhara Regional State Ethiopia. *South African Review of Sociology*. https://doi.org/10.1080/21528586.2015.1131191, https://www.researchgate.net/publication/299419063
Wenger, E. (1999). *Communities of practice. Learning, identity and meaning*. Cambridge University Press.
Wenger, E., et al. (2009). *Digital habitats: Stewarding technology for communities*. CPsquare.

Websites

https://www.sourcewatch.org/index.php?title=Mphatheleni_Makaulule
http://earthlorefoundation.org/about-us/history/
Macy, J. The work that connects website https://workthatreconnects.org/network/get-involved/
https://www.reuters.com/article/us-safrica-economy-unemployment-idUSKBN2AN0S1
Rogan, M., & Skinner, C. (2019). https://theconversation.com/south-africas-informal-sector-creates-jobs-but-shouldnt-be-romanticised-122745

Women, Education and Socialisation: Transforming Gender and Power in the MST, São Paulo, Brasil

Yvonne Corcoran-Nantes

Abstract The question of women and leadership and raising women's profile in institutions and organisations has been orientated from the outside through the investment of money in training for empowerment and leadership. Yet much of women's capacities comes from initiatives at the grass roots from within the female contingent itself. Women whose foremost role in the community, and the family is often ignored or taken for granted yet these are skills acquired through the course of their lives and carried on through the interactive and proactive socialisation of women as lifelong learning. This was certainly the case in the Landless Movement of Brasil (MST) whereby women's representation and leadership occurred outside of the local into state and national level. To the contrary of leadership profiles in other organisations in which women predominate at the local level and are less evident outside of that. From the outset women's role in farming for subsistence and the market was fundamental to the success of the movement. Their support for one another and the interchange of ideas formed a strong basis for their prominence in the movement and its success. This chapter focuses on women's role in the movement, its development and longevity rooted in a transformative praxis that underwrote their philosophy and action over many years (All the material in this article came from field work undertaken in Brasil between 2001 and 2005 unless otherwise stated).

Keywords Women · Sustainable agriculture · Landless movement · Awareness · Skills · Leadership · Personal relations

Y. Corcoran-Nantes (✉)
College of Humanities, Arts and Social Sciences, Flinders University, Adelaide, SA, Australia
e-mail: yvonne.corcoran-nantes@flinders.edu.au

J. J. McIntyre-Mills, Y. Corcoran-Nantes (eds.), *Transformative Education for Regeneration and Wellbeing*, Contemporary Systems Thinking,
https://doi.org/10.1007/978-981-19-3258-8_18

1 Introduction

In Brasil, and elsewhere in the world, across the 1970s witnessed a growing tide of rural to urban migration as agrarian capitalists, looking to expand their holdings and access credit, scoured the country for land removing peasant farmers rights and their means of survival in the process. In the 1980s there was an explosion of urban unrest across the major cities of Brasil from which popular social movements emerged with political demands ranging from land rights, access to water and electricity, health facilities to creches and housing (Da Gohn et al., 2012; Da Gohn & Marcondes, 1995) From this milieu of popular protest came the Movimento dos Trabalhadores Rurais Sem Terra (MST) – popularly known in English as the Landless Movement. It is generally recognised as one of the most dynamic political movements in the fight for land in Latin America and globally. Since 1985 the achievements of the movement are well known, settling hundreds of thousands of landless families on unused land, gaining title to that land and building a just and decent society within those settlements based on a philosophy bifurcated between Marxism and Liberation Theology (Fernandes, 2000). Raising the awareness of its members of the cultural politics and a culture of literacy which should be imbued across the generations.

Anyone visiting the MST settlements is rightly impressed by its struggle and philosophy which has borne fruit but not without cost. Members of the movement, both male and female, over the years have lost their lives in the fight for land usually at the hands of the Agro capitalists, that command their own land defence army and control of local politics and the police, with impunity. Dissenting voices are culled, intimidated and, in some cases assassinated. This has not deterred people from continuing to join the Movement in the struggle for land, most having worked as virtual slave labour either for the large landowners or in disadvantaged rentier farming arrangements. In the light of all this the settlements offer a new reality, traditional agriculture at its best and with not a square meter of cultivable land without something planted on it. Certainly, by the late 1990s a commitment to agroecology through what Mcintyre Mills[1] refers to as an 'ecological mindset' has been nurtured by a cultural politics of necessity whereby the individuals survival and wellbeing is inextricably linked to the specific ecological environment in which they live and on which they depend.[2]

The MST given its foremost position in the fight for land and land rights has come under greater scrutiny than many others with respect to its claims to intersectionality and equality. The MST claims that it is committed to the 'promotion of gender equality, Indigenous culture and a healthy and sustainable environment' the question is does the 'promotion' of these issues contribute much to the realisation of such a commitment. For research that focuses on this issue especially with respect

[1] See Chapter "Nurdles and Food Security: Education for Wellbeing and Implications for Pan-African Social and Environmental Justice" in this volume.

[2] See, Hernandez, 2020.

to gender the question centres on whether the experience in the MST has sustained or challenged 'unequal relationships' (Caldeira, 2009; Deslandes, 2009). The question itself seems simple enough but supporting the argument either way is far more complicated than it appears. It also subscribes to a 'top down' impetus for social change when the philosophy of the MST and its members subscribes to a 'history from below' as proscribed by EP Thompson (Thompson, 1966). This is most pertinent when considering the question of gender relations and social change and identifying gendered political practice on the settlements which are the focus of this and other studies in the field.

The complication imbued within the analysis is multi-fold as the settlers in different locations have different experiences, cultural mores, and geopolitical contexts. For example, settlements in Rio – where urban rural migrant settlers predominate and other settlements in the south with predominantly rural agricultural workers/peasant farmers. Expectations differ as do eco-political strategies and forms of co-existence. The women's experiences past and present and the nature of the traditional patriarchal gender relations that shape them are important in the women's analysis of how far they have come and what they have achieved. This chapter will focus on the question of women's participation in the MST from the women's standpoint in the socio-cultural contexts of the rural settlements in the region of Itapeva in the interior of São Paulo. The narrative presented here relies on insider testimony and analysis wherein aspirations of leadership often take second place to the question of independence and autonomy.

2 Rural Women Mobilising Since the 1970s

Rural women were noticeably present in the rural protest of the sixties for agrarian reform especially when, under the military dictatorship, a land reform law was passed in 1964 in which a rural underclass of sharecropper, agricultural wage labour and tenants were to be the beneficiaries (Deere, 2003). Then as now a landed oligarchy has proved to be the principal nemesis of those who work the land and Brasil remains the country with the largest disparity in land ownership in the world. The fight for access to land and the question of low agricultural wages placed women centrally in these protests because not only did they constitute unpaid family labour but as food producers sustaining family subsistence levels they were more severely affected (Guivant, 2003). The issues remained central for over 20 years where access to land became a unifying political objective in the face of rural 'modernisation' and production for export.

It was in the 1980s that rural women's mobilisation was most evident in a rising tide of social protest which brought together both rural and social movements for a single cause – the end of dictatorship and the full democratisation of Brasil. Rural women were building autonomous organisations and became the central membership constituency of rural and national trade union structures. As Sales notes:

> In 1985 alone rural women held 12 regional conventions of women agricultural workers in many regions across the country. The following year, in November 1986, rural women staged a National Convention of Rural Women Workers. In the years that followed, there was even greater involvement of women rural workers in conventions, seminars, campaigns about violence against women, maternal mortality and constitutional rights. These were just some of the isolated initiatives of groups and advisory panels existing around the work and organisation of women rural workers, which joined the new emerging social movements (de Sales, 2007: 438; *Translation mine*)

Within this context, women had begun to defend their place in the rural unions which, up to the mid-eighties, 'assumed' that only one person per household could be a member and behind this was the unmistakeable 'male head of household'. Such an assumption which was effectively disabused by the regular meetings of rural women union members, in the South of Brasil, pushing strongly to increase the numbers of women engaging with and joining the trade unions. The result being that by 1987 women constituted 29% of the union membership affiliated with the National Confederation of Agricultural Workers (CONTAG) (Deere, 2003) The concerns of women were initially concerned with discrimination and unequal pay and lack of benefits but this shifted to the imperative of gaining their rights to land, as set out in the 1988 Constitution, and most of all that of joint titled property (Guivant, 2003, Ribeiro Pessoa & Dal Ri, 2017) This background supported women's entry into the MST in the fight for land. Which by 2000 became one of the most powerful social movements in Brasil.

By the mid-nineties there was an apparent 50:50 ratio of delegates attending MST conferences and participating in the decision-making bodies. On the one hand it reflected the strong profile of women as members of the movement and active participants in the encampments and as members of the subsequent settlements while on the other it seemed a little contrived. This was a necessary public profile for a movement with a growing commitment to a more egalitarian political profile and an attempt to address the gender question. As Rodrigo an MST coordinator stated:

> Even from the early days of the moment women were turning up for the meetings, unaccompanied even when they had husbands or partners at home. When occupations were organised they would turn up with their partners or sometimes alone initially. The women have been a real driving force in this moment, and we have been pushed into creating space for their participation not only at the grass roots but also in the decision-making bodies. Women made it clear that their experiences were different and should be included in the political agenda. To be honest there were many men who did not accept that women should be part of the leadership or be present at meetings or conventions. Thankfully, there are fewer men who think that way in the movement now. But we are all clear that we need the women in this movement more than they need us. Without women this movement would have fallen long ago. (Itapeva)

In the 2000 elections of the MST nine women were elected to the 21-member national leadership. This level of representation was not reflected at the grass roots in the cooperatives and settlements which were able to exercise a certain relative autonomy in view of the cross regional cultural and geopolitical intersectionality. It was apparent from my earlier research that amongst the grass roots the idea of gender equality was far less imbued within existing gender relations, which prevailed within the rural villages and settlements, established by the movement itself. Nevertheless, this was a top down initiative, which was essentially to challenge the exclusion of women and children with respect to property ownership, access to credit and the political inclusion. A women's section was certainly formalised within the movement in the decade of the nineties with a clear political agenda to address gender inequality and ameliorate some of the injustices that women experienced in their personal relations and in their working and public lives. In some ways, while necessary, this seemed to be a disappointing development not least due to the fact that this reflects a modus operandi not necessarily adopted by popular movements previously but one which was very much a part of the modus operandi of political parties and trade unions (Corcoran-Nantes, 2002, 2008). Yet at grassroots level in the encampments and settlements women were frequently organising themselves into groups to address both practical solutions to subsistence issues, income generation as well as the discussion of gender specific issues within the movement and in their new lives.

The MST and its desire to shape an optimal agro-political community in which individualism and inequality did not hold centre of place in the settlements and where lived experience and co-existence with a protected ecological environment offered a meaningful example of possible alternatives to the exploitation and repression that its members had endured before. While the idea that it was possible within the new forms of socio-economic organization to produce what it called new men and women with a new society underpinned its political agenda. Idealistic as most political agendas are the practical implementation of such ideals was far more difficult to realise. If women wanted equality, then they were obliged to claim it and, in the process, assert their independence both individually and collectively. The narrative that follows focuses on the experience of women in the rural settlement of Pirituba II which was established in Itaberá and Itapeva, in the interior of the State of São Paulo.[3]

3 Entering the Movement and Participating in the Fight for Land

The rural settlement Pirituba 2 was the result of a protracted and incremental occupation of what was the Fazenda Pirituba. These occupations began in 1980 as part of a fight for land under the auspices of an emergent MST in the struggle for agrarian reform. This struggle brought together people from diverse social groups colonies/enclaves designated as 'Rural Settlement Projects' which came under the

[3] The women and men interviewed were settlers from the Agrovilas 3, 5 & 6 of the Pirituba 2 settlement. The names used as identifiers are not necessarily the actual names of the interviewees in accord with confidentiality.

jurisdiction of either federal or state government authority.[4] Women constituted over 45% of the total number of settlers and formed part of the initial encampments, land occupations, and the resultant settlements. The life histories and backgrounds of these women were similar and, despite claims from official quarters and others with respect to the background of those participating in the movement, all the women were from rural backgrounds and had been rural workers of peasant farmers up to entering the occupation. Only two women had come from urban community organisations affiliated with the MST and were part of the movement, but both had been brought up in the countryside and were from rural working families. Consistently women spoke of their lives as *bóias-frias*[5] and sharecroppers, the hardship, and the motivation for throwing in their lot with the MST to transform their lives.

> We had a small farm and worked on the rural estates. Before I worked on the farm, I worked on the famers land and whatever we produced he took 50%, we sometimes had to rent tools and we used donkeys for planting. We came from a sitio we couldn't plant anymore so we had nowhere to go…. I wasn't worried about joining the land occupation because we had nothing anyway. We understood what was happening through the other areas, so everything was clear. [Maria Elena, Agrovila 5]

> Being brought up in Alagoas we had a sitio [small farm] for twelve years. There were nine children, my father died and there was a lot of debt. We sold up to pay the debt to the banks when mother was still alive. We bought a house with the money left over in the town. Everyone had to work, some did shoeshine, I had to learn to clean house and I became a domestic because I was illiterate. [Nazare, Agrovila 5]

> I was a *bóia-fria*, there was no regular work and we suffered. We worked on a farm as casuals and between seasons I worked as a domestic. We grew beans, rice, corn, cotton and coffee. [Alin, Agrovila 3]

> I'm from Parana, we farmed a plot and we had to hand over 20% of what we earned from our production to the landlord. We wanted land for us…. I spent four months in a tent with my children. [Nilsa, Agrovila 6]

Most of these settlers were from the interior of São Paulo, already surviving through working on the land. Some were already members of the rural trade unions or active members of the Catholic Church and were the first to be targeted as a potential cohort for joining the movement. The numbers and potential squatters snowballed from those contacts. Early on meetings were held either at the local trade unions or even in people's homes to explain the aims of the movement and seek commitments from families who wished to be involved in the fight for land and acquiring 'a piece of land' of their own. Those who were interested in participating would attend regular orientation meetings for a period of up to 6 months. This was a period of preparation and planning. The participants were told what they had to do

---

[4] See chapter "Combining Focus and Circumspection: An Education in Natural Inclusion"in this volume.

[5] *Bóias-frias* is a Brasilian idiom which refers to casual agricultural wage labourers who often move from one rural estate to another for work and the term *bóias-frias* principally refers to the fact that they take their food to the fields and eat it cold.

and also the difficulties they might confront at the encampment and the dangers inherent in occupation. Preparation was important but on an individual level things did not always turn out as planned. Here women relate their experience of how they joined the movement:

> I lived in a neighbourhood near agrovila 1 and had relatives there. Then came agrovila 3 where I had an aunt and after that I wanted to join – there were lots of people from the family involved…. There was a motion to have an encampment near 3 so I decided to join. When 6 people came and asked me to join I decided to go to the meetings at the trade union…. I even held meetings at my house in the neighbourhood, but I didn't have coffee or anything. I had to borrow it….This way I had more information about what they had to do, to take what you could and what you would have to put with. [Zilda, Agrovila 6]

> I participated in the meeting as well as the rural trade union. They told us what the MST was about and I attended meetings for about six months. It was good preparation, we were told it could get violent and we might get evicted…I was frightened of getting shot but I was more frightened of my children going hungry. [Dulce, Agrovila 3]

All of the women we spoke to had joined the land occupations primarily at their own behest. It was they who attended the meeting and encouraged their husbands and partners to go with them. Partnered or not they went as a family –female-headed households would often take their mother with them. Single women usually went with their parents. The initial rules under the project were that preference was given to married couples and members could not have other means of income. One important feature of women's motivation was that they all argued that they had not chosen this path but that it had chosen them. They had absolutely nothing to lose but for some it meant going alone:

> I attended the initial meeting. I was a *bóia-fria* with no regular work we were really struggling. The rules of the movement at that time was that you need to be accompanied by a man – you go as a couple, a family. I was with my husband then but in the end, I decided to go without him. At first, he said no and then he said he needed another man to go. I had everything packed, and I left his stuff to one side. A truck came for us. It was a long distance, he didn't even help me carry my things. He did eventually agree to come but he left after a while. [Nilsa, Agrovila 6]

> We came just before Christmas. We attended the meetings went through the preparation then my husband said he wasn't going – damn him. My father helped me with money and food and my mother came with me and stayed for a while. [Ana, Agrovila 3]

> I became part of the MST because I was part of the organisation in the Bairro. We were selected to come here and exchange experiences. We came on our own and left the children. We lived in tents as part of the encampment. My husband was in São Paulo and after he arrived my mind opened up a lot…. Really, an awareness about a different life. I was also a person without land and without an education. [Nazare, Agrovila 5]

The encampments were hard places to be, makeshift tents and huts of canvas and plastic. They usually began on the side of the road in preparation for the invasion of occupying nearby land. Always shifting, the landlord would call the police or send in their henchmen to remove the encamped peasant farmers. The more famous and successful the movement became the more violent the response to the encampments. Moreover, without water or electricity those in them needed to generate income and work together to overcome the considerable difficulties this presented with men women and children surviving under challenging conditions.

Encampments were long term endeavours and when people were unable to access land to plant food they needed to work. Local farming estates always, on the lookout for casual labour, would send out trucks every morning to pick up the required labour. Mostly men, and some women would wait at the kerbside to be chosen for work. Some were lucky others not but each day people in the encampment would wait and hope. It was women who were expected to keep body and mind together setting up a collective kitchen, gathering water and taking care of the children. It was here that the basis of female solidarity and necessity led to the swift creation of women's groups. Not only to deal with the practical tasks within the encampment but as a source of mutual support and group cooperation. That laid the foundations for the gender groups and enterprise collectives which emerged in the settlements that followed.

> In the beginning extremely difficult – women got food, did washing and kitchen work. women's artisanal group to raise money. We made carpets and prepared food to sell by the roadside – but there were few customers and not that successful. We depended on the other settlements for support and food. [Jicelia, Agrovila 5)

> In the first encampment I was with my family. My father came first next and then we came afterwards we were there for one and a half years and then we were evicted by the police. At the second encampment there were many difficulties were considerable, we had nothing – just a mattress and beans. We had to look for water and find food we worked here and there to resolve problems – my family helped, We women went out and collected things from other areas and after six months we constructed the huts, we did this as a cooperative. It was a real achievement. [Lucia, Agrovila 3]

> Then there is the question of the problem of assault. Then we can't work on their land so we had to think. The coordinators of Agrovila 3 warned us to get out because the *jagunços*[6] would be arriving in five hours and they were going to be battering people - they weren't joking. Some people were really hurt. They wrecked the beans we planted, which had grown wonderfully, then they gave us time to leave… I just cried….it was so difficult. But we women picked ourselves up and started over – the women's group was really important to the survival of the camp. [Camila, Agrovila 5]

The women in the Agrovilas of Pirituba 2 spoke passionately of their time in the encampments, about what they had endured and the difficulties they faced. Women from the very beginning picked up and took responsibility for the day to day organisation of the encampment, the setting up of a communal kitchen, fund raising and the care of the community. Many spoke of how initially they felt alone, especially those who were 'unaccompanied' but coming together with other women strengthened their resolve as part of the process of adaptation. One of the female coordinators from Agrovila 6 spoke of the process as transformational:

> I was working with street people here in São Paulo and then heard there was a peasant body organising people to invade land and I had great expectations. I didn't actually know the MST or the leaders – nothing. We just came and camped here We felt that the struggle for

---

[6] *Jagunços were armed militias which large landholders used to 'protect' their interests.*

> Agrarian Reform was the only solution.... There were evictions in thundering rain, it felt bad but .... through this the other areas helped. I have already seen such misery in São Paulo, but for me it was worse than here. Our previous work with street people was basically assistance not transformation. What is happening here is transforming people's lives through experience and life education. [Lurdes, Agrovila 6]

While transforming the lives of rural families was often on the fast track social change with respect to traditional patriarchal family relations that shaped gender relations in the interior underpinned a new life in the settlements. Women were angered by the dismissal of their participation in the struggle for land and their ongoing equal participation in all areas of settlement life while being excluded from meetings. While the idea of equal input and cooperative work was attractive, men were less inclined to see equality extend to the decision-making body within the agrovilas and the settlement itself. The gender question was out and proud from the outset and women were unwilling to let the matter go; a change in gender relations would have to be a central part of the transformation and they were determined to make that happen.

> Initially, women were generally viewed as dependents. Yet women worked similar hours to men, earned the same wages but with no right to a voice.... Men didn't want women involved, in fact the running of the association/cooperative they referred to this as the 'men's group'...(laughs) It was as if the women had a group and the coordination was theirs.... Women were not allowed to enter. They didn't want women in the meetings – it was men who discussed the question of land tenure and the principals of the movement. Men formed the coordinating group – they were self-selecting and then people voted at an assembly of the cooperative.

> There were tractors for everyone and milking the same. You have to participate in the cooperative to vote. At the beginning there was a lot of arguments. As the women had a group we put it on the agenda in the assembly for this. There was only one person against the others, the man gave a speech said that a women's group was against the men – he left because of it. Very *machista*, he was so angry he left the next day. We wanted this a lot and we won, ... I have to say that it wasn't all the women that voted in favour though but now they are very much in the assembly. We work with the men, when they need help.... we do that but women are autonomous

Women in the various agrovilas spoke of the strong initial resistance form men with respect to their full participation in the Agrovilas, the question of a vote and the gender division of labour. Women formed a group to claim their rights within the settlement. Equality meant full equality. Women argued that if it was not for them there would be no struggle. For the most part women were the first to attend meetings and initiate the family's entry into the struggle, women were among the leadership of the encampment and the takeover of land. Attempts at their exclusion in the coordinating body of the cooperative was simply unacceptable.

> Once we established women's participation in the cooperative there was the question of a vote. We argued for parity in production. Voice, rights and income...Many men do not want the women to work on the land or go early with the men but women are going with their husbands to milk early. We have a right to vote, we deserve respect.

> When we arrived there no women leaders. But bit by bit women working together…doing alternate work, each area working with women's groups, although there are none in area 3, women are an integral part of production and equal partners. This isn't simply about ideology; the politics is in the practice. [Lurdes, Agrovila 6]

A further issue for the women was that the movement espoused equality and justice but in the 1990s there was still little of that for women. Despite the new constitution of 1988 giving men and women equal title over land the question of land rights was especially galling for the women of the movement who had actually fought for land with the men. The majority of families arrived leaving a highly exploitative rural environment in the interior of São Paulo where neither men nor women had rights over anything – including their labour. The irony of women being excluded from their right to land within a movement whose politics was based on that very premise was not lost on women. This proved to be one of the most important achievements by the women's groups on the settlement.

> The part of gender in the settlement was still not fronting here, women work on the land, deal with water and work in the house. The question of gender remains problematic. When we have a meeting, the women say that the main problem is having access to title, women are still considered dependents and only when the husband dies do they get title.…It was a hard battle but now women in the cooperative are signed up equally - both names on the title in this coordination. And we have women candidates that have entered in the coordination. [Nazare, Agrovila 5]

> We have had to work hard on this question. We are struggling because we don't have it in writing. It is very difficult working with this. Women are anguished by the problem. But until we come out of emergency status that problem of land rights cannot be solved. [Zilda, Agrovila 6]

> Once the women had gained a voice and representation in the movement the question of us having equal land registration followed fairly quickly. It wasn't a difficult battle when the MST has a political agenda of equal land rights. There was a real push from that point onwards for men to be more involved in domestic work in the home. Men left (laughs). [Dulce, Agrovila 5]

Women's determination to have some control over their lives and a real sense of independence both practically and financially became a driving impetus for the formation and strengthening of women's groups pursuing entrepreneurial initiatives to create an independent extra income.

4 Women's Collaborative Endeavours

In the very early days of the movement the MST had clear rules with respect to the operation and the organisation of settlements and the obligations of people within them. Settler families would be registered exclusively in cooperatives but by the late 1990s that rule was superseded by giving families an opportunity to be individual farmers or part of a cooperative. Quite simply because both kinds of farming had a very different gender division of labour and labour demand. Individual farms meant that women were faced with a much more intensive labour input both on the cultivable land and in the home. While in the cooperatives work sharing and a clearly agreed work roster governed activities and decision making which involved all its members. Women who were effectively from female headed households usually chose to be in a cooperative due to the breadth of support and shared responsibility for production.

> We have a market sector, agriculture, bee keeping, planting, animals. A group divides the work, making cream, milling and we have a roster for that. Hours are different. Everyone has a meeting with everyone, decide what we plant and that depends on the vote. [Zilda, Agrovila 6]

> We are like family in the cooperative and we are part of the association. There are also those working as individuals. At the beginning there was a coop, when we gained the land the decision was you could be in a coop or work individually. There are 24 families – 11 are in the cooperative. Some went to Agrovila 1 and others to 4 and others for 5. The coop adjusts itself and the individuals resolve themselves. Social activities leaders were self-selected, but decisions are made collectively. Coordinators are voted for but people put themselves up as candidates. [Lurdes, Agrovila 6]

> Now everything is fine, we are doing well. We plant corn and rice and raises chickens and cows and some people fatten pigs. The animals and everything are part of the cooperative. Men plant with the tractor. But everything men do women do too. [Anna, Agrovila 5]

> I work as an individual, we had a small farm before but here everything we plant or sell is ours. I get three alqueires, I plant rice, beans, corn and melon. We both plant the corn and I do the weeding. My husband and the children went on horticultural courses and I also go to the women's allotment to work where we share the produce. [Maria Eugenia, Agrovila 3]

> We are engaged in family work herding in a cooperative. We get children to help. We have 18 cattle and we plant rice – we have 3 alqueires[7] per family and we plant food….We engage in collective work doing everything together and we have parties. Other areas are very divided. [Nilsa, Agrovila 6]

While in the encampments women had already taken the initiative to form women's groups to work cooperatively in carrying out subsistence activities, setting up a community kitchen and where possible planting food. For one of the encampments two of the settled agrovilas gave them land to plant which helped considerably given that the particular encampment was there for 1 year and 8 months before the land claim was settled and a new agrovila emerged. There was, however, a need for women to engage in income generating activities within the location of the encampment. Men were more likely to go on the trucks to local farms as casual labour but the need to organise activities and tasks in the encampment, look after children and the vulnerable and undertake 'domestic' related work kept most women in the encampment. Any income generating activities needed to be established in the

[7] This is a Portuguese unit of land measurement dating back to the twelfth century, which people still use in the countryside. One alqueire paulista is equal to 2.42 hectares.

encampment, so the women took to organising. In the early days one group of women chose to make carpets while another set up a makeshift stall on the main road to sell handicrafts but both floundered. Carpet making fizzled out when they ran out of materials and had no money to acquire them while the temporary stall ended when the women found that there was insufficient traffic on the road to sustain it. While in the encampment the women's groups served a broader purpose, that of supporting one another and sharing experiences.

> I think forming the women's group was a survival strategy in itself. The stress, the conflict, the broken relationships and sick children… It all fell on the women. We formed a bond that has stayed with us, we grew and it gave us a raised awareness about ourselves and the new lives we were on the road to creating. [Nazare, Agrovila 5]

> When we were in the encampment, we formed a women's group and made carpets and food. The group was formed by women to resolve problems, exchange experiences. Once the Agrovila was set up, after a couple of years we started to look for a new enterprise for women. [Maria Elena, Agrovila 5]

Across the Pirtuba 2 settlement once Agrovilas were established a range of projects emerged although many were not gender specific. For example, there was a brick making project which was a joint effort when they were in the process of constructing homes and this went for approximately 2 years. After 4 years women set up a large fruit and vegetable allotment women and girls who worked on the allotment received an equal share of the produce. As with other collective activities on the settlement a roster was set up and the participants share was based on time dedicated to the work there. For many women, the produce gained from their work on the allotment made an important contribution to their family subsistence. Setting up additional women only enterprises had clear links to their previous lives in the interior where they were obliged to undertake other income generating activities to supplement family income. In the early years of the settlement and in the construction and establishment of Agrovilas as more land was won by the movement, the planting and development of each one took several years so the need for some supplement to family income fell onto the shoulders of women once again.

Following the MST rules that all income generating activities should occur on the settlements themselves[8]; but for women this presented no hardship because any activities that they engaged in needed to be located in their Agrovila whereby they could recalibrate their agricultural work and home obligations to accommodate their income generating activities. Some of the initiatives were specific to women's groups in particular areas while others were open to women from all over the settlement such as the women's allotment. Others were specific to a certain Agrovila. Agrovila 5 was an area that was extremely successful in establishing women's

[8] In the MST there were clear rules with respect to what settlers could and could not do. In view of the premise of the movements and its acquisition of rights of land through occupation the rules remained clear no one could undertake work outside of the settlements otherwise they would have to leave. Settlers had to derive income from activities within the settlement otherwise land could be taken back by the state or federal authorities for infringement of land titles agreements.

enterprises that were long standing and profitable. One of the enterprises was a small sweet factory that arose from an initiative by women with small children who needed something close by to work in, that was flexible and could generate the extra income that they needed.:

> We decided to set up a small, sweet factory which makes all kinds of sweets.... various types of *doce de leite* and *cocada*[9] which sells in the MST shop.[10] There's 16 women with young ones who work here. The profit is divided equally. When the women come in to make the sweets they have to sign in so we make sure that everyone has put in more or less the same hours. We are now applying for money from an International project to expand the factory. [Nazare, Agrovila 5]

> I work in the sweet factory a couple of days a week. How much we make and how much we sell depends on the season. In the summer we sell more than in the winter but Brasilians love sweets, so its been a good business. [Jicelia, Agrovila 5]

Perhaps the most successful venture was the founding of a pharmacy based on the production of natural and homeopathic medicines. This began when several years before a Swedish nurse, who was on a State project with the Central Trade Union (CUT), was sent to Itapeva. She was in Brasil to teach health education and gave first aid courses to the women and went from one Agrovila to the other but as she travelled around Pirituba 2 settlement it became apparent that working on medicine and medicines would benefit women practically and financially.

> It started with a nurse who offered us a first aid course. She noticed the women's allotment and particularly the bee keeping and growing medicinal plants – when she saw this the nurse asked if they want to do a course and expand their work with medicines. We were able to recognise the plants and in 3 years we bought books and learnt to do this. In November 1998 the nurse left. We make medicine for the pharmacy in our houses. Then we made a project and received money to construct a building to make this pharmacy and store. [Lucia,[11] Agrovila 5]

> There was a Swiss nurse that lived with the movement. She was 3 years with us and then went back. She gave courses in nursing and went from agrovila to agrovila. The course was in wound dressing, first aid and then she went along with the people to watch them and make sure they were doing everything right. [Maria Elena, Agrovila 5]

> The pharmacy has been a real turning point. There are 11 women who work there and all of them did a course with the nurse, so they are all trained. In the early days we didn't have enough time outside of the farm work to spend there, so it basically operated for two days a week, Tuesday and Thursday at that point... Now it's open every day and some of the women work full time while others work in shifts. In the cooperative we work on rosters, so women work in the pharmacy when they are off roster and some are excused from field work when they work in the pharmacy. [Nazare, Agrovila 5]

> We make creams, homeopathic medications to treat the heart, kidneys, and the liver we also make cough medicines.... we get some of the raw materials from the marketing sector of

---

[9] *Doce de leite* is a fudge sauce or spread made from milk and sugar while *cocada* is a very popular coconut candy made from fresh coconut and sugar that is usually homemade and both are very popular traditional Brasilian sweets.

[10] The MST has a shop in the centre of Sao Paulo selling produce from the settlements.

[11] Lucia is the pharmacy coordinator.

the settlement. Initially we would use the medicinal plants from the allotment and our own gardens. Now we have a large allotment next to the pharmacy where we grow everything, we need.... People on the settlement get the medications half price but outsiders have to pay full price. [Camila, Agrovila 5]

Women working together to generate income was a feature of their lives from encampment to settlement. There were other ventures in the Agrovilas of the Pirituba 2 settlement such as soap making in Agrovila 1 and bee keeping in Agrovila 6. The enterprises on Agrovila 5 especially included women who were heads of household they were also involved in the artisanal work. Both projects illustrated above expected women to put a certain number of hours in per week in order that the profits were shared fairly between the participants. The women's collectives were a means of empowering women, it offered women a sense of control and greater economic independence. Women also emphasised that it was a means of offering a real change from the previous arduous lifestyle of agricultural work.

The women groups met regularly, decisions were made in the groups with respect to the ventures they set up and to issues on the Agrovila and the settlement itself. They made decisions collectively, grievances would be brought to the group and put on the agenda of the settlement Coordinating body by their representatives. The solidarity and gender politics that came through all of this was in the word of the women themselves 'transforming'. As women found a voice in a traditionally patriarchal conservative environment and effectively were influencing a changed gender division of labour to one that was seen as more equitable, Decision making in the women's groups was transmitted to the Coordinating bodies at local and regional level. At regional level there were female coordinators who also represented locally, meetings were no longer bereft of women but witnessed significant attendance by women where before they were predominantly frequented by men.

Now in the cooperative women are signed up equally – both names on the title in this settlement. We have women candidates that enter the coordinating body. After all we do most of the work, we push the agenda forward, not backwards like the men. And we do this because we have a different vision. [Nazare, Agrovila 5]

We make them understand women, we want participation for women. There are now more women than ever at the meetings, before it was just men. We now have a group of women from different areas. Now I hold my head up, I am more valued as a person. Now I have respect, that's the word respect… .I never had that before. [Zilda, Agrovila 6]

5 Pathways Beyond

The changes occurring within the movement in Itapeva may not be universally applied and certainly are different than the situations described in research in other locations in Brasil.[12] In our research in the interior of São Paulo there were clear

[12] See, for example, Caldeira, 2008, 2009; Bereta da Silva, 2004.

indications the women's experience, tenacity and endeavour formed the basis for their collective struggle for social change. The MST as a movement had a clear political agenda for social change and the transformation of the agricultural sector through collective action, collective work in a move towards justice and equality. In addition, the adhesion and leading role of women in the fight for land and social change directly emanated from the movement's advocacy of a different philosophy of life and the rhetoric of a more egalitarian relationship. As Deere stated:

> Given the MST's commitment to social and gender equality, there is thus growing acceptance of the empowerment argument: that is, that gender relations must change which, in turn, requires increasing women's bargaining power within the household and community as complementary and interactive measures. [2003: 284]

In Itapeva, what was most important for women was the transformation of gender relations, a more equitable division of labour and a much-improved way of life. None of this could be fully realised in the present but the prospect of a different kind of socialisation constructed within the education given to the next generation – the children of the landless – moved beyond the ideological underpinnings of the movement and into the curriculum of schools established in the settlements across the country.[13] Thus, making education the locus of the fight for land and its politics of social transformation. As Schwendler and Thompson note:

> Regardless of the gender of the student, challenges will surface when entering back into the community. Being able to re-enter the school environment, discuss these issues and get feedback on how to address and confront these challenges means that the ability to question gender relations is continuously reinforced in the student. (2017: 111)

For women, the initial struggles themselves were part of a steep learning curve which they challenged incrementally at every step of their individual and collective journey. Together they were able to challenge the cultural conservatism of the rural interior and place the political promise and 'gender problematic', as they called it, on the movement's political agenda. Nevertheless, the practical tasks of encampments and the construction of settlements drove the women's groups need for collective work and decision making. It was in their own forums that the gender specific grievances emerged and the agenda for change which they pushed to the forefront of the political discussions within the assemblies and local meetings.

The construction of new realities in an evolving agricultural environment demanded alternative farming methods, the recalibration of gender relations, in both individual farms and collectives, and the economic and political empowerment of agricultural families. Women were a leading force in the achievements of the movement both in Itapeva and elsewhere and this was difficult to ignore. Women engaged in the dual struggle of first liberating land and then liberating themselves. Each being a protracted process. Moreover, not all transformational social movements are revolutionary and not all revolutionary movements are transformational but the experience of the political struggle and the changes that emanate from it are

---

[13] See chapter "Combining Focus and Circumspection: An Education in Natural Inclusion" in this volume.

without equal. For the women of the Pirituba 2 settlement there was little doubt that their lives had changed for the better and that this transformation would mature and be passed forward to the next generation and beyond.

References

Bereta da Silva, C. (2004). Relacoes de Genero e Subjetividades no Devir MST. *Revista Estudos Feministas, Florianopolis, 12*(1), 269–287.

Caldeira, R. (2008). 'My land, your social transformation': Conflicts within the landless people movement (MST), Rio de Janeiro, Brazil. *Journal of Rural Studies, 24*, 150–160.

Caldeira, R. (2009, May). The failed marriage between women and the landless People's movement (MST) in Brazil. *Journal of International Women's Studies, 10*(4), 237–258.

Corcoran-Nantes, Y. (2002). Women in popular urban movements, São Paulo, Brasil. In M. Pacione (Ed.), *The City: Critical perspectives in the social sciences* (pp. 636–654). Routledge.

Corcoran-Nantes, Y. (2008). Female consciousness or feminist consciousness?: Women's consciousness raising in community based struggles in Brazil. In T. D. Dickinson & R. K. Schaeffer (Eds.), *Transformations: Feminist pathways to global change*. Boulder, Colo.

Da Gohn, M., & Marcondes, G. (1995). *História dos movimentos e lutas sociais: A construção da cidadania dos brasileiros*. São Paulo.

Da Gohn, M., Marcondes, G., & Bringel, B. M. (2012). *Movimentos sociais na era global*. Petropolis.

de Sales, C. M. V. (2007). Mulheres Rurais: Tecendo Novas Relações E Reconhecendo Direitos. *Revista de Estudos Feministas, 15*(2, maio-agosto), 437–443.

Deere, C. (2003). Women's land rights and rural social movements in the Brazilian agrarian reform. *Journal of Agrarian Change, 3*(1/2), 257–288.

Deslandes, A. (2009). Giving way at the intersection. *Australian Feminist Studies, 24*(62), 421–437.

Fernandes, B. M. (2000). *A Formação do MST no Brasil*. Vozes.

Guivant, J. S. (2003). *Agrarian change, gender and land rights: A Brazilian case study*. Social Policy and Development Programme Paper No 14, June, UNRISD.

Hernandez, A. (2020). The emergence of agroecology as a political tool in the Brazilian landless movement. *Local Environment, 25*(3), 205–227.

Ribeiro Pessoa, J., & Dal Ri, N. M. (2017). Movimento dos Trabalhadores Rurais Sem Terra e a formação para a igualdade de gênero. *Revista Brasiliera de Educação do Campo, 2*(2), 581–606.

Schwendler, S. F., & Thompson, L. A. (2017). An education in gender and agroecology in Brazil's landless rural workers' Movement. *Gender and Education, 29*(1), 100–114.

Thompson, E. P. (1966, April). History from below. *Times Literary Supplement, 7*, 279–280.

Encounters and Mis-encounters: The Process of Political Pedagogy Between Educators and Families of the Landless Movement (MST) Settlement of Pirituba II, South-Eastern Sao Paulo, Brasil (1984–2006)

Edvaneide Barbosa da Silva and Yvonne Corcoran-Nantes

Abstract This research investigates the education process of children and youth from the MST settlement Pirituba II, located in the regions of Itapeva and Itaberá, in South-eastern São Paulo This chapter is about the Project of Rural Settlement (PRS) which comes under the jurisdiction of the Fundação ITESP which is linked to the São Paulo State Secretariat of Justice. The project arose from a land occupation instigated by peasant families, during the 1980s who had organized themselves as rural trades people and pastoralists in the Movimento dos Trabalhadores Rurais Sem Terra (MST) – popularly known in English as the Landless Movement. It was here that the question of basic education actively delivered in the settlement was applied in practice as a result of the fight for landless rights organised, especially, by the MST. Since the initial occupation of the Fazenda Pirituba, (A fazenda is a large landed estate of hundreds of hectares of arable land – they emerged from the period of colonialism between the 16th & 18th centuries when they were referred to as plantations owned by powerful colonisers and the church housing hundreds of thousands of slaves used for both agricultural, and household labour. Brasil was one of

Dr. Edvaneide Barbosa da Silva undertook the fieldwork research on which this chapter was based and wrote the principal draft in Brasilian Portuguese. Dr. Yvonne Corcoran-Nantes is a longstanding colleague and friend who has undertaken research, on occasion, with Edvaneide at the same settlement. Yvonne is responsible for the translation of the original draft into English, including, cited interviews and quotes from classic texts as well as for structural editing and additional material.

E. B. da Silva
Universidade de São Paulo, São Paulo, Brazil

Y. Corcoran-Nantes (✉)
Flinders University, Adelaide, Australia
e-mail: yvonne.corcoran-nantes@flinders.edu.au

J. J. McIntyre-Mills, Y. Corcoran-Nantes (eds.), *Transformative Education for Regeneration and Wellbeing*, Contemporary Systems Thinking,
https://doi.org/10.1007/978-981-19-3258-8_19

the last countries in the world to abolish slavery.) the settled families have fought for the development of written texts imbued with the philosophy of the movement and liberation theology to educate the young and upcoming generations of activists of the settlement. The qualitative participatory research offers clear insights into the process of transformative education in the grass roots settlements of the MST.

Keywords Education · Rural settlement · Movement of rural landless workers · Pedagogic projects · Liberation theology · Politics · Social action

1 Introduction

When we look at the post-cold war international scenario, at first glance, neoliberalism has left no possibility for existence of alternative societies. For Henri Lefebvre there is no such thing as absolute power because within any social order there are contradictions which open the possibility for political rupture, dissent and resistance.[1] In Brazil, this conflictual dissonance is manifest in the formation of social movements, from which the Landless Movement (MST) emerged. Movements of resistance against the course of neoliberal politics predominant within global capitalism continues to underwrite and reproduce social contradictions and socio-economic inequality between and within nations.

Against the entrenched interest of large landowners, supported by agrarian capitalism, the MST emerged leading a political struggle for agrarian reform which brought together diverse social groups of the country in colonies/enclaves designated as 'Rural Settlement Projects' which came under the jurisdiction of either federal or state government authority. It was withing this jurisdiction that the rural settlement of Pirituba II was established in Itaberá and Itapeva. A proportion of the peasant families on the settlement participated in the occupation of the lands of Fazenda Pirituba, through the rural workers trade unions of Itapeva, Itaberá e Itararé and the Solidarity Campaigns organised by part of the Catholic Church [Liberation Theology] during the process of forming the MST.

Through the secondary research consulted, and the interviews undertaken with respect to the origins of the families living on the settlement showed that a significant number derive from the South-eastern region of São Paulo where they worked as *bóias-frias* or sharecroppers according to those we interviewed. Moreover, it was not uncommon to hear settlers say that they had lived like 'slaves' on alien land. Thus, the appropriation and permanence of the settlement is associated with

[1] See, Lefebvre, Henri. *La presencia y la ausencia*. México: Fondo de Cultura Económica, 1983; *Lógica formal/lógica dialética*. Trad. Carlos Nelson Coutinho. 2ª ed. Rio de Janeiro: Civilização Brasileira. *O direito à cidade*. Trad. Rubens Eduardo Frias. São Paulo: Centauro, 2001b; idem. *A cidade do capital*. Trad. Maria Helena Rauta Ramos e Marilene Jamur. 2a ed. Rio de Janeiro: DP&A, 2001a.

winning one's freedom, this is in some ways possible to understand when witnessing the everyday lives of the peasant farmers on the Pirituba II settlement.

Since the earliest occupations of the Fazenda Pirituba, the peasant families did not have sufficient means to establish public schools on the settlement. In 2017 we were able to verify the establishment of five public schools, two were funded by the municipality and three by the state, which introduced a basic education for children and young people. Literacy education for young people and adults was envisioned by projects under the jurisdiction of the federal government.

For those families that participated in the MST this was not simply about having access to a form of cultural literacy through a mere instrumental form of education. After years of struggle, to have a public school established and maintained, the leaders of the MST felt the need for the development of educational curriculums by the teachers which was commensurate with the life experience and demands of the settlers. In this way, it would contribute to an understanding of the struggle for agrarian reform, increasing the challenge and critique of the politics of monoculture, environmentally destructive technology, food sovereignty and the practical experience imbued within agro-ecological production on the settlement. Lurdes, of the MST summed this up well:

> ....the pedagogical political project of the city is incompatible with that of a rural school. We don't simply want a rural school that is IN the countryside, we want a school that beyond being IN the countryside will be OF the countryside, that is truly a school OF the countryside because the political pedagogy of these schools really has to be linked, bound to the reality OF life in the countryside....so that the teachers that come to the school know what kind of community they are working in and increasingly mould that political pedagogical project to reflect our reality.[2]

In the conceptualisation elaborated by Paro (2001) a scholarly institution, in contradistinction to the dynamic of the capitalist mode of production, can contribute to the humanization of a person, when the way in which they educate is commensurate with the 'culturally historic present of the individual' (2001: 25). Thus, this research sought to analyse whether or not the school teachers in the public schools located in the settlement contributed to the cultural historicity of the settlement families or in other words in what way they incorporated within their educational political projects the lives and desires of the peasant settlers of Fazenda Pirituba?

We also applied the critique of Paulo Freire to traditional pedagogy, including his elaboration of a critical liberatory pedagogy. In relation to education and society Freire speaks of the 'banking' concept of education as a reflection of an oppressive society, creating a 'culture of silence' (2005: 28), whereby students become mere 'containers' to be 'filled' by the teacher, the students being the depository and the teacher the depositor:

> Education thus becomes an act of depositing, in which the students are the depositories and the teacher is the depositor. Instead of communicating, the teacher issues communiques and

---

[2] Lurdes Sanchez, is a settler in Area 6 and participates in the regional directorate of the MST/Itapeva and was part of the educational sector of the movement in 2006.

> makes deposits which the students patiently receive, memorize, and repeat. (Freire, 2005: 72)

In this way, Freire defines the limitations of the traditional school without any illusion that such an institution had any capacity to radically change Brasilian society. To focus on the possibilities a school presents, to contribute to the realisation that the oppressed are able to come together to effect a real transformation of an unjust society. As Freire states 'The adherence of the people is made possible by this unveiling of the world and of themselves, in authentic praxis' (2005: 169). As such the relationship between the teacher and teaching should be governed by 'dialogical action'. As Freire argues, 'the object of dialogical action is to make it possible for the oppressed, by perceiving their adhesion, to opt to transform an unjust reality' (2005: 174).

Our focus is on the nature of politico-pedagogical action between the teachers and the settler families. In this study we question whether the pedagogical curriculums developed by teachers engaged with the lived experience and challenges of the peasant settler families and in what way the relationships established between educators, the settlers and the leadership of the MST made a transformational pedagogical curriculum possible.

2 Methodology

The studies by Mainayo (1998) and Brandão (1982) influenced the initial methodological procedures that were contemplated to undertake this study based on qualitative research using participant observation. With respect to the interviews, it is necessary to be explicit that they were the primary, essential source for our analysis. In other words, this was not research that could be simply undertaken randomly. In construction the interview schedules we followed the influence of Meihy (2002) in making sure that the interviews were undertaken with the necessary rigour, interviewing 40 respondents consisting of teachers, ex students and parents.

Meihy (2002) presents three classifications of oral history: Oral life history, thematic oral history and oral traditions none of which are exclusionary. In the process of undertaking our research we did everything necessary to contemplate oral life history and thematic oral history given this was clearly what we wanted to do to obtain and analyse the life trajectories of the educators, ex-students and parents.

We also engaged in researching secondary source materials and a broad literature review. Using school documents and above all plans and the registration of teachers educational projects; publications of the MST which deal with education, publications of the *Fundação Instituto de Terras do Estado de São Paulo*/The Land Institute Foundation of São Paulo (ITESP), local newspapers, documents of the municipal secretaries of education of Itaberá e Itararé; locations and documents organised by the Regional Director of schools of Itararé as well as conversations, both formal and informal, with those responsible for education in schools in the region.

We accompanied and participated in the political, cultural and educational activities organised by the leadership of the MST/Itapeva as well as regional and state assemblies of the Movement; settler meetings in the Agrovilas 1 and 3; musical evenings, meetings of the educators of adults and young people, assemblies of the Landless Youth (*Sem terrinhas*), cultural activities organised by the Che Guevara group (formed by children and young people from Agrovila 3), birthdays in Agrovilas 1 and 3 as well as that of the MST at national level. With respect to schools, we accompanied and participated in pedagogical activities with the teachers and the community, for example the June festival (*Festa Junina*), end of year school celebrations and determinate celebrations organised by the Movement using school facilities. Although the research was undertaken within a certain time frame from 1984 to 2006, this also included the educational planning for 2007 presented by the Municipal Secretariat of Education of Itapeva referent to teachers within this municipality which was relevant to this research.

3 The Establishment of Public Schools in the Pirituba II Settlement

The establishment of the Pirituba II settlement cannot be analysed as complete. In the past, before the MST settlement existed, it was called Fazenda Pirituba and constituted a total area of 17,420 hectares, and located in the municipal areas of Itapeva and Itaberá in the south-eastern region of the State of São Paulo (ITESP, 2000). The fazenda passed to the State at the beginning of the 1950s, when a business called the Angatuba Company was unable to settle its mortgage debts. Initially the area was designated for the introduction of an incentive program for wheat growing to be administered by agricultural engineer Lino Vicenzi. In effect he basically leased or gave large extensions of land, pertaining to Fazenda Pirituba, to relatives and friends who were large cattle ranchers, called *Boiadeiros*[3] (Bombardi, 2005).

Research shows that the illicit acquisition of land from Fazenda Pirituba was not solely by the *Boiadeiros*. In the literature that we consulted we found that that there were two other large groups occupying that land – the Batagins[4] and the '*Holandeses*' – the Dutchmen. O ITESP informs us that in 1975, of the 142 contracts signed off by the State only 27 were actually made official:

> ….living permanently on the Fazenda were groups known *as the dutchmen, the batagins, furlans allotments and remnants of the Vicenzi family*….These groups did not meet the legal requirements of occupation and the State chose to exclude them by using its executive powers. At the same time, a Parliamentary Commission of Inquiry (CPI) of the Legislative Assembly concluded that Vicenzi, the engineer, should be dismissed from the public service without benefits. (ITESP, 2000: 68)[5]

[3] Roughly translated in English it means Cattlemen.

[4] The Batagins were well know cattle ranchers of the period.

[5] The italics inserted here are the Author's.

According to research undertaken by Iokoi (2005)

> In 1982, around 40 families of landless rural workers occupied one of the areas of the '*boiadeiros*', where they remained for four months. They were then expelled by the 'boiadeiros' armed guards. A second occupation occurred in April 1983, when families that participated in this action were ordered off the land by judicial process. One year later, in May 1984, there was another occupation, this time with the participation of 250 families that once again entered the area that was under the '*boiadeiros*' conrol. (2005: 36)

The families were settled on an emergency basis in the areas named '1' and '2' on the settlement project Pirituba II[6] in 1984, which resulted in a long trajectory of struggles. The peasant struggles of Itapeva and Itaberá were part of the resurgence of multiple social movements, in both the countryside and the city throughout the decade of the 1980s in Brasil. For Gohn (1994), the list of educational demands in the 1980s was inordinate. 'It was intrinsically linked to the actual political crisis of that time and the search for answers (or the equivalent) to structural problems created by the means and form of capitalist accumulation in the country' (Gohn, 1994: 58). Among the diverse demands enumerated by Gohn, it was the question of education which was an intrinsic part of the demands of various social movements in the country. In the case of the peasant settlers in areas 1 and 2 of Fazenda Pirituba, within the political project of the struggle for land it upheld the right to formal education for their children.

After the initial occupation of the land in Fazenda Pirituba, the peasant farmers spent some time in an encampment and from that point onward the construction of schools was integral to the demands of these settlers. Zeneide Moura, a settler in Agrovila 1 told us how this came about:

> Well, we entered here in the month of May, July is a holiday, but in August we had already erected a small school with two classrooms. From the outset we already had a teacher, so I went with her looking for students to see if it was possible to build a school, from there we took a list of students and went to the Secretariat (of Education of the Municipality of Itapeva) from that point we built it and when August came the lessons started.

According to Zeneide and the other settlers, the Terezinha M.R. Gomes Municipal School was established in Area 1 and, simultaneously, a State school was established in Agrovila 2, area 2. In this period Célia Ferraz, a teacher and resident in the municipality of Itaberá, was contracted by the State\ to teach classes in Terezinha M.R. Gomes Municipal School. After some time teaching at the school, Célia Ferraz had established a close relationship with the families. Little by little, Célia became close to the community and vice versa, through a dialectical educative process Célia learned from the women and they learned from her:

---

[6] Bombardi (2005) states that the roman algorithms distinguish between those that are known as 'lotistas', the first beneficiaries under the Rural Revision Law – which is 'Pirituba I, where the settlers occupied areas that had been illegally taken over griladas – 'Pirituba II'. Each new area conquered by a settlement was named according to Arabic algorithms: Área 1, Área 2 and so on. The occupation of the area that was undertaken by the first families was initiated because it was the portion of the fazenda that was taken by so called cattlemen, heirs and beneficiaries of Lino Vicenzi.

> The families were already in the settlement, they lived in houses that were made from planks of wood with cracks in the walls. There were no 'good' houses, they had no water, it was difficult and something completely new for me….After a year of working there I was very content, I made lots of friends, they would bring pots of *cocada,*[7] for me and I went to parties held in the Agrovila. And then it passed, I had to leave…. But I had this experience with them.

Between the end of 1980 and the beginning of 1990, the number of settler families increased because after the areas 1 (Itapeva) and 2 (Itaberá) were established, another peasant group had occupied other areas of the Fazenda and established settlements in area 3 (Itaberá) and area 4 (Itapeva).At the same time, a campaign was launched to establish a State School in Agrovila 3. The process to establish schools in areas 2 and 3 was similar to that situated in area 1. Thus, the public schools on the settlement were established based on collective action organized by groups of settler families This does not mean that there was a shared history between the political action of the individual groups.

In each school that was established, our research shows that the participation of the respective communities was differentiated according to the nature of the conflict, negotiation and their success in getting their demands met. At the beginning of 2000, the Governor Franco Montoro Municipal School was also established in area 1. In the same year, after various appeals by the settlers from two Agrovilas they managed to establish a high school. This was done through a partnership between the Municipal Secretariat for Education in Itapeva and the State Secretariat for Education of the State of Sao Paulo.

Officially the Bairro Engenheiro Maia State School was established in 1991, however, interviews with residents and settlers of the neighbourhood revealed a background to the story not officially registered but very much lived on in the memories of the families who succeeded, through considerable effort and resilience, for their children to have access to a culture of literacy. José Aparecido Ramos (Zezinho), a father and settler from Agrovila 3, spoke of when they fought for and gained area 3 which led to an increase in demand for education by both children and young people, and this led to the gradual expansion of the Bairro Engenheiro Maia State School. From early occupation until 2006 the settler families fought for and won five public schools. In the same period, all the children and young people had access to a basic education in the settlement.

During our research about the educational planning which occurred in 2007 in the public schools located in Agrovila 1, a proposal by the Municipal Secretariat of Education of Itapeva in conjunction with the Orsa Foundation –Itapeva Paper, came to our attention. The topics that would be developed and discussed with students was related to 'sustainable development'. According to Camargo (2015) this institution is part of the Orsa Group that develops social projects around infancy and adolescence in regions where it is possible to exploit land, as well as other benefits of nature through the monoculture of Eucalyptus. The interest of the Orsa Group in

[7] *Cocada* is a very popular traditional Brasilian candy that is usually homemade and is often sold by vendors on the streets of the cities.

a plantation of Eucalyptus is related to the Jari Project developed by Daniel Keith Ludwig:

> When north American billionaire Daniel Keith Ludwig decided to install a mega-enterprise in Amazônia, more precisely in the municipalities of Alemeirim in Pará, and Mazagão, in Amapá located specifically on the margins of river Jari, the affluent north of lower Amazona and those close to him described it as a 'project of a lifetime'….After receiving hundreds of millions in incentives in the wave of grand projects for Amazonia by the military government, in the face of mounting scandals and losses, the project was taken over by a group of national enterprises and , in 2000, it came under the control of the Orsa Group, a Paulista conglomerate in paper and packaging. The Orsa Group is known today as an example of a pioneer 'green company' receiving many awards recognising its social and environmental concern. Albeit that in Amazonia, the Orsa Group operates one of the largest landowning operations in the world which it has acquired through direct conflict with *posseiros*[8] and through judicial process pushing a politics of land *grilagem*[9] and exploiting the irregularities in land management plans in place. (Camargo, 2015: 61)

In 2006 the expansion in the cultivation of eucalyptus, pine and acacia affected the south of Bahia, São Paulo, Paraná, Santa Catarina e Rio Grande do Sul which had an impact on the soil in the regions where the absence of rain led to a lack of underground water leading to the degradation of soil fertility through an increase in the ratio of sand to soil (Oliveira, 2006). The peasant settlers that participated in the MST/Itapeva region at this time demonstrated significant concerns with respect to the advancement of this form of agro-cultivation in the region but most especially with respect to the operations of the Orsa Group with the acquiescence of ITESP. According to agronomist Francisco Feitosa,[10] in 2007, approximately 20 families used part of their plots to plant pine trees under a contract established between the settlers, the Orsa Group and ITESP. In this way the penetration of the settlement by the Orsa Group occurred deftly and gradually in this period was made viable with ITESP and the Municipal Secretariat of Education as accessories.

---

[8] *Posseiros,* are agricultural families/workers that occupy land which has been abandoned or not in productive use and engage in subsistence agriculture using the land productively thus operate what is commonly acknowledged as a 'right to use'. This would be defined as de fato land title. In view of this they can after so many years, productively using the land gain land title which legalises their rights becoming de jure land title. This land usually belongs to the government.

[9] *Grilagem* is the politics of land grabbing employed by companies and large landholders to expel *posseiros* from the land they live and work on either by using their armed hired hands to remove them through the threat of, or use of, violence or through the acquisition of 'legal title' via due process or other means to gain possession of the land.

[10] Interview was undertaken in April 2007 in his Itesp office in Itapeva where he worked as an agronomist.

4 Politico-Pedagogical Action Between the Educators and the Families in the Piritiuba II Settlement

These events cited above led to political action, combined or not, related to the political educational projects developed in the surrounding public schools in the settlement. In this way we can analyse the nature of the political educational projects between the teachers, students and the settler families that participated in the MST of that region. We are able to show the accounts and thoughts of the pedagogical practice of teachers Ruth Machado and Maria Olinda de Almeida Mazur (Marô), that in 2006 taught in the school of Agrovila 3. We analyse the different ideas of the school, the underlying respective teaching practice as well as trying to understand the ways in which they established a dialogue with the community of this Agrovila.

In October 2006, Márcia Ramos[11] e Zezinho together with other leaders of the MST and the community of Agrovila 3, organized an event called 'Education Week' in which they arranged diverse political, cultural, and educational activities. After invitations were sent out by the organisers and certain themes considered, we developed three workshops entitled 'Juvenile Sexuality'. The first workshop was developed in conjunction with the Grade 3 teacher Marô. The second, was developed with Grade 4 teacher Ruth Machado, both workshops were taught in the morning. In the afternoon we held a third workshop with the Che Guevara Group[12] from Agrovila 3 in the area reserved for the creche in the same Agrovila.

The participation of Grade 3 children of teacher Marô happened quite naturally which contributed significantly for students to be able to elaborate their questions relative to the subject without fear of writing something 'wrong' or asking for the teachers help. It was clear that Marô presented a teaching practice that went beyond the traditional ideas of education. We could also see that with respect to literacy she was a specialist because the way she spoke and her relationship with the students offered ideas far more profound with respect to the complex process of student learning.

In conversation with Marô, she told us that despite the difficulties, she had undertaken a specialist course on literacy at public university, in the State of Paraná, wherein she increased her knowledge in the field and was introduced to the theoretical contribution of Vigotsky. The teaching practice shown by Marô had an integral role in the educational mediation for the realization of the *Zone of proximal development* of the students:

> The zone of proximal development refers to those functions that have not yet developed ... These functions could be referred to as the 'shoots' or 'flowers' of development. The actual level of development is defined by retrospective mental development. While the zone of proximal development defines prospective mental development. (Vigotski, 2007: 98)

[11] Settler and member of the State Directorate of the MST in 2006.

[12] In the cited period, the Ernest Che Guevara Group was formed by children and adolescents of Agrovila 3, under the guidance of Marcia Ramos and Zezinho with the aims of developing cultural activities above all in music and sports activities.

Unaware of this contribution to educational science, the way in which the teacher, Ruth Machado, spoke with the students merely reinforced the sense of insecurity of the children which they tried to conceal in the teaching -learning relationship. When we asked the children to write down their questions/doubts with respect to the theme of the workshop practically half of the class stated that they did not know how to write the questions. The teacher did not seek to encourage them to try to write in their own way or indicate that she would be able to help them in their difficulty. To the contrary, she immediately proposed to write for them, highlighting the difficulties that the children had with written language, and, most forcibly, repeated it so that everyone could hear that they did not know how to write. However, the children showed they were extremely keen to learn about the issues presented in the workshop. From our point of view the children were very receptive, they required more attention because they were afraid of making mistakes, such as in writing or dancing. Nevertheless, they participated, and in the end, tried to overcome their shyness and fear, they danced and confided in both the teacher who developed the workshop and to their class teacher Ruth about their doubts and uncertainty.

In the end the students from Ruth's class and those in Marô's class and the Che Guevara group all raised many complex questions. The distinct traditional teaching practice of Ruth Machado did not prevent the children from expressing their doubts/questions about the subject in question. However, the objective of the activity, which consisted of encouraging the children to write down their questions in relation to the subject, was only partially satisfactory in the case of Ruth's class, because, as we mentioned earlier, the teacher undertook a writing exercise for the children.

5 Memory, Culture, and Place

We observed that the teaching practice of Marô transcended traditional teaching. Marô viewed the Education Week as an important moment to bring the community and the school closer together and vice versa. She informed us that she identified with educational approaches such as socio-constructivism and socio-interactionism. In both formal and informal conversations, accompanying the activities she organised in conjunction with the community of Agrovila 3 and our observations across the two phases of the educational workshops on 'Juvenile Sexuality' allowed us to confirm that this teacher uses a dialogical teaching praxis which engages with the culturally historic experience of the children.

The pedagogical praxis of Marô is also revealed in her records of her classes. One of the activities she recorded, was related to the research conducted with the class about the life histories of the students' relatives, having as a scenario the struggle for land and their permanence on the Fazenda Pirituba. The activity depended on the effective participation of the children. Zezinho was formally invited to visit the school in June 2006. On the scheduled date, Zezinho talked to the children about the process of land occupation until the development of a constitution by Agrovila 3.

Together with the class, Marô worked on a collective text based on the information presented by Zezinho in his talk to the class. Then, according to Marô's records, each child was encouraged to write about their families and something about their everyday lives. Each instant was planned for the children to recognize the trajectory of their families' struggles using appropriate methodologies for the attainment of literate culture.

Marô's efforts to encourage the participation of parents in school activities, is the result of an understanding that, in general, family interest in the schooling of the children left much to be desired:

> The community is welcoming and for this I am thankful. I just feel that parents could be more involved with the school, especially in taking an interest in their children's schooling....We are still determined to improve, and I am sure that every day we will increase these ties, to become a true collective, an active presence of the community in the school. The leadership of the Agrovila. we can always count on, because they support us every time, we need them and I believe that they will help us realize increased parental participation, and greater team and community spirit involving school activities. (Marô)

Another educational experience related to memory and place was at the Terezinha M.R. Gomes Municipal School situated in Agrovila 1. The Educational Coordinator, Eva Helena da Cunha, explained to us that the 'Local Memory Project' was an initiative of the partnership between the entities: "Museu da Pessoa", "Avisa lá" and the Votorantim company, Itapeva, which together with the Municipal Secretariat of Education in Itapeva chose a school located in the city and another in the countryside in which to implement the project.

Terezinha M.R. Gomes Municipal School was selected in 2006, the year in which the project was initiated by Valderli the 3rd grade teacher of elementary school. According to the project records, we observed that diverse historical sources were utilized such as: the testimonies of parents settled in Agrovilas 1, 4 and 6; newspapers in the municipality of Itapeva; pictorial sources which revealed the changes which occurred in the lives of settler families over more than two decades. On the first day, the children listened to stories from books, chosen in advance, with the aim of raising their awareness about the subject and recovering the memory of the neighborhood and the settlement. At the end of the activity, a detailed collective text was drawn up, enabling children to exercise conventional writing skills.

The activities performed by the children of the 3rd grade of Terezinha de M. R. Gomes Municipal School, was called 'The Memory Chest', and were also undertaken with their parents. Below are testimonials from two mothers. The first is Ivanilda Batista da Silva Aranha, a settler from Agrovila 4, and the second is Lurdes Sanchez, a settler from Area 6 and a participant in the Regional Directorate of the MST Itapeva and the educational sector of the Movement in 2006.

> To recall what was most significant in my life would be very difficult because there is no lack of scars....My most recent memory was coming here. I left a large city where I had lived for ten years, leaving everything behind, even my family, to live under canvas, where I learned many things, among them the meaning of struggle and humility. .... (Ivanilda Batista da Silva Aranha)

> I remember that on the night of the twenty first of December …. thirty families travelled towards Fazenda Pirituba. It was a night of heavy rain. What was the reason for the trip? It was to make a dream reality – to conquer a piece of land where we could build our lives. We arrived at … dawn. Soon we organized a community lunch with the communal supplies. We organized ourselves into a working group to guarantee our survival and endurance. It rained for 15 days straight. It was the first Christmas away from my family. But it was also the first Christmas passed simply around a campfire, discovering my new family – The Landless. This Christmas was the start of a new life for me and all the families in the encampment. To date, it has been eleven years of struggle, resistance, a lot of coexistence and learning. I remember those *companheiros* who donated food for us while we lived under canvas. I remember the celebrations in the encampment – Mother's Day, the first bean harvest, the celebrations under the bridge, the adult literacy classes, bathing in the stream, the shack lit with candles…. Over the years my three children were born – Iara, Cauê and Irai. Today we are to build a brick house, with electricity, with part of the land conquered and in production. My three children have been studying at the school on the settlement since the age of three. The community of families of Agrovila VI are united because our political struggle brought us closer together. (Lurdes Sanchez)

The stories and testimony of parents emphasized that, during the land occupation and the period of the encampment, they went through situations of shortages which they were able to overcome due to the solidarity they established in the times of considerable adversity: the lack of water; preparing food; protection from the cold; from the rain, even in the resistance to violence imposed by the threats of the henchmen sent at the behest of the alleged owners of Fazenda Pirituba.

The teaching coordinator Eva Helena reported that, in 2007, the project was ongoing in the school involving, as a priority, another 3rd grade class, and this time, included the 4th grade. We understand that the 'Memory Project' was an important experience developed by educators of the Terezinha M. R. Gomes Municipal School, enabling us to confirm that they were pedagogical practices, effectively dialogical among educators, students, and the settler community.

6 Schoolwork Rotation

According to the teaching coordinator Eva Helena, during the monitoring of teacher planning, in May 2006, schoolwork was generally developed in the same way as in other schools that deal with early childhood education. Each teacher was responsible for a particular class and tried to develop all the areas of knowledge satisfactorily with respect to early childhood education.

Throughout this conventional experience, the teachers realized that each of them were noticeably good in a particular area of knowledge, either because they liked one subject more than another or that they participated in undergraduate courses and had specialized in a particular topic. Informally, the teachers gradually developed a collective teaching experience which moved beyond the conventional teaching practice of early childhood education.

From a certain moment on, there was no longer one teacher for each class because all the teachers were committed to being responsible for all children enrolled in early childhood education. Thus, the coordinator Eva Helena explained that this experience culminated in a project called 'Schoolwork Rotation':

> This work started a year ago, this is the second year. We call it schoolwork rotation. It is the same as if it was the 5th to 8th grade, but the student is the one who moves and not the teacher, because we have an activity room, I mean, the student can have an activity here, then they go out and go to Arts, then they go to Movement, it is rotation, it is schoolwork rotation. The subjects are mathematics, spoken and written language, nature and society and the arts [according to] the PCNs. There is no nominated 'classroom' for each subject, there is only one activities room for games, a playroom. … The activity room has four cupboards and the material is collective, there is no material which belongs to only one teacher, the cupboard belonging to another, everything is collective, material is collective…. All the material is there and who needs it uses it. (Eva Helena da Cunha)[13]

According to the teaching coordinator, the central principle developed in early childhood education focuses on the integral development of the child and, although they did not have as their objective literacy itself, the children developed their writing skills through the activities promoted by the teachers. The same observation was made by the teachers involved in the 'schoolwork rotation' and who emphasized that this pedagogical project required discussions and partnerships with the community, and also with the other school teachers., because it was necessary to understand that play, fun, Arts, Movement, and group discussion are activities inherent to a child's cognitive development. Regarding the discussions with parents and teachers from the 1st to the 4th grade of elementary school (Phase I), the early childhood education teachers stated that they had experienced no difficulties, however it was not possible to say if they all understood the concept of 'schoolwork rotation' applied at the Terezinha M.R. Gomes Municipal School.

They added that 'schoolwork rotation' contributed to lessons having greater consistency which facilitated the implementation of interdisciplinarity in school. In 2007, the teachers concluded that it would be important to build a Municipal School of Early Childhood Education, because the Terezinha M.R. Gomes was designed for the delivery of the traditional basic education and as such presented structural problems. Early childhood education teachers did not mention the theoretical references that guided play-based activities they developed with the children. However, we believe that this politico-pedagogical action can be associated with the contributions of Vigotski (2007) when analyzing the role of the toy in the process of the humanization of a child:

> At preschool age, when desires emerge that cannot be immediately satisfied or forgotten, and there remains the characteristic of an earlier stage of a tendency towards the immediate satisfaction of these desires, the child's behavior changes. To resolve this tension, the preschool child is involved in an illusory and imaginary world where unachievable desires can be fulfilled, and this world is what we call a toy. …. (Vigotski, 2007: 108–118)

[13] This follows the central line of work presented by the Referential National Curriculum for Early Childhood Education. At the time of the research this consisted of: Movement, Visual Arts, Music, Oral language and Writing, Nature and Society, Mathematics.

7 The Struggle to Maintain Bairro Engenheiro Maia State School

After the achievement of establishing schools, the settler and surrounding community faced serious problems related to the infrastructure in the Bairro Engenheiro Maia State School. The expansion of the school building was important to the settlement community, but this did not guarantee the autonomy and freedom imbued within the new way of teaching that the MST desired. Moreover, it also could not be considered a sufficient guarantee of the quality of care for children and adolescents of the settlement and those students living in the Engenheiro Maia neighbourhood.

As of May 1998, there were diverse forms of participation, in order that the school could maintain minimal function. The parents organized themselves, formed commissions, participated in meetings with officials in the municipalities of Itapeva, Itaberá and Itararé. The minutes of meetings and assemblies, as well as all documents sent to the education authorities of the region, reveal that the school's problems were directly related to the toilets not working, the lack of transport for children and young people, no security; a lack of employees; a precarious building structure; lack of equipment for the kitchen, and the low SARESP exam results.[14]

The case of the toilets seems to have been one of the most pressing problems. So much so that the families of the settlement and the Engenheiro Maia community held a marathon of meetings, culminating in assemblies and other forms of mobilization to pressure the local authorities responsible for school equipment. Clearly, the parents tried to solve the problems of the school following the hierarchy of responsibility, that is, going first to the school directorate, then to the educational agencies, and lastly to the Guardianship Council. This conglomeration of problems prevented the teachers from developing teaching and learning within reasonable conditions. Furthermore, this situation highlights the indeterminate nature of the achievements of the settlers, resulting in the interminable mobilization of peasant families in the struggle for rights.

Faced with numerous unsuccessful attempts to resolve the prejudicial situation to the lives of school children, the parents met and decided not to send their children to that particular school. This act of defiance by the peasant families, although meeting with the disapproval of the State, was a most effective form of pressure, whereby they managed to elicit a commitment from the Regional director to introduce solutions for part of the problems.

The forbearance of settler families that we witnessed at every stage of attempts to engage in discussions with the institutions responsible for education in Itaberá was noteworthy. In the minutes/reports, the records show that at each scheduled meeting with a certain regional body, the parents committee meticulously repeated the entire mobilization process, demonstrating a willingness to contribute to the

[14] School Performance Assessment System of the State of São Paulo [SARESP]. This is an evaluation of basic education developed by the Department of Education of the State of São Paulo since 1996.

resolution of problems, so that their children could return to study in an environment suitable for the dissemination of knowledge.

About a week after the suspension of classes, as a result of parental action, they were able to reach an agreement with the bodies responsible for school infrastructure, with respect to having a functioning water supply system and transportation for students. This was not a result of the independent volition of the educational entities responsible that basic education became a reality for the children of the settlers, quite the reverse, it was entirely the result of the concerted action of the settler families who managed to establish successful negotiations with these entities and secure tangible outcomes.

The fact that parents decided that their children should return to class does not mean that all the problems had been resolved. The outcome of the stoppage also reveals that, concerned about the slow response of the educational bodies of Itaberá, the parents of the Association of Parents and Teachers of the Bairro Engenheiro Maia State School resorted to the public power of another municipality, which responded, as an emergency measure, to the parents' request. It is apparent, therefore, that the process of establishing schools within the settlement was a result of the collective political struggle of settler families. Parallel to the conquest of and permanence on the land, they undertook incessant and diverse struggles to gain access to a school education for their children.

The intense participation of families in daily school life was not restricted to external issues. In 1999, conflict regarding the political pedagogical project of Bairro Engenheiro Maia State School was manifest in the relationship between director Maria Elidia, settler families and leaders of the MST. Maria Elidia stated that she did not agree with the way in which the parents, members of the MST, tried to engage in the pedagogical discussions of the school. According to her, on planning day, the MST proposed that the teachers select some life stories for the students to be familiar with.

The parents and former students of the school offer an alternative perspective to that of Maria Elidia, stating that there was no pedagogical discussion whatsoever during her administration, and they had experienced significant difficulties in establishing a dialogical relationship with the director. In their view, the solution to the problem was the removal of Maria Elidia from the school directorate. To this end, they organized and forwarded a petition to the Regional Board of Education, culminating in the episode that became known as the 'departure of Director Maria Elídia'.

The parents and former students we interviewed, who participated in the MST, reported that they had difficulties in establishing dialogue from the outset of the management of Maria Elídia, due to her 'authoritarian' and 'prejudiced' practices:

> I think she [Maria Elídia] had contempt for the students, she talked about whether the students would want to be dumb like their father and work on the farm all their lives, as if farming was inferior to any other job, you know? Perhaps when she expressed herself in this way, she spoke without thinking, but it was something that tainted the school a lot, it affected the students a lot. (Alessandra da Silva Carvalho, former student of Bairro Engenheiro Maia State School)

During the interviews with mothers and former students that participated in the Movement, we identified a perspective of professional training that could contribute to better living conditions in the settlement. From this perspective, settler Ana Lúcia explained:

> I think the study is important because for those of us here who work with the land, ….if we had an agronomist, a technician, who was from the settlement it would be great, because we don't you have to use agronomists that do not even come here, they only come to know how much we have harvested and sometimes, when you call them, it takes a lot of work for them to come, and usually the agronomists merely 'sell poison', those who come here say that here you have to pass such poison,[15] if you are not smart, you pass the poison without needing to on the crop. You have to be very smart to fight these people, so if we had, a person, a technician, an agronomist of our own, it would be different, if we had a veterinarian of our own it would be different.... we need to study so that those who graduate come back to work in the settlements [...]. (Ana Lúcia, settler, Agrovila 5)

We believe that Maria Elídia's teaching practice can be identified with Traditional Pedagogy. One of the issues she highlighted, in interview, concerned her attachment to educational planning which is elaborated exclusively by the State, reducing the community and students to mere recipients of a certain unquestionable and complete knowledge, and the role of school management was only to establish a few expeditious suggestions of the families.

In designing a plan centered on state guidelines that, supposedly, is inalterable, Maria Elídia reduced not only students and the community to the condition of non-subjects of education, but also that of her role and the educators was reduced to passive professionals, uncritical in relation to the possibilities of elaborating educational curriculums consistent with the reality of the socio-geographical place. A traditional concept of education does not enable teachers to know, effectively, the lived reality of the students.

By requesting the democratic management of the Bairro Engenheiro Maia State School, the parents and student settlers of the Movement, presented a conflict of a pedagogical nature. On the one hand, we identify the presence of traditional pedagogy, underlying the educational practices of the then director. On the other, the desire of certain parents and students to try to inscribe, in school, tolerant practices related to the daily lives of the settlers.

> Those who use the banking approach, knowingly or unknowingly (for there are innumerable well-intentioned bank-clerk teachers who do not realize that they are serving only to dehumanize), fail to perceive that the deposits themselves contain contradictions about reality. But, sooner or later, these contradictions may lead formerly passive students to turn against their domestication and the attempt to domesticate reality. They may discover through existential experience that their present way of life is irreconcilable with their vocation to become fully human. They may perceive through their relations with reality that reality is really a *process*, undergoing constant transformation. (Freire, 2005: 73)

[15] Poison referred to here are usually broad spectrum weed killers which the MST refuses to use because they are committed to organic farming.

Reports indicate that, after the petition for the departure of Maria Elídia, there was no delay, on the part of the actual Director of Teaching, in delivering an affirmative response to the settler's request. In October 2000, Deputy Director Celia Ferraz officially took over from Maria Elidia:

> It got better because the Deputy Director was totally unlike the Director, you know? She spoke really well with the students, she got along very well with the students, that was Director Celia…. (Alessandra da Silva Carvalho, former student of Bairro Engenheiro Maia State School)

The activities developed with the community, during the period when Celia Ferraz was Director of the school, did not culminate in formal moments of pedagogical discussion. However, the relationship between families, school management and other teachers improved significantly:

> Look, I don't remember what the educational curriculum of the school was, I don't remember, I know there was a curriculum as all schools have but within this curriculum , as I said to you, there was this rapport, this desire that a student should feel good to become citizens, to have citizenship…. (Célia Ferráz, former director of Bairro Engenheiro Maia State School)

> …. No, it was not discussed [the educational curriculum of the school], it was more a relationship. (Zezinho, father and settler)

> Not that I remember. The educational curriculum was not discussed with the community, it was a standard kind that you have in all schools…. (Alessandra da Silva Carvalho, former student of Bairro Engenheiro Maia State School)

However, the dialogical practice of Celia Ferraz facilitated educational activities in the school with the participation of the students and community, as well as the participation of Celia Ferraz in the MST events in the settlement. Among the school documents we find records and photographs referring to the projects developed, these are: The Democracy at School Project, with the participation of the student João Antônio, who gave lectures on the theme 'External Debt' 'Community Project goes to School' which promoted the formation of a cleaning and painting collective at the school in Agrovila 2 and a gardening and painting collective at Bairro Engenheiro Maia State School.

Celia Ferraz pointed out that the parents and students of the MST educated other parents and students with respect to their participation in meetings, parties and work collectives. Interviews at the school, with parents, ex-students and Celia Ferraz, left a strong impression that the educational subjects - teachers, students and family were educated in unison:

> For the dialogical, problem-posing teacher-student, the program content of education is neither a gift nor an imposition—bits of information to be deposited in the students—but rather the organized, systematized, and developed "re-presentation" to individuals of the things about which they want to know more. (Freire, 2005: 91)

During the management of Celia Ferraz, although the community had not formally participated in the school planning, the participation of the students in specific school events, in our view, made it possible to include, to a certain extent, the life experiences and needs of the settled community. It seems to us, that the substitution of Maria Elidia by Celia Ferraz represented, for the settled community identified with the MST, a shift from a 'banking education' (Traditional Pedagogy) to a dialogical education (Liberation Pedagogy).

Officially, Celia's permanence in the position of Director was only 1 year and 1 month. Celia stated that when she accepted the invitation to be Director of the school, she had health problems and that was the reason she did not remain in the position longer. Interviews with parents and former students revealed that, for them, the reasons for Celia's departure were far from the truth. The departure of Celia Ferraz seems to have led to a loss of momentum in the continuity of schoolwork articulated through a prism of liberational concepts of education.

7.1 Final Considerations

The agrarian reform conceived by the MST was not restricted to the struggle for land. Parallel to the formation of the rural settlement, they organized collective struggles for the implementation of cooperatives, basic health centres, improvements in public transport, designated areas for cultural and leisure activities, as well as public schools and other demands in the context of the struggle for basic rights.

The experiences highlighted and analyzed showed that, especially since 2000, some teachers were willing to develop politico-pedagogical projects consistent with the lived reality of the students and their respective families. According to the experiences systematized in our research, they dedicate considerable effort in the cultural and historical actualization of the children and young people of their community.

Since 1990, the Movement, at national level, has developed a politico-pedagogical project around a proposal called 'Rural Education'. The MST, to its credit, brought together several researchers and entities around this project which culminated in the publication of the *Rural Education Notebooks* (Caldart, 1997). In the Pirituba II settlement, the commitment of the Leaders of the Movement who organized activities related to 'Rural Education' was noticeable with the teachers at the public schools established in and around the settlement.

One of the most cited theoretical references in the literature of the MST, and present in the Education Notebooks elaborated by the national education sector of the Movement, is Paulo Freire. The important and extensive study of this philosopher of education leaves us in no doubt as to his concept of an emancipatory education for the exploited and oppressed of the countryside and the city. Although he understood the limits of the school under the domain of capitalist interests, he dedicated his whole life to thinking about a dialogical education (Freire, 2005).

Another theme presented in the MST's Rural Education Notebooks, considered a categorical imperative for the formulation of an educational curriculum in schools established in rural encampments and settlements, concerns 'education for work'. Moisey Pistrak is one of the main authors cited by the national education sector of the Movement, which contributes to the concept of a school of labour (Pistrak,

2000). In his words: 'It is not a question of studying any kind of human work, any kind of expenditure of muscular and nervous energies, but of studying only socially useful work [...]' (Pistrak, 2000: 50).

We verified that the centrality of the theme of "work" in school, according to the MST's Education Notebooks, is based on the work of Marx, whose influence is noticeable in Pistrak's work. In Capital, Marx's contribution to thinking about an education with the perspective of 'work in general', 'concrete', 'creator of use values', indispensable for the satisfaction of 'spiritual and social' needs of man, 'whose number and extent are determined by the general level of civilization' (Marx, 1984: 261–262). In this same study, Marx leaves us reflections that allow us to realize that the concept of 'work' is not an end in itself, because the goal is humanization. Thus, work becomes a means or, as Marx himself stated: work is a means to an end (Marx, 1984: 261–262).

The young people interviewed who participate in the MST showed a clear commitment to their studies. They spoke of their intention to contribute to the betterment of the demands of the settlement and the Movement itself:

> (...) I don't think about myself, that's a personal thing, because I am part of the Movement and the Movement is part of me, so it is a collective ideal that I have....So these doctors who are going to come here are going to strengthen the Movement even more.... (João Antonio, former student of Bairro Engenheiro Maia)[16]

The materials published by the Movement, a study of the cited literature, the interviews conducted with parents and former students, who participated in the MST/ Itapeva, allowed us to understand that they wished to study 'socially useful work' (Pistrak, 2000: 50). This is not an education for labor/work alienation, that is, educating for a limited labour market to reproduce the *status quo*. It is an education for life where work is an integral and fundamental part of that life, one of socio-political change which synchronises with memory, culture, and place.

The choice of this emancipatory perspective in relation to work is also present during the formation of the schools. In this research, we observed that the origin of the conflicts narrated by parents and former students in relation to the process of implementation of the Bairro Engenheiro Maia State School concerns the pedagogical concepts that permeated the practices of teachers, especially that of the former director Maria Elidia. By affirming that the teaching practice of the former director was consistent with traditional pedagogy, means that she was unaware of the life experience and demands of the settlers. If, on the one hand, we observe traditional pedagogical practice on the part of the former director, on the other hand, we understand that the her replacement by Celia Ferraz led to a fleeting attempt, by the parents and students of the MST, to implement dialogical pedagogical practices.

The management style of Celia Ferraz is highlighted in our study because it was an innovative experience. Although it remained short, we can see that it was a time when parents and students, that identified with the MST, felt that the school belonged to the community. Celia Ferraz did not offer any thoughts about her educational

[16] In 2006, João Antônio, the son of settlers of Agrovila 3, was studying medicine in Cuba.

philosophy, she informed us that under her management there was no formal moment for the elaboration of school planning in conjunction with the community. However, through the affective relationships that she established with both students and parents we were able to see that this educator was very keen to learn about the daily life of the settled community.

We observed that from 2005 the intentional formulation of educational political projects occurred in these schools, according to the records. In the State School at Agrovila 3 we observed different teaching practices between teachers Marô and Ruth Machado. The latter demonstrated some awareness of the educational objectives of the MST, as well as the theoretical contribution of Paulo Freire in the Movement:

> … They are very keen to follow Paulo Freire's psychology. Of course, we use it, we apply some of it, but we can't do everything. It's not possible, because then we move away from the curriculum here. What we can do, according to what they want, is extra classes with the children. (Ruth Machado, Teacher at the State School, Agrovila 3)

We observed that, for this teacher, the centrality of the teaching-learning process lies in the execution of a supposed elaborate plan *a priori,* which bears no relation to the experience of the students. By assuming that her teaching practice does not envisage the entire proposal of the Movement, or even to state that it does so 'in part', Ruth Machado revealed that she is not familiar with the complex proposal of the liberation pedagogy. Because, in order to implement dialogical practice, it is necessary to have as a starting point student experience, that is, an investigative methodology of everyday life and, from a collective construct, carry out the teaching and learning process. Thus, themes such as 'agrarian reform', 'work on the farm' and 'pesticides' presented by Ruth as secondary, are, in fact, central to the main life scenario of the students.

By failing to conceive the importance of getting involved in the dilemmas of their students and not understanding them as historical subjects, during the development of the workshop in the 'Education Week' held with the class of Ruth Machado, we were able to confirm the children's difficulty in exercising their autonomy in writing. The teacher did not encourage them, to write in their own way .

Concomitantly in the same school, during the teaching activities developed by the teacher Marô to present a memory of place, we were able to observe the development of the oral and writing skills of the children, as well as other language forms. Marô informed us that he noticed similarities between his theoretical and methodological points of reference of teaching and the pedagogical project of the MST. Due to the importance of everyday life in the teaching-learning relationship, the resourcefulness of the writing of the students of Marô acquired meaning. The different teaching practices developed in the State School of Agrovila 3 demonstrated that the act of writing, when developed together with life in the settlement, has meaning for the children, that is, their involvement in the teaching-learning process occurs as subjects.

Terezinha M. R. Gomes State School also developed activities which highlighted the life history of settled families. The 'Local Memory Project' project focused on, as a priority, the 3rd grade of elementary school in the years 2006 and 2007 and, during this period, its continuity was not confirmed. Nevertheless, it was an important experience, making it possible for us to affirm that this constituted dialogical pedagogical practices between educators, students and the settled community. In relation to educators, we observed that the intention was to build an educational political project that we consider to be consistent with liberation pedagogy.[17]

Regarding early childhood education in this same school, 'schoolwork rotation' drew our attention. We consider this form of school planning an innovative experience that is part of the liberation concept of education, because it has developed children's play activities with the purpose of promoting interdisciplinary knowledge. We had the opportunity to participate in the end of year party of 2007, in which the entire educational team of the school participated. Using theatre, singing and dance performances, among other forms of artistic expression, it was possible to observe the collective work of the early childhood education teachers.

The presence of a dialogical pedagogical practice highlighted in this research underpinned a substantive effort to (re)construct peasant life in the settlement. However, it is important to state that these practices were carried out in a timely manner and were not linked to a political pedagogical project developed by teachers and the settler community. We determined that the participation of the settler community in activities developed in the schools did not extend to the possible process of elaboration of an education pedagogical political project. Regarding this issue, Hélio a settler of Agrovila 5 stated: "[...] we conquered schools, but we do not engage in teaching."

This mis-encounter between the formation of schools and the absence of a politico-pedagogical project elaborated by a set of teachers, together with, and about, the settled community, constitutes the teaching-learning process in the Pirituba II settlement. Concomitantly we found that in view of its profitable interests the Orsa Group, represented by its Foundation, took advantage of this lacuna to forge proposals from its standpoint which were contrary to the complex (re)construction of peasant life in the region. It is important to highlight that the Orsa Group offensive was made possible through facilitation by state entities: Municipal Department of Education of Itapeva in the scope of teaching and Itesp with regard to the lease of land for a pine tree plantation.

In relation to the Orsa Foundation's proposals in schools in 2007, teachers were in favor of the development of educational projects related thematically to problems observed in the community, but not necessarily with the participation of entities where they were unable to identify the real objectives of the 'supposed partners'. The MST leaders who were observing this process challenged the involvement of the Foundation and its interest in schools in the region.

In relation to school education, our position is based on theoretical contributions that have no illusion about the reproductive nature of classes that hold power. The contradictions observed in capitalist societies are also produced and reproduced in

[17] See Freire, Paulo (2005) *Pedagogy of the Oppressed*, Continuum Press: New York & London.

schools. At the heart of these contradictions arise the "residues", as Lefebvre (1979) called them by analyzing more broadly the possibilities of insurgencies to power, both at the macro societal level and interpersonal relations. In this respect, Martins interpretations (1997) emphasize Lefebvre's contributions to our study:

> Utopia is in the residual, it is in what cannot be captured by power and those who have power. [...] The tendency for the apparent perpetuation of capitalism, its exploitation and its relations of power is not repeated without reproducing with it what denies it and that proposes, at the very heart of exploitation and oppression - liberation. The absolute power of capitalism does not exist, it is a fraud. It is a capitalist's dream. (Martins, 1997: 73–74)

The research presented here has demonstrated that given progressive management and the commitment of teachers, individually and collectively, to synchronise curricula with the surroundings and lives of students and their families a transformational educational curriculum is possible.

References

Bombardi, Larissa Mies. (2005). *Campesinato, Luta de Classe e Reforma Agrária (A Lei de Revisão Agrária em São Paulo). Tese de doutorado.* Programa de Pós Graduação em Geografia Humana, Departamento de Geografia, Faculdade de Filosofia, Letras e Ciências Humanas da Universidade de São Paulo.

Brandão, Carlos Rodrigues. Diário de Campo. (1982). *A Antropologia como Alegoria*. Brasiliense.

Caldart, Roseli Salete. (1997). *Educação em movimento. Formação de educadoras e educadores no MST*. Vozes.

Camargo, Maria Luíza Gutierrez de. (2015). *O latifúndio do Projeto Jari e a propriedade da terra na Amazônia brasileira.* Dissertação (Mestrado em Geografia Humana) – Faculdade de Filosofia, Letras e Ciências Humanas da Universidade de São Paulo.

Freire, P. (2005). *Pedagogy of the oppressed* (30th Anniversary ed.). Continuum Press.

Instituto De Terras De São Paulo. (2000). *Mediação no campo: Estratégias de Ação em Situações de Conflito Fundiário*. Cadernos Itesp. 2ª ed. São Paulo, n. 06, Dez. 2000.

Gohn, Maria da Glória Marcondes. (1994). *Movimentos sociais e educação*. Cortez.

Iokoi, Zilda Márcia Grícoli. et. al (coord.). (2005). *Vozes da Terra dos assentados rurais de São Paulo*. Fundação Itesp.

Lefebvre, H. (2001a). *A cidade do capital. Trad. Maria Helena Rauta Ramos e Marilene Jamur.* 2a ed. DP&A.

Lefebvre, Henri. (2001b). *O direito à cidade*. Trad. Rubens Eduardo Frias. Centauro.

Lefebvre, H. (1983). *La presencia y la ausencia*. Fondo de Cultura Económica.

Lefebvre, H. (1979). *Lógica formal/lógica dialética*. Trad. Carlos Nelson Coutinho. 2ª ed. Rio de Janeiro: Civilização Brasileira.

Martins, José de Souza. (1997). A questão agrária brasileira e o papel do MST. In: Stédile. João Pedro. (org.). *A reforma agrária e a luta do MST*. Vozes.

Marx, Karl. O Capital. (1984). *Crítica da Economia Política. O Processo de Produção do Capital.* Tradução: Reginaldo Sant'Anna. DIFEL.

Meihy, José Carlos Sebe Bom. (2002). Manual de História Oral. 4ª ed. Edições Loyola.

Minayo, Maria Cecília de Souza. (1998). *O Desafio do Conhecimento: pesquisa qualitativa em saúde*. Hucitec/Abrasco.

Oliveira, Ariovaldo Umbelino de. (2006). O latifúndio do eucalipto: a humanidade é nada frente à história da natureza. In: *Movimento de Mulheres Camponesas/Rio Grande do Sul. Entrevista especial*, 23 jun.

Paro, V. H. (2001). *Escritos sobre Educação*. Xamã.

Pistrak, M. (2000). *Fundamentos da Escola do Trabalho. Tradução: Daniel Aarão Reis Filho*. Expressão Popular.

Vigotski, Lev Semennovich. (2007). *A Formação Social da Mente*: o desenvolvimento dos processos psicológicos superiores. Orgs. Michael Cole...[et al.]; Trad. José Cipolla Neto, Luis Silveira, Solange Castro. 7.ed. Martins Fontes.

Combining Focus and Circumspection: An Education in Natural Inclusion

Alan Rayner

Abstract This chapter draws on my experience of presenting a trans-disciplinary course, entitled 'Life, Environment and People', at the University of Bath, UK, from 2001 to 2011. The course was available to final year undergraduate students studying for degrees in biological sciences, natural sciences, management and psychology. It was based on my understanding of the fundamental evolutionary principle of 'natural inclusion'. This principle expands beyond the abstract Darwinian concept of 'natural selection', which is founded in distanced perception. It enables us to understand self-identity as a local expression of habitat, not an independent entity required to compete successfully with others within a predefined arena in order to survive. It hence offers the possibility of learning to understand life in a more compassionate, creative and environmentally sustainable way than has come to predominate in modern human culture. This possibility is served by a cross-disciplinary educational approach that encourages intellectual discernment, emotional awareness, imaginative enquiry and a truly participatory learning process that is receptive and responsive to diverse viewpoints and human qualities.

Keywords Nature · Inclusion · Education · Creativity · Compassion · Sustainability

1 Introduction: The Need for Inclusive Perception in a Domesticated Predator (Photo 1)

The painting symbolizes the bringing together of unidirectional and all-round views of reality into a comprehensive situational awareness of individual self-identity as a dynamic natural inclusion and expression of its contextual neighbourhood.

A. Rayner (✉)
Bath and North East Somerset, Bathford, UK
e-mail: alan@admrayner.plus.com

J. J. McIntyre-Mills, Y. Corcoran-Nantes (eds.), *Transformative Education for Regeneration and Wellbeing*, Contemporary Systems Thinking,
https://doi.org/10.1007/978-981-19-3258-8_20

Photo 1 'Honeysuckle Sharing Circle'. (Oil painting on canvas by Alan Rayner, 2003)

A candelabra of honeysuckle flowers at different stages of development emerges from a central focus in receptive-responsive relationship with an outer circle of flowers representing a spectrum of colours differentiated from 'white' light.

Focus and Circumspection
We have eyes on the front of our faces
Giving us binocular vision
We have fingers and thumbs
Enabling us to grasp and throw
We have massive frontal brain lobes
Helping us to focus our attention
.
All of which make us great
At sorting wheat from chaff
And singling out the prey and fruit and vegetables
That keep us alive
And hunger at bay
As we make our way
Through the wildness of life
…

But there comes a day
When we feel the need
To be at home
As well as away
To be settled
Not always on the move
.
And for that we need
A different kind of vision
A circumspection
Including ourselves
Within what's all around us
A panorama
That can only come circumferentially
In sharing circles of view
And deep introspection
From me into you
And you into me
Until we can see
Both inwards and outwards
Radially and tangentially
Quite naturally
In very good company
.

As bipedal apes lacking the ability to digest cellulose and having a hunter-gatherer ancestry, it is always important to recall that as individuals we are predisposed to view the world about us in a narrowly focused way.

In itself, such focused attention is not a bad thing. If not accompanied by other modes of perception, it can, however have the detrimental effect of mentally dislocating us as external observers from what we observe in a detached frame of reference or 'perspective' as 'objects'. It goes with a life style on the move as individuals or small groups searching out and being able to distinguish and gather, catch and kill the sources of food that we need to sustain our lives. And so long as this life style persists, we remain in a dynamically balanced relationship with our natural environment, never more than temporarily exceeding in population density what can be sustained by the organic and inorganic resources within our immediate vicinity. It is a well-known ecological truth that wild individual predators cannot outnumber, and indeed are often much rarer than the prey on which they depend.

Problems begin, however, when we begin to settle and socialise in large numbers, culminating in the huge and complex societies and conurbations of modern civilization, which are served by increasingly intensive systems of agriculture and international trade. Under such circumstances the persistence of a separative – and hence competitive – world view will be prone to put us at odds with one another, other life forms and our surroundings. This causes psychological, social and environmental harm by way of conflict, depredation and deprivation. Moreover, these problems can't simply be solved by switching to an entirely unified lifestyle in which individual uniqueness is suppressed in monotonous cultures of collaborative conformity. Such cultures are incapable of responding creatively to changing circumstances. We haven't evolved to live that way and it is no more ecologically sustainable in the long run than its competitive alternative.

What is needed in this situation is a way of understanding our human place in the natural world that combines and balances – rather than polarizes – separative and unifying world views into a more comprehensive situational awareness that neither view can provide on its own. In other words we need to combine our ability to concentrate our attention unidirectionally on particular localities outside of ourselves with an awareness of what occurs all around and within ourselves. We need to combine focus and circumspection as a naturally balanced and co-creative way in which to live and relate to one another and our surroundings.

By the turn of the millennium, as an evolutionary ecologist, mycologist and university educator, I sensed that this need had become urgent, both for my own psychological wellness and for all my companion human and other life forms on planet Earth. The twentieth century had brought unprecedented scientific and technological advances, yet these had been turned to misuse in two devastating world wars and numerous smaller-scale wars. An 'Iron Curtain' had been drawn between rampant western capitalism and stifling eastern communism. Meanwhile a so-called 'third world' experienced terrible poverty and exploitation. Dreadful racial and religious prejudices burgeoned as the natural world we inhabit became an increasingly depleted and polluted place. Fearful depression and anxiety marred my own and many others' mental health.

So, what were our academic institutions, and especially our scientific institutions doing to alleviate this growing human and planetary crisis? Precious little, so far as I could tell, beyond attempting to design and impose short-term technological and legalistic fixes. And in many ways they were aggravating the crisis by promulgating the hierarchical, separative and monopolizing world views that had done so much to precipitate it. Notions of 'selfish genes 'prevailing in a Darwinian 'struggle for life' gained prominence alongside 'Game Theory', 'monetarism' and the materialistic belief that no reality exists other than that which can be quantified. Research and education succumbed to political and economic expedience in place of truth-seeking. Far from being 'led out into wider awareness', students were required to assimilate and pass tests affirming their adherence to the status quo.

Eventually, in the soul-searching wake of a personal crisis of confidence in my ability to do what I felt expected to do as an academic, I resolved instead to do whatever I could to assist in a process of educational transformation based on combining depth of focus with all-round circumspection.

2 'Life, Environment and People' – An Education in Natural Inclusion

For much of my working life I had been a teacher and researcher in biology at the University of Bath, UK. I wanted to share the enthusiasm for the natural world that had grown within me during my childhood and student days.

My educational intention had therefore always been to help students to acquire scientific knowledge and understanding in a memorable, pleasurable, discerning and imaginatively open-minded way. But in the prevailing competitive Darwinian climate of Academia I felt compromised by the pressure to provide and acquire technical knowledge without any deep appreciation of its meaning, origins, interrelatedness and implications for the way we live. Such pressure limits possibilities for innovation, synergy, versatility and wise practice in the long run. I had therefore sought to inspire delight in the diversity of the natural world and a common-spirited, questioning approach to understanding and learning *from* (not just *about*) it, which takes nothing for granted.

Two questions guided both my teaching and my research regarding any scientific proposition: is this consistent with sensory experience (i.e. 'evidence') and does it make consistent sense? I invited students to ask these questions themselves as a rigorous way of exploring the foundations of current scientific praxis and its application to real world problems:

- How well does this correspond with actual findings?
- How readily can it be integrated with understandings from other viewpoints?

Those questions reflected my concern that strongly hierarchical and commercially motivated education systems may not adequately help students become aware of the influence and underlying assumptions of scientific knowledge and praxis in a social and environmental context. As will become evident, I had good reason to feel this this concern.

I wanted to awaken possibilities in students for more realistic, imaginative and empathic forms of scientific understanding and enquiry. I worked towards this by explicitly including philosophical and aesthetic considerations within basic courses on fungal and plant diversity, plant symbiosis and pathology and biological ethics. But I also took the opportunity to innovate more radically by developing a transdisciplinary final year undergraduate course entitled 'Life, Environment and People' ('LEP').

I presented LEP from 2001 to 2011. I included mathematical, artistic, social, psychological, cosmological, educational and technological as well as specifically biological and ecological topics. It was available as an option to students studying for Biology, Natural Sciences, Psychology and Management degrees. Between 60 and 80 students typically attended each year.

I have never felt more strongly committed to my work as an educator than I did during this course. It epitomised all that I feel a true University education, indeed *anyeducation*, should strive for by way of a 'leading out into wider awareness in preparation for life yet to come'. I hoped it might provide a model for a new educational approach of the kind so desperately needed in what has come to be known as a volatile, unpredictable, complex and ambiguous (VUCA) world. The students themselves mostly loved the course and some still write to me saying how influential it has been in their lives.

But my fears of not being understood were to be born out, all the more so now in 2021 when so much of what I foresaw at the turn of the millennium is coming to pass. Presenting this course wasn't easy for me. It was one of the hardest, most harrowing endeavours of my working life. I presented it in the teeth of profound resistance, if not outright antipathy and misrepresentation from colleagues and external examiners. In retrospect I feel it was a small miracle that I kept it going for so long until the mental strain became too much. Had it not been for the support of the students themselves and some colleagues in other Departments, it would have been curtailed within the first three years.

What was so challenging about it to the existing academy? To put it bluntly, I wasn't preaching the Gospel of reductionist Science as set down by prior authority. I was told that I was confusing the students and undermining their faith in what they had been taught to accept as unquestionable. I was criticized for introducing them to what at the time were quite novel concepts, such as fractal geometry and chaos theory, and for casting doubt on genetic determinism and the legitimacy of definitive logic. I was accused of saying things I would never say, of being anti-scientific, of lacking intellectual rigour and clarity, and of engaging in 'free-fall philosophy'. I was also castigated for introducing students to my own heterodox thinking while not offering its orthodox alternative. But what did I *actually* do?

2.1 Four Fundamental Pedagogical Principles Underpinned LEP

1. The explicit intention to learn *from* life science about our human relationships with our natural neighbourhood in ways that enhance quality of life.
2. Participatory and improvisational enquiry. Through a questioning and conversational approach, I encouraged students to express their personal views and knowledge for others to consider and learn from. I expressed my own emerging understandings of 'natural inclusion' (see below). Nonetheless, I strictly regarded my role *only* as an experienced guide to the terrain, *not* an expert instructor. I worked to a guideline 'agenda', but allowed the conversation to evolve its own course within this.
3. Willingness to question and invite questioning of all kinds of assumptions and beliefs to assess their validity and open up possibilities for new ways of thinking and working.
4. Encouragement of diverse modes of enquiry and communication, including artistic and poetic expression to complement conventional analytical methodology and language. As well as formal essays, students were encouraged to produce one piece of informal or 'artistic' coursework. Every year this was a source of great excitement, co-creativity and pleasure, with students displaying a wide array of 'hidden' talents that they put considerable effort into bringing out into the open. See: http://www.youtube.com/watch?v=0RjNYfVVAmg and http://www.youtube.com/watch?v=9empiCQHLVA

Given this varied and participatory approach, I needed to devise suitably flexible but fair and consistent assessment criteria. I found the following were effective:

Reflective Quality:- does the work accurately and thoughtfully reflect themes emerging during the course? Are the scientific ideas that are conveyed and/or challenged fairly represented, in a way that demonstrates sound discernment/understanding/scholarship in your own learning?
Creativity:- does the work display imaginative thought and (where applicable) practical resourcefulness in relation to the theme/subject matter addressed?
Communicative Quality:- does the work communicate a clear message and/or evoke imagination and thought?
Quality of Execution:- is there evidence of skilful work?
Endeavour:- is there evidence of care and effort?

Students were generally very enthusiastic. Once they got used to the freedom to think creatively and express their own knowledge and understanding – and feel valued for so doing – their enjoyment of co-inquiring with one another became palpable. One typical but especially comprehensive student comment was as follows:

> The advantages of this module to the students who attended it are beyond describing for those who did not and have not attended it. In fact, it is exactly what I expected, but never received, from a University education until this module. It challenges the way you think, begs that you broaden your horizons, opens your eyes to new ways of viewing the world and of being more in tune with what is happening in the world today while offering the best possible experience of a lecturer-led class I have ever had. This is complemented by a fantastic lecturer who had more impact on my time at University than all my other lecturers combined and the stimulation to young minds of such a sort is something only to be encouraged. This module challenges and does away with all of the staid and repetitive approaches to learning and teaching which I and the majority of my friends have found to be commonplace (and most disappointing) ... instead taking a fresh and wholly worthwhile approach which not only engages the student in the idea but, lo and behold, actually attempts to involve us and value what we think and say rather than encouraging us to hold no opinion other than that which can be "backed up" by those who have come before, thus generating an unhealthy attitude towards learning by promulgating the advance of only that which is already thought. Such an approach kills fresh ideas, dulls our intellect (not sharpens!) and devalues what we, as young students, have to offer as well as devaluing us in ourselves. This module should not be underestimated in its ability to create positive diversity of thought, of views and opinions and it was, to my mind, attended by the type of students that Bath University should be trying to cultivate and promote: conscientious, responsible, outside-the-box thinkers, debaters, fresh and brilliant new minds with a deep social and human conscience. I believe those who attended this course will be the leaders and thinkers of tomorrow, and will do all the better, and BE all the better as people, for having attended this module. Again, I cannot stress its importance and the impact it had on all those who attended, and the closeness generated among the group of people who did attend….
>
> This unit was the best one I took during what was already a very rigorous and developmental course at Bath. I believe the only thing that could make this unit better would be to either make it compulsory to all lecturers and all 4th year students or to promote it more broadly.

In the following sections I will discuss some of the issues that the course raised, how these have been developed in my subsequent work, and how they relate to current interest in regenerative development. I often introduced these issues by asking a simple key question that opened the door to deep and far-reaching exploration and sometimes unexpected findings.

Throughout you may notice the recurrence of the central theme of what I call 'natural inclusion' (e.g. Rayner, 2011, 2017, 2018, 2021). One way in which I describe natural inclusion is *the mutually inclusive, co-creative, receptive-responsive relationship between intangible spatial stillness and energetic motion in the being, becoming and evolutionary diversification of all material bodies, including our own*. Another way is 'evolution in response to receptive invitation'. A third is 'the mutual inclusion of love in life and life in love'.

You may also notice me using diverse modes of formal and informal linguistic and artistic expression, just as I encouraged the LEP students to do. And hopefully my educational intent in doing this will also clarify. You may also find it helpful to view this recorded conversation:- (167) "Mutual natural inclusion of receptive space and informative flux" – A conversation with Alan Rayner – YouTube (https://www.youtube.com/watch?v=IKBlGtzEJoo).

2.1.1 Self-in Neighbourhood; Neighbourhood in Self: The Comprehensive Situational Awareness of Natural Inclusionality

Here, I want to ask – as in this short video presentation https://www.youtube.com/watch?v=es6eg8e_0aY:- "Who or what is your neighbourhood?"

Why do I ask this question? Why does it deserve your special attention when you are already so bombarded with so many other concerns in your modern everyday life? What difference could seeking an answer to it make to your life, my life, human life, life on Earth?

What thoughts, feelings, sensations and intuitions are aroused in you as you contemplate this question? Does it arouse your curiosity? Does it puzzle you? Do you even want to contemplate it? What benefit could it bring you and your loved ones to do so?

To contemplate this question, I feel, helps us to develop and enhance what may be described as our 'situational awareness'. This is the naturally inclusive humility of understanding that we are not solely responsible for what we think and do – and what really is, by way of our spatial and energetic circumstance, and how we need to relate to it.

I say this because the question helps us to explore the relationship between the uniqueness of our individual self-identity and situational context. Deep tensions arise whenever we think that the former either excludes the latter (as in Einstein's view that 'the environment is everything that isn't me) or is indistinguishable from the latter (as in non-dual mysticism).

When I used this question to introduce LEP, the initial responses I received from students generally fell into two distinct categories. They were either individual-focused (my neighbourhood is what locally surrounds but doesn't necessarily include me) or group-focused (my neighbourhood is the local community of people of which I am one).

We were then able to explore together how and why such different views could arise and how the tension of being pulled in seemingly opposite ways by them could actually prove to be creative. I have been through this process myself, both emotionally and in my scientific research. I was attracted at first by the difference between seeking personal (individual) benefit and social benefit. Then I realized how and why each necessarily includes the other.

This is why the neighbourhood question is ultimately so deep and far-reaching in its implications. It invites us to reflect on the natural relationship between what physical science calls 'space, time, energy and matter', and where, if anywhere, 'self' and 'neighbourhood' are perceived to begin and end. Hence we were drawn into the philosophy of 'natural inclusionality', based on awareness of natural inclusion (see above) as a fundamental evolutionary and ecological principle.

Have you ever thought about the crucial role of space everywhere as an invisible, frictionless presence in Nature? I didn't until the turn of the millennium. Then I suddenly realized that although space is not a substance, it's not possible for any material form, let alone any kind of life, to exist without it.

And with that simple realization, much naturally 'fell into place' (or, actually, fell into 'place-time') in a way that was new for me. Moreover, much that many of us human beings have, for millennia, assumed or been taught is *definitively* true fell out of place as inconsistent both with our actual life experience and realistic reasoning. *This is because definitive thought isolates or equates matter from or with space instead of recognising that these two natural occurrences are distinct but mutually inclusive.*

Did you, or do you notice what I am doing in the above video? You see me first of all in a 'selfie' view, speaking words. Then I stop speaking and spin myself around. You see me not just objectively 'head on', but at the centre of a panoramic circulation in which I am bodily included. Then I walk forward leaving the place where I was standing behind me. Now, I spin around again in the other direction. Finally, I speak words of gratitude.

What do you make of this? Why did I make this short video? Let me now try to explain, with the help of this illustration, entitled 'Floral Awareness' (Photo 2):-

The painting illustrates different kinds of situational awareness, making allegorical use of *Narcissus* flowers. Please view from top to bottom and left to right, and refer to text for explication

Please imagine being within a solitary *Narcissus* flower.
Pause.

Are you finding this difficult? Are you questioning what's the point, because, quite clearly, you are *not* a *Narcissus* flower? Many of us would, because we've become so used to looking *at* what we see with our eyesight as being outside of our selves, separated from us by gaps of space and by both their own and our own bodily boundaries. Do you notice how this perspective view has the effect of excluding ourselves from our neighbours and neighbourhood?

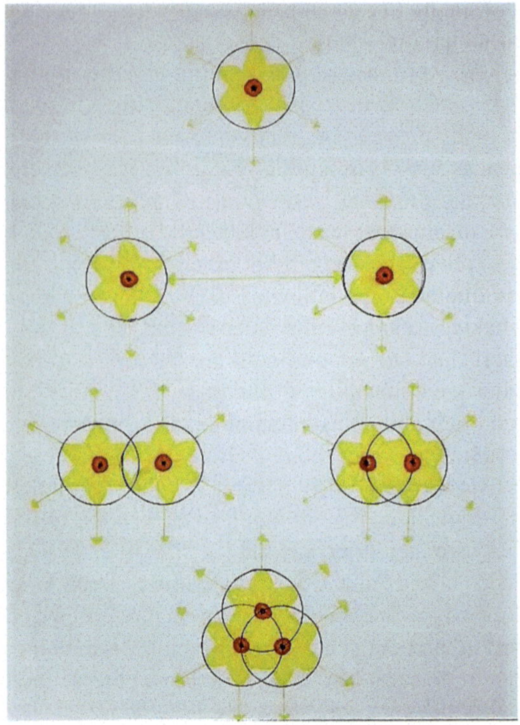

Photo 2 'Floral Awareness'. (Pen, ink and watercolor on paper by Alan Rayner, 2021)

Those of us who, like me, have been trained as natural scientists have even been taught that this is the *only* way in which we can impartially view and understand reality. We are urged to view nature as 'third persons' situated outside what we are observing and to analyse it anatomically into its constituent 'parts'.

And there's no doubt that we have learned an enormous amount via this 'reductive' route. For example, we would first see a *Narcissus* flower from outside-inwards as a star-formation of six radiating tepals (a technical name for petal-like sepals). These surround a cup called a corona. Within this is a whorl of 'male' pollen-bearing stamens around a central 'female' ovary surmounted by a receptive stigma and style through which fertilization takes place leading to fruit and seed-formation. Then, focusing down further we would recognise the tissues and cells that make up these structures. Then we would recognise the organelles, molecules and ultimately atoms and subatomic 'particles' within these cells.

But how much can we learn this way concerning how it really is to be a *Narcissus* flower? And how much would we learn concerning how it really is to be human using this same analytical process on ourselves? What would we learn about our emotions, our relationships, our place in the world? And, when it comes right down to the ultimate depth of focus at subatomic scale, how can we understand the findings of quantum mechanics?

Clearly this third-person 'distanced learning' from outside-inwards is not as comprehensively impartial as it is often made out to be! In fact in some ways it is profoundly biased and can be very misleading. It can even be harmful to our health, through deliberately ignoring the most deeply personal aspects of our being. Here, by way of illustration, is what Charles Darwin had to say about it:-

> A scientific man ought to have no wishes, no affections, – a mere heart of stone.

That is why I invite you – and invite myself – to imagine *being* within a solitary *Narcissus* flower.

I do this because it helps us to move f*rom reductive to receptive awareness*. Receptive awareness enables a radically different, more intimate kind of understanding to come into play. We can *empathise with* – not distance ourselves from – what we are observing. We can begin to appreciate what we are observing – and also appreciate our selves – from inside-outwards as well as from outside-inwards. And when we do that, our situational awareness expands. We include both ourselves and our neighbours dynamically within the common space of our natural neighbourhood.

But how can we do this? Here's how I do it:-

I look at the flower shown at the top of 'Floral Awareness'. Now I close my eyes and reach down imaginatively into the cavity of my heart. Here I feel, often achingly, my deepest joys and sorrows. I open my eyes and take another look at the flower. Now I allow the receptive, feeling awareness in my heart cavity to be present within the central corona, where the ovary is located (shown as a dark circle in the illustration). I retain that feeling awareness as I close my eyes again.

I notice how this feeling awareness radiates out in all directions from my receptive heart-centre into an endless open field of intangible spatial stillness. I reach out to a circular perimeter within this openness from within the inner sanctum of my grail-like corona to the tips of my tepals. But my awareness doesn't stop at this perimeter. Instead it continues infinitely outwards. I have no immediate awareness of past or future. I am simply here, now, somewhere in communion with space everywhere. All is quiet. All is calm. Blissful. Unconditional. Loving.

This is my primary, first-person, receptive awareness. I can recall it from infancy. I experience it when deep in meditation or even whilst sunbathing or floating on still water with eyes closed. My intellect is at rest. I have no thoughts or sense of my individual self-identity. I am not aware of my external appearance, or even that I have an external appearance.

Please notice that this does not mean I have no awareness. It means that my awareness is of the very deepest, dreamiest and most far-reaching kind. It is pregnant with creative potential. But I can easily forget it. When I do forget it, or even deliberately ignore it like Darwin's scientific man, I know from personal experience that the consequences are distressing. In itself it cannot sustain my life. But without it, my life becomes insufferable in many ways. I need to hold onto it, for dear life!

Suddenly, as in the second row of 'Floral Awareness', my blissful first-person awareness is interrupted some distance away by another *Narcissus* flower. How do I feel about it? Remember that at this stage I have no awareness of my individuality. So, what am I to make of this intruder? Is it real or illusory? Is it within or outside my awareness? Is it aware of me?

Now, please imagine with me that despite being a radiant expression of plant life, I am curious and I can move bodily from one locality to another. The other *Narcissus* intrigues me. I want to find out more 'about' it. All my first-person awareness may now be withdrawn as I focus my attention on its appearance from outside-inwards. I feel a desire to analyse it from third-person distance. Since I feel no affinity with it, I can harm it.

Now I find that if I touch it, it resists the pressure of my touch, as if it has a built-in inertia that wants to stay where it is. In that touching experience I realize that I too have a resistive bodily surface, even though I may not yet be fully aware of it. If the other flower moves, I realize that it can be no longer where it was and it could also be no longer where it currently is. It has both a historical 'past' and a yet-to-become 'future', as I do, too.

I wonder what makes it move. I perceive this as an external 'force' – a 'pull' or a 'push'. This force accelerates it across a measurable distance within a measurable duration. I have entered into pure 'third-person' awareness. This is what led Isaac Newton to formulate his mechanical laws of motion. It has become idealized by the analytical science and logic that isolates the 'observer' from the 'observed', and 'self' from 'neighbourhood'.

I notice a lurch in the cavity of my heart as my awareness shifts from inside-outwards all-round receptivity to 'straightforward', narrowly focused 'tunnel vision'. My awareness has been hugely embellished by the knowledge of tangible material form, including my own body. I may catch fascinating sight of my body in a reflective surface or photograph. But I believe that it is quite separate from other bodies, even if, like the other *Narcissus*, they resemble my own.

It is, quite literally, a shattering transition, like falling out of love. I leave the idyll of my primary, uninterrupted, receptive first-person awareness behind and walk straight into the grip of a mental trap. This trap fixes the position of my own and other bodies in a static box-frame of distance and duration. The latter are divided up into standard lengths, such as metres and hours, as if they are tangible substances. Now I really do feel all alone, isolated from, not all at one with everywhere. I am simultaneously an independent agent and a boxed-in definitive singleness at odds with others and my surroundings. The other *Narcissus* is a competitor, who I am pitted against in a life-or-death struggle for existence. So the Darwinian story of evolution tells me.

This trap can be very difficult to get out of, especially if it is constantly reinforced by your peers and teachers. Yet there is a way out, and this becomes apparent to me as soon as I allow my first-person awareness to combine fluidly with my third-person awareness in a mutually inclusive rather than mutually exclusive way.

I begin to enter the increasing intimacy of 'second-person' awareness if I and the other *Narcissus* can recognise one another as having in common both a distinctive external form and a receptive awareness radiating from within into overlap with each other. Space no longer seems to distance us but is the continuous receptive, silent, frictionless stillness everywhere that includes both of us, and that we both include in our bodies. Our bodily boundaries no longer seem to isolate us, because they are freely permeable to space as a continuous immaterial presence that cannot be cut or displaced. And because this is so, they cannot be fixed in place. They can only be formed temporarily. Ultimately, at subatomic scale, all material bodies are 'flow-forms', tangible manifestations of natural energy flow around local receptive centres of space.

Notice how the third person view of material bodies as inert objects that are forced into motion from outside has been turned inside-out through its mutual inclusion with first person receptive awareness. *Matter is now understood to be an energetic configuration of space somewhere within space everywhere*, not a rigid exclusion of space. Receptive influence is the primary source of natural motivation, not executive force!

Holding the receptive openness of space dynamically within our hearts, I and the other *Narcissus* feel called reciprocally into one another's awareness We each respond energetically to this calling of *mutual attraction* – the love in our hearts – by drawing closer. Then, *if* we can *attune* compatibly, a binding relationship begins to develop through the reciprocal flow of energy around and between us. We each feel enlivened by this flow as we become a 'loving couple'! But if our energy flows are not compatible, we will dissociate.

Not every loving couple lives 'happily ever after', however! There is a possibility that in our second-person reciprocity, I and my partner may become so intimate that we withdraw our first-person awareness from the wider-world. As a couple, we now become just as isolated from our neighbourhood as might a single individual.

Another possibility is that as a single individual aware of my own bodily form, I may form a second-person bond exclusively with myself. This is a psychological condition known as 'narcissism', which can result in vanity and a will to power over others.

These possibilities for 'self-isolation' or 'couple-isolation' dissipate as soon as we regain first-person awareness. Now the possibility for second-person relationship extends to groups of three or more. But again, if first-person awareness is lost, cults and cliques may form as 'bunches of Narcissi'. This is why I remarked earlier, 'I need to hold onto it, for dear life!'

So what does this all mean with regard to my introductory question: who or what is your neighbourhood?

What I have sought to bring out is that first-person (subjective), third-person (objective) and second-person (reciprocal) awareness each reveal a distinct aspect of *our contextual situation as dynamic local inclusions and expressions of our natural neighbourhood, in variable spatial and temporal proximity to one another*:-

First-person awareness reveals the receptive continuity of our beingness, radiating from inside-outwards within the infinite, intangible stillness of space everywhere.

Third-person awareness reveals the occurrence of material form and movement, as sensed from outside-inwards

Second person awareness reveals the receptive-responsive relationship between spatial stillness and energetic motion that manifests in material forms and their association and dissociation.

None of these in isolation is sufficient to yield comprehensive situational awareness of our natural neighbourhood. Each is prone to be misleading in a different way. Comprehensive situational awareness – what might aptly be called '4th person awareness – requires the distinctive contribution of all three in mutual correspondence. This is what I have been seeking in my explorations and explications of natural inclusionality.

As a further illustration, here is how I might draw simultaneously on my first, second and third-person awareness as a first-aider attending you in an emergency. I would draw on my first person awareness to bring a feeling of calm. I would draw on my second-person awareness to treat you with compassionate consideration as a companion human being in need of care. I would draw on my third person awareness to identify what's wrong and what's needed to remedy it.

Now, I want to consider some of the profound implications of natural inclusionality for life, environment and people, which came to light during our enquiries.

2.1.2 Embodied Water Flows and Sentient Evolution: Why Organisms Are Not Machines

The idea that living organisms are sophisticated machines has long been with us as a product of third-person perception. Before evolutionary theory became generally accepted, organisms were widely regarded as the creations of an external maker, in much the same way that we regard the products of our own manufacturing industry. And to this day it remains common for the structure and function of living organisms to be described in terms of machine and construction metaphors such as 'building blocks', 'blueprints', 'component parts', 'wiring', 'plumbing' etc.

There is something deeply troubling about regarding human and other organisms as reproducible living machines, which Mary Shelley's depiction of Frankenstein's Monster was intended to evoke. Some *vital quality* is missing from this mechanistic picture. The Romantic Movement sought to point out this missing quality. But it was overlooked in the rush for technological progress. Meanwhile, far from using an understanding of evolutionary process to expose the myth of the externally created machine-organism, adherents of natural selection theory developed an even more emotionally alienating metaphor.

During the twentieth century, the widespread acceptance of Darwinian theory combined with the discoveries of genetics and development of information technology to convert the natural organism into a fully autonomous form of artificial intelligence, complete with hardware and software. The organism became a set of designer genes in a robotic body whose only function was to convey digital information from one generation to the next as exclusively and prolifically as possible. This selfish replicator came fully equipped to exterminate its opposition and calculate the costs and benefits of its actions in order to make more of itself in what, apart from random accidents, is a fully definable future.

What could be missing from this picture? I put it to LEP students that when scientists explore the possibility of biological life evolving on other planets, the first substance they look for isn't DNA, but an altogether much simpler chemical compound: water. Why? What is it about water that is so *vital* to biological evolution?

We recognised that in carbon-based biological life, water takes on the role of *receptive medium* that space has universally. Life, when active, is wet, not dry! Underlying this role is the receptive-responsive relational chemistry of what happens when electrons flow around and between atomic centres. Here are some of the details:-

Water is a surface-tense *liquid* at atmospheric pressure and between 273 and 373 °K (freezing 'point' and boiling 'point'). This is amazing in view of its low molecular weight, but can be understood in terms of the receptive-responsive relationship of oxygen and hydrogen atoms, which gives rise to a phenomenon known as 'hydrogen-bonding'. The inductive space within an oxygen atom is *receptive* to what is conventionally labelled 'negative charge', conveyed by electrons, whereas a hydrogen atom *responsively releases* negative charge from the environs of the 'positively charged' proton in its nucleus.

Correspondingly, water molecules are *dipolar*, with electrical charge being distributed unevenly so as to be relatively negative in the vicinity of oxygen and relatively positive in the vicinity of the two hydrogen atoms. Complementary charges attract, like male and female, and so water molecules *couple* with one another to variable degrees to form complex associations. These associations become more fluid as temperature increases towards boiling point, and stiffer as temperature decreases towards freezing point. Significantly, the densest configurations occur 4 °K *above* freezing point, so that ice floats on water – and life can continue to move around in seas, lakes, ponds and puddles whose surface is frozen.

This electrical dipolarity is also very significant in enabling water to participate in the dynamic relational flow form of organisms as a *differential solvent* – a medium that dissolves and so fluidises materials to varying degrees depending on their composition. 'Polar' materials are 'hydrophilic' ('water-loving'), and dissolve readily. These include salts, sugars, acids, alkalis, alcohol and *some* parts of proteins. 'Non-polar' (neutrally charged) materials are 'hydrophobic' (water-repellent), and don't dissolve readily. They include fats, oils and *other* parts of proteins. Whereas hydrophilic materials are permeable to water flow, hydrophobic materials are impermeable.

By bringing together a suitable mix of hydrophilic and hydrophobic materials in a watery context, it is possible for living organisms to assume an enormous variety of flow-forms, based on the development of what is known as 'osmotic pressure' within a partially permeable bodily boundary. This variety corresponds in many ways with the forms that water itself assumes within, upon and above Earth's surface. Bodies of water can dissociate into variably sized individual droplets and branches and associate into streams, rivers, pools, clouds and labyrinthine networks, depending on how and where they gather in, distribute and conserve sources of energy in diverse circumstances. The same kinds of river-like pattern can be found in slime bacteria swarms, slime mould plasmodia, fungal colonies, army ant swarms, wildebeest herds, trees, leaf venation, blood systems and nervous systems. The same kinds of currents, ripples and vortices as those found in ocean and atmosphere occur both within individual bodies and collective groupings of organisms ranging in scale from minuscule to global.

All in all, organic life as an embodied water flow is a far cry from a digital computer. It is also a far cry from the suspended animation of dormant life in what is truly 'survival' mode (see below). The latter conserves energy in seeds, spores, eggs, cysts, sclerotia, pupae etc. It does not 'compete to survive', but 'survives to revive'.

There is, however, an even deeper way in which organisms differ from machines. And to appreciate this, it is necessary to bring to mind our first person awareness of innermost receptivity. Living organisms are *internally motivated* manifestations of biological evolution whereas machines are both designed and moved by external agency. In other words, organisms possess sentience – some form of receptive and responsive awareness of their natural situation and future possibilities – whereas machines do not.

Attempts to explain the origin and evolution of life as if organisms are machines designed, driven and controlled by external agency therefore fall fundamentally short of addressing the pivotal question of life: whence and how does sentience arise? Even Charles Darwin recognised this problem when he commented, 'one might as well speculate about the origin of matter'.

Darwin's comment, made at a time when atoms were still believed to be hard, massy, indivisible particles, was unwittingly prescient. For, as Doug Marman (2018) has shown convincingly, all the mechanistically inexplicable findings of quantum mechanics that have been made subsequently have equivalents in the receptive-responsive relationships formed by living organisms with one another and their natural surroundings. This suggests that the origin of life and the origin of matter really do share a common ancestry.

Moreover, it is possible to work out from simple first principles what is essential both for the existence of matter and for the existence of life, but has been overlooked by all scientific theory and methodology that views reality solely from the standpoint of an external observer who discounts internal agency. Quite simply, as I have clarified here and elsewhere (e.g. Rayner , 2017, 2018), this is the mutually inclusive relationship between omnipresent intangible spatial stillness and locally informative energetic flux as receptive and responsive presences in all material bodies, including our own living human bodies.

Here I will focus on the central role of intangible receptive spatial influence as a source of internal agency in the origin and evolution of tangible material and living form. No sooner is this intangible (i.e. massless and frictionless) influence recognised in the core of all living bodies as well as atomic and subatomic form then our perception of the reality of nature shifts radically from definitively objective to dynamically inclusive.

We begin to appreciate that natural space and boundaries are sources of receptive continuity and dynamic distinction, not definitive discontinuity as in the abstract instantaneous geometry of dimensionless points, breadth-less lines, depthless surfaces and isolated solids. Matter is not ultimately assembled by external force from pre-existing and separable discrete entities ('building blocks') and neither is life.

Both arise dynamically from energetic flow around and between receptive centres of space somewhere within space everywhere. The role of these receptive centres is invitational and catalytic, not executive. They call energy into circulatory flow-form around themselves, they do not control it. *Evolution occurs primarily in response to receptive invitation, not mechanical causation.*

2.1.3 From Automaton to Sharer: The Inclusional Nature of Individuals

What Is an Individual?

That was the research question that ultimately led me to question everything I'd been told by conventional analytical science in general and evolutionary biology in particular. *Because it has no definitive answer.* As William Wordsworth put it:- 'In nature, everything is distinct, yet nothing defined into absolute, independent singleness'.

The very idea that reality can be subdivided or aggregated into discrete numerical units that can be treated as independent 'wholes' acted upon by external 'force' is inconsistent with evidence and incapable of making consistent natural sense. It is an abstraction from nature, not a truthful representation of nature. And while it can be useful in the short term for certain navigational purposes, it is capable of causing great psychological, social and environmental harm.

The assumption of definitive completeness, which underlies modern analytical science and artificial intelligence is simply false. Nature is intrinsically fluid, not digital. There is no definitive separation and

There are no wholes in Nature
Not even one.

There are only dynamically distinct localities in place-time. And the reason for this is simple. Natural space is continuous. Natural boundaries are dynamically informed.

Material bodies, from sub-atomic scale outwards are *flow-forms* – co-creations of space and energy in receptive-responsive relationship. Every body dwells within every other body's receptive-responsive gravitational and energetic influence. Natural boundaries are intrinsically dynamic distinctions between inner and outer space, not rigid partitions. Natural space is an infinite, intangible, continuum, not a 3-dimensional box. Evolution occurs in response to receptive influence, *natural inclusion*, not selective external force. Individuals inhabit and give local expression to the shared space they all inhabit, dynamically associating and dissociating, never self-isolating.

2.1.4 Dissipative and Conservative Organization: The Essence of Regenerative Sustainability

The very term applied to an 'organism' implies that it is not just a haphazard assortment of odds and ends, but something that is 'organized' in some way. What is natural organization, how does it arise and how is it sustained or lost?

You might imagine that this is a question to which modern evolutionary, thermodynamic, social and complex systems theories would have the answer well sussed. There is, however, a simple and perhaps by now familiar life lesson – readily available to any of us from our personal experience – that these abstract theories have not learned:-

All falls naturally into place when the receptive influence of continuous space and informative influence of dynamic local boundaries are taken into consideration.

All falls paradoxically out of place when space and boundaries are treated objectively as sources of definitive separation or unification.

We can thereby recognise that there are *two* radically different yet complementary modes of natural organisation.

Naturally dissipative organisation maximises boundary surface exposure in response to *external* receptive influence through proliferation of particulate, folded or branching structure.

Naturally conservative organisation minimizes boundary surface exposure in response to *internal* receptive influence through integration into packaged or networked structure.

In simple terms, a rough, leaky surface can both absorb and lose more energy from and to its outside than a smooth, impermeable one.

This distinction has especial relevance to understanding the evolution and ecology of living systems, and how this relates in turn to human organisations. To put it very basically, as summarised in Photo 3, *lifeforms both thrive and survive by dynamically balancing between dissipative and conservative organisation as internal and external energy availability varies* (Rayner, 1997).

Dissipative organisation is *generative*, associated with assimilation and distribution of external energy sources into growth and reproduction. For example, a flowering plant assimilates the energy of sunlight into the proliferation of roots, shoots, branches, leaves, flowers, seeds and vegetative propagules.

Conservative organisation is *protective*, associated with the storage and redistribution of internal energy sources within survival and exploratory structures. A flowering plant reduces loss by means of impermeable coatings and reduction of surface area to volume ratio within enlarged storage organs and elongated, sparsely branched or unbranched roots and stems that pioneer growth into new territory.

The ability to *combine* and *convert* between these modes enables life forms to vary their pattern of development to suit changing environmental circumstances in different dynamic localities ('place-times'). Conversion from dissipative to conservative organisation is associated with boundary integration and sealing. Conversion

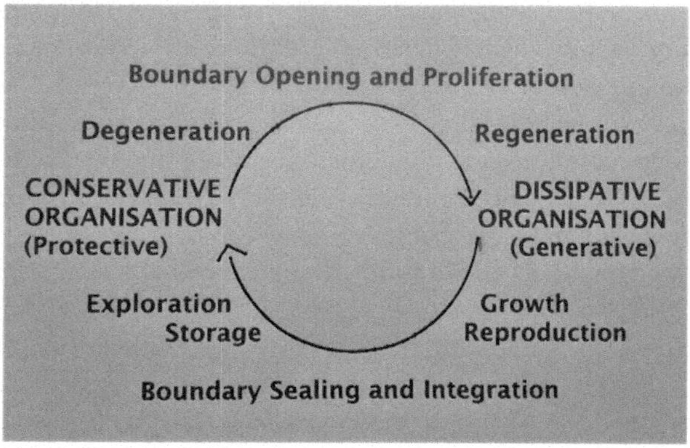

Photo 3 The relationship between conservative and dissipative organization in living systems

from conservative to dissipative organisation is associated with boundary opening and proliferation.

This is the essence of ecological and evolutionary sustainability as a creative, degenerative and regenerative process. A simple lesson, you might think, and one that any of us can learn simply by observing how life forms – including us human beings – naturally attune their development and behaviour to the circumstances we find ourselves in. And yet a lesson that we miss as soon as we mentally isolate ourselves and others, by *definition*, from our spatial and energetic context. For we then dislocate our thinking from our awareness of how it feels, both inwardly and outwardly to be alive and vulnerable, in need of sustenance and care.

2.1.5 The Natural Mathematical Physics of Time as Current and Space as Stillness

There is a radical difference between the abstract binary logic of mutual exclusion or unification upon which conventional mathematics and physics is founded and the natural discernment/inference from inner receptive awareness of mutual inclusion and dynamical distinction (natural inclusionality). The latter is a relational 'Logos', which includes the intangible beingness/stillness of space within the dynamic becomingness of motion, and the dynamic becomingness of motion within the intangible beingness of space.

With natural inclusionality comes an understanding of the reality of the natural flow geometry of *current time* locally within and around the continuous stillness of openly receptive space everywhere. Intangible future possibility precedes and invites tangible current actuality, just as the tangible flow of a river both creates and follows spatial paths of least resistance through landscape under the receptive influence of gravity.

To disregard intangible reality is to create the perception that material past mechanically *causes* material future. To disregard tangible reality is to create the perception that 'all is one continuum' and tangible phenomena exist only as 'appearances' in the mind. Both perceptions in different ways are unworkable.

In Nature, mechanical causation is subsidiary to situational causation. Clockwork time is subsidiary to energetic current. Material distance is subsidiary to spatial omnipresence.

Newtonian mechanics, which *does* disregard – as an absence of measurable/tangible presence – the intangible receptive influence of space – as a frictionless presence – overlooks this Logos. Which is why quantum mechanics cannot be understood by Newtonian mechanics. The latter overlooks 'the vitality of the intangible' (Rayner, 2018) from a third-person standpoint. The Logos IS mathematical physics, but not as we know it, Mr Spock! It is a new, and much more fundamental kind of physics, which traditional physicists resist like the plague it isn't.

Infinite Quality
Everybody is a dynamic inclusion of infinity
Say that again
Every body is a dynamic inclusion of infinity
.
What does this mean for you
And what does this mean for me?
.
How does this affect our calculations?
.
Quantitatively, not in the least byte
Qualitatively, immeasurably
.
Every body is immeasurably more than
What it's counted to be
.
That includes you
That includes me
So let's learn to live within
Each other's loving reach
Most carefully
.
Infinity is not a whole number
And neither are we
Infinity is the limitless welcome
In grace
The immaterial space
From which we all emerge
To which we all recede
In droplets of light
To live and find rest
In stirrings and peace
.

Unbroken Movement
So, you want to know
The meaning of 'Flow'
As the meaning of 'Life'
Transcending all strife
Around and between all bodily form
As standard deviations from idealized norm

.
Well, the answer's quite clear
If you don't hold so dear
To the notion
That motion
Is something forced from outside
The swell of the tide
In infinitesimal units of length
Mistaken for strength
By calculus mind
Unprepared to be kind
In a world it perceives
To be nothing but leaves
In a book made to measure
The size of our treasure
.
Flow is unbroken movement
In unbroken time
In unbroken space
Around and between receptive centres of grace -
This limitless, soft stillness,
Which we cannot sever
Then, now, or forever
With any hard edginess
That seeks to rule and divide
The place where we hide
From where we belong
In the heart of our song
.

2.1.6 The Natural Confluence of Art, Science and Soulful Spirituality in an Aesthetic Appreciation of the World About Us

I have long felt that a deep and truthful understanding of our human place in the natural world should *combine* rather than *oppose* scientific, artistic and spiritual perceptions. I think that natural inclusionality offers us the comprehensive situational awareness to make this possible. Since words alone are, however, inadequate to convey this awareness (see below), I have repeatedly asked myself 'how can the mutually inclusive relationship between scientific, artistic and spiritual perceptions of Nature be shown pictorially?' Photo 4 shows one of my more recent attempts to do this, and you may notice in it also some resemblance to Photo 3.

2.1.7 The Natural Language of Flow: From Static Definition to Fluid Distinction

The verbal language that we use inescapably both reflects and reinforces the way we think. Given the cultural predominance of definitive thought, many of us have been brought up to use language in a very rigid way, believing that this enhances clarity, certainty and precision and removes ambiguity. Moreover the structure of the English language itself separates subject from object, uses ‘nouns’ as ‘thing’ words and verbs as ‘doing’ words and defines word meanings in dictionaries. This has the effect of consolidating definitive thought at the expense of recognising the intrinsic fluidity of natural phenomena that arises from the receptive-responsive relationship between space and energy. By seeking to ‘define things’ we immediately fall into a language trap that can give quite the wrong impression of what we truly mean. This is especially a problem for anyone who recognises and wants to communicate the reality of natural inclusion. This is why I am hyper-aware of language that implies one kind or other of definitive isolation of ‘oneness’ from ‘otherness’. I try to avoid it as best I can by finding more fluidly descriptive words and using poetic and metaphorical form – even if this is liable to annoy or puzzle literally-minded people. Hence, for example, I reserve the use of ‘connection’ literally to describe tangible linkage, as when two electrical wires are placed in contact with one another. ‘Space’

Photo 4 ‘Confluence’ (Oil painting on board by Alan Rayner, 2021) Here combined are the local tangible elements of Earth (solid), Water (liquid) and Air (gas) formed from the intangible elements of Fire/light and Space/darkness, symbolized further by the two wings, two tail streamers and head and body of an arctic tern (which migrates annually between Arctic and Antarctic circles). We also see a transition from radial/spherical symmetry associated with stillness and bilateral symmetry associated with movement. Correspondingly, the imagery symbolizes the vitality of the mutually inclusive bodily and soulful/spiritual middle in bringing seemingly opposite but actually complementary viewpoints (as per left and right wings) into confluence

as an intangible presence cannot literally do this. This is why when speaking of 'mutual influence' I tend to use the language of 'relationship', 'reciprocity', 'attunement', 'confluence', 'resonance' etc, not 'connection'.

2.1.8 Nature as a Source of Moral Guidance: From Selfishness and Altruism to Needfulness

The idea that Nature is at best amoral and at worst totally immoral – an indiscriminate rag bag of good and bad, or cesspit of bad behaviour – has long been embedded in the human psyche. It is epitomized by this comment made by Richard Dawkins:

> If you wish, as I do, to build a society in which individuals co-operate generously and unselfishly towards a common good, you can expect little help from biological nature. Let us teach generosity and altruism, because we are born selfish.

Here, I suggest that the converse is true: an awareness of how we naturally are as needful, living human beings in the world as it actually is, full of life, is a source of moral guidance that enables us to value honesty, reasonableness and kindness as vital qualities of loving care for self and neighbourhood. What causes us, on the other hand, to be selfishly deceitful, unreasonable and cruel is the product of a profound misconception of reality, which isolates or conflates human identity from or with the remainder of Nature. What, then, does Nature's moral guidance have to say to us?

The 'Golden Rule of Nature' is to live, love and be loved in accordance with how you naturally are in the world as it actually is. This can also be framed as 'love your natural neighbours and neighbourhood as you love your self', or, even more simply as 'Love other(s) as you love your self'. This is very different from the pursuit of hegemonic or coercive power – through the ideological abstraction or conflation of human self-identity from or with Nature.

What comes into your mind when you hear me speak of 'Nature'. Perhaps a tranquil rural idyll? Maybe a riot of sex and violence? Somewhere you yearn to be or want to 'connect' with? Somewhere you want to escape from or avoid like the plague? If you look 'Nature' up in a dictionary you'll find any number of definitions and shades of meaning. These depend upon whether and where Nature is thought to begin and end and whether human beings and artefacts are included or excluded.

What I personally have in mind when I speak of Nature is all actual occurrence. Notice that this three-word description has no physical limits in extent or duration and includes both tangible substance and intangible space and energy. As an occurrence it comprises more dynamically than what is present instantaneously and more spatially than exists within local material boundaries. Hence Nature is envisaged to be truly continuous and non-definable, encompassing all transient material form and eternal receptive space.

In these terms it is inconceivable that anywhere or anyone separate from Nature can actually exist other than in the human imagination, and neither can the human imagination itself. Yes, material bodies can be distinguished from one another and their surroundings as dynamically distinct identities but they cannot be defined as separate entities within discrete boundary limits that isolate them from one another and space.

Calls to 'connect', 'reconnect' and 'work with Nature' are all indicative of thinking that Nature is a separate entity from human self- and group-identity, instead of recognising that human identity is a local dynamic inclusion and expression of Nature. By the same token, claims that 'We are all One' or 'Part of Nature as a Unified Whole' implies thinking of Nature either as formless or as a definable object in which our individual uniqueness is either denied or subordinated.

With natural inclusional awareness, comes true humility. Without it comes the arrogance or humiliation of situating one's self-identity hierarchically above or beneath others in a 'Great Chain of Being'. Alternatively all sense of the vitality of individual uniqueness and freedom of movement is lost in a web or sea of all-oneness. In other words, instead of being understood and appreciated in context, the vitality of our 'ego' as a centre of receptive-responsive awareness becomes either isolated from its natural situation or lost altogether in passive self-negation.

With true humility comes the realization that *all tangible content is contextually* sourced. And so the influence of our contextual situation – our 'natural neighbourhood' – on our lives is so comprehensive that we have very little individual responsibility for what happens. We can only respond receptively to current circumstances in which we find ourselves in order to live, love and be loved as well as we can. If we individually or collectively seek to deny this reality one way or another, we make life personally and culturally difficult for our selves.

This realization actually frees us from the propensity to take individual credit or shoulder blame for what is perceived to go well or badly, and instead to understand how it arises primarily from our contextual situation.

Where we recognise that our contextual situation is adverse, we can recognise why this is and seek with others' help collectively to change it and individually to cope with and not exacerbate it. To try to do this alone is a Sisyphean task.

Not least, we can recognise how a lack of situational awareness in one way or another contributes to a hostile culture in which we truly do 'struggle for life'.

2.1.9 Inclusive Democracy – Caring, Sharing and Co-creativity

The way I designed LEP itself epitomizes natural inclusion as an evolutionary learning process in which all kinds combine in mutual communion to make an ever-changing world, full of invention and re-invention. As such it also epitomizes a mode of social governance, following the principle of the 'sharing circle' illustrated in Photo 1, that I consider to be truly democratic – for all, through all, by all – rather than hierarchical or partisan. Sadly, such true democracy is virtually non-existent in the modern human world. In its place we have systems of governance that are founded either on an overarching sovereign or dictatorial power over other, or opposition between alternative ideologies. Such systems are inescapably oppressive and confrontational and hence detrimental to community spirit and natural truth-seeking. They exacerbate fear, intolerance and dishonesty in place of love, open-mindedness and wisdom. They restrict and subvert the kind of inventiveness that enables us to enhance our human quality of life in tune with our natural neighbourhood. So much that we take pride in as purely human invention *what* has been discovered and invented by natural biological evolution, as biomimeticists are increasingly recognising. We need only to become more aware of it by attending carefully to what Nature has to teach us far beyond the bounds of abstract human rationality.

> *Care-Taking*
> Neglect arises from a lack of humility
> An assumption of superiority
> That does not care to know
> How it feels
> To be in need of care
> And instead perceives that need
> To be infra dig
> A hole to climb out from
> Not call for help in
> .
> Neglect arises from ignorance of what it takes
> To be alive and loving
> It takes care
> That's why we need to be care-takers
> Not controllers, stewards or custodians of Nature
> Truly to be our selves
> In companionship with all and sundry

2.1.10 No Yoke – Liberating Our Selves From False Perceptions of Space and Boundaries

For millennia, much of humanity has been living and suffering under the yoke of a logical fallacy that treats space and boundaries as causes of definitive separation between one material thing and another:

> The **whole** philosophy of Hell rests on a recognition of the axiom **that one thing is not another thing**, and, specifically, **that one self is not another self**. My good is my good and your good is yours. What one gains another loses. Even an inanimate object is what it is by excluding all other objects from the space it occupies; as it expands, it does so by pushing all other objects aside or by absorbing them. A self does the same… 'To be' means 'to be in competition'. (C.S. Lewis's 'Screwtape Letters' from a senior to an apprentice devil)

This fallacy has become deeply entrenched in the abstract philosophy, science, mathematics and governance of modern culture. And yet its falsehood becomes apparent as soon as we contemplate such simple questions as:- 'how can a material body have volume if it contains no space?' and 'how can a material boundary cut the inside of a body off from its outside, unless it has no thickness and hence doesn't exist?'

Remnant
All that remains
When all's been said and done
Is what's always been here
And always will be
.
The vast presence of absence
Endlessly calling for attention
Waiting to burst into life
At the strike of a light
.
Endlessly forgotten
By the clamour of inattention to innermost receptivity
Frothing over hidden depth
Carving initials
Into stone and bark
.
Claiming
'I was here'
Meaning
'I really mattered, didn't I?'
Which could never have happened
In the first place
Without what remains
When all's been said and done
.
We are but traces in eternity
Flowing lines that gather around receptive cores
Before passing on from here to there
We are not building blocks or machine parts
Neither are we made from building blocks or machine parts
We are receptive-responsive relationships
All the way in and all the way out
The music of silence wrapped in sound
And sound wrapped in silence
The play of light enveloping darkness
As darkness envelops the play of light
Solvent and solute in natural solution,
Not one or other at odds
That's what really matters
•

Admission

If you truly wish to unlock the door
To a wide-open new paradigm of love in life
And life in love
Stop looking for the key!
.
First find the keyhole-
The innermost receptivity
Around which the key revolves
To reveal those different aspects
That come into life when admitted

.
Receptive love
Invites and welcomes in;
Responsive love
Moves towards and flows around;
Reciprocal love
Dynamically embodies each in the other
Receptive stillness within responsive circulation,
Hearts within minds
Within the wide-openness
The ever-present grace of endless space
.

There's no admission without admission

Innermost Calling
Within every living soul
Is an innermost calling
To include and be included
In the life and love of all
Around its self
.
A yearning to belong
In natural communion
Reaching inwards
Reaching outwards
Simultaneously
.
A deep receptivity
That cannot be seen
It can only be felt
As it calls into life
What cannot be seen
Until it comes to life
In radiant form
.
The zero within one within all
The stillness included in motion in stillness
Making music together
In every living body
.
A calling that needs to be heard
In the silence
That enshrouds and permeates us
Neither shut in nor shut out
From the realm of our imaginations
.
For when it's ignored
In rational pursuit
To know what can't be known
To control what can't be controlled
Our souls will rebel
And give us Hell
Until and unless

We learn to listen
Carefully

.

NaturalBelonging
Please say, after me
'I belong to Nature,
Nature does not belong to me'
That makes all the world of difference
Does it not?

References

Marman, D. (2018). The lenses of perception interpretation of quantum mechanics. *Integral Review, 14*, 5–143.

Rayner, A. D. M. (1997). *Degrees of freedom – Living in dynamic boundaries*. Imperial College Press.

Rayner, A. (2011). *NaturesScope – Unlocking our natural empathy and creativity*. O Books.

Rayner, A. (2017). *The origin of life patterns in the natural inclusion of space in flux*. Springer.

Rayner, A. (2018). The vitality of the intangible: Crossing the threshold from abstract materialism to natural reality. *Human Arenas, 1*, 9–20.

Rayner, A. (2021). The natural inclusion of diversity in community: Swirls within swirls and variations around a central theme. In J. J. McIntyre-Mills & Y. Corcoran-Nantes (Eds.), *From polarization to multi-species relationships, contemporary systems thinking* (pp. 729–747). Springer.

The γ – Principle of Natural Inclusion: From Competition Versus Cooperation to Heartfelt Relay

Alan Rayner

Abstract The age-old contradiction between competition and cooperation as alternative perceptions of life and evolution is reconciled by the γ-principle of natural inclusion. This principle has three fundamental tenets, all of which need to be recognised if it is to be understood comprehensively: (1) Natural bodily boundaries are intrinsically dynamic, not fixed statically in place – i.e. they are dynamically informed and sustained by continuous energy flow and dissipate if the latter ceases; (2) natural void space is a continuous intangible (i.e. frictionless) presence that simultaneously invites energy flow both inwardly (towards and around zero) and outwardly (towards infinity); (3) all material bodies, from sub-atomic to galactic, are mutual inclusions of void space and energetic motion in receptive γ responsive relationship. Hence Nature is understood to be intrinsically self-organizing in response to receptive influence, not governed by external force. The implications of this for individual, social and environmental understanding are profound.

Keywords Competition · Co-operation · Natural inclusion · Non-linearity · Self-organization

1 Competition Versus Cooperation: A False Dichotomy

For millennia, human thought and culture has been dominated by two alternative ways of relating to one another and other life forms: competition and cooperation. Moreover, these two alternatives have themselves been perceived to be in competition with one another as diametric opposites: to cooperate is perceived to limit

A. Rayner (✉)
Bath and North East Somerset, Bathford, UK
e-mail: alan@admrayner.plus.com

J. J. McIntyre-Mills, Y. Corcoran-Nantes (eds.), *Transformative Education for Regeneration and Wellbeing*, Contemporary Systems Thinking,
https://doi.org/10.1007/978-981-19-3258-8_21

individual freedom, whereas to be independent is to preclude cooperation. Consequently these alternatives reside at the root of much psychological, social and environmental conflict, which adversely affects our ability simply to live, love and be loved as we naturally are and can be.

This adversarial way of thinking can be traced back through recorded human history at least as far as ancient Greece, where it emerged in the 'atomism' of Democritus and Aristotle's divisive Laws of Contradiction and the Excluded Middle. Underlying these Laws is the following premise:-

Tangible matter is entirely separate from intangible space.

This premise is known as 'dualism'. It arises from the 'third person' perception of a material body exclusively from outside-inwards as a 3-dimensional solid object frozen in a snapshot of space and time and with no awareness of its interior space or dynamics. The existence of the body is hence considered to begin and end definitively at its surface. Matter and space are regarded as mutually exclusive – 'something' that exists physically and can be quantified in discrete numerical units, and 'nothing', which doesn't.

Dualism is at the root of the 'Cartesian Split' between 'Matter' and 'Mind' and is culturally institutionalized in objective scientific method, mathematical abstraction, linguistic definition, theological exclusion, governmental control and educational authority. It is the foundation of the binary logic of '1' (numerical presence) or '0' (numerical absence) upon which digital computing relies. In treating a numerical figure as independent from its environment (as per Einstein's contention that 'the environment is everything that isn't me'), it also underlies the idea that any 'one' must either 'compete' or 'cooperate' with any other one. The possibility for any kind of relationship in which both individual distinction and collective togetherness exist co-creatively is excluded from consideration.

The alternative, 'non-dual' or 'unitary' view of reality that is often expressed in favour of cooperation is, by contrast, based on this premise: 'All is one continuous whole' The corollary that no actual distinction exists between material presence and absence – and hence that any perceived distinction is illusory – arises from 'first person' awareness from inside-outwards. This awareness 'sees through and beyond' surface appearances to the continuous 'oneness' of all existence. It can give rise to a belief that 'self' or 'ego' does not or should not exist.

Which of these premises – dualistic or non-dual – makes most sense of your actual life experience?

Do they relate to how you feel in your heart? How do they influence your perception of 'right' and 'wrong', 'true' and 'false'?

Do you recognise how each contradicts the other as 'false' or 'wrong' in a different way?

Is there any way in which each could be recognised to be true in one way, yet false in another way?

2 Reconciliation: The γ -Principle of Natural Inclusion

Can the contradiction between competition and cooperation, dualism and non-dualism, be reconciled by a more comprehensive and heartfelt view that accepts the truths of both while rejecting their respective falsehoods?

I both think and feel that the answer is yes! There is a third way of perceiving reality that co-creatively combines the truths of both binary and unitary conceptions into a comprehensive understanding of how Nature actually works. This can be traced within and applied equally well to scientific findings, engineering, psychotherapy, spiritual wisdom traditions, democratic governance, need-oriented business, philosophy and, not least, the arts – in other words all aspects of human endeavour.

I think this third way is not only possible but urgently needed. One way it could be appreciated is through what I call here the γ – principle of natural inclusion. (Rayner, 2011, 2017, 2018, 2021) Gamma (γ) is the third letter of the Greek alphabet and can hence be thought of numerically as the creative coming together of alpha and beta (1 and 2; individual and coupled) ways of being, just as the symbol itself visually displays. Moreover this coming together is both fluidly dynamical and open-ended, like the 'stitching in time' symbolized in Photo 1, which depicts the flow of growth, maturation, degeneration and regeneration through the seasons in natural communities.

Here the inflows and outflows from locally embodied loops of life represent a continuous reconfiguration or relay from each into other, not a stand-still monopoly or confrontation. This corresponds with the pulse and circulation of lifeblood out from and into our responsive γ receptive human hearts. Life is perceived in a

Photo 1 'Stitching in Time' (Oil painting on board by Alan Rayner, 2020)

heartfelt way as a gift of natural energy flow, to be received, sustained and passed on, all in good time.

Natural inclusion arouses deep feelings of joy, passion and compassion in my heart. Indeed I would go so far as to say that this awareness needs primarily to be heartfelt as a soft, warm, dark, receptive love of life before it can be fully understood and appreciated intellectually. Intellect alone tends to preclude awareness of natural inclusion due to the predominance of dualistic thought, which isolates reason from emotion.

There are three fundamental aspects of natural inclusion that I regard as vital to recognise. (1) Natural bodily boundaries are intrinsically dynamic, not fixed statically in place. i.e. they are dynamically informed and sustained by continuous energy flow and disappear if the latter ceases. (2) Natural void space is a continuous intangible (i.e. frictionless) presence that simultaneously invites energy flow both inwardly (towards and around zero) and outwardly (towards infinity). (3) All material bodies, from sub-atomic to galactic, are mutual inclusions of void space and energetic motion in receptive γ responsive relationship. These bodies, as dynamic local configurations of omnipresent space, can both associate with one another into couples or collectives with reduced surface: volume ratios and dissociate into individual identities with increased surface exposure, depending on energetic context and relative flux.

Whereas these principles can be traced within many existing avenues of thought and feeling, too many for me to mention specifically here, I have been hard-pushed to find any that comprehensively, explicitly and unambiguously recognise all three – even though some do come very close. Neither have any of my companion researchers, despite extensive enquiries. This may be because the tendency to retain some adherence to dualism or non-dualism is very strong. The closest philosophical approach I feel we have encountered is that of phenomenology, and indeed I would be happy to describe natural inclusionality as a natural phenomenology based upon the receptive γ responsive relationship between space and energetic flux.

So why is this of such fundamental importance to the way we live our lives as they naturally are in the world as it actually is? In a nutshell, I think it comes down to no less than how we regard or disregard the relationship between life and love.

3 What Truly Motivates Us – The Natural Inclusion of Life in Love and Love in Life

Objectivistic views of reality give rise to the Newtonian conception that the movement of material bodies is driven or dragged by mechanical force situated outside the object in question. A ball driven upwards by being kicked is dragged back to Earth by gravity. It doesn't move of its own accord – i.e. it has no inner motivation or internal agency. This notion, when applied to life forms results in a forceful, command & control view of management and leadership, in which it is assumed that the individual needs to be compelled to work by external push or pull because it has no source of internal inspiration that will induce it to do so. Sadly, many of us have experienced the associated 'top-down', hierarchical attitude of mind all-too painfully. Many are the times when I have heard fellow educators say of a failing student that they 'need a kick up the backside'! Even the most hard-line objectivist will, however, be hard pushed to deny that there is a literally vital difference between the behaviour of life forms, not least human beings, and that of dead bodies and machines.

Only dead fish go with the flow, as the saying goes. Live fish swim. This difference implies that some kind of self-motivation does exist within life forms. But where and how does this self-motivation arise? For purely objectivistic thought this is a huge and ultimately insoluble conundrum because an object that depends on some kind of external motive force to move it simply cannot move itself!

Given the predominance of objectivistic thought resulting from dualism, however, this hasn't prevented people from trying to imagine or discover some kind of internal mechanism in charge of operations – a central controller, administrator, administration or process that governs the individual material elements – atoms, cells, organs, individuals etc – that comprise the life form.

So, what form might this forceful internal mechanism take? Who, or what is 'boss'. A wide variety of possibilities have been considered. Maybe it is some kind of invisible 'little man' or 'homunculus' within the man. Maybe it is a brain telling the body what to do, or maybe it is a 'selfish gene' asserting its right to be selectively preserved at others' cost. Maybe it is some kind of internal government or 'parliament'. But the objective objection to such internal agencies remains the same:- who or what is managing the manager? It's a good question and it has inspired some significant efforts to circumvent the issue.

One approach that has gathered momentum over recent decades suggests that living systems, as gatherings of what may be huge numbers of individual elements, organise themselves as variably coherent wholes. This kind of explanation falls under the general heading of what have been called 'complex system' and 'self-organisation' theories. Complex system theories have been greatly assisted by advances in computing technology. This enables coexisting individual and collective dynamic tendencies to be modelled mathematically using non-linear equations, and compared with observed behaviour patterns. And there is no doubt that the theoretical modelling can correspond remarkably closely with what is observed, as well as what might be expected from the γ – principle.

While close correspondence between theoretical model and actual observation cannot, in itself, account for why this correspondence exists, it does at least go some way towards showing that in reality individuality and collective coherence are not the mutually exclusive polarities that binary ('either/or') logic presupposes. Nor are they indistinguishable or illusory, as unitary 'all-is-oneness' insists. Rather, they are co-creative tendencies from which a vast variety of natural patterns of behaviour arise.

A particularly clear exposition of this difference between contradictory polarities and complementary tendencies has been provided by Kelso and Engstrøm (2006), who make use of the tilde or 'squiggle' character (~) to signify this relationship. My use of the gamma (γ) symbol takes this a crucial step further, for reasons I will now clarify.

When we do carefully question *why* the correspondence between model and reality exists, we begin to recognise the shortcomings of purely objectivistic explanations. This is because the modelling itself makes a variety of assumptions that are inconsistent with objective assumptions. For example, intrinsic movement, cyclicity, spontaneity and heterogeneity are all vital ingredients of the models. While it is valid to suggest that patterns of variably coordinated behaviour can arise spontaneously from these qualities, the question of how and whence all these qualities arise cannot be answered if it is assumed that an object can only be moved by external force and is by default static. And while it is often suggested that the answer resides in quantum mechanics, if one thing is clear from the findings of quantum mechanics, this is that they cannot be understood objectively.

Notice now that objectivistic theories of reality are all deliberately or unwittingly 'emotion-free', due to their reliance on third person perception alone. They have nothing to say about and do not recognise how it feels to inhabit a living body, why it feels that way or where the internal source or inspiration for movement resides. They all comply with Charles Darwin's dictum:- "a scientific man ought to have no wishes, no affections – a mere heart of stone".

So, we might ask, how would subjective perception account for the behaviour of life-forms? The answer is implicit in Darwin's statement. Motivation is emotional. It arises from a feeling of receptive internal need or desire that induces energetic responsiveness. So where might such need or desire arise? Have you ever felt hungry or thirsty? How would you explain such feelings to an objective observer interested only in what moves you from outside? Tricky? But you might find more understanding if you were to describe to a compassionate fellow human being what you experience subjectively as a painful sense of inner hollowness – a 'presence of absence' – needing to be satiated if you are to thrive.

Hunger and thirst are examples of sources of inner motivation that induce responsive inward flow and outward movement from and towards sources of food and water (and hence energy). And we channel that flow both instinctively and through learning from experience and those in our neighbourhood with whom we form receptive γ responsive relationships.

Moreover, hunger and thirst are not the only inner motivations that we human beings have if we are to thrive and enjoy life. Humans need more than bread and water alone. We seek out sources of clothing, housing, material goods, entertainment, interest, companionship, beauty and so on. And we seek love in the form of soul mates and sexual partners, without whom in the latter case we cannot procreate.

All these motivations spring from a receptive sense of intangible lack, a 'presence of absence' – which needs to be satiated. But if we either ignore or deny the existence of our own and others' dynamic bodily boundaries, this receptivity can become transformed into greed or passivity.

Now consider what happens if we combine the heartfelt subjective view and the discerning objective view in mutually inclusive dynamic relationship, not either/or confrontation. What do we get? We get the γ – principle of the natural inclusion of life in love and love in life! The innermost receptivity, represented by the space within the loop of the γ symbol calls energy into and out from circulation around its intangible centre, in continuous relay. Nature is understood to be *intrinsically* self-organizing in response to *receptive influence*, not governed by external force. The implications of this for individual, social and environmental understanding are profound.

Acknowledgement I would like to thank Roy Reynolds and Doug Marman for insightful commentary on drafts of this chapter, and all those who have helped me in one way and another over the years.

References

Kelso, J. A. S., & Engstrøm, D. A. (2006). *The complementary nature*. MIT Press.
Rayner, A. (2011). *NaturesScope – Unlocking our natural empathy and creativity*. O Books.
Rayner, A. (2017). *The origin of life patterns in the natural inclusion of space in flux*. Springer.
Rayner, A. (2018). The vitality of the intangible: Crossing the threshold from abstract materialism to natural reality. *Human Arenas, 1*, 9–20.
Rayner, A. (2021). The natural inclusion of diversity in community: Swirls within swirls and variations around a central theme. In J. J. McIntyre-Mills & Y. Corcoran-Nantes (Eds.), *From polarization to multi-species relationships, contemporary systems thinking* (pp. 729–747). Springer.

Systemic Praxis and Education to Protect the Commons

Janet J. McIntyre, R. Wirawan, T. Flanagan, and D. Finlayson

Abstract The chapter stresses the need to protect the commons through transformative education and practice that helps to reframe the way in which human beings see themselves in relation to other living systems. It also stresses that technocratic approaches to poverty and climate change – without caring for sentient beings and nature – could lead to new forms of authoritarianism and exploitation. All decision making needs to be guided by ethical decision making on how to protect all the interdependent users of a shared habitat. The chapter makes the case that education, design and the ethical interventions that support wellbeing – and prevent pathology or the undermining of ecosystems – should be sanctioned in law.

An early version of this paper was presented at the 65th Annual Meeting of the International Society for the Systems Sciences in 2021. The virtual conference was titled: The Art and Science of the Impossible: The Human Experience. This chapter is an updated version of the paper presented and cites asynchronous conversations with Rudolf Wirawan, Tom Flanagan and Dennis Finlayson to explore areas of concern raised in this paper in the form of reflections and a metalogue.

J. J. McIntyre-Mills, in conversation with R. Wirawan, T. Flanagan, and D. Finlayson.

J. J. McIntyre (✉)
College of Education, University of South Africa, Pretoria Central, Gauteng, South Africa

College of Professions, Adelaide University, Adelaide, Australia
e-mail: janet.mcintyre@adelaide.edu.au

R. Wirawan
Adelaide University, Adelaide, Australia
e-mail: rudolf.wirawan@adelaide.edu.au

T. Flanagan
Wesleyan University, Middletown, CT, USA
e-mail: trflanagan@aol.com

D. Finlayson
International Systems Sciences, Ashland, TN, USA
e-mail: dennis.finlayson@myphone.coop

J. J. McIntyre-Mills, Y. Corcoran-Nantes (eds.), *Transformative Education for Regeneration and Wellbeing*, Contemporary Systems Thinking,
https://doi.org/10.1007/978-981-19-3258-8_22

Keywords Education · Conversation · Feedback · Transformation · Multispecies relationships · Regeneration

1 Introduction: The Art of Systemic Vision

The annual Ramsay art exhibition "Close Contact" awards prizes to a winning artist and the people's choice. It provides food for free public education on systemic issues on invasion, displacement and risk and for all those with eyes to see. I was reminded as I scanned my Covid QR code for entry to the gallery, lodged my votes for the people's choice, after having a coffee and logging into a news website (that no doubt also tracked my news preferences) that in postmodern society we are also *the product* (Zuboff & Schwandt, 2019). Children were asked by the gallery guide to reflect on what the art gallery provides to South Australians.

The first child answered that he thought it provided a place for people to "sell paintings." The guide explained that this did not happen directly through the gallery. The child persisted confidently that nevertheless the gallery's main purpose is to enable artists to "make a name" and to "sell their paintings".

At this point the guide conceded that this was indeed correct. A young girl said that the gallery "provides entertainment." "Oh, good point," said the female guide warmly ...." and anything else?" she prompted, but the group were keen to be allowed to move on through the gallery which is rich in colour, texture, movement and sound. Of course it *also* provides a context for critical thinking and education which no doubt was the reason for the guided tour.

The winner, Kate Bohunnis, created a sculpture "Edge of Excess" of a pendulum swinging above a swathe of pink which could represent flesh as the scythe hovered just above the flesh, symbolic of ambivalence and risk, a subject on every one's minds during a pandemic. My attention was drawn to the work by Juanelle McKenzie who wove a sun (Yurndu) out of emu's feathers during the Covid epidemic. Surrounding the sun is a mandala of quandong seeds. The emu plays a role in regenerating the quandong fruit, thus closing the circle. The painting gives a lesson in systems thinking which to my mind is both elegant and profound.

A brave and profound installation by Hoda Ashar is in a gallery devoted to surveillance and violence. Hoda's images of whistle-blowers summarise the importance of enabling people to *speak out* about the abuse of the frail, the disabled and asylum seekers. She draws attention to the price paid by the whistle-blowers who had been silenced and threatened with imprisonment for speaking out. Her exhibit is placed near to the installation of watch towers by Sam Cranstoun who according to the gallery notes had read Foucault's 1977, "Discipline and Punish" which inspired the creation of the sculptures for the installation. The commodification of the powerless could have included not only the colonised, women, young people, the frail – but also animals who were left out of the frame, except in their referencing by Aboriginal artist Juanelle McKenzie.

This morning I checked the world news and noted that the whistle-blower journalist from Belarus had been captured (hijacked mid-air) and that the EU had decided to block flights from Belarus and to avoid their air space.[1] In Hong Kong[2] those who have spoken out or protested are barred from participating in local government.

Surveillance and power need to be balanced by global citizens who speak out as whistle blowers and who show alternative ways of engaging with the environment.

In the next gallery a video "Rising Tide" played softly in the background. Daniel McKewan made a video showing yachts floating amongst ruined city blocks, without providing any suggestions of an alternative future to the apocalypse.

The close sense of connection with the land is evident in the work of Kieren Karritpul called "Weaving Myself: the landscape and the land". According to the gallery notes he chose to paint the woven connection he felt for county as he is not permitted to weave as he is male and this is *women's business*. His painting represents weaving together the many strands of life. Each tiny strand represents the cyclical breathing that supports life as we know it. This spiritual painting is about breathing. It reminds me of Abram's (2020) epiphany when he heard Macey explaining *interbeing* in terms of the way humans breathe in oxygen and breathe out carbon dioxide which plants breathe in and photosynthesise. They then 'breathe out' oxygen for us to breathe in. Plants also produce the food on which we depend. He explains that *spiritus* is Latin for breath which animates our being and stresses that consciousness is non local (Damery & Abram, 2013: 109). Damery & Abram (2013: 109) which Jung (1959) considers as a possibility as it also helps people to believe in continuity when they face the inevitability of death. It also helps people to think that they are part of nature and return to it.

The notion of learning from nature is perhaps the lesson that the school children could have taken away with them as the message of hope depicted in the regenerative cycle played by the emu, the quandong fruit, and the sun as shown by McKenzie. To me her work illustrates our hybridity and interconnectedness with other species and with the forces of nature. We can indeed learn from a flightless emu, because, like Icarus human beings have flown too close to the sun. Hubris has prompted us to try to control people and nature.

A palimpsest is a document on which we write and re-write. Our history is written in the land and carried by the stories children tell their grandchildren.

Greta Thunberg[3] congratulated the group of eight young people in Australia who have successfully taken the federal environment minister to court in a bid to hold

---

[1] Beake, N. 2021 26th May Belarus plane: 'Perfect storm' prompts EU to act fast – BBC News, Brussels https://www.bbc.com/news/world-europe-57239522

[2] From protests to 'patriots': Why China is crushing Hong Kong dissent https://www.bbc.com/news/av/world-asia-57225142

them accountable for the impact that extending a coal mine in New South Wales would have on climate change. The court supported a ruling that the minister had a:

> "duty of care to protect children from future personal injury caused by climate change."

2 Statement of the Design Problem: Balancing Individualism and Collectivism Through Standing Together

If education means to lead out, the question which we need to consider is what sort of future do we want for future generations of living systems?[4]

Despite power imbalances, human beings throughout history have been able to make design choices with consequences for ourselves, others, and the environment.

As Buckminster Fuller explained we are all part of one system and thus all nations need to protect spaceship earth (1978). Understanding systems thinking is of central relevance to education, because it shapes the content, structure and process of education. Human beings need to take urgent action to protect our planet Greta Thunberg (2019)[5] and other young advocates have underlined that world leaders must make better choices.

Chief Raoni, Nobel-peace prize nominee (2020)[6] stresses the need to protect the forests as these are the lungs of the world as detailed by Forsetto (2021)[7] and in the Piaraçu Manifesto which makes a plea for the peoples of the world to stand together.

The Dalai Lama (2005) in his book, "The universe in an atom" underlines the importance of thinking and behaviour as all causes have consequences. Lama[8] (2020) Systems thinking, quantum physics and critical analytical ability are of vital importance to transformative education.

---

[3] Greta Thunberg hails 'amazing' Australian court ruling as 'a huge win for the whole climate movement' The climate activist hailed the Federal Court's decision on Twitter and offered a" big congratulations" to the teenagers who achieved it. Read in SBS News: https://apple.news/ACYjrnsz2QBGGja_u-YzA-w

[4] Systems thinking and practice needs to be" more than spaghetti and meatballs", to cite the dismissive 'putdown' given to a draft policy document produced by a policy advisor to the Australian labour government.

[5] https://www.google.com/search?q=thunberg+how+dare+you+speech&oq=thunberg+and+how+dare+&aqs=chrome.2.69i57j0i22i30l2j0i22i30i395l4j0i390i395l2.24047j1j7&sourceid=chrome&ie=UTF-8

[6] https://news.mongabay.com/2020/02/chief-raoni-nobel-peace-prize-nominee-presides-historic-meeting-with-over-600-indigenous-leaders-in-brazil/

[7] Forsetto cites the landmark event with the Mebengokre Peoples of Brazil and indigenous peoples and leaders as follows: "*The landmark debates lasted for four days and, after extensive review, resulted in the Piaraçu Manifesto, a four-page document denouncing the Brazilian government for its role in putting indigenous peoples at risk. ..*"

[8] https://www.theguardian.com/world/2020/nov/11/buddha-would-be-green-dalai-lama-calls-for-urgent-climate-action

Pope John Paul's (2015) encyclical begins by quoting St Francis of Assisi and stressing that the earth is "our common home."[9]

All these leaders stress the importance of solidarity and understanding what we share in common.

3 A Community of Practice for Fostering Solidarity and Engaging Participants

A community of practice is a concept coined by Etienne Wenger (1999), a social anthropologist who found that informal communications based on reciprocity enabled people to achieve their goals, because when people identify with an area of concern, they were prepared to help one another to achieve a specific goal. He found that reciprocity and task orientation helped communities to self-organize. This concept can be applied to a range of contexts, particularly public education.

The ongoing community of practice (spanning disciplines and cultures) applied by members of the team that contributed to the volume (and particularly chapters "Learning from Nature's Classroom: Reframing Economics, Accounting and Accountability" and "Women, Education and Socialisation: Transforming Gender and Power in the MST, São Paulo, Brasil") developed around *the need for transformationineducationand economics*. It includes academics and members of the community who learn from one another and who pass on what they have gleaned.

In this volume on transformative education some of the chapters propose a decentralized approach to education based on working with the marginalized other chapters have stressed the importance of extending rights to the disabled and to recognize the sentience of living systems. This has implications for *ethics, governance, andeducation*.

3.1 Area of Concern

Wellbeing stocks (Stiglitz et al., 2010) need to be supported by *social, economic, and environmental* indicators developed within a community context and supported locally, nationally, and internationally in line with the UN (2030) Sustainability agenda. These arguments have been spelled out elsewhere in Planetary Passport (McIntyre-Mills, 2017a), From Wall Street to Wellbeing (2014). The latest IPCC (2021) report stresses that climate change is accelerating. It is time to apply laws to support positive feedback to mitigate and adapt to risks.

[9] Encyclical letter ***laudato*** *si'*of the holy father francison care for our common home https://www.vatican.va/content/francesco/en/encyclicals/documents/papafrancesco_20150524_enciclica-laudato-si.html#_ftn11

What kinds of re-enforcement encourage people to move away from business as usual to support wellbeing?

- **Positive** re-enforcement may include rewards such as subsidies and access to resources.
- **Negative** re-enforcements may include barriers to subsidies and resources or penalties involving community service, fines, or incarceration.

The challenge is to create the will to protect the commons and to try out an alternative form of accounting and accountability at the local, national and post national regional level.[10]

The contributions made by Ann Florini (2003) on scaling up the Aarhus Convention could be applied more widely. Currently the convention ensures that local people with concerns about the environment have the right to information, the right to be heard locally within the European Union and to take their concerns to the European parliament or European court (if they are not satisfied). The Aarhus Convention (1998) provides three policy pillars to enable this everyday engagement to occur. The policy pillars include:

- The right of all residents in the EU to access information
- The right to be heard and the right to take the areas of concern to the European Parliament and then to the European Court if the issues are not satisfactorily addressed.

Florini (2003) stresses that the policy provides a valuable potential platform for extending democratic rights to residents within and beyond a nation state so that social and environmental justice concerns could be addressed at a scaled up post-national regional level.

Other potential approaches that could contribute to a new, reframed architecture to protect social and environmental justice include the suggestion by Geoffrey Robertson (2021), a human rights lawyer who stresses that organisations, governments, and individuals need to be held to account by *blocking their assets and rightto travel* – if they commit crimes against humanity.

The late Polly Higgins (2013, a barrister advocated for an Ecocide Law for crimes against the environment, whilst Vandana Shiva (2012a, b, 2019, 2020a, b),a nuclear scientist and ecofeminist, has advocated for seed and water security and taken Monsanto to court, leading to her winning the Sydney Peace Prize.

David Held (2004, Global Covenant) has stressed that local interventions to support social and environmental justice need to be buttressed in law. These above-mentioned policies, covenants, and bills could be applied using a scaled up form of engagement from the local household and neighbourhood level, to the local

---

[10]The prototype software to achieve this goal *could* be applied and scaled up as detailed in *Wall Street to Wellbeing*, *Planetary Passport* and the edited collection *From Polarisation to Multispecies Relationships*(McIntyre-Mills and Corcoran Nantes, 2021). The approach could provide a means for local social, economic and environmental accounting, accountability and transparency.

government, national and post national regionalist level to ensure that living systems are protected.

The challenge is to create the will to engage. One of the ways forward is to enable people to see the benefits of social, economic, and environmental accounting and accountability by addressing food, energy, and water security in relation to employment creation. These three aspects are common concerns that could be a basis for scaling up engagement and support for protecting the global commons which could be supported by giving *positivefeedback* and rewards to those who support wellbeing stocks and conversely giving *negativefeedback* and sanctions to those who undermine wellbeing stocks.

The proposed governance process could be supported through local engagement at the local government level where subsidies and supports provide incentives to those who support the common good.

It is arguable that democracy has lost its way. Rights need to be balanced by responsibility and policy decisions need to be informed by critical systemic engagement.

Those who place their rights above the rights of current and future generations ought to be held to account (see McIntyre-Mills, 2014, 2017a, b). We need to be guided by axioms such as: *We can be free and diverse to the extent that freedom and diversity does not undermine therightsof current and future generations.*

A form of 'Magnitsky Law' (Robertson, 2021) could be used to bolster a planetary passport. Those who support wellbeing stocks are given positive sanctions and those who undermine wellbeing stocks and who commit crimes that undermine social or environmental justice could be sanctioned, using a form of law that freezes assets and halts the ability to travel (Robertson, 2021). This would address similar concerns raised by Bruce (2016) who stresses that whilst an international court dedicated to addressing Climate Change issues would be a way forward, it is insufficient. An Ecocide Law (Higgins et al., 2013) would be helpful as a way to implement sanctions against those who render young people homeless in the future. This is an action that needs to be taken to ensure that those who do not show a duty of care towards young people are sanctioned.

3.2 Community of Practice Approach to Explore What Motivates*Transformation*

This requires exploring *feelings* and these of course cannot be explored by means of artificial intelligence (AI) so the research combines exploring qualitative insights by means of zoom interviews and in depth case studies. As Solms (2021) stresses drawing on the work of Chalmers (1995) and Oliver Sacks (see Draaisma, 2015), consciousness is that part of human function which refers to feelings about personal experiences. Consciousness is therefore *more than brain function.*

The aim of the community of practice is to inspire the marginalised, particularly women and young people, to use their own talents and resources by adapting the Indonesian decree that every village should foster enterprise.

Indigenous and local knowledge is the basis for creating local green enterprises in line with the UN Sustainable Development Goals. Members of the community of practice are engaging in zoom conversations with local facilitators to *train the trainers* and inspire their engagement. Our hope that the webinar engagements organized around three modules (together with local leaders who have been nominated by local chiefs/local community) will be able to form co-operatives with a green focus. We provide them with links and information. Some of the small pilots are in the early stages of scoping, some are at ethics approval stage, whilst some have been completed.

Reconsidering boundaries for social and environmental justice has implications for our understanding of what Hannah Arendt (1963) called "the banality of evil" with reference to the everyday decisions by ordinary people that led to the holocaust in Germany. The need to extend the social contract to protect human rights (not only the rights of citizens), requires a rethinking of governance and global citizenship. Social justice in turn needs to be buttressed by supporting environmental justice through supporting an international law of Ecocide (Higgins et al., 2013) that stresses the need to protect the wellbeing of an ecosystem and *all the inhabitants* of a region. It is couched in deliberately non- anthropocentric language and goes on to stress that organisations and companies need to be held to account. We need to stop ecocide. New forms of governance and democracy could be based on protecting habitat for a range of species. Education and governance systems need to foster an ecology of mind (Bateson, 1972) drawing on a respect for multiple species and Indigenous Knowledge Systems (Odora Hoppers, 2013) and an eco-fracturing approach (Pauli, 2010). This has implications for the production, distribution, and consumption chain and how we view our rights and responsibilities. Aspects of our life such as the nature of food, the nature of work and the nature of governance are all constructs. We need to ensure that economic systems support diverse species and the habitats on which they rely.

Social, economic, and environmental transformation at the local, national, and post national regional level is required to support human security.

Human security is in turn dependent upon protecting vulnerable multispecies relationships spanning organic and inorganiclife (Diagram 1).

3.3 Food, Energy and Water: A Way to Resource the Commons and Address Existential Risks

Elinor Ostrom's research (2008, 2010, 2014) demonstrates that people can and do share resources under certain circumstances, these eight principles relate to setting clear boundaries and rules, appropriate decision-making processes and monitoring linked with appropriately applied sanctions and processes. She also stresses that the local community rights need some recognition by wider government and that the local systems need to be nested within layers of governance. The challenge which we face is how to motivates people to move from business as usual

Diagram 1 What motivates people to move from business as usual towards re-generative living?. (Sources: McIntyre-Mills (2008) and Wirawan (2021), see chapter "Learning from Nature's Classroom: Reframing Economics, Accounting and Accountability" of this volume)

towards supporting wellbeing stocks and enthusiasm for managing the commons (Flanagan, 2014).

Forsyth and Johnson (2014: 8) cite Ostrom (1990: 43) and explain that rules matter, and that people are more likely to follow them when they understand 'the costs and benefits' and have a stake in their formulation.

Vandana Shiva (2018, 2020a, b) highlights that protecting water and seeds, for example requires not only good local practice but national and international action[11] whilst Polly Higgins et al. (2013) propose the ecocide law as a way to protect the commons. So, we are looking at community dynamics including the material and non-material aspects that motivate people to protect the commons.

- What reinforcing dynamics in social enterprises could help to protect wellbeing stocks?
- What social, economic and environmental indicators are meaningful to local people, i.e., what are the perceived pathways to wellbeing? These provide the basis for measures or indicators of wellbeing
- What works, why and how in learning organisations/communities to support social enterprise, because people are able to earn, learn and grow a future together?
- What are participants perceptions of an engagement process to track, trace and hold markets to account?

[11] It is worth noting that Elinor Ostrom who won a Nobel Peace Prize in 2009 and Vandana Shiva won the Sydney Peace Prize in 2010.

4 Humanity Can Only Be Protected by Protecting *Biodiversity*

The big issues of the day are poverty, climate change and pandemics and the root causes can be said to be linked with a lack of representation, accountability, regeneration, and sustainability of the environment.

Better design is the way to address the big issues of the day – this requires positive feedback or reinforcement to prompt current generations to think about their everyday decisions. Once we lose participation in the agora and critical engagement we are ruled by the market (Baumann, 2000, 2011).

I have been inspired by John P Van Gigch's notion of meta-design (1991) and was introduced to West Churchman's *Design of Inquiring systems*. This set of questions for inquiry prompts us to think in terms of 12 questions (critical heuristics) to consider the implications of "what is the case" and "what ought to be the case."

The area of concern for a proposed Global Covenant (Held, 2004) and proposed *Planetary Passport to Protect People and the Planet* is to address poverty, climate change, displacement of people and destruction of habitat. The PP could strive to balance individual and collective needs in line with Held's notion of a so-called 'Global Covenant' and sanctions could be applied to those who undermine wellbeing stocks. Stiglitz et al.'s (2010) wellbeing stocks could be supported by enabling people to 'be the change' on a daily basis through the way they choose to live their lives and making social contracts through the on-line system to protect local resource systems. Their footprint can be monitored locally, and they can generate transformation locally by rewarding sustainable living through the provision of positive supports.

5 Reframing Ethical Frameworks and Key Concepts: Education, Interbeing, and Teleonomy to Support the Wellbeing of Living Systems

Each generation of life writes their history on the landscape which in turn shapes the limits within which we live. We are interbeings shaped by many species and the environment on which we rely. Human beings are just one strand within a living system. *Interbeing* is explained by David Abram (2020) as human beings breathing in the oxygen that plants breathe out and vice versa. Abram explains that the concept was coined by the Vietnamese religious scholar Thich Nhat Hanh, who translated *pratityasamutpada* as 'interbeing'. We (like other species) are reliant on many systems for our survival. This is understood by many religious faiths, particularly Buddhism[12] and Indigenous knowledge systems, that see the land as our mother, for

[12] Saka Dawu, the 26th of May is an auspicious day on a Buddhist calendar. A key precept for those who espouse the dharma is the equality of all sentient beings. This is a hopeful way forward to assist regeneration.

instance. It is time to re-write it and to re-generate hope. Politically, critical engagement informed by praxis needs to ensure that local people retain their rights and responsibilities to write their futures.

This is also the case in Australia where lessons were shared by mentors whilst undertaking research in the Northern Territory (McIntyre-Mills, 2003) and South Australia (McIntyre-Mills, 2008). Sundanese wisdom from West Java and South Africa (McIntyre-Mills et al., 2018, 2019a, b, Quan-Baffour et al, 2019) underline similar lessons on living in ways that support people and in nature.

One of the most profound lessons I learned in 2020 was from watching the documentary "My Octopus Teacher" about regeneration and an ecological mindset. The documentary directed by Pippa Ehrlich and James Reed (released on Netflix) was filmed off the cost of South Africa. where I spent three months from October to December, 2020.[13]

The documentary shows Craig Foster's underwater journey to learn more about a kelp forest which he visits every day. He befriends an octopus who teaches him about transformation and resilience.

The sense of connection and trust once the curious female learned to trust another species is poignant. As I watched I thought about the life chances of sea creatures on this coastline that are increasingly threatened by sewage and plastic, including nurdles that have washed up from container ships that have lost cargo destined for factories to produce molded plastic.

Currently ethical frameworks are inadequate (see McIntyre-Mills, 2014) as they are inherently anthropocentric and misplaced; humanity can only be protected by protecting the biodiversity in which we are embedded. This could be progressed through implementing this design *foreducation* curricula, structures and processes that extend the 'Frontiers of Justice'[14] (Nussbaum, 2006) to enable capabilities.

5.1 Assumptions and Values

Sentience is not the preserve of human animals; it is shared by all animals and perhaps some would argue that machines have a form of sentience. Once this is accepted, it raises questions about rights, responsibilities and where we draw the

[13] https://www.youtube.com/watch?v=3s0LTDhqe5A

[14] In "Frontiers of Justice", Nussbaum (2006) stresses the need for the right to a life worth living to be extended to all sentient beings. Nussbaum's (2011) ten capabilities are directly concerned with wellbeing, a life worth living and the extent to which these can be addressed. These capabilities, include: (1) living a life that is not cut short prematurely (2) bodily health, (3) bodily integrity, (4) sense, imagination, and thoughts,(5) emotions, (6) practical reason, (7) affiliation, (8) links to other species, (9) play, (10) control over one's environment. In Nussbaum's version, the Capabilities approach focus on" the protection of areas of freedom so central that their removal makes a life not worthy of human dignity." (p31)

line of inclusion and exclusion. It has implications for ethics, law, governance and democracy and how we perceive the notion of rights and respect for living systems.

Once we appreciate our hybridity and interconnectedness (McIntyre-Mills, 2017) we have to accept that current architectures of governance are obsolete. We need to accept that rights cannot be exerted at the expense of other generations or other species in order to maintain the current carbon economy. What will the future of governance entail?

- Perhaps a Zoopolis (Donaldson & Kymlicka, 2011) managed by benevolent stewards to ensure that sentient users are protected within a suitable habitat?
- Perhaps a planetary passport (McIntyre-Mills, 2017) could be developed on the basis of balancing individual and collective needs governed by the axiom: we can be free and diverse to the extent that our freedom and diversity does not undermine the rights of others (including all sentient beings)?

5.1.1 Systemic Ethics Underpinned by a Transformation in Values Detailed in the Uluru Statement of the Heart

The Uluru statement of the heart suggests that as a nation we need to respect our dependency on the land and that Aboriginal voices have not yet been heard as the land and waters have not been placed at the heart of policy decisions:

> We, gathered at the 2017 National Constitutional Convention, coming from all points of the southern sky, make this statement from the heart: Our Aboriginal and Torres Strait Islander tribes were the first sovereign Nations of the Australian continent and its adjacent islands, and possessed it under our own laws and customs. This our ancestors did, according to the reckoning of our culture, from the Creation, according to the common law from 'time immemorial', and according to science more than 60,000 years ago. This sovereignty is a spiritual notion: the ancestral tie between the land, or 'mother nature', and the Aboriginal and Torres Strait Islander peoples who were born therefrom, remain attached thereto, and must one-day return thither to be united with our ancestors. This link is the basis of the ownership of the soil, or better, of sovereignty. It has never been ceded or extinguished and co-exists with the sovereignty of the Crown.

Ethical guidelines for human researchers, practitioners, and members of the public need to recognise our biological kinship and interdependency which can be challenging given the lack of respect for the rights of the powerless which has been evident

The implications of interbeing have been made apparent by Covid. Abram (2020) uses *interbeing* and *inter-breathing* as synonyms that illustrate the notion of *dependent co-arising* or mutual causation within one system comprising organic and inorganic life . Covid -19 can be seen as feedback in response to the anthropocentric social, economic, and environmental decision making that disrespects other living systems (McIntyre-Mills, 2021a, b, c). As stressed in "From Polarisation to

Multispecies Relationships" and a related article: "Communication and culture: a multispecies endeavour within a shared habitat" – throughout the history of colonisation (Strang, 2017) and imperialism, the environment and people have been commodified.

Respect for multiple species needs to be based on recognition of our hybridity and interconnectedness informed by a sense of awe and supported by norms rooted in a belief in our precarious interdependency and the implications of dispossession (see Butler, 2011; Butler and Athanasiou, 2013). A moral compass to protect diversity needs to be buttressed by a Global Covenant (Held, 2004) and the Ecocide Law (Higgins et al., 2013, Higgins, 2018) whilst mindful that stressing animal rights to Indigenous people who have themselves been denied rights by colonists who have dispossessed and displaced their ancestors can be hard to accept (Strang, 2017), particularly as the environmental legacy left by First Nations bears better scrutiny!

I would prefer the design of limits to be informed by ethics derived from structured dialogues (Flanagan & Christakis, 2009, 2010, forthcoming; Christakis & Bausch, 2006; McIntyre-Mills et al., 2008), rather than a stack (Bratton, 2015) managed solely by algorithms spanning every layer, although I doubt that this is what Bratton intends. The process of making meaning together with people is a cornerstone of design according to Alexander Christakis, not only to test out ideas, but in order to enable people to have a stake in the decision-making process. This sense of 'stake' is vital when addressing the big issues of the day that require valuetransformations. Christakis has explained that one of the problems with the approach of the *Club of Rome* was that it did not enable people to have a say in the modelling of climate change.

We live in a world where human beings are abused on a regular basis and in a world where we think it is our right to treat workers, animals, plants, and nature itself as commodities or as a means of production and profit. It is time to revise the ways we see ourselves in relation to others whilst preserving the sacred human right to be creative.

Top down approaches to "One Village One Product" were deemed to be unsuccessful in Japan (Claymore, 2007) which provides instructive lessons for those wishing to emulate the idea in Indonesia. This means that the commons need to be protected in ways that do not universalize governance in ways that undermine local agency and leadership through top-down approaches.

The scope for human beings to have a say and to overturn structures seems to me a basic human capability that *should not be lost*. Mindfulness and the choice to act and think in terms of chosen values is a basic notion linked with the human spirit or soul. The brave spirits who rise above internment in camps (such as Viktor Frankl[15]) based on their inner vision and whose hopes and beliefs sustain them would argue

[15] " Between stimulus and response there is a space. In that space is our power to choose our response. In our response lies our growth and our freedom". *Viktor E. Frankl*

that this human soul is not the same as brain functioning or an algorithm. This is what Chalmers (1995) called the *hard question*: What are the qualities that make up the mind that experiences and chooses to perceive in a particular way? We ought to be free and diverse to the extent that freedom and diversity does not undermine the rights of this generation of living systems or the next. It thus behoves us to think in terms of stewardship for multiple species, but not to lose the ability to think critically and mindfully.

The notion that interventions to manage the commons – through a form of post national governance that could span users and their immediate environment – is problematic, if it is not guided by conscious, critical inter-subjective decision making which supports a new form of democracy based on *'if / then'* scenarios. Bratton (2015) suggests so-called stack sovereignty[16] (Bratton, 2015) which to my mind could lead to dehumanizing governance. In a post anthropocentric world, some may say this is acceptable, but the risk is that a worse option could be created.

Martha Nussbaum (2006) has stressed the need to extend the boundaries of rights and responsibilities as the social contract is too limited. Many Indigenous first nations extend the boundary to protect the environment as do ecological humanists and feminists such as Haraway (1991, 2016) or anthropologists such as Bird Rose (2008, 2015) and physicists turned eco-feminists such as Vandana Shiva(2020a, b), whilst the ecocide movement stresses that rights and responsibilities need to be buttressed by international law.

Protecting the global commons has been theorised by Ostrom who won a Nobel Peace Prize, but Shiva who stridently points out that local governance does not go far enough – did not! She won the Sydney Peace Prize (usually given to those who are less mainstream in their thinking). The work of Vandana Shiva or the hopeful views expressed by Litfin (2009, 2013) are grounded in the hope that social movements from below will help to influence change by providing some of the balancing from below to offset the decisions from above. Whilst reading about Bretton's concept of "The Stack" I asked some colleagues about some of my concerns about how liquid boundaries governed by algorithms could prevail without ongoing monitoring *from below* as well as *above*. Their responses are summarized in the following vignette extracted with their permission from a series of emails over a few days provided a basis for this edited exchange that helped crystalize some issues.

[16] To cite the summary of Bratton's (2016) book:: "Bratton explores six layers of The Stack: Earth, Cloud, City, Address, Interface, User. Each is mapped on its own terms and understood as a component within the larger whole built from hard and soft systems intermingling – … This model, informed by the logic of the multi-layered structure of protocol" stacks," in which network technologies operate within a modular and vertical order, offers a comprehensive image of our emerging infrastructure and a platform for its ongoing reinvention. The Stack is an interdisciplinary design brief for a new geopolitics that works with and for planetary-scale computation. Interweaving the continental, urban, and perceptual scales, it shows how we can better build, dwell within, communicate with, and govern our worlds.thestack.org}

6 Re-membering Boundaries and Exploring the Motivation to Expand Our Solidarity with All Living Systems

What potential pathways can promote educational opportunities and redress the food and water insecurity associated with a growing population and environmental destruction? (also see McIntyre-Mills et al., 2019b). What motivates people to transform their thinking and practice at a personal and political level? These are two central concerns for a critical systemic approach (Midgley, 2000; Midgley & Lindhult, 2021; Churchman, 1971) to assessing our rights and responsibilities. The following table sets out design considerations for transformative educational praxis. Lines of inclusion and exclusion needs to be informed by considering where to draw boundaries on the basis of questioning (Table 1):

Table 1 Design considerations for transformative education and research: multispecies relationships

Boundaries The links across loss of habitat and pandemics[a] and implications for multispecies relationships stressed by Jane Goodall,[b] Donaldson and Kymlicka (2011) in 'Zoopolis'.	Where do we draw the line of inclusion and exclusion when we make policy? In what contexts should global control be permissible to balance individual and collective needs? The line needs to be drawn where excessive misuse of resources by one species places other living systems at risk.[c]
Perspectives Need for re-framing what constitutes ethical thinking and practice.[d]	Why are some species and people are prioritized? This has implications for our perspectives on rights to commodify others and this in turn has implications for property (ownership) and food (what constitutes food) Power imbalances needs to be addressed to ensure that resources are not captured, stockpiled and misused at the expense of the majority of living systems in this generation and the next.
Relationships Frans De Waal – animals have a sense of empathy, fairness and ability to reciprocate Eva Meier – animals communicate Susanne Simard, Monica Gagliano, Peter Wohlleben and Allen Rayner – plants communicate	What are our relationships with others (human, animal, environment)? What can we learn from the teleonomy of Covid-19 about climate change, habitat loss, trafficking and containment of sentient beings? How can sentient beings be assured that they can express their capabilities so that they live a life worth living (Nussbaum, 2006, 2011)
Systems Social and environmental justice needs to underpin designs non -anthropocentric designs	How can we design our systems, our infrastructure and technology in ways that ensure we do not exceed our boundaries?

[a]https://www.cnn.com/videos/us/2020/03/19/coronavirus-jane-goodall-acfc-full-episode-vpx.cnn
[b]Representation of multiple species depends on improving social, economic and environmental accountability, sustainability and regeneration. Rapid growth in human populations, rapid development of sprawling urban environments that are not designed to create balance across species

(continued)

Table 1 (continued)

impact food, water and energy insecurity. 'From Polarisation to Multispecies Relationships' (McIntyre-Mills et al., 2021, chapters 15, 17, 18 and McIntyre-Mills, chapter 26) explains that living systems are defined as a continuum from organic to inorganic life. The Covid-19 pandemic has provided Goodall (2020) the platform to speak out about how cross-species infections are caused by the way human beings are encroaching on the habitat of wild animals, containing them in inappropriate spaces, trafficking and transporting them in ways that are cruel that undermine their health and result in cross species infections. The case for a new form of eco-systemic governance to protect sentient beings and their habitat is developed (McIntyre-Mills, 2021, chapters 5 and 8, McIntyre-Mills and Christakis, chapter 14)

[c]An 'Ecocide Law'(Higgins et al., 2013) is needed to protect the capabilities of all sentient beings, in order to address the systemic challenges that can together be called a 'Risk Society. Diverse species (that are normally not in contact with one another) are thrust together when they are displaced farmed, transported, contained and marketed without considering their rights or our responsibility as stewards to balance and protect diverse species rights. Viruses have the potential to jump across species, because of the way in which many host species have lost both their freedom and their habitat

[d]In some cultures, rocks and plants, as well as a range of creatures are seen as kin that need to be protected (see McIntyre-Mills, 2003, Romm and Lethole, 2021 and chapters "Regenerative Education and Safe Habitats for Diverse Species: Caterpillar Dreaming Butterfly Being", "AA!!Venda Women and Social Enterprise: Stepwise Process to Regenerative and Sustainable Living" and "Women, Education and Socialisation: Transforming Gender and Power in the MST, São Paulo, Brasil" this volume)

Box 1

Metalogue to address the Area of Concern

Janet: What do you think of the so-called "sovereignty of the stack?"

Dennis: The key is in the title "governance of" … It is the key area to research and perhaps apply some ethical considerations too....as in critical systems......or stakeholder preference etc.. Eventually the whole of the governance might be a co-op of co-ops at every level of the stack.....i.e. governance by members and users etc…But I am not well informed in the field...but I will try to get a response from some people who might be...a co-op of co-ops could work ?"

Janet: *Coop of Coop* versus *Co-op of Co-op* makes me think of chicken coops – a worry indeed if users are treated by a great designer no better than chickens cooped up.... actually you and Tom have the nub of it...*the designer has to have a soul*, but there is no guarantee....which is why systems must be open and critical....But if we have *co-ops* of co-ops then perhaps we could try out another way of working? In a meeting held by P.L in Venda (who has set up a co-op) asked the women to consider the capability requirements for the wellbeing of chickens and asked them what they think *the chickens want....* the women gave a very insightful responses to the question: "*How can chicken farming be combined with bamboo farming?*" (see chapter "AA!!Venda Women and Social Enterprise: Stepwise Process to Regenerative and Sustainable Living" in this volume). They are in favour of free-range chickens in co-operatives, because the chickens like to 'run around', 'scratch', 'perch'

(continued)

Box 1 (continued)

and 'sing'. A good metaphor, because we need to avoid totalitarian top-down approaches that limit our *species specific*capabilities…

P.L. indicated that there are many free-range chicken in their community and that most households have chicken, and how those chickens live lives that are worth living by having large spaces in which they can express the full range of their capabilities (Nussbaum, 2006, 2011)

Janet continues: " …. Actually, I thought you were being very profound by mentioning chicken coops, because it made me think of invisible boundaries by which we are governed. These not so 'liquid boundaries' are the future.

I did not know about clathrates, until Tom Flanagan mentioned them, so I looked up their meaning and I see that they can be an invisible cage like structure in liquids. This is a good analogy, because in the stack each space could be mapped in the same way as clathrate. I had to look this up (thanks Tom). It is defined as follows:

> "Clathrate is a structure in which water molecules under certain conditions bond to form complex networks of molecules forming cage-like structures that encapsulate a guest molecule, which is a gas" cited from: Investigating Seafloors and Oceans, 2017[18]"

Tom: Janet, my take on Ben Bratton's often explicitly poetic notion of "the stack" is organic. By this I mean that he takes a view of the built and natural world as a thinking entity. He speaks of "platforms" as constructs of institutional thoughts, both enabling and constraining imagined possibilities much as does language(s) in human systems. Platforms can be concrete (much as cultures imprint their beliefs in the artifacts of human traditions) and also as abstract echoes of those artifacts. The abstract, cloud-carried echoes are, then, shadows of past experiences offered forward in combinations that represent new possibilities. He mentions roads as platforms that enable and constrain possible flows of traffic. Roads make us all drivers. In similar fashion, telecommunication makes all callers, and cities make us all inhabitants. I see this as defining us as agents of and for "the stack." As a self-designated "designer of possibilities," Bratton sees the crucible of creativity as the imagined overlay of "real(?)" platforms. What can this mean, given that I haven't read his forthcoming book? He conflates the notion of a creatively sapient superorganism that is subconsciously testing computational combinations of platforms seeking acceptably felt new hybrid platforms with a notion of cultural evolution….This dynamic feels like language building, which can mean quite

(continued)

[17] www.sciencedirect.com

Box 1 (continued)

different things to those familiar only with the expressive ambivalence of romantic language, the encased precision of complex words in the Germanic tradition, or the deeply layered experience of polytonal, Asian expression... for possible examples... all augmented to a digital platform of programmers lingua franca. But I digress. When Bretton speaks of "data that slides through the glass wires and ambient electric fields" I sense that he is alluding to the slippery substance of meaning.... carried in" words"... to which the Bee Gees sang: "I have only words to steal your heart away." So, in responding to a needed change in global culture in the face of global climate change, I think that Bretton might be saying that only the spiritual power of poetry can save us... saving us only when it touches, resonates with, and echoes beyond the platforms that sustain and constrain us in so many deeply individual ways.

Is this a response to your question? The " stack" is the collective being. It rules with sovereignty.... but may rule well or poorly... depending on our enculturated attitudes to our poets?

.... I feel a great deal pivots on our concept of HOW sentient entities think. Is sentience computational or is it an analogy of the emotio-cognitive mind? Under what conditions can a sentient system see beyond what IS to select what MAY BE. Hasan Özbekhan wrote that a system's values locks a system into a vectoral projection of the present into a changing future (where the system may or may not survive). There can be no real change without a deep change in the configuration of value systems.

A subconscious awareness computationally chugging away seems too mechanical to accommodate creative adjustments in an envisioned need to change course. Is the stack more than this? Yes, system feedback loops can affect system changes, yet only within the organic boundaries of systemic capacities. If systems are to evolve new capacities, they must be less of themselves (that is less conscribed to homeostatic operation) in order to more than themselves (that is to be open to autopoietic dynamics with effectors that are beyond the dominating influence of an existing sovereign system). Systems must be both consistent yet also open to change.... imperfectly perfect, if you will. Like the imagined clathrate structures in fluid water.[18]

Because systems bend toward optimization in resource-constrained realities, their operations are pulled into a gravity well... this is their core value of optimization.., driving the system into a pit centred always upon a foregone

(continued)

[18] See Dillon (2002)" Clathrate is a structure in which water molecules under certain conditions bond to form complex networks of molecules forming cage-like structures that encapsulate a guest molecule, which is a gas" cited from: Investigating Seafloors and Oceans, 2017 www.sciencedirect.com

Box 1 (continued)

notion that what HAD been best for the system remains best for the system. Western culture offers scant praise for optimization. No?

Thus, only in a resource-free domain of imagination can a sapient system choose to reimagine itself, if and only if it were to consciously choose to do so. As a core value, a system must hold sacred its capacity to be capable of imagining itself in new ways... of discovering new core values... of being more than" aware" by being reflectively" self-aware." To hold core values securely yet gently.

Maybe?

Let me add a humorous note taken from a mind[19] far wiser than my own:

> A reasonable man seeks to fit into the reality that surrounds him. An unreasonable man seeks to alter the world that surrounds him. Therefore, all progress is dependent upon the unreasonable man.

We may have given ourselves over to riding upon the shoulders of unreasonable men far too long. Perhaps our salvation now rests in the strength of unreasonable women ????

Janet: Thank you for taking the time to engage. Your emails are poetic and I like the quote! (ha , but women, like all genders along the gender spectrum can be very human – more is the pity) …. But yes, gender is a key factor in shaping life chances and agency (McIntyre-Mills et al., 2018, 2019a, b, 2020a, b forthcoming). *Women* need to be designers (Haraway, 1991) who address challenges or they risk being *disempowered* by the designs made by others. Women need to exert agency in the political design space to protect their rights and to speak out for the voiceless The tragedy of the enlightenment (western and eastern style) is that human beings have overestimated our capabilities and now (by concentrating on a market driven state that honours technical solutions to deep systemic failures) we face the hubris of a military industrial state, a tiger of many stripes (no offense to the non-metaphorical tigers)….

I am still working my way through "The stack." It is worth a careful read as I fear for democracy and governance – it could take a step backwards towards totalitarianism and it shows every sign of doing just that as the drums of war bang on justifying the overreach by every stripe of the military complex….. Poetry is vital along with the values that support 'wonder' and 'wilderness' to draw on Gerald Manly Hopkins.

By the way, I had an interesting time listening to a city planner who was giving a master class on a ferry in Sydney to a group of year 12 students last

(continued)

[19] George Bernard Shaw https://www.goodreads.com/quotes/536961-the-reasonable-man-adapts-himself-to-the-world-the-unreasonable accessed 14/05/2021.

Box 1 (continued)

Friday. As we went past Barangaroo (on Sydney harbour) he explained that the tower blocks were completely sustainable and that they had a negative carbon footprint. They use sea water for air-conditioning which is commendable, but I could not help regard them as ugly and unwelcoming....some trees were planted around the new model precinct above which the infamous Packer Casino towers. The giant city scape to my eyes lacks the intimacy of the quaint small scale Historical Rocks section which was rescued from demolition by Green Bans and the action of Indigenous people who actually asked the builders association to support them. They said that this area should be retained. Luckily a small slice of Sydney remains – the tiny cottages and cobble streets where the ex-convicts plied their trade alongside the locals who got the worst deal from everyone.....Today the area is for tourists but it retains something of the past. A sense of community. Strangers talk to one another because of the human scale. I met a colleague of Jack Munday because we were both watching the same documentary in the local Rocks Museum. The local shop owner on the corner offered to let me buy an orange juice (and pay later) as I went into the shop without my purse during an early morning evacuation as a result of a faulty fire alarm.

Can a sense of community and mateship be socialised in the cultural history of a small-scale place? Small communities foster strong emotional attachments through face-to-face engagement and emotional connections. These connections can be lost when we demolish and re-invent places. These are not entirely idle musings but are based on my recollection of the sadness of the people who were re-located tragically from District 6 in Cape Town as part of the terrible apartheid 'gentrification' which was of course quite the opposite. The area remains a scar on the side of Table Mountain. Years ago (not by chance as I was doing research on how people felt about the relocation), I met some of the people in the relocated area of Bonteheuwel and they told me that the reason why the windswept areas of the bleak Flats were covered in papers and broken bottles was that no one felt 'at home' there.....

So how can the notion of sustainable habitats that balance rural and urban areas achieve a sense of being at home – not just for human beings but for other species?

The notion of urban forests on rooftops and sides of buildings appeals, but we will need to have places that create small communities.....it also makes sense to limit the size of urban sprawl which has been said time and time again....

Dennis: I like the idea of poetry, but Janet – to get to your concern on critical engagement – the clue is in governance, users, and the address, it needs stakeholder involvement .....

(continued)

Box 1 (continued)

Janet: Tom and Dennis, yes, indeed. I have not read "Terraforming", but have been listening to some of Bratton's talks, reflecting on sections of "The Stack", and reading papers that are available. It seems that he has embraced a technocratic approach and that he follows the work of Srnicek and Williams (2015) "Inventing the Future: Post capitalism and a World Without Work" in which they too embrace technology. I think that the socialism at the expense of soul could be problematic. I much prefer the work of Deborah Bird Rose (a social anthropologist) who worked in the area of ecological humanities in Australia. She introduced the notion of *ethnography spanning species* and the need to face up to the era of extinctions; she posed the need for an ethic of '*Love in a time of extinctions*' (2008) and the need to care for others. This lack of care and commodification is the problem we face today. I believe *the human mind* is more that the brain and that consciousness is a continuum of living systems. Many species have consciousness, we need to tap into that and realise that human beings are the problem and the solution, paradoxically! We need to re-learn our place within living systems.

Janet to Dennis: Yes interbeing. teleonomy and living systems would be a good title for a special integration group for the International Society for Systems Sciences. It is the area that needs to be addressed, in order to address living systems. I think that relationships across social, economic, and environmental systems need to be discussed at ISSS – The issue is how relationships shape evolution – hence the importance of values through which we gendered human beings 'write' our evolutionary journey – which is heading for disaster.

I think that Alexander Christakis (2006, Christakis and Kakoulaki, 2021) was right that dialogue is indeed vital for democracy and the process of drawing out key themes will be central for our designs. He understood the first principle of cybernetics, namely that diversity needs to inform all designs (Ashby, 1956). We can add that biodiversity is equally important as Scott (1998) pointed out so profoundly in "Seeing like a state," in which he illustrates a range of examples in which monocultures result in losses to the environment and to society. Vandana Shiva (2012a) wrote a book with the title "Monocultures of the Mind" in which she made a similar argument.

Conversations in my opinion are the key to design. They are both a means and an end, because in the process of engaging in conversation we build friendship and trust. As these bonds grow, we extend our 'tolerance' and if we combine the conversations with 'hospitable' engagement then we are likely to make progress in maintaining and progressing the democratic project. Some of these ideas were discussed some years ago in "Philosophy in a time of terror" by Jurgen Habermas and Jacques Derrida in conversation with Borrodori (2003).

(continued)

Box 1 (continued)

It is interesting that the human brain has evolved in such a way that the talents needed for design are not carried by one person, but by groups of people with *different experiences andvalues*. Finding a way to work together across the divides remains the challenge. Perhaps focusing on shared needs, namely water, food, energy, and habitat security could make the difference.

Some of the pathways to addressing complexity have been shown by the paths walked by Stafford Beer, Alexander Christakis, Ken Bausch, Tom Flanagan and others who focus on dialogue and engagement. I think that Sue Gabriel's round table approach is another, whilst my take on mixed methods is to weave narratives based on listening, observing, and trying to understand views, in order to try to find contextual patterns….

Tom to Janet: The topics that you raised are critically important to talk about and yet also equally difficult to write about, as should be the case for all things which are written. We write about things and talk about things so that we might then think about things. Often, we think it's the other way around. But if we don't talk about something, we don't focus on it sufficiently to really think about it at all.

I have been thinking about the sacred quite a bit recently, largely because Craig Lindell (a board member for the Institute for 21st Century Agoras) and I have been involved in a writing effort on the topic of a secular church in a sacred world. Neither Craig nor I are deists. However, we are both deeply spiritual in our experience of life. I see no inconsistency with this. And I feel the tragic loss of religious practices which do not genuinely cultivate a sense of spiritual presence. This brings me to a few comments on the front piece of the article.

Interbeing, as a state of mind, was conceptually introduced to me by Daniel Christian Whal, author of the Regenerative Economy. In this sense interbeing is a state of mind that parallels interdependency as a state of body. Interbeing is a state of spiritual presence where interdependency is a state of physical presence. Being present is never as simple as we might think it is. In my simple, frequently three-part philosophy, Sentience as a form a Feeling is secondary to Sapience as a form of Thinking and Presence as a form of Interbeing. Presencing, as a verb, is, for me, living to the highest capacity. None of this contradicts your thoughts as expressed in the paper, and I sincerely hope that they do weave together well with your themes. I like to feel that my understanding of ethics and morals follows the writing of Hasan Özbekhan. I also like to feel that my understanding of ethics flows from an appreciation of the centrality and the frontiers of identities. And my appreciation of identity reflects the unavoidably subjective understandings of individuals as they seek to become the negotiated objective understandings of groups. A truly voiced

(continued)

Box 1 (continued)

subjective individual understanding, whether accurate as a measure of a "reality" or not, is an objective data point in any inclusive social understanding. First-phase science as an objective practice recognizes its "objectivity" as a reflection of an observation that, while subjectively experienced in multiple potential ways, remains unchanged by observers. The science, as a community, negotiates the aggregation and balance among the objective events of reported understanding in the form of some unitive, if not universally coherent "objective social truth." Second-phase and third-phase scientific observations, harvesting subjective and intersubjective observations, likewise are resolved into objective social truths through negotiated understandings. Any participant's sense of identity is defined by their feeling of centrality with respect to the negotiating process.... if they are not part of that process, then the objective social truth from the negotiation is not their truth. My point is that truth value is only relevant to the means within which individuals are included in the truth negotiation. There is no way that an outcome of any Community Thinking can be imposed upon another community without forcefully colonizing that Community with its tyranny. So then, from my perspective, the path towards broad inclusion must include an invitation to deeper thinking. This thinking must be driven by inquiry and must be preceded by a passionate desire to inquire. Because answers are the endpoint of thought, presenting an answer as a catalyst for thinking is like presenting a tombstone as a catalyst for life! Hence my passion for poets... artists who can paint familiarly mysterious visions with their broadly felt words.

Janet to Tom: So, the process of engaging is actually one of the most important aspects. It is a means to an end and an end in itself .....

Tom to Janet:: We cannot interconnect our minds without first finding an experiential way of interconnecting our hearts... we cannot find a way of interconnecting our hearts without sharing a hope for such interconnection... and, perhaps most essentially, we cannot share a hope to connect our hopes without a belief that such inclusive sharing is actually possible. It is not so much that we must speak for the earth as, I think, it is that we must feel the presence of the earth when we speak. In this necessity it is a tragic consequence that religions that draw our spirits beyond the world carry our spirits away from the world. A new sensation, a new experience of religion is emerging, though. It will run in parallel with older traditions if allowed but will cultivate a renewed sense of a unitive oneness that now will include the presence of all "creations" on earth. This new experience will weave it's way into culture as a secular church in a spiritual world. Maybe. Social contract, regardless of how well rights and responsibilities are designed to balance, cannot endure outside of a social covenant. We, the people, must come to terms with a hope,

(continued)

Box 1 (continued)

promise, and belief that we will become one people... infinitely complex along the fractal of our frontiers yet also immutable – secure in the essence of our imperfect and evolving humanity. In this sense, a breech in social contract is dwarfed by a breech in our social covenant.

What pending force of planetary reality will shatter our comforts and allow us to feel the presence of the possibility of a new covenant within which our humanity once again finds itself to be inclusive and interdependent? I think that as the old ways converge upon their event horizon, we will be left with little else than to imagine the unimaginable, make peace with the mysteries, and find safety in the otherness that we each carry as both our flag and our cross. But I digress…

Janet to Tom: So, perhaps humanity will realise that we are all members of one (threatened) species and that we need to respect our place within the web of life. Humility supported by beliefs that support this view, whether they are ecological, religious, or based on an expanded form of pragmatism, will help to support this goal. Narrow idealism or pragmatism could be our downfall. This is why we need a non anthropocentric approach to design informed by systemic ethics (McIntyre-Mills, 2014).

Interestingly Alan Rayner – whose work on inclusionality grew out of his work on fungi and communication – introduced me to the work of Daniel Christian Wahl who has engaged in dialogues with Alan as detailed in Transformative Education (2021).

Box 2
Conversation on Artificial Intelligence

Janet: The question that comes to mind is why do people want new technology that is designed to support competition and the old neo-liberal economy?

Rudolf: No, we do not want all this technology, the same as we do not want to become slaves for the 1% Elite, Therefore, we also need to master the technology, so we can counterbalance it.

Janet: Perhaps the way technology is designed needs to be considered, this is why Donna Haraway has stressed the need for women and the marginalised to become designers, rather than objects of other people's designs. These designs must ensure that we balance individual and collective needs. Designs that stress only competition could be very dangerous, because Bots could create designs where human beings are superfluous. This may be good for some species and the environment (for a while) but perhaps the bots will consider the environment and other species superfluous too, because they do not need food or water, but they will need an energy supply of some sort, not that that is much comfort!

(continued)

Box 2 (continued)

Rudolf: Nothing is useless or useful, it all depends on how you use it. Horsepower is replaced with mechanical power. Human power by mechanical mind power. …we need to do is be prepared for change.

Janet: Seriously, the notion that people will be without jobs will mean they will need to create their own jobs by eco-facturing (Gunter Pauli, 2010) and marketing their own goods. Power is a central concern, who decides and why is central to design (see Flood, 1999, Flood and Romm, 2018).

Rudolf: I believe, when we apply all this technology appropriately, it will help us in implementing eco-facturing and make the world a better place for all.

Janet: Shiva argues that technology is never value free, so designs need to be very carefully considered (Swer, 2020). I like the idea of designing algorithms to work for people and nature to support "wellbeing stocks". These seem to be a secular version of what many Sundanese or Balinese believed to be important in Indonesia.

Rudolf: The algorithm is already working in nature, with or without human knowledge or understanding. Unfortunately, humans like to change nature to satisfy their ego. That is the reason for pollution, war, human made disaster.

Janet: Technological change is not inevitable it is driven by design and design is driven by choice and choices are in turn driven by values that are driven by either collaboration or conflict (these are the two evolutionary principles)

What form of reinforcement is supportive of social and environmental justice? Policy and law that rewards regeneration and sustainability. But how? We need local government support and post national governance as per the architectures mentioned in my slide on the need for biodiversity to be protected at multiple levels of governance, from below and above. My concern is Big brother doing the wrong thing! How does the open-source lever help to hold big brother algorithms to account so that local and post national wellbeing stocks are protected?

This is the focus of the effort to inspire alternative forms of technology to enable people to earn, while they learn and grow a future together.

Rudolf: Drawing on Taoist wisdom, we do the opposite to balance the imbalance. This is Wu Wei.

Janet: How does a farmer use wu wei ?

Rudolf: The farmer works co-operatively with nature.

Janet: I have shared with you a paper that appeared some years ago on wuli, shili renli which could be helpful? How will we apply the wuli (rules), shili (sensing) renli (relational) approach (Attwater, 2002, Gu and Zhu, 2000) that addresses areas of concern in a practical manner?

(continued)

Box 2 (continued)

Perhaps the shili (sensing) will be promoted by renli (relations) and wuli (rules) generated by the score card and the tracking system you propose for organic products?

By focusing on the basic ingredients for life, water, nutrients, energy, and a viable niche/habitat/home we could move a step closer to fostering a will to protect the commons. If participants in the dialogue realise that survival is a joint endeavour, perhaps they will understand?

Rudolf: By building in measures associated with social enterprise we could guide decisions that make sense to people.

Janet: Yes, the design is based on working with people to create relationships that are meaningful so that the decisions are informed by their perceptions and values. Indicators of wellbeing need to be meaningful so the design needs to reflect the diversity of the participants and their diverse life experiences, I also think that striving for harmony should not prevent people also engaging critically (thus testing out ideas) when they talk about what matters to them?

Rudolf: The system can be demonstrated diagrammatically (see Diagram 2).

Janet: So, by asking people to tell stories about living well we can learn from one another and share what matters to us? We could explore how our

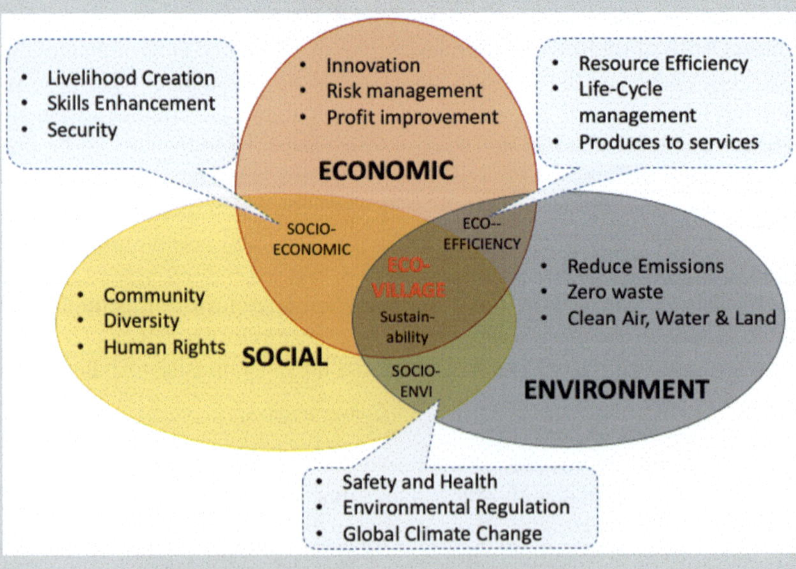

Diagram 2 Measuring the effectiveness of social enterprise

(continued)

Box 2 (continued)

opinions also differ from one another. Participants could be encouraged to explore the social, economic, and environmental factors that shape their lives through prompts to the conversation. For example, in the pathways to wellbeing prototype (McIntyre-Mills et al., 2014) people were asked to explore what they have, what they need, what they are willing to add or discard in order to move away from *Business as Usual* and to take small (regular and progressive steps) towards supporting what Joseph Stiglitz et al. (2010) call '*Wellbeing Stocks.*'[20]

The wellbeing of living systems can only be protected through extending the capabilities of human beings to protect organic and inorganic life, through redrawing the boundaries beyond the narrow social contract to protect all living systems:

Re-enforcement or positive feedback could help people to move towards more re-generative approaches. The challenge will be to get provincial governments to reward regions that are showing progress and to discuss what 'so-called 'progress entails, namely protection of biodiversity as inherently linked with social, economic, and environmental wellbeing.

Rudolf: Through engaging in online discussions and telling stories about what wellbeing means to them in everyday life, they could generate a lot of data, which can be used to measure what they value in terms of wellbeing indicators. All data generated could be recorded in a blockchain system to ensure that the recorded data is immutable.

Janet: How could block chain assist in supporting the circular economy?

Diagram 3 depicts the design for Pathways to Regenerate Living systems using Natural Language Processing (NLP) and Machine Learning (ML) to classify the data from stories to produce the Influencer Map, which contains the following three bar charts: Business as Usual, Small Changes, and Re-generating Living.

Janet: I am worried about over reliance on machine learning,[21] because people informed by ethical dilemmas need to make the final decisions, but a

(continued)

[20] The definition is as follows: '1. Material living standards (income, consumption and wealth), 2. Health, 3. Education, 4. Personal activities including work, 5. Political voice and governance, 6. Social connections and relationships, 7. Environment (present and future conditions), 8. Insecurity, of an economy as well as a physical nature'. This definition of wellbeing stocks fits well with the way in which both Indigenous and Non-Indigenous Australians connect with Country in Australia and elsewhere and the way in which critical systems thinkers and complexity theorists understand.

[21] https://www.youtube.com/watch?v=d4rBh6DBHywFei-Fei Li & Yuval Noah Harari (2019) in Conversation – The Coming AI Upheaval.

Box 2 (continued)

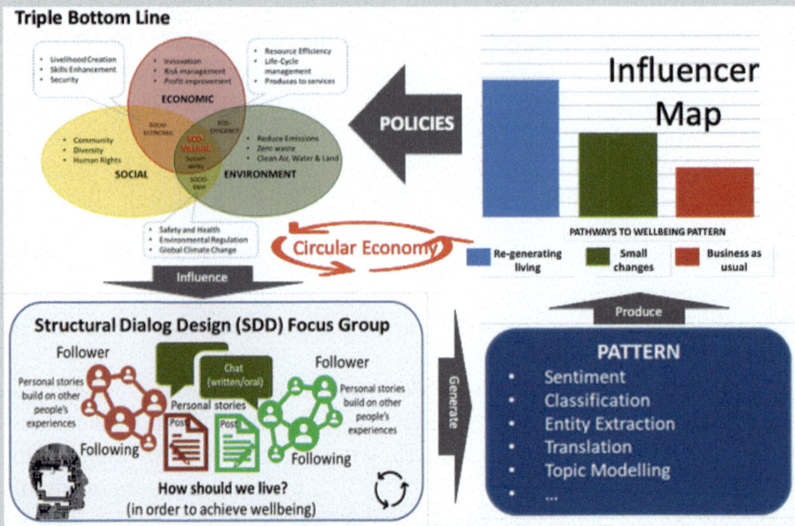

Diagram 3 The cycle to support a circular economy. (Source: Details are discussed in the chapter "Transformative Education: Employing the Balanced Scorecard for Regenerative Development" of this volume: Earn while they learn and grow a future together: exploring the potential of Eco-villages –to support a web-based economy (The Web site on Eco Village is located at http://wirasoftfoundation.org/en_GB/web/bic-sig))

form of engagement which shows how well (or badly) a local community is doing – -as individuals and as a collective --could help to inform better critical decisions and it could also be helpful to ensure that subsidies and incentives along with status and recognition are given to those who are supporting the common good.

This could help to introduce a cultural shift supported by incentives to engage in activities that support regeneration measured by wellbeing indicators.

Rudolf: Yes, the Influencer Map could be used to demonstrate to what extent a Circular Economy is being achieved. Members of social enterprises could also use smart contracts to support marketing green organic products that can be tracked and traced using the system.

Janet: I like the idea of being able to market products by demonstrating their organic origins. The design could also support Gunther Pauli's (2010) notion of no waste.

So perhaps the challenge that we will face is HOW to motivate people to engage with this system in a way that balances the needs of multiple species – without being top down AND to ensure that the process is not highjacked for

(continued)

Box 2 (continued)

authoritarian purposes! In Indonesia, there is a chance that this initiative could be scaled up, because it is a presidential decree[22] that all villages subscribe to the "One Village, One Product approach". This decree could be adapted to foster engagement to support the needs of the marginalised – but in order to encourage local governments internationally to engage in mapping and modelling their carbon footprint would require a combination of both positive and negative sanctions managed by international law.

I find it encouraging that young people are taking out lawsuits against the environment minister, as reported by Morton (2021):

> The federal court of Australia has found the environment minister, Susan Ley, has a duty of care to protect young people from the climate crisis in a judgment hailed by lawyers and teenagers who brought the case as a world first. Eight teenagers and an octogenarian nun had sought an injunction to prevent Ley approving a proposal by Whitehaven Coal to expand the Vickery coalmine in northern New South Wales, arguing the minister had a common law duty of care to protect younger people against future harm from climate change.

This is not an isolated incidence. Taylor et al. (2021) report how children are taking on countries in Europe: "Imagine, 33 of the biggest economies in the world saying, 'We don't want this' and six youths saying, 'We are going to do this.'"

7 Action Learning and Action Research: Gender Sensitive Youth Agency in the Green Economy – Capacity Enhancement Approaches

In order to address the problem of social exclusion, unemployment, and habitat degradation, place-based research focuses on empowering local people. Education and governance systems need to address the production, consumption, and redistribution cycle and the potential to adapt and scale up the 'One village, one /many enterprises ' to support resilience (see chapters "Collective Action for Regeneration of the Web of Life in the Face of Disruptive Injustice" and "Women, Education and Socialisation: Transforming Gender and Power in the MST, São Paulo, Brasil" this volume) in which we strive to explore behaviour and valuetransformation in co-operatives and social enterprise groups and in the process learn more about the potential of regenerative enterprises that address food, energy, and water

[22] One village, one product' (OVAP, Morihiko Hiramatsu – Governor of Oita prefecture, 1979, Yogyakarta, 2014) was applied by President Jokowi in 2008–2009.

insecurity. Currently unemployment and lack of access to tertiary or vocational training are major human security challenges in a degraded environment. Highly urbanised, environmentally affected regions face the cascading social, economic, and environmental challenges that impact on the habitat across the continuum from domestic, liminal, agricultural, and wild animal life, to draw on Donaldson and Kymlicka (2011).

Transformation from business as usual requires taking steps towards protecting "well-being stocks" (Stiglitz et al., 2010) by exploring[23] the extent to which:

(a) salience, trust, and engagement, the key underlying principles of participatory democracy (McIntyre-Mills & Christakis, 2021) and intersubjectivity (Christakis and Kakoulaki, 2021), are indeed supported by a community of practice to support agricultural activities.
(b) social, economic, and environmental accounting and accountability can be supported by means of the community of practice teaching the implications of decisions by means of scenarios.

The multi-sit symposium in 2017 held in South Australia and West Java addressed the production, consumption and redistribution cycle and the potential to adapt and scale up the 'One village, one /many enterprises' to support resilience. The Africa-Indonesia Bandung Conference in 1955 on development stressed the importance of de-colonisation and local voices, but indigenous people and many of the rural and urban poor feel left behind in South Africa and West Java. In line with these principles, we explored behaviour and valuetransformation in co-operatives and social enterprise groups to ascertain what works why and how to inform, adapt, and map regenerative business opportunities. According to South Africa's implementation of the 2030 agenda for sustainable development 'Solving Complex Challenges Together', SDG 10 addressing inequality will need more attention to give opportunities to young people.

In line with The Indigenous Knowledge Systems (IKS) Policy (November 2004) the project aims to create a community of practice to enable IKS and to enhance institutional capacity building in line with Odessa Hopper's (2013) vision for the positive protection and development of IKS. The approach draws on Sen and in particular Nussbaum's (2011) capabilities approach for human development.

[23]This volume builds on previous work (McIntyre-Mills et al. 2019a, b, McIntyre-Mills and Corcoran Nantes, 2020) in which we expressed our concerns that highly urbanised regions face food and water insecurity and are at risk of becoming food deserts unless everyday strategies are explored with service users and providers to find better pathways to resilience and wellbeing for the most vulnerable members of the population. Most of the global urban population *will be located in Asia (52%) and Africa (21%)*, according to the UN (2014: 11). This has implications for food, water and energy security. High rate of urbanization by 2050 the majority of the world's population will be living in cities in Africa and Asia.

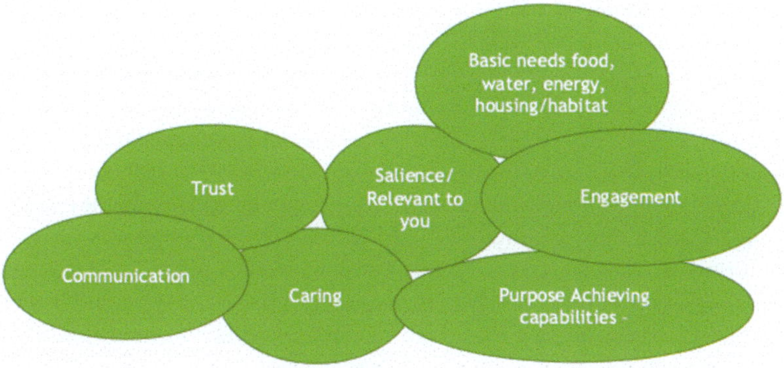

Diagram 4 Key considerations for engagement

It draws inspiration from Indonesia's" One village, one /many product(s) approach" to development.[24] This Participatory Action Research aims to:

- Foster a demonstration pilot as a first step towards scaling up regional development hubs to create local employment opportunities and thus reduce rural to urban migration and the associated risks.
- Study co-operative social engagement on what works, why and how in Indonesia could provide insights for South Africa using a community of practice[25] (Wenger, 1999)
- Extend and deepen an understanding of democratic and governance opportunities, whilst recognizing that Ostrom's research has revealed principles and patterns amongst co-operative groups but the place-based research on what works, why, and how also reveals the limitations of local engagement and the need for advocacy for post national, post regionalist governance.

Each of the case studies to date reveal limitations and to address the social, economic, and environmental challenges at the local level, but this is not enough, we also need to address challenges by achieving post national solidarity (Diagram 4).

The hypothesis is that the greater the level of participation in social, economic, and environmental decision making based on 'if / then scenarios' linked to green

[24] Alam Endah, Cibodas and Tarumajaya in West Java where local residents are actively engaged in the 'One Village, One enterprise approach decreed by President Jakowi and Kediri City where Wirawan engaged in training and helped to inspire their activities related to a Computer Training Co-operative

[25] Manyeledi Border of Botswana NE Province, community study and Tiger Kloof Educational Institute, Vryberg set up an Organic Co-op with the Disabled, scoping visit. Ocean View Organic Farm (Fish Hoek)linked with Ocean View High School and the Co-op board, exemplar for action learning for Bamboo Co-operative, Venda, South Africa.

social enterprise and supported by a community of practice – the better the public education outcomes in line with UN Sustainable Development Goals.

The research[26] explores whether groups can manage shared resources in such a way that the commons (habitat) for multiple diverse species can be maintained whilst simultaneously setting up a viable circular economy that enables those –who are currently marginalised and at risk of displacement – to establish their own income streams within the co-operatively run social enterprises that are supported as learning organisations/communities.

New forms of education, economics, democracy, and governance are needed. The potential of eco-villages needs to be explored. Imagine if we could have rural-urban campuses based on an eco-village model? Ecovillages could link products, markets and de-centralised systems of learning through caring and growing a future together.[27]

Core design principles are suggested, namely salience, trust, and engagement to protect living systems and the people who are affected.

These principles need to underpin the decision-making process. These principles are discussed in the paper together with the importance of 'being the change' and 'interbeing' through expanding pragmatism to consider the social, economic and environmental implications of choices. Systemic Ethical decisions honour 'freedom and diversity' to the extent that freedom and diversity are not undermined by power imbalances and find a way to counterbalance:

- *Educationas usual* or neo-liberal *factoryeducation* that *does not* address the convergent social, economic and environmental challenges
- *Steps towards alternative forms ofeducation* at the primary, secondary, tertiary level linked at each level with Vocational Education and training that addresses the social, economic and environmental challenges to support
- *Transformative educationnodes and networks* that balances individual and collective social, economic and environmental protection across rural and urban areas that draws on the local tacit knowledge and is supported by local learning hubs where people could meet, greet, learn new skills in a multi-generational way with facilitators spanning places of formal and informal learning.

[26] The research draws on interdisciplinary contributions, in particular the work of Elinor Ostrom on managing the commons, Vandana Shiva on sustainable agriculture, Gunter Pauli on circular green economies, Polly Higgins and colleagues who founded the movement to introduce an Ecocide Law and Peter Senge's work on organisational learning.

[27] The pilots on eco villages propose an alternative cyclical economy based on eco-villages supporting urban hubs to re-generate rural-urban balance based on ecofacturing, to use Gunter Pauli's concept. The development of eco-facturing using local products such as cassava for bioplastics, bamboo for biochar, fair trade coffee, mushroom production using the coffee grains and honey are discussed as examples of ecofacturing that are currently being developed in Indonesia as well as pilot projects in South Africa.

7.1 Postscript

The future is shaped by our thinking and practice. It is possible for praxis to be regenerative; it is a matter of choice. As we finalised the volume, people waiting outside the airport in Kabul were bombed and women face increasing risks.

We need a large dose of humility and an equally large dose of courage to take a practical leap towards living differently every day. Please help us to scale up the eco village: ***Community of practice to inspire agencyand hope to addresswellbeing stocks:***

janet.mcintyre@adelaide.edu.au
rudolf.wirawan@adelaide.edu.au

The Web site on Eco Village is located at http://wirasoftfoundation.org/en_GB/web/bic-sig[28]

Bibliography

Aarhus Convention. (1998). *On access to information. Public participation and access to justice in environmental matters*, Denmark, 25 June. http://ec.europa.eu/environment/aarhus/

Abram, D. (2020). Foreword in Kaza, S. 2020. In *A wild love for the world: Joanna Macy and the work of our time*. Shambhala.

Arendt, H. (1963). *Eichmann in Jerusalem: A report on the Banality of Evil*. Viking Press.

Ashby, W. R. (1956). *An introduction to cybernetics*. Chapman and Hall.

Attwater, R. (2002). Mixing meta-methodologies and philosophies: Wuli-shili-renli, pragmatic and practical philosophy. In *International Systems Sciences Annual Conference*, Shanghai.

Bateson, G. (1972). *Steps to an ecology of mind*. Ballantine.

Baumann, Z. (2000). *Liquid modernity*. Polity.

Baumann, Z. (2011). From agora to the market place, and where to from here? *Journal of Globalisation Studies, 21*(1), 3–14.

Borrodori, G. (2003). *Philosophy in a time of terror. Dialogues with Habermas and Derrida*. University of Chicago Press.

Bostrom, N. (2011). *Existential risk prevention as the most important task for humanity*, www.existential-risk.org

Bratton, B. H. (2018). *The whole age of computer has made it where —*. https://www.youtube.com/watch?v=ZwfQde5WWjA

Bratton, B. H. (2015). *The stack: On software and sovereignty*. MIT Press.

Bratton, B. H. (2020). *Terra forming the world order*. https://palladiummag.com/2021/01/11/benjamin-h-bratton-on-terraforming-the-world-order/

Bruce, S. (2016). The project of a world environment court. In C. Tomuschat (Ed.), *The OSCE court of conciliation and arbitration*. Brill.

Buckminster Fuller, R. (1978). *Operating manual for space ship earth*. Dutton.

[28] Community engagement to address : UN Sustainable Dev Goals, Gender Mainstreaming

Critical Systemic thinking, Indigenous knowledge systems, webinars on engagement to empower the marginalised and Circular economy

Butler, J. (2011, May 28). *Precarious life: The obligations of proximity*. The Neale Wheeler Watson Lecture, Nobel Museum, Svenska, 28 May. Available at: www.youtube.com/watch?v=KJT69AQtDtg. MIT Press.

Butler, J., & Athanasiou, A. (2013). *Dispossession: The Performative in the Political*. Polity.

Butler, J., & Taylor, S. (2010). *Taylor, A. Examined Life. Excursions with contemporary thinkers*. The New Press. https://cjgl.cdrs.columbia.edu/article/performing-interdependence-judith-butler-and-sunaura-taylor-in-the-examined-life-2/

Chalmers, D. J. (1995). Facing up to the problem of consciousness. *Journal of Consciousness Studies, 2*, 200–219.

Christakis, A., & Bausch, K. (2006). *How people harness their collective wisdom and power to construct the future in co-laboratories of democracy*. Information Age.

Christakis, A., & Kakoulaki, M. (2021). Chapter 35: Objectifying intersubjectivity for a scientific (re)volution through inclusion. In J. J. McIntyre-Mills & Y. Corcoran-Nantes (Eds.), *From polarisation to multi species relationships: Re-generation of the commons in the era of mass extinctions*. Springer.

Churchman, C. W. (1971). *The design of inquiring systems: Basic concepts of systems and organisations*. Basic Books.

Claymore, Y (2007). A study for Sustainable Local Development through One Town One Product: An Overview of OTOP in Thailand. In *Paper was presented in "International OVOP Policy Association (IOPA)" Conference*, at Zhejiang University, Shanghai, China.

Dalai Lama. (2005). *The Universe in a single atom. The convergence of science and spirituality*. Harmony.

Dalai Lama, His Holiness. (2020). https://www.theguardian.com/world/2020/nov/11/buddha-would-be-green-dalai-lama-calls-for-urgent-climate-action

Damery, P., & Abram, D. (2013). The environmental crisis and the Psyche. *Jung Journal: Culture & Psyche, WINTER 2013, 7*(1), 104–119. (Taylor & Francis, on behalf of C.G. Jung Institute of San Francisco Stable): https://www.jstor.org/stable/10.2307/26596525

De Waal, F. (2009). *The age of empathy: Nature's lessons for a Kinder society*. Harmony Books.

Dillon, W. P. (2002). Gas Hydrate in the Ocean Environment. In R. A. Meyers (Ed.), *Encyclopedia of physical science and technology* (3rd ed., pp. 473–486). Academic Press, ISBN 9780122274107. https://doi.org/10.1016/B0-12-227410-5/00276-3. https://www.sciencedirect.com/science/article/pii/B0122274105002763

Donaldson, S., & Kymlicka, W. (2011). *Zoopolis: A political theory of animal rights*. Oxford University Press.

Draaisma, D. (2015). Oliver Sacks (1933–2015). *Nature, 525*, 188. https://doi.org/10.1038/525188a

Flanagan, T. R. (2014). Designing the means for governing the commons. In G. S. Metcalf (Ed.), *Social systems and design* (pp. 147–166). Springer.

Flanagan, T., & Christakis, A. (2010). *The talking point: Creating an environment for exploring complex meaning*. Information Age Publishing.

Flanagan, T.R. & Christakis, A. N. (2009, 2022). *The talking point: Creating an environment for exploring complex meaning*. 2, : Information Age Publishing. 2n edition forthcoming

Flood, R. (1999). *Rethinking the fifth discipline: Learning within the unknowable*. Routledge.

Flood, R. L., & Romm, N. R. A. (2018). A systemic approach to processes of power in learning organisations: Part 1 – Literature, theory, and methodology of triple loop learning. *The Learning Organization, 25*(4), 260–272.

Florini, A. (2003). *The coming democracy*. Island Press.

Forsetto, R. (2021). 5 February 2020 news.mongabay.com cites the Manifest of Piaraçu (20/Jan/2020). https://apiboficial.org/2020/01/20/manifest-of-piaracu/?lang=en

Forsyth, T., & Johnson, C. (2014). Elinor Ostrom's legacy: Governing the commons and the rational choice controversy. *Development and Change, 45*, 1093–1110. https://doi.org/10.1111/dech.12110

Frankl, V. (1992). *Man's search for meaning: An introduction to logotherapy* (4th ed.). Beacon Press.

Gibson, K., Bird Rose, D., & Fincher, R. (2015). *Manifesto for living in the Anthropocene*. Punctum Books.
Gildea, F. (2020). Accelerating down a road to nowhere: On inventing the future by Nick Srnicek and Alex Williams. *The Political Quarterly, 91*, 359–363. https://doi.org/10.1111/1467-923X.12850
Goodall. (2020). https://www.cnn.com/videos/us/2020/03/19/coronavirus-jane-goodall-acfc-full-episode-vpx.cnn
Gu, J., & Zhu, Z. (2000). Knowing wuli, sensing shili, caring for renli: Methodology of the WSR approach. *Systemic Practice and Action Research,**13***(1), 21–58.
Harari, Y. N. (2021). The World after Covid, *FT Weekend Digital Festival*. https://www.youtube.com/watch?v=iWo4OrGhGxI
Haraway, D. (1991). *The reinvention of nature*. Free Association Books.
Haraway, D. J. (2016). *Staying with the Trouble: Making Kin in the Chthulucene*. Duke University Press.
Hardin, G. (1968). The tragedy of the commons. *Science, 162*(3859), 1243–1248.
Held, D. (2004). *Global covenant: The social democratic alternative to the Washington Consensus*. Polity.
Higgins, P. (2012). *Ecocide, the 5th Crime Against Peace*. https://www.youtube.com/watch?v=8EuxYzQ65H4 see the link in this lecture to Stop Ecocide, Change the Law https://www.youtube.com/channel/UCCYoIf880oaM419JO8XUhEA
Higgins, P. (2018). *The 2018 Hague freedom lecture by Polly Higgins*. https://www.youtube.com/watch?v=kQn8oA6e9To
Higgins, P., Short, D., & South, N. (2013). Protecting the planet: A proposal for a law of ecocide. *Crime, Law and Social Change, 59*, 251–266. https://doi.org/10.1007/s10611-013-9413-6
IPCC. (2021). *IPCC sixth assessment repORT climate change 2022: Mitigation of climate change*. https://www.ipcc.ch/report/ar6/wg3/
Jackson, M. (2000). *Systems approaches to management*. Kluwer.
Jacobsen, M. H. (2013). From Modern Utopia to Liquid Modern Anti-Utopia? *Indigo: Humanities Magazine for Young People, 8*, 24–49.
Jung, C. V. interview with Freeman, J. (969) Face To Face | Carl Gustav Jung. (1959). *HQ Professor Jung is interviewed at his home in Switzerland by John Freeman*. https://youtu.be/2AMu-G51yTY
Koesler, A. (1968). *The ghost in the machine*. Macmillan.
Litfin, K. T. (2009). Reinventing the future: The global ecovillage movement as a holistic knowledge community. In G. Kütting & R. Lipschutz (Eds.), *Environmental governance: Knowledge and power in a local-global world*. Routledge.
Litfin, K. (2013). *Ecovillages: Lessons for sustainable community*. Polity Press.
McIntyre-Mills, J. (2003). *Critical systemic praxis for social and environmental justice: Participatory policy design for a global age* (The contemporary systems series). Kluwer.
McIntyre-Mills, J. (2008). *User-centric design to meet complex needs*. Nova Science. 423 page book. Appendix on software design by Denise de Vries. Output from Linkage LP0560406 Nova Science.
McIntyre-Mills, J. (2014). *Systemic ethics and non-anthropocentric stewardship* (p. 270). Springer.
McIntyre-Mills, J. (2017a). *Planetary Passport for representation, accountability and re-generation* (Contemporary Systems Series). Springer.
McIntyre-Mills, J. (2017b). Recognising our hybridity and interconnectedness: Implications for social and environmental justice. *Current Sociology*. https://uk.sagepub.com/en-gb/journals-permissions
McIntyre-Mills, J. (2018). Policy design for non-anthropocentric pathways to protect biodiversity and regenerate the land. *International Journal of Educational Development in Africa*. https://doi.org/10.25159/2312-3540/2865

McIntyre-Mills, J. (2021a). Chapter 3 'From old to new taxonomies of rights, relationships and responsibilities to protect habitat'. In *From polarisation to multispecies relationships in the age of mass extinctions* (pp. 31–54). Springer.
McIntyre-Mills, J. (2021b). Chapter 26: McIntyre-Mills, J. Biopolitics and food security to protect social and environmental justice. In *From polarisation to multispecies relationships in the age of mass extinctions* (pp. 575–602). Springer.
McIntyre-Mills, J. (2021c). Chapter 26: Biopolitics and food security to protect social and environmental justice. In *From polarisation to multispecies relationships* (pp. 575–602). Springer.
McIntyre-Mills, J. J. (2021a). Chapter 5: Consciousness for balancing individualism and collectivism. In J. McIntyre-Mills & Corcoran-Nantes (Eds.), *From polarisation to multispecies relationships* (pp. 67–86). Springer.
McIntyre-Mills, J. J. (2021b). Chapter 8: Stewardship: An anthropocentric misnomer? Rights, responsibilities and multispecies relationships. In J. McIntyre-Mills & Corcoran-Nantes (Eds.), *From polarisation to multispecies relationships* (pp. 119–140). Springer.
McIntyre-Mills, J., & Christakis, A. N. (2021). Chapter 14: Social and environmental justice. In J. McIntyre-Mills & Corcoran-Nantes (Eds.), *From polarisation to multispecies relationships* (pp. 283–308). Springer.
McIntyre-Mills, J., & Corcoran-Nantes. (2021). *From polarisation to multispecies relationships* (p. 777). Springer.
McIntyre-Mills, J. J., & Corcoran-Nantes, Y. (Eds.). (2021). *From polarisation to multispecies relationships in the age of mass extinctions*. Springer.
McIntyre-Mills, J., & de Vries, D. (2012). Transformation from Wall Street to Well-being. *Systems Research and Behavioral Science*. First published online: 10 OCT 2012. https://doi.org/10.1002/sres.2133
McIntyre-Mills, J., & with contributions by Wirawan, R., Widiansingh, I., & Riswanda. (2021). Chapter 15: Social engagement to protect multispecies habitat: implications for re-generation and food security. In *From polarisation to multispecies relationships in the age of mass extinctions* (pp. 309–368). Springer.
McIntyre-Mills, J. with Corcoran-Nantes, Y., Wirawan, R. and Widiansingh, I. (2021) Chapter 17: The potential of eco-facturing: Towards social and environmental justice through vocational education and training *From polarisation to multispecies relationships in the age of mass extinctions,* Springer pp. 391–424.
McIntyre-Mills, J., & Wirawan, R. (2021). Chapter 18: From education as usual to creating a post national learning community. In *From polarisation to multispecies relationships in the age of mass extinctions* (pp. 425–450). Springer.
McIntyre-Mills, J. with contributions by Wirawan, R and Widiansingh, I and Riswanda. (2021). Chapter 15: Social engagement to protect multispecies habitat: Implications for re-generation and food security. In J. McIntyre-Mills & Corcoran-Nantes (Eds.), *From polarisation to multispecies relationships* (pp. 309–368). Springer.
McIntyre-Mills, J. with Wirawan, R. (2021). Chapter 17: The potential of eco-facturing: Towards social and environmental justice through vocational education and training. In J. McIntyre-Mills & Corcoran-Nantes (Eds.), *From polarisation to multispecies relationships* (pp. 391–424). Springer.
McIntyre-Mills, J., & Wirawan, R. (2021). Chapter 18: Creating a post national learning community. In J. McIntyre-Mills & Corcoran-Nantes (Eds.), *From polarisation to multispecies relationships* (pp. 425–450). Springer.
McIntyre-Mills, J., Bausch, K., Christakis, A., & de Vries, D. (2008). How can we break the mould: democracy, semiotics and regional governance beyond the nation state? *Systems Research and Behavioural Science, 25*(2), 305–322.
McIntyre-Mills, J., Romm, N. R. A., & Corcoran Nantes, Y. (2019). *Democracy and governance for resourcing the commons: Theory and practice on rural-urban balance*. Springer.
Meijer, E. (2016, translated, 2019). *Animal languages: The secret conversations of the living world* John Murray. London

McIntyre-Mills, J., Corcoran-Nantes, Y., Widianingsih, I., & Wirawan, R. (2019a). Chapter 7. Alam Endah: Rural Camelot in West Java—A case study of empowerment and integrated rural development. In J. McIntyre-Mills, R. NRA, & Y. Corcoran Nantes (Eds.), *Democracy and governance for resourcing the commons: Theory and practice on rural-urban balance* (pp. 179–192). Springer. https://doi.org/10.1007/978-3-030-04891-4. Print ISBN: 978-3-030-04890-7 Human Research Ethics LIVING VIRTUOUSLY AND WELL 7213.

McIntyre-Mills, J., Corcoran-Nantes, Y., Wirawan, R., Widianingsih, I., Saptenno, I., & Sari, N. H. (2019b). Chapter 8. Freeing the Luwaks and escaping the iron cage: Vignettes on surviving in the anthropocene by joining up the social, economic and environmental policy dots—an application of critical systemic thinking to areas of concern. In J. McIntyre-Mills, N. Romm, & Y. Corcoran-Nantes (Eds.), *Democracy and governance for resourcing the commons. Contemporary systems thinking* (pp. 195–232). Springer. https://doi.org/10.1007/978-3-030-04891-4. Print ISBN: 978-3-030-04890- Human Research Ethics LIVING VIRTUOUSLY AND WELL 7213.

Midgley, G. (2000). *Systemic intervention. Contemporary systems thinking*. Springer.

Midgley, G., & Lindhult, E. (2021). A systems perspective on systemic innovation. *Systems Research and Behavioral Science, 38*(5), 635–670. https://doi.org/10.1002/sres.2819

Midgley, G., & Rajagopalan, R. (2021). *Critical systems thinking, systemic intervention, and beyond*. https://doi.org/10.1007/978-981-13-0370-8_7-1

Morton, A. (2021). Australian court finds government has duty to protect young people from climate crisis. *The Guardian*. https://www.theguardian.com/australia-news/2021/may/27/australian-court-finds-government-has-duty-to-protect-young-people-from-climate-crisis

Nussbaum, M. (2006). *Frontiers of justice: Disability, nationality, species membership*. Harvard University Press.

Nussbaum, M. (2011). *Creating Capabilities: The Human Development Approach*. The Belknap Press.

Odora Hoppers, C. (2013). Community engagement, globalisation, and restorative action: Approaching systems and research in the universities. *Journal of Adult and Continuing Education, 19*(2)) Autumn 2013: 94–102. https://doi.org/10.7227/JACE.19.2.7

Ostrom, E. (1990). *Governing the commons: The evolution of institutions for collective action*. Cambridge University Press.

Ostrom, E. (2008). *Design principles of robust property-rights institutions: What have we learned?* Elinor Ostrom Workshop in Political Theory and Policy Analysis Indiana University Center for the Study of Institutional Diversity Arizona State University. http://www.indiana.edu/~workshop

Ostrom, E. (2010). Elinor Ostrom Nobel prize in economics lecture. https://www.youtube.com/watch?v=T6OgRki5SgM gives an update on the 2009 lecture.

Ostrom, E. (2014, Jun 25). Elinor Ostrom Nobel prize in economics lecture. https://www.youtube.com/watch?v=T6OgRki5SgM

Ozbekhan, H. (1968). *The triumph of technology: "Can implies ought."* System Development Corporation paper. http://redesignresearch.com/dialogues/wp-content/uploads/2013/08/Ozbekhan-The-Triumph-Of-Technology-_Can-Implies-Ought_.pdf

Ozbekhan, H. (1970). *The predicament of mankind: A quest for structured responses to growing world-wide complexities and uncertainties*. http://web.archive.org/web/20070928083748/http://www.cwaltd.com/pdf/clubrome

Palese, E. (2013). Zygmunt Bauman. Individual and society in the liquid modernity. *SpringerPlus, 191*. https://doi.org/10.1186/2193-1801-2-191

Paul, P. J. (2015). Encyclical letter laudato si'of the holy father francis on care for our common home. https://www.vatican.va/content/francesco/en/encyclicals/documents/papafrancesco_20150524_enciclica-

Pauli, G. (2010). *The Blue economy: Report to the Club of Rome*. Paradigm Publications.

Quan-Baffour, K. P., McIntyre-Mills, J., & Romm, N. R. A. (2019). Ubuntu: Dialogue on connectedness, environmental protection and education. In J. McIntyre-Mills & Romm. NRA (Eds.),

Mixed methods and cross disciplinary research: Towards cultivating eco-systemic living (pp. 221–250). Springer.

Rayner, A. (2010). Inclusionality and sustainability – Attuning with the currency of natural energy flow and how this contrasts with abstract economic rationality. *Environmental Economics, 1*(1), 98–108.

Robertson, G. (2021). *Bad people – And how to be Rid of them*. Random House.

Romm, N. R. A., & Lethole, P. V. (2021). Chapter 6: Prospects for sustainable living with focus on interrelatedness, interdependence and mutuality: Some African perspectives. In *From polarisation to multispecies relationships in an era of mass extinctions*. Springer.

Rose, D. B. (2008). Love in the time of extinctions. *The Australian Journal of Anthropology, 19*, 81–84. https://doi.org/10.1111/j.1835-9310.2008.tb00112.x

Rose, D. B. (2015). *Love at the edge of extinction*. http://deborahbirdrose.com/links/

Scott, J. (1998). *Seeing like a State*. Yale University Press.

Shiva, V. (2012a). *Monocultures of the mind*. Third World Network.

Shiva, V. (2012b). *Making peace with the earth*. Fernwood Publishing.

Shiva. (2018). *Oneness vs the 1%: Shattering illusions, seeding freedom*. Spinifex Press.

Shiva, V. (2019). *We must fight back against the 1 percent to stop the sixth mass extinction*. https://www.democracynow.org/2019/2/22/vandana_shiva_we_must_fight_back

Shiva, V. (2020a). *Two paths to the future of food and farming – EcoFarm Keynote*. https://www.youtube.com/watch?v=9vNPGwqR4TQ

Shiva, V. (2020b, March 25). *Ecofeminism and the decolonization of women, nature and the future*. https://www.youtube.com/watch?v=hVbbov9Rfjg – Uploaded by San Telmo Museoa.

Simard, S. (2016). *How trees talk to each other*. https://www.youtube.com/watch?v=Un2yBgIAxYs

Simard and Defrenne. (2019). *The secret language of trees – Camille Defrenne and Suzanne Simard*. https://www.youtube.com/watch?v=V4m9SefyRjg

Solms, M. (2021). *The Hidden Spring: A Journey to the Source of Consciousness*. https://www.youtube.com/watch?v=CmuYrnOVmfk

Srnicek, N. and Williams (2015) A. Inventing the future: Postcapitalism and a world without work, .

Stiglitz, J., Sen, A., & Fitoussi, J. P. (2010). *Mis-measuring our Lives: Why the GDP doesn't add up*. The New Press.

Strang, V. (2017). Justice for All: Inconvenient truths and reconciliation in human-non-human relations. In H. Kopnina & E. Shoreman-Ouimet (Eds.), *Routledge international handbook of environmental anthropology* (pp. 259–275). Routledge.

Sutton, R. (1992). *Machine learning. Introduction: The challenge of reinforcement learning* (Vol. 8, pp. 225–227). Kluwer Academic Publishers.

Swer, G. M. (2020). How technology drives the history of the green revolution. *Capitalism Nature Socialism*, https://doi.org/10.1080/10455752.2020.1862889

Taylor, M., Holden, E., Collyns, D., Standaert, M., & Kassam, A. (2021). *The young people taking their countries to court over climate inaction*. https://www.theguardian.com/environment/2021/may/07/the-young-people-taking-their-countries-to-court-over-climate-inaction

Thunberg, G. (2019). *Greta Thunberg to world leaders: 'How dare you? You have stolen my dreams and my childhood'* https://www.google.com/search?q=thunberg+how+dare+you+speech & oq=thunberg+and+how+dare+ & aqs=chrome.2.69i57j0i22i30l2j0i22i30i395l4j0i390i395l2.24047j1j7 & sourceid=chrome & ie=UTF-8

Wahl, C. (2016). *Designing regenerative cultures*. Triarchy Press.

Wahl, C. (2019). Sustainability is not enough in. In G. D. Dabelko & K. Conca (Eds.), *Green Planet Blues: Critical perspectives on global environmental politics*. Taylor & Francis Group, 2019. ProQuest Ebook Central. http://ebookcentral.proquest.com/lib/flinders/detail.action?docID=5892383

Wenger, E. (1999). *Communities of practice. Learning, identity and meaning*. Cambridge University Press.

Wenger, E., & Smith, J. (2009). *Digital habitats: Stewarding technology for communities*. CP Square.

Wirawan. (2021). *Chapter 6 this volume Learning from nature's classroom: Reframing economics, accounting and accountability.*

Wirawan, R., Finlayson, D., & Flanagan, T. (2021). Architectures to protect the commons: A community of practice to inspire agency and hope. In *International systems sciences, paper delivered 65th annual meeting of the international society for the systems sciences, "The Art and Science of the Impossible: The Human Experience"*. https://www.isss.org/online-2021

Wohlleben, P. (2016). *The hidden life of trees: What they feel, how they communicate: Discoveries from a secret world.* Greystone Books.

Wohlleben, P., & Simard, S. (2016). https://www.intelligent-trees.com/; https://vimeo.com/ondemand/intelligenttreesvimeor

Zuboff, S., & Schwandt, K. (2019). *The age of surveillance capitalism: The fight for a human future at the new frontier of power.* Profile Books. https://www.youtube.com/watch?v=GwxOxQ1AOEghttps://www.youtube.com/watch?v=GwxOxQ1AOEg

Postscript: A Meta Logue with Reynaldo

Janet J. McIntyre-Mills and Yvonne Corcoran-Nantes

Janet: Brother Reynaldo[1] when I looked at your paintings and the video of your work,[2] I was struck by one of the paintings in particular. It spoke to my frame of reference and when I looked at it the following came to mind:

Cascade for tears wept,
webs for support and
feathers of hope to take flight

Then you told me it is called 'Peacock Cascades.'
Please tell me a bit about yourself as a systemic thinker and practitioner?

Reynaldo: Sister Janet the creative human being is doing something that delights the spirit so much that any effort involved seems to be a trifle, as they enter into a state of fluid creative joy from the pleasure of attaining a beautiful result for many, rather than just for oneself.

The magic of a work of art is its transcendence and its potential to communicate with others.

A piece of art usually persists over time unless a contingent situation destroys it before its creator dies, although few creators (artists) ever think about that when applying his (her) abilities to bringing their art into existence!

[1] rtrevino8522@yahoo.com.mx

[2] Reynaldo Treviño Cisneros – Revelaciones.mp4 – Google Drive.

J. J. McIntyre-Mills
College of Education, University of South Africa, Pretoria Central, Gauteng, South Africa

College of Professions, Adelaide University, Adelaide, Australia

Y. Corcoran-Nantes
College of Humanities, Arts and Social Sciences, Flinders University, Adelaide, Australia

J. J. McIntyre-Mills, Y. Corcoran-Nantes (eds.), *Transformative Education for Regeneration and Wellbeing*, Contemporary Systems Thinking,
https://doi.org/10.1007/978-981-19-3258-8

Reynaldo Trevino *Cascades Peacock*

Reynaldo: On our planet there are many types of cascades. Some of them are very beautiful and are born from natural springs flowing down the mountains and embellishing the sites where they develop harmony all around the flora and the unevenness of the rocks where they fall.

These kind of streams usually bring peace to the souls of wanderers. They know that it is pure water; the most important source for the development of life.

However, there are other cascades that come from artificial sources like industrial waste waters very strongly contaminated with chemical toxic residues- resulting from uncontrolled processes of transformation.

It is our time to decide which cascades we want to maintain and which ones should be avoided by correcting the processes that produce them.

A solution is to apply from the beginning regenerative processes that eliminate any possible damage of the original natural conditions and that support the life of all native multiple species.

Intelligent and responsible foresight is always needed. And this kind of foresight must be an indispensable purpose of transformative education. There are more questions to be addressed by learners than the ones their teachers could imagine to be pertinent.

The resemblance of feathers of a peacock flying upwards in the painting intends to animate the set of questions to be addressed whenever transcendental actions are to be launched for the betterment of our world living conditions. Any intended project needs to be reflected over with questions such as:

- why? what for?, how?, where?
- for whom? by whom?,
- when?, with what resources?,
- what are the departing points?,
- what resources might be added in intermediate stages?,
- what indicators and monitoring systems might be employed to evaluate and ponder progress and possible achievements?,
- what subproducts or collateral effects emerge?
- what values are shared all along the implementation processes?,
- how often the monitoring is required? and
- what emergent properties might become usable to improve, accelerate or decelerate, according to new necessities, or possibly, to change the route?

This is systemic thinking—it is not an individual responsibility but a collective one.

Good questions and answers can be represented through artistic creations to prompt *thinking through* the taken for granted—as a step towards transformation towards ways to support living systems.

Afterword

We would like to thank the contributors to this volume on re-generative education and the importance of safe habitats for diverse species. Together we aim to make a contribution to theory and practice by making suggestions as to how to re-frame the content, structure and process of education for transformation. The case is made for a more integrated curriculum where learners across primary, secondary and tertiary levels are given the opportunity to do applied learning and where unemployment is a family issue to enable intergenerational learning through vocational training that enables people to earn, learn and grow a future through green co-operatives linked with a learning community.

This book was conceived of before the world was plunged into a global pandemic which in the end has, in its own way, added a nuanced gravitas to the subject matter and pathways into the future. Initially, we had considered our experiences in the field with respect to both formal and informal education and the changes that had occurred over the last decade most especially with formal education – the construction of curriculum and the opening up of flexible delivery and the experiment with 'new ways 'of learning. Outside of the formal structure we noted there is a centuries old history of learning the accumulation and dissemination of knowledges and the way they are passed on to individuals and communities and their importance both within and across cultures. We considered how impactful those learning environments were in the creation of new ways of thinking which in many cultures has shaped and transformed the way they lived. The authors in this book have all had experience in the field and their contributions illustrate the rich tapestry of learning and ways of transmitting knowledge/s which enriches the world we live in and gives up hope for change that is positive, dynamic and transformational.

In the process of bringing together the authors and their knowledge and experiences along came COVID-19 presenting one of the most devastating challenges we could ever have anticipated. Everyone was now required to think on their feet, with respect to their teaching which was being transformed as we wrote – schools

J. J. McIntyre-Mills, Y. Corcoran-Nantes (eds.), *Transformative Education for Regeneration and Wellbeing*, Contemporary Systems Thinking,
https://doi.org/10.1007/978-981-19-3258-8

closing, distance learning in many cases hurriedly put together without precedent and maintaining communication through social media and across the internet highway that we had only used socially and occasionally for professional purpose was now to be the lynchpin of our existence as academics and social practitioners. As the formal structure of teaching collapsed through the closure of the 'live venues' such as schools colleges, universities and community centres those engaged in knowledge transmission and enrolled students went through a legal separation. At that point the renegotiation of space, the rebuilding of a learning community, the retention of contact and support all of which was occurring 'on the run' as it were. No time for contemplation, for testing out our ideas or hoping for a swift ending to what remains a serious epidemical crisis. As in any other crisis necessity is the mother of invention as they say and nowhere more so than in the eye of the storm that is COVID-19. It has forced people to confront their relationships outside of formal structures, relationships with one another, with the environment and the greening of livelihoods and productive processes.

The mantra for Covid-19 has been 'not until everyone is protected are, we all protected' as it should be for vaccination so should the way forward be. Socio-political change underwrites a transformation of a society both local and global. At no other time in history, has action 'from below' been more important than now. This volume offers many examples and lessons of ways in which our approach to life, to education and to the very idea of transformation itself can be negotiated. Learning from one another outside of formal structures is equally if not more important than the traditional formal educational structures rarefied in our global world. This is a change not necessarily of our choosing either in time or place but it is a change we had to have. Moving us out of and beyond our comfort zone has not been a welcome change but an abrupt about face with respect to our attitudes to present and future realities. We often hear people say how they cannot wait for things to get back to normal, when the reality is it never will. We have moved into an epoch of a new normal, new ways of doing things and ways of creative thinking. Learning is a process; the term lifelong learning has never been so relevant as now. We hope that this volume contributes to this process and may we all learn together moving forward as a caring, forward thinking global community.

Glossary of Key Concepts

- **Anthropocentric** approaches are people centred and this volume advocates for an approach that is focused on multiple species
- **Boundaries** need to be considered in a conceptual and spatial sense when considering ethical decisions. This volume argues that we need to extend our sense of solidarity with other species and recognise our relationality and interdependency as living systems. By extending recognition and understanding our relationships with others it has implications for ethics, democracy, governance and education policy.
- **Circular economy** as applied in this volume is based on the work of Vandana Shiva and others who support working with nature in ways that regenerate it, rather than depleting the environment or exploiting the people.
- **Critical systemic thinking and practice** are based on working with people in learning organisations and learning relationships that support triple loop learning (Flood & Romm, 2018) with people who test out ideas to ensure that appropriate, contextual decisions are made with participants to protect multispecies relationships (McIntyre-Mills & Corcoran Nantes, 2021)
- **Ecofacturing** means working with the sun, the wind, the soil, the earth to support living in ways that protect wellbeing stocks rather than striving for short term profit.
- **Inclusionality, solidarity and gender mainstreaming** concepts are extended to include solidarity with the environment and the need to re-frame rights and responsibilities towards the voiceless and differently abled.
- **Multispecies relationships** refer to the many species that depend on one another for their shared survival.
- **Natural** inclusion according to Alan Rayner (2021, pers. comm) is more than a 'concept', it is fundamental evolutionary principle and 'percept' of ecological and evolutionary philosophy that can be described as "a flow of receiving, sustaining and passing on."

J. J. McIntyre-Mills, Y. Corcoran-Nantes (eds.), *Transformative Education for Regeneration and Wellbeing*, Contemporary Systems Thinking,
https://doi.org/10.1007/978-981-19-3258-8

- **Power with**, not power over is a core tenet of a relational approach that respects multiple species.
- **Relationality** is an approach that recognises that all nature is part of one system
- **Regeneration** refers to education that supports social and environmental wellbeing as a life-long practice in learning organisations and communities that support social and environmental justice.
- **Systemic** is based on the idea that relationships are iterative, not linear. Cause and effect results in ongoing cycles that result in positive or negative cascades that feed forward or deplete living systems
- **Transformative** education combines face to face learning in the community that draws on the local tacit knowledge and is supported by local e-learning hubs where people could meet, greet, learn new skills in a multi-generational way with paid facilitators who work to support on line learning that could support education and vocational training through a range of alterative options to support nodes and networks that enable people to earn, while they learn and grow a future together. The focus is on moving from dualism between human beings and other species to supporting relationships that protect living systems.
- **Wellbeing stocks**, a concept coined by Stiglitz et al. (2010) refers to multidimensional measures of wellbeing to protect current and future generations.

Index

J. J. McIntyre-Mills, Y. Corcoran-Nantes (eds.), *Transformative Education for Regeneration and Wellbeing*, Contemporary Systems Thinking,
https://doi.org/10.1007/978-981-19-3258-8

The manufacturer's authorised representative in the EU is Springer Nature Customer Service Centre GmbH, Europaplatz 3, 69115 Heidelberg, Germany. If you have any concerns regarding our products, please contact ProductSafety@springernature.com

Printed and bound by CPI Group (UK) Ltd, Croydon, CR0 4YY
30/09/2026
02225315-0016